Developing New Functional Food and Nutraceutical Products

Developing New Functional Food and Nutraceutical Products

Edited by

Debasis Bagchi
University of Houston College of Pharmacy
Houston, TX, United States

Sreejayan Nair
University of Wyoming, School of Pharmacy
Laramie, WY, United States

ELSEVIER

AMSTERDAM • BOSTON • HEIDELBERG • LONDON
NEW YORK • OXFORD • PARIS • SAN DIEGO
SAN FRANCISCO • SINGAPORE • SYDNEY • TOKYO

Academic Press is an imprint of Elsevier

Academic Press is an imprint of Elsevier
125 London Wall, London EC2Y 5AS, United Kingdom
525 B Street, Suite 1800, San Diego, CA 92101-4495, United States
50 Hampshire Street, 5th Floor, Cambridge, MA 02139, United States
The Boulevard, Langford Lane, Kidlington, Oxford OX5 1GB, United Kingdom

Library of Congress Cataloging-in-Publication Data
A catalog record for this book is available from the Library of Congress

British Library Cataloguing-in-Publication Data
A catalogue record for this book is available from the British Library

ISBN: 978-0-12-802780-6

Publisher: Nikki Levy
Acquisition Editor: Megan Ball
Editorial Project Manager: Billie Jean Fernandez
Production Project Manager: Nicky Carter
Designer: Greg Harris

Typeset by Thomson Digital

Dedicated to my beloved and respected
Professor Sidney J. Stohs, PhD, FACN,
CNS, ATS, FAPhA, FASAHP

Contents

Contributors

Kristy Appelhans
Herbalife Nutrition, Los Angeles, CA, United States

Margie Atwater
Mind Genomics Associates, White Plains, NY, United States

Kartik Baruah
Laboratory of Aquaculture & Artemia Reference Center, Department of Animal Production, Faculty of Bioscience Engineering, Ghent University, Belgium

Brent Batzer
Amin Talati & Upadhye, LLC, Chicago, IL, United States

Vladimir Bessonov
Laboratory of Food Chemistry, Federal Research Centre of Nutrition and Biotechnology, Moscow, Russia

Joe Bogue
Department of Food Business and Development, University College Cork, Cork, Ireland

Nicola Luigi Bragazzi
Department of Health Sciences (DISSAL), School of Public Health, University of Genoa, Genoa, Italy

Joyce Cao
Herbalife Nutrition, Los Angeles, CA, United States

Annamaria Di Capua
Operative Unit of Hygiene and Public Health, Local Health Service, La Spezia, Italy

George Carrera, Jr.
Amin Talati & Upadhye, LLC, Chicago, IL, United States

Scott Chaffee
Department of Surgery, Comprehensive Wound Center, Dorothy M. Davis Heart & Lung Research Institute, The Ohio State University Wexner Medical Center, Columbus, OH, United States

Leighton K. Chong
Udell Associates, Castro Valley, CA, United States

Orla Collins
Department of Food Business and Development, University College Cork, Cork, Ireland

Joseph Cwik
Amin Talati & Upadhye, LLC, Chicago, IL, United States

Amitava Das
Department of Surgery, Comprehensive Wound Center, Dorothy M. Davis Heart & Lung Research Institute, The Ohio State University Wexner Medical Center, Columbus, OH, United States

Sourya Datta
Supply Chain Manager at eBay, San Jose, CA, United States

Ryan Dickerson
Department of Surgery, Comprehensive Wound Center, Dorothy M. Davis Heart & Lung Research Institute, The Ohio State University Wexner Medical Center, Columbus, OH, United States

Bernard W. Downs
Udell Associates, Castro Valley, CA, United States

Antonia Erz
Department of Marketing, Copenhagen Business School, Copenhagen, Denmark

Surya P. Gautam
CT Institute of Pharmaceutical Sciences, Sahapur Campus, Jalandhar, India

Mahua Ghosh
Department of Chemical Technology, University of Calcutta, Kolkata, India

Vincenza Gianfredi
Department of Experimental Medicine, School of Specialization in Hygiene and Preventive Medicine, University of Perugia, Perugia, Italy

Surashree Sen Gupta
Department of Chemical Technology, University of Calcutta, Kolkata, India

Abhishek Gurnani
Amin Talati & Upadhye, LLC, Chicago, IL, United States

Jeff Hilton
BrandHive, Salt Lake City, UT, United States

James E. Hoadley
EAS Consulting Group, Alexandria, VA, United States

Chun Hu
Nutrilite Health Institute, Buena Park, CA, United States

John Hudson
Department of Economics, University of Bath, Bath, United Kingdom

Vasily Isakov
Department of Gastroenterology and Hepatology, Federal Research Centre of Nutrition and Biotechnology, Moscow, Russia

Michelle Jackson
Venable LLP, Washington, DC, United States

Raj K. Keservani
School of Pharmaceutical Sciences, Rajiv Gandhi Proudyogiki Vishwavidyalaya, Bhopal, India

Rajesh K. Kesharwani
Department of Biotechnology, NIET, NIMS University, Shobha Nagar, Jaipur, India

Kristen Klesh
Venable LLP, Washington, DC, United States

Alla Kochetkova
Laboratory of Food Biotechnology and Foods for Special Uses, Federal Research Centre of Nutrition and Biotechnology, Moscow, Russia

Jacqueline Kuler
Amin Talati & Upadhye, LLC, Chicago, IL, United States

Claudia Lewis
Venable LLP, Washington, DC, United States

Francesco Maddalo
Operative Unit of Hygiene and Public Health, Local Health Service, La Spezia, Italy

Lorenzo Marensi
Hygiene and Public Health Unit, Local Health Unit 3 of Genoa, Genoa, Italy

Palma Ann Marone
Toxicology and Pathology Associates, Department of Pharmacology and Toxicology, Medical College of Virginia, Virginia Commonwealth University, Richmond, VA, United States

Mariano Martini
Department of Health Sciences (DISSAL), Section of Bioethics, University of Genoa, Genoa, Italy

Shane T. McDonald
Principal Flavor Chemist, Kalsec Inc., Kalamazoo, MI, United States

David Moskowitz
Mind Genomics Associates, White Plains, NY, United States

Howard Moskowitz
Mind Genomics Associates, White Plains, NY, United States

Parisa Norouzitallab
Laboratory of Aquaculture & *Artemia* Reference Center, Department of Animal Production, Faculty of Bioscience Engineering, Ghent University; Laboratory of Immunology and Animal Biotechnology, Department of Animal Production, Faculty of Bioscience Engineering, Ghent University, Belgium

Daniele Nucci
Department of Experimental Medicine, Experimental Centre for Health Promotion and education, University of Perugia, Perugia, Italy

Asim Kumar Pal
APC Nutrients, Secunderabad, India

Matthew Poliner
Venable LLP, Washington, DC, United States

Sebastiano Porretta
Experimental Station for the Food Preserving Industry, Parma, Italy

Stephen Rappaport
Stephen D. Rappaport Consulting LLC, Norwalk, CT, United States

Livia Rossi
Mind Genomics Associates, White Plains, NY, United States

Sashwati Roy
Department of Surgery, Comprehensive Wound Center, Dorothy M. Davis Heart & Lung Research Institute, The Ohio State University Wexner Medical Center, Columbus, OH, United States

Teresa Concetta Saporita
Hygiene and Public Health Unit, Local Health Unit 3 of Genoa, Genoa, Italy

Varuzhan Sarkisyan
Laboratory of Food Biotechnology and Foods for Special Uses, Federal Research Centre of Nutrition and Biotechnology, Moscow, Russia

Chandan K. Sen
Department of Surgery, Comprehensive Wound Center, Dorothy M. Davis Heart & Lung Research Institute, The Ohio State University Wexner Medical Center, Columbus, OH, United States

Andrew Shao
Herbalife Nutrition, Los Angeles, CA, United States

Anil K. Sharma
Department of Pharmaceutics, Delhi Institute of Pharmaceutical Sciences and Research, New Delhi, India

Jennifer Shield
Amin Talati & Upadhye, LLC, Chicago, IL, United States

Elena Smirnova
Laboratory of Food Biotechnology and Foods for Special Uses, Federal Research Centre of Nutrition and Biotechnology, Moscow, Russia

Karin Tollin
Department of Marketing, Copenhagen Business School, Copenhagen, Denmark

Flavio Tovani
Sant'Anna School of Advanced Studies, Pisa, Italy

Amy Jane Troy
Department of Food Business and Development, University College Cork, Cork, Ireland

Victor Tutelyan
Laboratory of Enzymology of Nutrition, Federal Research Centre of Nutrition and Biotechnology, Moscow, Russia

Lawrence J. Udell
Udell Associates, Castro Valley, CA, United States

Jesper Vej
Department of Marketing, Copenhagen Business School, Copenhagen, Denmark

Brigitte Velema
Mind Genomics Associates, White Plains, NY, United States

Santosh K. Verma
CT Institute of Pharmaceutical Sciences, Sahapur Campus, Jalandhar, India

Tina Vukasović
DOBA Faculty, Prešernova, Maribor; International School for Social and Business Studies, Mariborska cesta, Celje; Faculty of Mathematics, Natural Sciences and Information Technologies, University of Primorska, Glagoljaška, Koper, Slovenia

Jerzy Zawistowski
Food, Nutrition and Health, Faculty of Land and Food Systems, University of British Columbia, Vancouver, BC, Canada

Preface

Let food be thy medicine, and let medicine be thy food.
Hippocrates, Father of Medicine (460–377 BC)

Nutraceuticals and functional foods are becoming increasingly popular worldwide. The "Ayurvedic Medicine" from India and the Chinese Traditional Medicine have demonstrated the therapeutic efficacy of a number of botanical or herbal medicines for thousands of years. The Unani (or Yunani) system of medicine is a Persian Arabic traditional medicine practiced in India (1526–1707 AD) and South-Central Asia, which was based on the teachings of the Greek physician Hippocrates. In recent years, there have been significant advances in the development of pharmaceuticals and drugs aimed at treating a variety of disease conditions. However, most of these pharmaceuticals or drugs are expensive and are associated with a number of adverse side effects. More recently, a significant number of nutraceuticals and functional/medical foods have been introduced to the worldwide marketplace, backed by broad-based safety and efficacy studies. These natural products are often less expensive, and are available as over-the-counter products and have therefore gained popularity. To keep up with these changes medical and allied-health schools are training physicians and health-care professionals on the use and benefits of alternative and complementary medicine.

This book was conceptualized with a vision to convey to our eminent readers the diverse nature of nutraceuticals and functional and medical foods and the challenges and opportunities associated with developing, characterizing, and marketing them. The rationale for the development of these products is the documented long history of their use. It is important to recognize however that some of the nutraceuticals and functional and medical foods may exhibit untoward adverse effects, and therefore must undergo rigorous safety, toxicity, and mutagenicity tests. Especially, long-term safety studies are warranted to rule out potential harm caused by these agents. There are several important considerations that need to be addressed before developing nutraceuticals and functional foods, including (1) availability, (2) cost-effectiveness, (3) safety, (4) efficacy in clinical settings, (5) regulatory constraints and approval, (6) marketability, (7) supply chain and effective distribution, and (8) customers' acceptance.

The introductory chapter in this book is by Jeff Hilton, a 25 year veteran of the healthy lifestyles category and brand strategist. This chapter describes the growth pattern and emerging opportunities in the nutraceuticals and functional foods market. The chapter that follows, by Joe Bogue and coworkers from the Department of Food Business and Development, University College Cork, Ireland, highlights market analysis and concept development of functional foods. The third chapter by

Sourya Datta, from eBay, focuses on the diverse marketing challenges and strategies, while the fourth chapter describes the strategic viewpoints of innovation and marketing teams in the development of novel functional foods. Staying with the theme of marketing, the subsequent chapter by Dr. Palma Ann Marone alludes to marketing trends in the context of food safety and consumer awareness.

Amin Talati & Upadhye, LLC, a well-respected legal team from Chicago, dwell on intellectual property in relation to nutraceuticals and functional foods, in Chapter 6. An additional chapter by this group extends the discussion to regulatory aspects of new product development and product-approval procedures and guidelines in the United States. Formulation of nutraceuticals and challenges associated with formulation development, with special emphasis on stability and safety evaluations, form the basis of the next two chapters designed individually by Dr. Ghosh and Dr. Keservani and their eminent teams. The chapter by Sourya Datta, eBay, which alludes to the challenges associated with sourcing, supply chain, and manufacturing of nutraceuticals and functional foods. Next we include a chapter that describes the analysis of raw materials and their quality control. The following chapter describes the strategies for preclinical and clinical evaluations of nutraceuticals. Given the growing importance of flavors, we chose to include a chapter by Dr. Shane T. McDonald, a renowned flavor chemist, that deals with the concepts of flavor creation in novel nutraceutical and functional food formulations. Howard Moskowitz and his team discuss issues related to mind genomics and cognitive economics in Chapter 14.

Chapters 15–18 focus on the regulatory aspects relating to nutraceuticals and functional foods. This includes a chapter by Dr. Andrew Shao (Herbalife Inc., CA), which alludes to the regulations governing global market entry for nutraceuticals and functional foods. Another chapter by James E. Hoadley, who has years of experience with the US Food and Drug Administration (FDA), discusses the salient features of new dietary ingredient (NDI) and generally recognized as safe (GRAS) aspects of regulations related to nutraceuticals. Trials and tribulations associated with nutraceutical development in Europe, the Asian subcontinent, and China are discussed in chapters by Drs Bragazzi, Zawistowski, and Hu, respectively. The challenges associated with the functional foods marketplace in Slovenia, especially among young customers, is described in Chapter 20 by Dr. Tina Vukasović. In the subsequent chapter, Claudia A. Lewis et al. discuss the overall regulatory status of medical food. The concluding Chapter 22 by Bernard W. Downs highlights the intricate issues associated with regulations, including intellectual property, branding, and trademark of nutraceuticals and functional foods. We have also included two chapters as examples for the cutting-edge research on nutraceuticals and functional foods. Dr. Chandan K. Sen (Associate Dean and Director, Ohio State University Wound Healing Clinic) and coworkers discuss the value of nutritional supplements in diabetic wound healing, which is a rapidly emerging problem, given the rise in global obesity. Dr. Kartik Baruah from Ghent University, Belgium describes the development of novel aquaculture for fisheries.

Dr. Kristy Applehans from Herbalife Inc., CA, discusses the strategies on adverse event reporting by the nutraceutical industries worldwide in Chapter 25. Genetically modified products, also known as GMO foods, are gaining popularity globally, and Prof. John Hudson provides a compelling analysis of GMO in Chapter 26.

We sincerely thank our eminent authors for their invaluable contribution to this book and we expect that the book will be useful to our readers. The editors sincerely thank Namrata Bagchi, BE, for her valuable inputs and suggestions.

Debasis Bagchi, PhD, MACN, CNS, MAIChE
Sreejayan Nair, PhD, FACN

CHAPTER

1 Growth patterns and emerging opportunities in nutraceutical and functional food categories: market overview

Jeff Hilton

BrandHive, Salt Lake City, UT, United States

1.1 INTRODUCTION

This chapter is perhaps unique in that it is written by a 25-year veteran of the natural products industry. My background includes positions with Nature's Way, Nutraceutical Corporation, and consulting work with many of the top United States supplement and food companies including Solgar, NBTY, Natrol, Schiff, Dupont, and LESAFFRE. So I do come at this from somewhat of an insider's perspective. It's what I do every day. In that way, I hope my perspective is insightful for you. You will probably also note a slightly instructional tone as I tend to write as if I am speaking to a client.

Where do I begin to dissect trends in this $150+ billion sector? It's growing overall at over 10% per year with functional food and beverages leading the way, followed by nutraceuticals and supplements at a somewhat lower growth rate. It is an industry going through almost constant change and evolution, partly due to the changing regulatory environment, and partly due to the ever-changing attitudes of everyday consumers regarding health and wellness.

As an overview to the content for this chapter, I have identified *eight major market trends* that I think are changing the face of the functional food, beverage, and dietary supplement business from a macro perspective. Those eight trends are:

1. Marketplace convergence of categories, channels, technology and consumers,
2. Accelerated growth of functional foods and beverages,
3. Cobranding partnerships between ingredient suppliers and manufacturers,
4. Increased focus on science and claims validation,
5. Growing demand for sustainable and eco-friendly products,
6. Explosion in active nutrition as consumers become more health conscious,
7. Emergence and popularity of innovative dosage and delivery forms, and
8. Rise of the millennials and a new paradigm on health and wellness.

Developing New Functional Food and Nutraceutical Products. http://dx.doi.org/10.1016/B978-0-12-802780-6.00001-8

1.2 TREND ONE: THE POWER OF CONVERGENCE TO CHANGE A MARKETPLACE

There's something in the air these days. It's the sound of change and evolution in the healthy lifestyles sector. It's a movement with momentum and it is driving the marketplace. I like to call it *convergence*. And it manifests itself in a number of ways that provide insight into where we are headed as an industry. Any savvy industry observer can feel the winds of change uprooting "business as usual" and turning the paradigms we have all operated under for decades on their heads, frankly. I have identified four specific types of convergence that I think are changing the way consumers think and buy healthy or better-for-you products and services.

1.2.1 CATEGORY CONVERGENCE

The lines are blurring between supplements, foods, beverages, and pharmaceuticals. As consumers become aware of validated nutrients that they want to incorporate into their diets, they want to consume them in the most convenient and nonintrusive manner possible. So they are increasingly relying on enhanced and fortified foods and beverages to supply the dietary supplementation traditionally and historically delivered through tablets and capsules. Smart dietary supplement manufacturers and suppliers should already be developing plans to make sure they have a primary or co-branded presence in the burgeoning functional food and beverage categories. In addition, convenience and portability are driving consumers toward alternative dosage forms that are more easily consumed and transported including liquids, gels, shots, gummies, and stick packs. Smart dietary supplement manufacturers are already testing alternative delivery forms or partnerships for their nutrients. Actual and virtual retail shelves are changing, and for good reason. Pill fatigue is not just a seniors concern.... it's a people concern.

1.2.2 CHANNEL CONVERGENCE

The paradigm of control in the marketplace has been turned upside down, with the consumer now in control and wielding all the power. They can quickly and easily price shop and compare retail offerings with direct-to-consumer websites. They can dissect and research claimed features and benefits and access product reviews in seconds. Today's consumer is all about transparency and authenticity, so there is no place to hide. Plus increasingly more integrative health providers are recommending and even dispensing dietary supplements to their patients. So as these various distribution channels continue to overlap and merge, the competition for consumer mind share is intensifying. Smart marketers are establishing revenue streams in each of these growing channels to tap into consumer trends in self-care that show no signs of diminishing. The name of the game in today's environment is to be wherever the customer is, and that's not easy.

1.2.3 TECHNOLOGY CONVERGENCE

If the 23 & Me, Fit Bits, Apple watches, and other innovations in wearable technology have taught us anything, it is that consumers don't need (or want) a doctor to tell them what's up with their day-to-day health status. For catastrophic illness, MDs are still the first resort. However, for more general maladies and aging/prevention strategies, consumers are self-diagnosing in droves and using technology to monitor their cholesterol, blood pressure, sleep patterns, and even genetic propensities. Who knew? And that's just the beginning. According to thought leaders and professionals, the "quantified self" is only just off in the distance as consumers take charge and proactively work to monitor their own health and wellness needs. Smart marketers are working on plans now to integrate their products with these emerging self-assessment technologies or get left behind.

1.2.4 CONSUMER CONVERGENCE

So in my opinion, where all of this is leading is to a significant marketplace opportunity that rivals any we have seen in the past few decades. Consumers I think are empowered but they are also confused. They are armed with tactics, but don't have an overall integrated plan of action for improved or sustained health as they age. They have their enhanced beverages, nutritional gummies, nutricosmetics, and their digital devices, but they don't understand how to bring it all together in a cohesive and efficient and effective way. Where this is heading is toward a more customized, personalized healthcare delivery that takes into consideration the whole patient or consumer. What is lacking is an approach that integrates these elements into an actionable, credible, and easy-to-follow plan or regimen that can be tracked, customized, and replicated across a broad population.

Smart industry players are working to capitalize on this explosive megatrend resulting from the convergence of multiple categories, channels, technologies, and consumer demographics. Smart marketers recognize that this multifaceted convergence is one of the single biggest market opportunities we've seen in this nation of aging baby boomers and entitled millennials.

1.3 TREND TWO: FUNCTIONAL FOODS AND BEVERAGES

I remember back in the 1990s sitting in meetings talking about the potential of the functional food and beverage sector, which at that point was in its infancy. Now, over a decade later, we are in the ramp-up to significant expansion of the category with new products entering the market on an almost daily basis. A simple walk down the beverage aisle of Whole Foods provides an excellent example of the proliferation that has occurred. The sheer volume of functional beverage entries is both staggering and intimidating. Even more sobering, probably 50% of those brands won't be on the shelf one year from now. The beverage aisle is a brutal retail battleground where brands either thrive or die. Functional food brands have not yet proliferated to the extent that functional beverages have, but based upon a review of marketplace activity and trends, many more functional food product introductions are on the horizon. And the key driver here is the consumer, believe it or not. In survey after survey, consumers have stated that they prefer their health and wellness to result primarily from what they eat and drink on an ongoing daily basis, so it is not a huge leap to expect expanded growth for these once vertically niched market segments.

The following seeks to identify, review, and discuss some significant and relevant ingredient and formulation trends in these growing categories.

1.3.1 WHAT IS A FUNCTIONAL FOOD?

As defined by the Institute of Medicine in Washington, functional foods are "those foods that encompass potentially healthful products, including any modified food or ingredient that may provide a health benefit beyond the traditional nutrients it contains." Functional foods can include foods like cereals, breads, yogurts, snacks, and beverages that are fortified with vitamins, herbs, or other specialty ingredients. A specialty ingredient may be a naturally nutrient-rich whole food source such as spirulina, garlic, or soy, or a specific component of a food, like Omega-3 fatty acids from salmon.

1.3.2 HOW BIG IS THIS SECTOR?

Functional foods and beverages represent roughly a $50 billion category in the United States retail market. Recent years have seen the discontinuation of many slow-moving entries particularly in the beverage sector. That said, the functional category is still outpacing traditional food and beverage sales in the United States. So the first trend of note is brand proliferation followed by a shakeout of slow movers. We are already seeing this happen within the beverage sector, and foods will follow suit as the number of market entries expands.

1.3.3 WHAT SALES CHANNELS ARE PERFORMING BEST?

Not surprisingly, nearly 90% of functional food sales occur in mass-market channels including food, drug, mass volume stores (Walmart/Target), and club stores (Costco/Sam's). Large CPG (consumer packaged goods) companies like General Mills, Kraft, Nestlé, Tropicana, and Coca Cola have led the charge in this sector over the past several years, and they continue to own the category in the minds of consumers looking for added functionality. It is also interesting to note that almost 60% of functional product sales are in the beverage sector, which has taken the lead in both product development and consumer education.

1.3.4 SEARCHING FOR NATURALLY FUNCTIONAL INGREDIENTS

Increasingly consumers are looking for foods and ingredients that are "naturally functional" such as blueberries, pomegranates, whole grains, protein, fiber, and soy.

These types of ingredients have formed the basis of the functional food offerings we have seen in the marketplace to date as they already have some consumer awareness and acceptance. Finally, these ingredients have ODI (old dietary ingredient) status, and are also generally recognized as safe (GRAS) for food and beverage formulations. That means manufacturers can make nutritive value-based health claims that are less likely to be challenged by regulators.

1.3.5 FUNCTIONAL VERSUS INHERENTLY FUNCTIONAL

At a basic level, there are two primary types of functional food/beverage products:

1. "*Better for you*" foods contain less or none of certain ingredients perceived by consumers to be less desirable. Examples include:
   a. low-sodium soups
   b. reduced fat-, mayonnaise-, cholesterol-, and or cholesterol-free baked goods
   c. no sugar added desserts.

   These products are not truly "functional" in my opinion, but do offer consumers a way to manage their nutrient intake and improve their health.
2. "*Added functionality*" foods offer nutrients in efficacious doses added to food products for improved consumer health and wellness. Examples include:
   a. high-fiber products,
   b. soy products,
   c. juices with added calcium or vitamins,
   d. fortified waters, and
   e. naturally antioxidant-rich juices.

   These types of products fit the classic profile of a functional food.

While both categories are growing, added functionality foods have captured the interest and imagination of those consumers looking for the convenience of eating foods that contain supplement-level doses of health-improving nutrients.

1.3.6 REVISITING FOOD/SUPPLEMENT CONVERGENCE

Again, one trend in product development and consumer demand that is hard to ignore is the merging of foods and dietary supplements. It's no secret that consumers would prefer to get their nutrients primarily through the foods and beverages they consume, rather than taking additional supplements to fill in the gaps. Not surprisingly, manufacturers on both sides of the equation are working to find ways to deliver efficacious doses of nutrients that deliver on taste and performance expectations when formulated into a food or beverage product.

Several factors are driving this convergence, including:

- reduced discretionary income,
- 2-for-1 mentality,
- pill fatigue among older baby boomer consumers,

- rise in overall health awareness among Americans,
- convenience, and
- trend toward simplifying life and nutritional intake.

Another clear factor in the convergence of foods and supplements is the emergence of "gateway nutrients." Consumers are becoming increasingly aware of and perceive certain nutrients as having added value. As their perception of added value increases, they become more willing to pay a higher price for foods that include these nutrients. Examples of gateway nutrients include probiotics, Omega-3's, antioxidants, soy, and green tea.

1.3.7 DOCTOR, DOCTOR

Whether you are talking foods, beverages, or supplements, the product development and marketing focus has turned to condition-specific applications. Increasingly, consumers are shopping to meet their specific and identified health needs. Today's consumers expect more targeted health solutions and are honing in on potential solutions that meet their condition-specific needs. Some of this targeted shopping behavior is recession-driven, and it helps that retailers are increasingly merchandising products by health needs. But even beyond that, consumers are looking for a more focused approach to wellness that allows them to assess the results of taking a product. They look for specific health improvements, or lack thereof, and modify their shopping accordingly. Some of the most dominant health conditions include energy/performance, healthy aging, immunity, digestion, cardiovascular, joint support, weight loss, and diabetes. See Table 1.1 for ingredients that extend the promise of a condition-specific health benefit.

1.3.8 NOT SO EASY

It is estimated that roughly 80% of new functional products fail in the marketplace. This is largely due to inadequate promotion, inadequate consumer and trade education (just because it is a food, don't assume that consumers will readily see the added value or be willing to pay more for it), and the fatal flaw of introducing what is basically an undifferentiated me-too product (how many energy drinks can the market support?). Despite increased creativity and innovation in the functional products sector, don't expect this failure rate to improve anytime soon.

1.3.9 WHY DO NEW PRODUCTS FAIL?

The majority of new product failures have to do with what companies didn't do enough of, specifically promotion, education, and market differentiation. As a result, the brand failed to make a connection with the consumer and sales fell short of projections.

Other issues have emerged which have more to do with product development than with marketing. One concern is the lack of efficacious doses offered in functional

Table 1.1 Condition-specific Functional Foods

Conditions	Ingredients	Examples
Energy	Green tea, caffeine, guarana, phytochemicals	Perky Jerky, Dynapep, Guru Energy Drink, 180°
Performance	Electrolytes, amino acids, protein	O.N.E. Coconut Water, Nasoya Tofu, Silk Soy Milk
Healthy aging	Omega-3s, antioxidants, trans resveratrol, sea buckthorn	Tofu Plus, Vitamin water, Welch's Healthy Start, Fruit2Day, Sibu Beauty, Gummy Pandas, Drench water
Heart health	Soy, fiber, cocona, phytosterols	Fiber One, Soy Joy, Barilla Plus Pasta, Progresso (w/added fiber), Kashi Heart to Heart Oatmeal
Joint & bone health	Turmeric, soy, calcium, antioxidants (inflammation management)	Yoplait (w/added calcium), Joint Juice, POM Wonderful
Cognitive health	Citicoline, resveratrol, selenium, L-theanine (in tea leaves)	Nawgan, red wine extracts, green teas, Kaia Foods Sprouted Sunflower Seeds
Digestion	Fiber, probiotics, whole grains, oats, enzymes	Barilla Plus Pasta, Mojo Milk, GT's Kombucha Teas, Activia, Tropicana juice w/fiber
Immunity	Camu camu, echinacea	Wildly Nutritious Immunity Blend, Praeventia, Yogi Immunity Tea, Immune Water
Weight management	Turmeric, soy, capsaicin, green tea, caffeine	Lychee Wasabi *Capsaicin* Spiced Elixir, Naga chocolate bar (w/Indian curry)
Diabetes	Stevia, yacon	Zevia Natural Sodas, Yerbamate Royale Tea (w/stevia), Terramazon Yacon Slices, LovingEarth Organic Yacon Syrup
Stress management	Omega-3s, magnesium, nuts, oats	
Sleep support	Melatonin	Tru2U Tart Cherry Juice

food and beverage products. If the products lack enough of the active ingredient to provide results for the consumer, the repeat purchase and word-of-mouth referrals will not happen. The consumer is trying to justify the added cost of a value-added product and they are definitely taking delivery of promised health benefits into consideration as they assess whether the benefits delivered are worth the price paid.

Another critical issue is the lack of legitimate consumer research on both the ingredients added to the products as well as the composite functional product itself. Increasingly consumers know how to do their homework to see if the added ingredients are clinically proven to be effective.

Finally, the consumer is open to information but needs to be educated about new ingredients with which he or she is not familiar. Ingredients such as fiber, protein, and calcium bring with them a cachet of awareness and other attributes, while natural ingredients, such as antioxidants, probiotics, and Omega-3's generally require a more thorough job of education and promotion to get seated with the consumer. Product education takes time and money, something that most companies underestimate when planning for a product introduction.

1.3.10 REGULATORY ROADBLOCKS

One emerging problem within the category, which definitely impacts the consumer, is the increased use of exaggerated or inaccurate health claims for functional products. Not only does this represent a significant area of risk for manufacturers with regulatory agencies, but it also erodes trust and credibility for all products at the consumer level. This type of offence occurs across the board, but is particularly rampant in the immunity, heart, and cognitive and weight loss categories. Both the FTC (Federal Trade Commission) and FDA (Food and Drug Administration) have stepped up their policing of these types of violations and FTC legal actions with Coca Cola for Vitamin Water, Kellogg's for Rice Krispies, and Frosted Mini-Wheats, and also POM Wonderful demonstrate their desire to clearly define and separate conventional foods and beverages from dietary supplements. That said, the marketplace battle is far from over and as consumer demand and education increase, the temptation for companies to make aggressive health claims for their functional offerings will only grow stronger. Meanwhile, both sides show no signs of slowing in the race both to meet consumer needs and to protect the consumer from misleading or exaggerated claims.

1.3.11 A SWEET STORY

Probably one of the most significant events in the functional food and beverage sector over the past few years has been the proliferation of functional sweeteners, specifically stevia and agave. These sweetener alternatives offer the sweetness of sugar without the calories, plus they have a low glycemic index and comparable taste. This has spurred a bonanza of opportunities for companies to reformulate with one of these alternatives, offering less of what consumers don't want and additional health benefits at the same time.

1.3.12 OH TO BE GREEN AND SUSTAINABLE

While consumer demand and awareness are growing slowly, they are increasingly interested in earth-friendly packaging options that offer ecological benefits in addition to standard product benefits. Particularly as highly visible companies like Coca Cola and Pepsi announce bottles made mostly from plants or other postconsumer material, the expectations of consumers will rise as they begin to consider the origins, footprint, and postconsumer life of various packaging alternatives. Green attributes and sustainability are certainly not compelling reasons to buy at any price, but when other factors are equal they do carry increased weight in the purchase decision. This is particularly true with single-use convenience products that are consumed and discarded promptly after purchase.

1.3.13 SUPER FRUITS STILL GOING STRONG

There are literally hundreds of undiscovered fruits hidden in the rain forests of the world. Don't expect this trend to disappear anytime soon. Americans are enamored with these healthy superfoods, at least for now. But the category is showing some signs of fatigue, particularly as mainstream fruits such as blueberries, grapes, cranberries, and pomegranates spend heavily to secure their rightful piece of the superfruit market pie. Meanwhile, keep your eyes peeled for newcomers sea buckthorn berries (beauty), yacon (diabetes), and camu camu (immunity).

1.3.14 JUST ADD SPICE

One of the newest (and oldest) players in the functional ingredient space is spices and seasonings, which have long provided flavor to food but are now being positioned as providing health benefits in the proper dosage. Examples of this trend include rosemary, red peppers, oregano, turmeric (curry), cinnamon, and ginger.

Spices support numerous conditions according to scientific literature. Turmeric is widely recognized as an antiinflammatory agent, making it a top choice for joint health formulations. Red peppers jumpstart metabolic activity and aid in weight loss. Ginger has a calming effect on the stomach and is a favorite in digestive health products. Cinnamon helps manage blood sugar and rosemary boosts liver function. In short, spices make a valuable addition in nearly every condition-specific category.

1.3.15 INGREDIENT POWER PLAYERS

Antioxidants continue to be the belle of the ball, primarily due to increased consumer awareness and demand. Shoppers may not know what antioxidants are, but they do know they are good for health and wellness. Likewise, the other power players in functional foods and beverages continue to be fiber, soy or whey protein, Omega-3's, and probiotics/prebiotics.

1.3.16 KEEP IT SIMPLE FOR THE CONSUMER

Consumer research supports what we all suspect. Moms are totally overwhelmed with food choice and making the right decisions for themselves and their families. Generally, the fewer ingredients on the nutrition facts panel the better. Keeping it simple is their new mantra. In particular, moms are keeping an eye out for overly processed food products and trying to avoid them whenever possible. Trans fats, high fructose corn syrup, preservatives, and artificial flavors and colors are securely on the radar of things to watch out for. Particularly when it comes to staples for the household, such as milk, bread, and eggs, they are attempting to change entrenched habits and begin buying whole grains and organic alternatives. In addition, mothers are increasingly concerned about incorporating more fresh items in the meals they serve. There is tremendous opportunity for functional food and beverage products to take advantage of this desire for foods that provide a dose of medicine. Research shows that the healthier a product is perceived to be, the less processed it is also perceived to be.

1.3.17 TASTE COMES FIRST

Don't go thinking that because consumers are seeking greater functionality in the foods and beverages they consume, taste is a secondary consideration to performance. Successful brands must deliver on both great performance and great taste. Delivery on those two key benefits has been the driver behind the success of functional brands like Corazonas, Fiber One, and Special K in recent years. The old adage "if it doesn't taste good, nothing else matters" is still true.

1.3.18 CLOSING THE GAP

It's important to remember that product education in this category is in its infancy. Consumers may talk about eating healthy and making better food choices, but that doesn't always translate to purchase behavior. The more your customers know about your products' health benefits and clinical support, the more likely they are to embrace your brand and act on their convictions.

Without a doubt, functional foods and beverages is definitely a category to watch over the coming years. Consumers are clamoring for new products and are looking to food first for enhanced nutrition and functionality. Eventually demand will exceed supply.

1.4 TREND THREE: GROWTH IN COBRANDING ACTIVITY

The business of suppliers cobranding ingredients with manufacturers has steadily expanded over the past 10 years, and for good reason. It's basically a win–win proposition driven by joint research, promotion, and education at both the trade and consumer levels. With the escalating cost manufacturers pay for research-backed, branded raw ingredients, they are increasingly looking to suppliers to share the burden of funding the education and promotional activities required to launch a successful ingredient brand. In short, suppliers are being asked to step up and participate more aggressively in terms of both effort and dollars. And that seems fair. Why should manufacturers be required to foot the branding bill alone? After all, the comarketing concept is to share both the risks and the rewards.

1.4.1 IS BRANDING ALWAYS THE RIGHT STRATEGY?

Branding is the ongoing process of creating and maintaining a unique and compelling brand image in the minds of current and potential customers. Building brand awareness, acceptance, preference, and most important loyalty for an ingredient is not a simple or inexpensive process, and companies should carefully consider whether or not branding and/or eventually cobranding is the best choice.

In general, companies should brand when they:

- hold valuable patents or trade secrets,
- have scientific support specific to their ingredient,
- will be doing business in substantial, active categories, and
- have potential cobranding partners to share risk and invest in the brand.

In general, companies should not brand when they:

- are selling in undifferentiated commodity categories,
- have no compelling scientific support relating to their specific ingredient,
- have minimal funds to devote to promotion and education, and
- have no meaningful intellectual property.

1.4.2 SURVEY THE MARKET AND ACT ACCORDINGLY

As companies make branding decisions, they should be sensitive to what is happening in the marketplace. It is easy to see as we survey the consumer landscape that confidence in dietary supplements has been eroding for some time. Consumers are more skeptical, confused, and unsure about dietary supplements and functional foods. They are looking for ways to assess and understand product quality and make better purchase decisions. As a result, we find in our marketing research with consumers that science and technical superiority are increasingly driving the brand message. And that has important implications for marketers in both dietary supplements and functional foods. It requires manufacturers and suppliers to:

- create brand value through science and other intellectual property,
- integrate science into the brand value proposition,
- take a more education-focused approach to marketing,
- expand their intellectual property portfolio, and
- look for research-driven marketing partners.

1.4.3 CONSISTENTLY COMMUNICATE THE KEY BRAND MESSAGES

Once the brand value proposition has been defined and honed, it is important to integrate that message through all marketing communications. Effective branding means consistently and persuasively communicating the product positioning at all points of customer contact.

1.4.4 WORK TO CREATE BRAND VALUE

There are a number of potential strategies for creating and sustaining brand value for an ingredient. Outlined briefly next are several ideas to consider:

- Launch a third-party website to promote the health benefits of a product category.
- Sponsor an annual symposium or gathering of scientists working in a category.
- Conduct consistent retailer training.
- Plan and execute trade show seminars and events.
- Create a distinctive new logo for a brand to differentiate it in the marketplace.
- Consider distributing a matte story focusing on a newsworthy aspect of the ingredient story.
- Contract with a celebrity spokesperson.
- Find an author to write a book promoting an ingredient.
- Sponsor an athlete who benefits from taking the product.

The future growth of the natural products industry lies in smart manufacturers and savvy suppliers partnering to bring innovative and science-based ingredients to market. Consumers are demanding it. Retailers will support it. And it can drive business to new heights if it's done right.

1.5 TREND FOUR: INCREASED IMPORTANCE OF SCIENCE AND CLINICAL VALIDATION

1.5.1 MARKETING SCIENCE TO TRADE AND CONSUMER AUDIENCES

As an industry, finished goods manufacturers and raw material suppliers need to improve the way in which they talk about science in their marketing materials. Much of what circulates today is overwritten, needlessly complicated, and often irrelevant to its intended audience. Here are some basic guidelines that can help to keep a science-based message both focused and relevant.

1.5.2 INTEGRATE SCIENCE INTO THE BRAND POSITIONING

The surest way to accomplish this is to make the scientific validation and support generated for the product central to the product's positioning and value proposition. That means more than just adding a line of copy to ads and brochures about the clinical research that has been conducted. Consider the following options:

- Revise the product name or identity/logo to communicate a more clinical or science-focused attitude.
- Revise the packaging to reflect a more science-driven image.
- Consider putting the product in a box with a consumer-friendly product insert or booklet summarizing and visualizing relevant research findings. Also consider a standing fifth panel on the box highlighting the product's technical superiority.
- Make science more central to consumer seminars, retailer training, trade show activity, and product literature.

- Consider using a health care practitioner or other technical expert to endorse or speak on behalf of the product.
- Examine the tone and attitude of the brand's advertising. Make it more indication-specific. Use science-oriented visuals and illustrations. Then carry that over to ancillary marketing tools including website, direct mail, and press materials.

The typical consumer's attention span is shrinking faster than the Brazilian rain forest. Given the flighty nature of attention spans today, how can you convey complex scientific topics so that consumers understand and remember them? Is it simply a matter of chunking the material into smaller bites? How do you take our increasingly visual culture into account? While smaller bites and a visual or two can help, brands that integrate more than one of the following tactics into their copy can help consumers read, understand, and remember their science.

1.5.2.1 Tell a story

While we like to think we are motivated strictly by facts, we receive and store information more easily in story form. Which is easier and more interesting for you: memorizing the factual sequence of information on the left or remembering the story on the right?

Stories are easy to process and remember because they string data together in a sequence. The consumer follows the sequence in order to recall the data and stores a single data set, rather than separate data bits, allowing for easier retrieval.

1.5.2.2 Use metaphors

Stories are full of metaphors. Metaphors can be imagery and narrative that tell a story (including similes and analogies), or they can be simple phrases, such as "feeling blue." Metaphors let you relate a complex idea to something your consumer already understands. This helps them grasp the new idea quicker and increases trust in the brand.

For example, to explain the relationship between endothelial health and arterial plaque buildup, we compared an unhealthy endothelial gastrointestinal (GI) lining (which is rough and sticky) to flypaper. Cholesterol and fats passing through the bloodstream stick to this flypaper, building up over time and blocking arteries. Restoring the health of the endothelial lining is like replacing flypaper with smooth silk. It was easy for consumers to understand that cholesterol and fats are less likely to stick to a silky smooth endothelium.

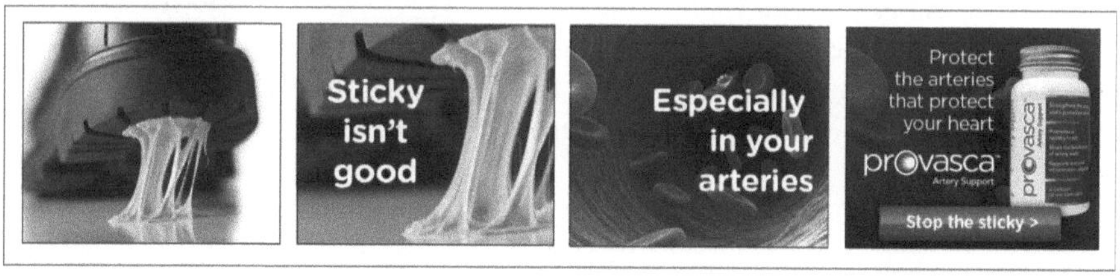

This ad campaign uses metaphor to help consumers understand the relationship between endothelial health and arterial plaque buildup.

1.5.2.3 Reach different parts of the brain

Present technical information in different forms—such as music, sound effects, rhymes, and images—to activate different areas of the consumer's brain. This activation helps draw them into the story.

The big three are:

- visuals,
- audio, and
- video.

1.5.2.3.1 Visuals

Anything you can show the consumer reduces the amount of energy needed to process your information, making you and your brand appear straightforward, helpful, and truthful. The bottom line: make messaging as visual as possible, as often as you can.

Good visuals are expensive and take time to create. Still, a single infographic can make a bigger impression on consumers than an entire database of written facts. Visuals have been shown to:

- increase time spent with the material,
- increase understanding, and
- increase retention.

1.5.2.3.1.1 The connection between seeing and remembering. A recent study on retention of visual information describes pictures as having a "direct route" to long-term memory storage. In fact, the study found that "each image stores its own information as a coherent 'chunk' or concept."[a]

1.5.2.3.1.2 Making words visual. Adding visuals can be as simple as formatting a passage as a callout or adding color to a list. In fact, adding color visuals increases the willingness to read by 80%.[b]

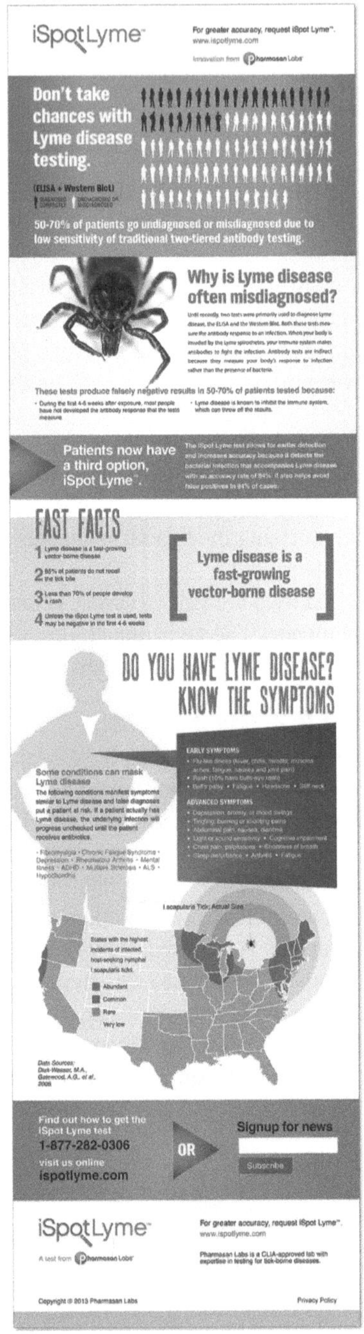

This infographic for a new Lyme disease test gives consumers multiple connection points to the copy, greatly increasing engagement.

[a]Brady, T.F., Konkle, T., Alvarez, G.A., Oliva, A., 2008. Visual long-term memory has a massive storage capacity for object details. PNAS 105 (38), 14325–14329.

[b]Green, R.E., 1989. The persuasive properties of color. Marketing Communications, October, 50–54.

1.5.2.3.2 Audio

While audio has always played a vital role in radio and TV, the growth of videos and other online marketing methods are extending its reach. Use audio elements (voices, music, sounds) to convey tone and activate an emotional response to a message. For example, a voice can sound authoritative, friendly, or matter-of-fact, whatever the marketing campaign requires. A human voice can engender trust, sounds can provide context, and music can reach emotional centers that written words can't touch.

1.5.2.3.2.1 The brain remembers what it hears and sees. Audio alone is not as helpful as audio combined with print or images. As psychologist Jerome Bruner of New York University describes, people only remember 10% of what they hear and 20% of what they read, but they retain 80% of information they both hear and see.

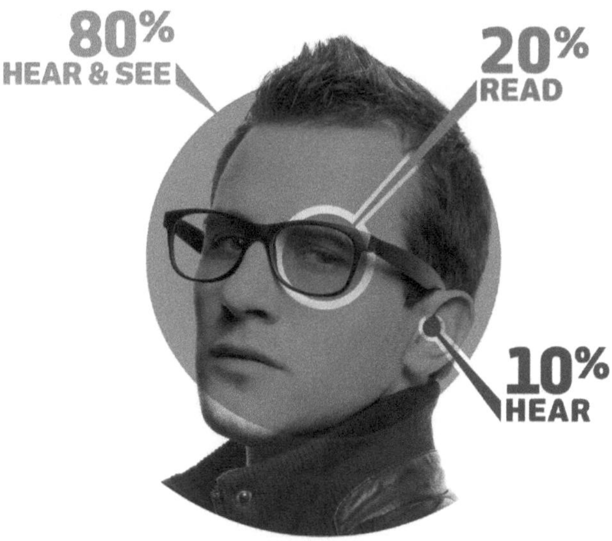

1.5.2.4 Video

Walk through an airport or sit in a lobby waiting to get your tires rotated and you're sure to see more people watching videos on more devices today than ever before. YouTube is up to 1.3 million views a minute, and Adobe released data in Apr. 2013 that shows mobile video consumption grew by 300% in a two-year span.

How can videos help explain science? Videos can accelerate scientific learning in three ways:

1. video allows you to combine your written and audio messages for greater impact,
2. video helps consumers visualize novel concepts, and
3. video allows you to break the learning sequence into chunks to ease consumers into more technical topics.

The high cost of video production. Animated videos are expensive. However, you can reuse short animation sequences to produce a sequence of videos on a similar topic that help consumers drill down into the science.

1.5.3 BREAK IT UP

Would you want to read this chapter if it weren't broken up into seven small segments? In a similar fashion, your readers will appreciate you finding ways to break your topic up into bite-sized pieces.

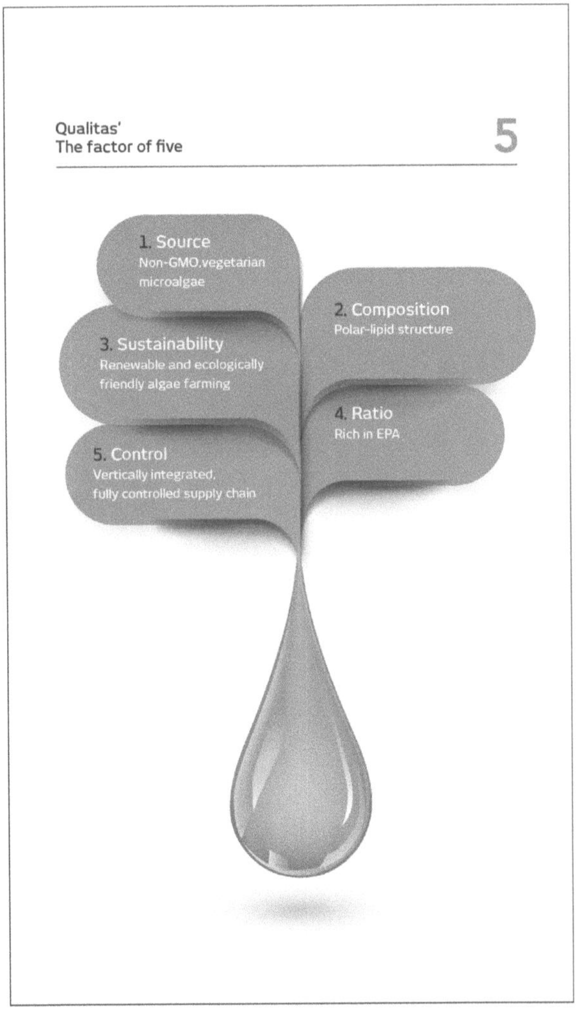

1.5.3.1 Use repetition

Remember writing your first college paper? You were probably told to: (1) tell your reader what you're going to say, (2) say it, and (3) summarize what you just said. This seemingly redundant device helps readers learn new information. It actually pushes new information through short-term memory into long-term storage through memory consolidation.

The brain stores short-term memories for about 20–30 seconds. To transfer memories from short-term memory to long-term storage, brain signals need to fire in sequence more than once. The saying "neurons that fire together, wire together" is referring to this memory consolidation process. When you repeat information (introduce, explain, summarize), you've enabled the repeated firing of neurons to form a new memory.

1.5.4 NAME THAT STRUCTURE

When organizing complex information, we start with a foundation of tightly integrated ideas, insights, and messages. Once we know what those are, we organize those elements into a structure and then name the structure to help the reader remember what the different structural elements are. The name provides the reader with a frame for grouping the following four or five or six points. It also personalizes the information for the consumer, making it easier to understand.

1.5.5 SUMMARIZE THE FINDINGS

The FDA makes it difficult for dietary supplement brands to summarize their research findings without getting into FDA hot water. Here's a quick summary of what you can and can't do on either your marketing collateral or your website:

- *You can't* summarize your study or a third-party study in consumer-friendly terms, leaving out all the disease references.
- *You can* include the complete abstract, a link to a complete abstract or a link to the entire study.

So, given these limitations, how do you summarize the research? We often write an article that includes a summary. This article does not appear on the branded website or on branded materials, but exists somewhere else that publishes scientific information for consumers, like the online version of *Psychology Today* or *WebMD's* online publication. In the summary, we remove using statistical terms (such as variation from the mean) and lengthy method analyses to make the results more interesting and easier to understand. For dietary supplement brands, the article should not include references or photos of branded products or have links to the brand website. Such a connection should be at least "two clicks" away from the article. (Note: two clicks is a suggestion, not a surefire way to avoid FDA scrutiny.)

1.5.6 REACH THEM AT THE RIGHT PLACE AND TIME

1.5.6.1 The right place (the doctor's office)

Even today's busy consumers have places where they are forced to sit still, such as the doctor's office. Brands can develop a range of marketing collateral for the doctor's office:

- *A short video that would run on a loop* in the doctor's waiting room. Most doctor waiting rooms now show the news, which offers no benefit to the doctor and provides no new health-related information for the patient.
- *A video on a specific procedure or test (such as, neurotransmitter testing) or a specific condition*, as a way of informing the patient of basic information that the practitioner would otherwise have to repeat over and over again to new patients.
- *An engaging poster or infographic* that informs the consumer about a topic.

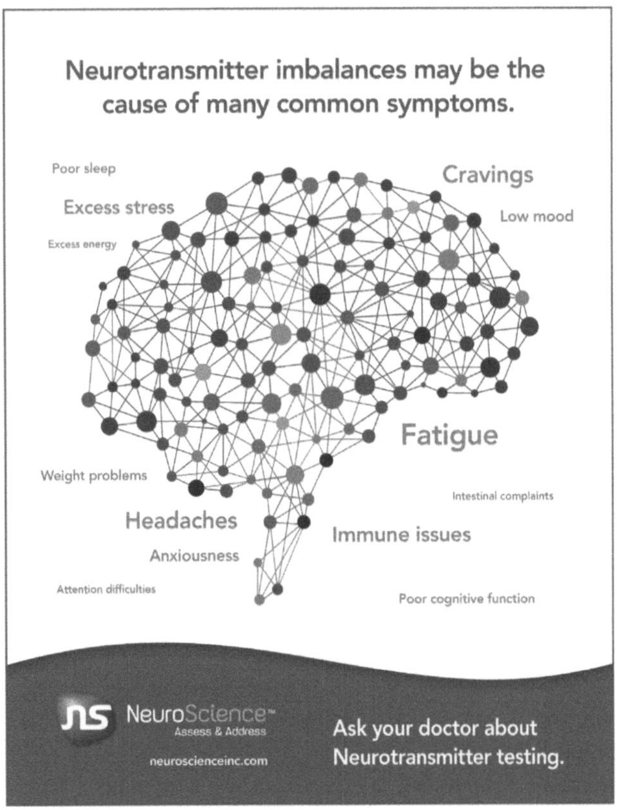

1.5.6.2 The right time (prerecorded podcasts and webinars)

Let your consumers pick the best time to view information by supplying pre-recorded podcasts and webinars. When you make this information available 24*7, the consumer is more likely to find and digest the science when they are actively seeking out information on a topic, such as following a diagnosis or after experiencing a life event.

1.6 TREND FIVE: MORE BRANDS TOUTING SUSTAINABILITY AND ECO-FRIENDLY MESSAGING

Marketers in the health products category are getting a lot of peer pressure lately to hop on the green bandwagon. After all, green is the new black. On a surface level for consumers, it has become a type of status symbol, and for manufacturers it is occasionally just a half-hearted public relations attempt. Whether the motives of consumers and manufacturers are truly altruistic, or simply it's a matter of being politically correct, the marketplace is showing an increased interest in and demand for green, sustainable products and services. No longer is this an isolated fad or trend targeting the hemp-wearing, sandal-footed, tie-dye counterculture. Finally "going green" is going mainstream. Like any new market trend, there are challenges in positioning product/service benefits in a meaningful and relevant way to consumers.

1.6.1 WHAT IS GREEN OR SUSTAINABLE?

Not only are there economic challenges (being "green" is costly for both manufacturers and consumers) to being eco-friendly, but there are barriers to positioning as well. Consumers are confused about what green means. They think that green products are too expensive. They are skeptical about whether a product or a company is truly green. And they are concerned that green products won't perform as well as conventional products. There are studies that identify both positives and negatives in marketing green products. Some say consumers perceive green products as too costly, while others insist that the added cost is insignificant when compared to the positive long-term result. A recent survey found that almost 40% of Americans prefer to buy products they believe to be "environmentally friendly." But on the flip side, a Gallup poll recently asked whether the environment should take a priority over the economy. In 2000, 70% of Americans held this to be true. But as the economy began to weaken, a Mar. 2008 poll of the same question revealed that the environment as a priority had dropped by 39%.

"Green" and even "sustainable" are fairly ambiguous terms. They are used loosely across the marketplace. Sometimes green is used purely as a PR tool. Faster than you can say "carbon footprint," a myriad of companies are pumping out reusable grocery bags with their logo and proudly announcing that they're suddenly a "green company." Nice beginning. However, try to grow a conscience if you're one of "them" and look at the concept of green marketing from a broader perspective.

There's much more to green than green. And most consumers are much more informed and skeptical than you might think. Detailing negative fallout to companies that "green wash" or "guilt wash," an Ipsos Reid study reported that 70% of Americans "strongly agree" that companies selling green products are doing so merely as a marketing ploy.

In a perfect world, "green" essentially means better for the environment. This takes into consideration the entire manufacturing process from facilities and ingredients to packaging materials used. As the term "green" is so broadly used, and at risk of becoming innocuous to the ever-vigilant consumer, there's a definite need for greater definition and regulation of terms and categories in green marketing. Could green be the next "natural," a term that has lost all consumer credibility or relevance? Time will tell. The United States government has stepped in to protect and maintain truth in advertising. Since 1992, the FTC has been issuing guidelines called the Guides for the Use of Environmental Marketing Claims, or in a more sustainable and less bureaucratic use of language, The Green Guides. These guides are intended to help companies avoid claims that may be untrue or deceptive to consumers, not unlike the development of FDA guidelines for health claims.

Aside from a consensus on the definition of green, it's important, as a company, to first ask if it makes business sense to go green. There, of course, is that nagging moral dilemma. Pesky morals aside, take a good look at the bottom line and the target audience. Do they really care? Sure, we wish everyone cared about the planet becoming a noxious landfill, but the reality is that there are people who want good, inexpensive, and reliable products, regardless of their impact on the planet. The positioning of a product as green may have no relevance to the customer being targeted. The other core consideration is whether being socially responsible is one of a company's core values. If it doesn't flow from an honest desire and commitment to help the environment, maybe it's not the best alternative.

While the category of green marketing is relatively new, the marketing challenge is nothing new. It's all about positioning, relevance, and overcoming consumer objections to get a desired result. And the same is true for the economy. We've seen this happen before and it will most likely happen again. The challenge is to differentiate the brand and position it in a relevant and compelling way, building value that will appeal to cost-conscious consumers.

Green marketing is a balancing act of short-term and long-term benefits. Depending on a company's level of commitment to green, it can have a positive or negative outcome. If you're going to be a green company, make sure you're honestly green. If you claim that your product is certified organic, but then use nonrecyclable materials or environmentally insolvent manufacturing processes that leave a hole in the ozone the size of Texas, then why bother? You are in business to make a profit, but short-term gains may be outweighed by long-term perception of your brand, assuming that you're completely on the green bandwagon. Sure, you can fool some of the people some of the time. But genuine goodwill and altruism toward people and the planet will endure longer than the facade created by less-than-genuine green companies.

1.7 TREND SIX: THE EMERGENCE OF THE ACTIVE NUTRITION CATEGORY

The sports nutrition segment has always been an active one, but has historically represented only a sliver of overall supplement sales. And while bodybuilders and elite athletes have traditionally been the target audience focus for sports supplements, these individuals actually account for only about 5% of total sports nutrition category sales. This highlights the importance of the mainstream consumer and what performance supplementation means for everyday health and active lifestyle needs. Indeed, the sports nutrition category is morphing into the active nutrition category.

Industry analyst firm Euromonitor reported that from 2008–2014, the sports nutrition market grew 8% annually to $8.8 billion, and the market continues to grow at approximately 10% annually. To leverage this market, it is critical for product manufacturers to pay attention to a number of unique and emerging consumer segments in the space, specifically:

- actively tuned-in,
- performance driven,
- gracefully maturing, and
- driven to lose.

While trends in consumer education and access to information continue to fuel mainstream sports nutrition growth broadly, each of these user segments has its own goals, drivers, and motivators for seeking out and using sports nutrition products. There is no "typical" sports nutrition consumer any longer, so category segmentation is key.

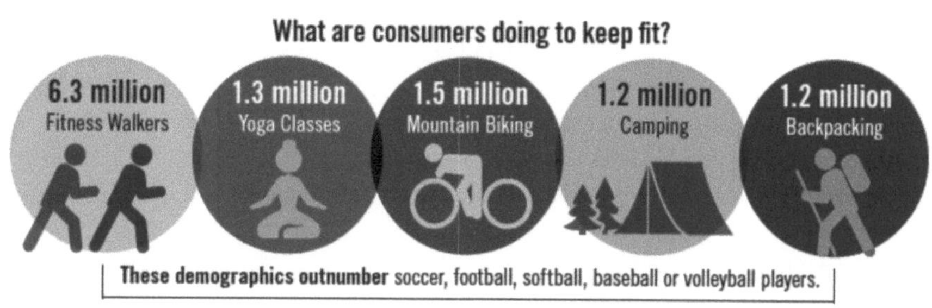

1.7.1 ACTIVELY TUNED-IN

The Actively Tuned-in segment includes informed, active consumers who strive for ways to feel good, look good, and enjoy long-term health. Fitness and exercise are a way of life. While they do seek specific foods and nutrition products to support their active lifestyles, performance improvement is *not* a key motivator. Women tend to make up a significant portion of this group. They view sports nutrition as an integrated part of their active lives, but don't believe it should require complicated or special supplementation regimens. They are on the go and in the know. They research products before purchasing, but are also looking for convenience. They expect products they buy to fit naturally into their healthy and active lifestyle, and believe foods, beverages, and snacks should not only taste good, but also should be healthy and help them reach their goals.

1.7.2 PERFORMANCE DRIVEN

The Performance Driven group tends to be young and very active. They seek sports nutrition solutions that give them a performance advantage and a winning edge. They strive for more power and energy to support strenuous exercise. They work hard and play hard. They exercise regularly and have sports nutrition regimens that include both pre- and postworkout supplementation, as well as hydration. They tend to be early adopters of new products and unique benefits. When they do find products that give them proven results and noticeable outcomes, they remain loyal to these brands. They want to feel the effects. They put the functional benefits delivered above taste and cost when it comes to sports nutrition products.

1.7.3 GRACEFULLY MATURING

The Gracefully Maturing segment is motivated to maintain endurance longer, and individuals want to have the energy to continue with daily activities as they age. Many in this category tend to think of themselves as 10–13 years younger than their actual age. This more mature group includes both men and women and its key motivators include energy, youthfulness, and injury prevention. They are interested in bone and joint health, and also want to slow the process of muscle wasting (sarcopenia), which happens naturally as we age. They look primarily to protein, as well as other products for bone and joint health. Feeling good and staying active are more important to this group than improving appearance or achieving weight loss. Also, many individuals in this segment have difficulty chewing and swallowing, and want products delivered in alternative forms that are easy to take. Many prefer RTDs and fortified beverages as opposed to big supplement tablets and hard or chewy snack bars.

1.7.4 DRIVEN TO LOSE

This segment is primarily driven by the need for weight management. This group includes both men and women, and is not constrained to a particular age group. Because they are overweight or obese, they not only want to feel better about themselves, but may also be driven by medical issues. They may have been advised by healthcare professionals to incorporate diet and lifestyle changes to help them lose weight, but are often confused by the options. They tend to have many demands on their time and energy that prevent them from eating healthily and exercising. They look for products that provide more motivation and energy, as well as thermogenic and satiety support. However, it is often difficult for them to change their routines or implement new regimens

Three marketing tips for reaching active nutrition consumers:

1. Deliver solid research and easy-to-swallow science.

   Because consumers in all segments continue to grow more and more savvy, many are looking for evidence and scientific backing. Any sports nutrition supplement should have ingredients that have been the subject of human clinical studies. This will help in building a solid foundation for making strong structure/function claims and will also ensure consumer loyalty and repeat sales. With sports nutrition supplements in particular, consumers expect to see results.
2. Emphasize convenience and alternative delivery forms.
   A large part of the population is incapable of, or resistant to swallowing pills. Additionally, many consumers are looking for convenient options, so consider alternative delivery forms including powders, liquids, RTDs, and fortified foods and bars to increase product adoption and continued use.
3. Provide clear messages, clean labels.

   The increase in mainstream consumer awareness is a major growth opportunity for companies to expand their sports nutrition offerings that cater to specific preferences and requirements, including products that are gluten-free, dairy-free, vegetarian, and non-GMO.

No doubt the active nutrition sector will continue to expand as Americans become more active based on concerns regarding lifestyle diseases such as obesity, diabetes, and heart disease.

1.8 TREND SEVEN: INNOVATION IN DOSAGE AND DELIVERY FORMATS

The retail shelves across the United States, both physical and virtual, are changing. What has historically been dominated by boxes, canisters, and bottles is morphing into something quite distinctive. And it's no accident. Consumers are changing the way they manage and consume supplemental nutrition, and smart manufacturers and suppliers are paying attention. Or at least they should be.

The move toward more innovative, convenient dosage and delivery forms for functional foods, beverages, and supplements is in full swing. And it only takes a brief visit to Whole Foods or GNC to observe it in all its glory. Single-serve tear pouches, gummies, shots, strips, liquids, rapid-dissolve tablets, sprays, gels, and chewables are invading retail shelves and consumers are rushing to purchase.

1.8.1 PILL FATIGUE—NOT JUST FOR SENIORS

When it comes to dietary supplements, I have long held the belief, and research confirms, that pill fatigue is not just a concern for aging seniors, but also for both baby boomers and millennials. The trend among all demographic groups is to take fewer "pills" and supplement their food intake with more convenient and less "intrusive" delivery forms. The stereotypical supplement junkie who takes 15–20 capsules and tablets every morning is becoming more the exception than the rule. Today's consumer is supplementing their diet with a range of nutritional solutions delivered in a myriad of ways. Specifically, through fortified foods and beverages, basic daily nutrition in the form of a multivitamin, plus occasion-specific or need-specific products dictated by their daily schedule or personal health profile.

So the consumer trend overall, in my opinion, is toward a more diverse selection of dosage forms based on needs and lifestyle, accompanied by lifestyle changes. This has important implications for both manufacturers of supplements and suppliers of raw materials.

1.8.2 WHAT ARE THE KEY CONSUMER DRIVERS?

So what is driving consumers toward these innovative delivery forms? Well, most consumers are:

- looking for convenience and flexibility,
- searching for new and innovative health solutions,
- enjoying newfound empowerment to proactively manage their own health and wellness,
- attracted to bright, shiny new things,
- increasingly receptive to the recommendations of peers, friends, and family, and
- two words: social media.

1.8.3 WHAT'S SOCIAL GOT TO DO WITH IT?

It's hard to overstate the impact that social forums have had on changing supplementation habits and practices. People talk about health on these social platforms. It's a topic that we all have in common. We think about it every day. The media harp on about it. So health awareness is high, and consumers generally don't honestly feel they are doing enough to safeguard their own health, even if they eat right and exercise. Millennials are all about maximizing and intensifying their daily life experience, and baby boomers are all about living longer and healthier until they are 90 years old. This makes for a fertile and vibrant dialogue online and new dosage and delivery forms lend themselves to sharing and exchange and recommendations.

1.8.4 TARGETING SPECIFIC CONSUMER SEGMENTS WITH DELIVERY TECHNOLOGY

Manufacturers, suppliers, and marketers are all trying to target consumers with messaging and presentation elements that intrigue, compel, and facilitate engagement. One often-overlooked way to do this is to deliver the formula or ingredient in an innovative delivery form. Senior adults can effectively be reached with powdered mix-in or sprinkle-on ingredients, liquids, and sublinguals. Millennials are partial to gummies, shots, and chewables. Gen X is all about gels and strong flavor delivery. Smart marketers are taking advantage of these delivery innovations to bring excitement, energy, and word-of-mouth benefits to their products.

1.8.5 SO WHAT?

What I think we are all learning about the new emerging consumer, or Consumer 2.0 as I like to call them, is that they are smarter, more skeptical, and much more selective than consumers of old. They have access to multiple venues to shop, compare, and purchase. They basically hold all the cards. Yet for all their power and control over access, they remain confused about supplements generally and rely heavily on user-generated content (read: peer influence) to make buying decisions. They are also shunning traditional rituals like the "three meals per day" paradigm and moving to new eating patterns and habits around snacking. What does that have to do with supplements, you ask? Well, successful supplements of the future will have to compete for the limited attention of a busy and convenience-driven consumer who wants your potential health solutions to be available on demand and on their terms. So delivering the product/ingredients in a dosage and a form that are interesting, flexible, and conform to their lifestyle is vital and quickly becoming a cost of entry into the category.

1.9 TREND EIGHT: HOW MILLENNIALS ARE CHANGING THE CONVERSATION AND THE MARKETPLACE

1.9.1 SO WHAT'S ALL THE FUSS ABOUT MILLENNIALS?

It seems like everyone is talking about the millennial generation. Why is that? Most likely because these 80+ million consumers are consistently breaking the rules established by their baby boomer parents and siblings to set their own course in almost every aspect of their lives, from the workplace, to how they shop, to how they buy, to how they consume. And of particular importance to many reading this chapter, their attitudes about health and wellness are no exception.

1.9.2 MILLENNIALS ARE THE NEXT CONSUMER WAVE FOR HEALTHY LIFESTYLE PRODUCTS

For the past decade or so, marketing healthy foods and beverages has primarily been focused on the baby boomers. And they are a dominant demographic force to be sure. But not enough attention is paid, in my opinion, to the emerging, up-and-coming consumers for the healthy lifestyle products many of us manufacture and sell.

Millennials will soon be the single largest global demographic segment, and these 18–33 year olds are indeed out to change the world. Make no mistake about that.

But for today's functional food and beverage marketer who has long targeted the aforementioned baby boomer, making the transition to successfully reach millennials is not as easy as it seems. Marketing healthy products to a boomer and marketing healthy products to a millennial couldn't be more different tasks; yet our inclination as marketers is to pitch them the same way and approach them with similar strategies and tactics. And that's just plain wrong.

1.9.3 MILLENNIALS WANT AND DEMAND BOTH CHOICE AND ENGAGEMENT

As a base premise, it's important to understand that millennials seek and embrace choice. In fact, they take pride in the choices available to them and like to have a wide range of product options they can select from. They want to buy products that fit, and even better enhance, their chosen lifestyle. So for manufacturers, their products must demonstrate both relevance and value to this demographic in terms of what they are about, and what they want to accomplish. They will not build their daily routine around a product. Manufacturers must earn that spot and actively work to maintain it.

Another aspect of my premise is that millennials want to connect with brands on both a rational and an emotional level. And unless brands foster that type of relationship, brand loyalty will never flourish. The kind of give and take relationship that all marketers seek to create between brand and consumer is founded on trust, and that trust is built by allowing millennials to actively participate in and interact with the brand. They want to customize what you offer to their own particular needs and lifestyle. They may be open to listening to your messaging or viewing your content, but it will be on their own terms and on their own timeline. If you as a brand are not willing to open your doors to them, they will most likely not become loyal customers. It's really that simple. It is not easy for established brands, seeking to cultivate millennial customers, to readily give them that type of unfettered access. I get that. However, we must seek to understand that the rules of engagement have shifted in favor of the consumer, and millennials relish that control and transparency. They are not about to give it up. So, to be frank, if you want to play, you go by their playbook.

1.9.4 MILLENNIALS HAVE A PECULIAR RELATIONSHIP WITH BRANDS

To put it succinctly, millennials speak brand. They have been inundated with brands and messaging since they were children. They get brands. I like to say that they "see behind the marketing curtain." They also have a healthy mistrust of brands generally, and are looking for transparency and authenticity in the brands they purchase and tell their friends about. They are somewhat big brand averse. There are exceptions of course, and I'm talking about Apple and Patagonia, for example. But most often they prefer smaller niche brands that they can adopt, adapt, and customize to meet their specific needs. So if you are not willing to pay that price, then millennials may not be your ideal target customer.

1.9.5 MILLENNIALS LOOK AT HEALTH AND WELLNESS DIFFERENTLY

They are not really all that worried about aging, or prevention or health preservation. They are more focused on living their current life to the fullest and maintaining good health to run as fast and as hard as possible NOW, not 40 years from now. Their path through life is much less sequential and linear than more traditional boomers. They move along a more personalized and random timeline of events including marriage, children, travel, schooling, and career. They take a similar approach to achieving health and wellness, by mixing and matching a variety of strategies that work for them. They are very comfortable mixing improved diet, exercise, supplements, pharmaceutical drugs, functional beverages, and over-the-counter medicines to meet their individual daily health and performance needs. Natural as a marketing term or descriptor is meaningless to most of them. They prefer "authentic" to natural. It speaks to what they are looking for in the brands they trust. But before you rush to hang out a "We're Authentic" sign, make sure that it honestly comes from inside the company and can be validated. Millennials can smell hypocrisy a mile away, and they will hunt you down if you try to pull one over on them. It won't be pretty.

1.9.6 MILLENNIALS VALUE USER-GENERATED CONTENT (UGC)

In terms of pitching a brand to millennials, what matters most to them is not advertising or overt sales promotion, but rather what their friends, peers, and families think about a particular product or service. They would prefer to see or hear testimonials or read reviews from peers or siblings or close friends. They don't want to be sold. They want to know how what you have to offer can improve the way they live their life. And they use social media outlets, websites, and smart phones to access that user-generated content. It holds more weight than any other single source. If you are not incorporating user-generated content into your marketing outreach, you are passing by a significant opportunity to open a powerful channel of communication. And don't forget that the most important rule about your brand content is that it needs to be easily shared with others. So video is ideal, followed by still graphics or a combination of words and graphics. When someone shares your content with a friend, it's more about their friendship with that person than about your brand content, but you have been allowed to participate in that exchange, which is where the value comes in for the brand.

1.9.7 WHY DOES SOCIAL MEDIA MATTER?

Social media plays a critical role in the exchange of UGC. Research results show that 65% of millennials are disconnected from their devices for less than 1 hour per day. And much of that time is spent online with peers and friends and family in social settings. So if you are looking for millennials, you will inevitably find them on social media. But tackling social media is easier said than done. Clients approach me saying, "Hey, I think I need a Facebook page." Or "Hey I need a blog." I smile to myself because it is a common approach to begin with the tactics rather than backing up and asking, "What can social media do for my brand, and which platforms would be best for me and what I want to accomplish?" So my point is that it is always wise to plan social media from a strategic perspective before jumping to specific tactics. You will get better results that way. Then once you decide which platforms make sense, the basic steps are as follows:

- develop platform strategy,
- establish social presence and map out content,
- produce and publish content,
- measure ROI, and
- analyze and report.

And if anyone tells you that social media is difficult to track in a meaningful and accurate way, don't believe it. It can and has to be done.

1.9.8 STORYTELLING IS KEY TO REACHING AND PERSUADING MILLENNIALS

No doubt you have heard this before, but communicating the product value proposition in story form is the most compelling way to appeal to a millennial consumer. But remember that they want to know the WHOLE story, not just what you choose to tell them. They want to know how your brand came to be, what you stand for, features and benefits, and also what happens to the package after it is used up. They want to know what your company values and supports. They want to know the company behind the product. Prime communication tools for this type of engaging storytelling include both video and infographics. This demographic is plugged in at all times and wants information about your brand that they can easily access, assimilate, and share with peers.

1.9.9 A WORD OF CAUTION

So on a final note, it is important to recognize that although there are key traits or characteristics which most millennials have in common, they are still a diverse group of individuals, not some monolithic block. A number of hypotheses have been put forward recently regarding a more detailed segmentation of this group, but there is really nothing definitive at this point. But like any population, individuals fall on all sides of the median or norm. Nonetheless, these tenets that I have covered should assist you in better understanding the general orientation, interests, and attitudes of this burgeoning population.

CHAPTER

Market analysis and concept development of functional foods

2

Joe Bogue, Orla Collins, Amy Jane Troy
Department of Food Business and Development, University College Cork, Cork, Ireland

2.1 INTRODUCTION

There have been many changes in the consumer package goods (CPG) market over the last number of years as consumers seek novel health benefits from food and beverage products. As the functional food industry continues to experience new innovations and sales growth, the development of new functional foods and beverages with added ingredients remains a continued focus for global food firms. Benefits that are linked to energy, aging, satiety, and fatigue represent lucrative opportunities for food and beverage firms. However, firms often struggle to gain product success due to various factors such as consumer resistance, legislative barriers, and technological hurdles.

Health and wellness has been one of the most important drivers of innovation in food and beverage markets. However, the development of new functional foods presents many marketing and technological challenges to product developers with high reported failure rates in this category. Reasons for failure in the functional foods market include: too many benefits from a single brand, benefits that are often not relevant to the consumer, relying on the selling power of the ingredient rather than the benefit, and using nonrelevant carriers (Mellentin, 2009). In many cases, firms have not adequately addressed consumer needs or have struggled to market science and technology to consumers. This chapter will look at market trends and the marketing and technological challenges in the new product development (NPD) process. A case study will then focus on using a market-oriented technique, user-centered design (UCD), to develop new product concepts for the aging consumer segment. Aging consumers provide much greater challenges when using user-design methodologies than traditional groups and need to be fully integrated into the design process (Newell et al., 2007).

Developing New Functional Food and Nutraceutical Products. http://dx.doi.org/10.1016/B978-0-12-802780-6.00002-X

2.2 FUNCTIONAL FOODS: KEY MARKETS, CONSUMER TRENDS, AND NEW PRODUCT TRENDS

The functional foods market offers new product opportunities for food firms that develop new products that meet consumer expectations, through understanding consumer choice motives and the trade-offs consumers make when choosing these foods. The functional foods market experienced a global increase in value terms of 26.7% from 2009 to 2013 (Leatherhead Food Research, 2014). With an annual average growth rate of about 8.5%, the global market for functional foods is expected to exceed US$ 305.4 billion by 2020. The trends driving growth in the functional food market include: consumers interested in the prevention of health issues, the increasing cost of healthcare, the steady increase in life expectancy, and the desire of the aging for improved quality of life in their later years (Bigliardi and Galati, 2013).

The key functional foods markets globally are the United States and Japan, followed by Asia Pacific and European markets. In 2013, China had the highest expenditure on health and wellness retail products, followed by Brazil, the United States, Russia, and Mexico. Global food firms account for more than 75% of the functional foods market and include key players such as: Coca Cola Co., Dean Foods, General Mills Inc., Kellogg Co., Kraft Foods, Nestle S.A., and Pepsi Co. Key growth functional foods categories include health snacks and breakfast products that include cereals and grains (Industry Arc, 2014).

With sales of US$ 43.9 billion in 2012, a 6.9% increase from 2011, the United States dominates the functional foods market, and is expected to reach a compound annual growth rate (CAGR) of 8.7% globally by 2020 (Euromonitor, 2014; Nutrition Business Journal, 2013). This growth is driven by the rising popularity of functional foods, with 6 in 10 US adults consuming "at least occasionally" one of these products. The most common consumed items were yogurts for digestive health and cereals for heart health. Other key categories include: cholesterol-lowering margarine and butters, important for heart health; shakes and bars to reduce hunger or increase energy; and health and energy beverages (MSI, 2012). Europe accounts for an additional 20.2% of the global market value, with European regional sales dominated by the United Kingdom, France, and Germany (Nunes, 2014).

The Chinese functional foods market was valued at US$ 24.6 billion in 2012, and is expected to continue growing, but by a reduced percentage by 2017. Among the reasons for this increased demand for functional products are increased consumer purchasing power and higher concerns about health and food safety. Increased interest from Chinese consumers in dairy products is driving purchases of functional products, coupled to a lesser extent with incremental product improvements and innovations (Agriculture and Agri-food Canada, 2014).

Chinese consumers are very concerned with health, and believe that functional foods are a great way to take care of the body. The main consumer segments that purchase functional foods are married couples and households with children, with variations in product popularity by region. Products containing "friendly" bacteria are more common in Beijing, traditional cooling herbs in Guangzhou, and tonic herbs in Guangzhou and Chengdu (Agriculture and Agri-food Canada, 2014). From a strategic perspective, Chinese firms want to launch more premium products as consumers become more sophisticated in their purchasing habits. This will increase purchases of products offering either very specific or multiple health benefits, as well as items that incorporate ingredients commonly found in traditional Chinese medicine.

The Brazilian market is also an important functional foods market, worth US$ 8.7 billion in 2014 and expected to increase by 12% by 2019 (Euromonitor, 2015; Martins, 2015). The success of this market can be attributed to the fact that consumers want to live healthier lives, added to the increase in purchasing power by the middle class in Brazil, that now want to buy premium products. Most innovations are expected to take place in the dairy and bakery categories. Among the new products launched in Brazil in 2015 was Grano Brasilis Panqueca, which contained linseed, a functional ingredient that claimed to prevent cardiac and immune diseases, with antiinflammatory and antioxidant properties.

Euromonitor (2014) found that consumers were more concerned with health, especially with general wellbeing, weight management, digestive health, energy boosting, and endurance; and that the fastest-growing health products were related to: energy boosting, food intolerance, general wellbeing, digestive health, and beauty. For consumers' healthfulness is now a more complete concept, including attributes such as "fresh," "real," "avoidance of certain substances," "inclusion of positives," "high quality," "close to the farm," and even "ethical practices" (Hartman Group, 2013).

Consumers' interest in functional foods has increased and according to an International Food Information Council Survey, almost 9 out of 10 American consumers are interested in learning more about functional foods and 89% of the interviewed consumers could name at least one functional food and its associated health benefit. The consumer groups that benefit most from functional foods include: infants, toddlers, followed by children from 4 to 12 years old, and aging consumers (those 60 or older) (IFIC, 2013a). Functional foods consumption is linked to consumers' knowledge and the perceived benefits and factors that influence purchase include: cost, taste, and the product positioning platform used to market a functional food such as foods that are perceived as "pure," "basic," and "natural" (IFIC, 2013b).

Datamonitor (2014) found that the majority of European consumers believed in positive nutrition, with 56% consuming food and beverages to improve health. A total of 19% of European consumers indicated they consumed functional foods at least a few times a week while 16% consumed functional beverages a number of times during the week (Datamonitor, 2014). Functional beverages were the fastest-growing segment of the functional foods market with a CAGR of 9% in 2013. Dairy was also a major volume growth area with growing penetration. It is expected to account for 40% of overall consumer product demand by 2025 (D'Costa, 2014).

2.3 MARKETING AND TECHNOLOGICAL CHALLENGES FOR THE CONCEPTUALIZATION OF NEW PRODUCTS

Although consumer interest in functional foods is increasing, the development and marketing of functional foods presents food firms with a range of key challenges, not only from a technological perspective but also in the marketing of science and technology to consumers. The high reported failure rates (70–90%) suggest that many new functional foods have not met with acceptance or consumers have not fully understood the product concept on offer to them.

One of the key NPD questions to ask is: how can firms develop more market-oriented functional foods that gain consumer acceptance? For new product personnel an important part of developing successful functional foods is optimizing the intrinsic (sensory) and extrinsic (marketing) attributes for specific consumer segments, such as the aging consumer segment. This illustrates the multidisciplinary knowledge-intensive process that characterizes NPD and the importance of integrating the consumer with the early stages of the NPD process to develop more market-oriented products.

2.3.1 NPD TRENDS AND CHALLENGES WITHIN THE FUNCTIONAL FOODS INDUSTRY

A drive toward sustainable living and prophylactic medication has made health and wellness food and drinks move more into mainstream markets. Consumer trends play an important role in guiding product development in this market. Euromonitor (2014) outlined the top five functional food and drink trends in 2014: natural energy boosting, enhanced natural protein, power of digestion, rise of greater prevention, and deficiency alert. For European consumers the top health concern was tiredness/fatigue (41%), followed by stress (39%), insomnia (27%), general sight degeneration (23%), overwork (23%), high/low blood pressure (23%), nearsightedness (23%), snoring (23%), depression (21%), and restlessness (20%) (Datamonitor, 2014; D'Costa, 2014). Functional foods targeted at the categories outlined earlier, with their bioactive constituents, provide a means of protecting consumers from certain chronic diseases (Vella et al., 2014). Nutrition and health information, and the source of this information, is an important factor in influencing acceptance of functional foods by communicating the health benefits in a clear and consistent manner that targets the specific needs of consumers. Unlike the sensory characteristics of a food product, the health benefits of functional foods cannot be seen directly by consumers, hence information pertaining to health benefits and the ways in which this information is communicated can influence perceptions of functional foods (Vella et al., 2014).

An understanding of consumer attitudes toward functional foods and natural health products, including factors that can act as a barrier to consumer acceptance, may be found by generating consumer insights at the earlier stages of the

NPD process (Malla et al., 2007). The long-term marketplace success of innovative functional foods is dependent on consumer acceptance, and attitudes and perceptions related to these products (Vella et al., 2014). A market-oriented approach to NPD significantly helps to ascertain the feasibility and level of market acceptance of new product concepts, define the target groups, and identify the optimal extrinsic and intrinsic attributes driving consumers' preferences and acceptance of innovative foods and beverages (Jakobsen et al., 2014; Bogue et al., 2005; van Kleef et al., 2002).

The advances of the internet, including the role of consumers in disseminating information on social media, offer new possibilities for involving consumers in the NPD process (Prahalad and Ramaswamy, 2004). The role of the consumer has changed from isolated to connected, from unaware to connected, from passive to active. Food and beverage firms need to consider customizing their NPD practices to take into consideration the changing market environment where increasing health awareness, improved technology, and concern among consumers are raising a number of challenges facing product developers (Khan et al., 2014). This is particularly true in the context of developing new functional foods and beverages that gain consumer acceptance.

2.3.2 NPD PROCESS FOR FUNCTIONAL FOODS AND BEVERAGES

The central goal of the NPD process has been identified as the creation of products that create superior value for consumers (Slater and Narver, 2000). A consumer-led NPD process consists of four stages: (1) opportunity identification, (2) product design and development, (3) testing, and (4) introduction and launch (Costa and Jongen, 2006; van Kleef et al., 2005; Urban and Hauser, 1993). Innovative food products account for less than 2% of all NPD launches, and of this 2%, it is likely that 75% would experience product failure (Khan et al., 2014, Hardy, 2010).

Central to the development of successful functional food and beverages is the management of the front-end phase, which crucially determines the eventual fate of a new product or service idea (Akbar and Tzokas, 2013). The early stages of the NPD process are called the "fuzzy" front-end stages as the direction that the product development will take is not very clear at this point. The awareness that decisions made early in the NPD process can have a significant impact on a project in terms of energy, cost, and sustainability has resulted in the need to focus more attention on the early concept development stages (Chandrasegaran et al., 2013). A more market-oriented process integrates the consumer as codesigner into the early conceptualization stages of the NPD process. A variety of multinational food firms, including Danone, Unilever, and Procter and Gamble, has invested significant financial and time resources into the earlier NPD stages when developing new products for the health and wellness market. In doing so, they have also invested significant resources into consumer insight research that highlights consumers' current and unmet needs (Datamonitor, 2011; Bleiel, 2010; Danone, 2008).

There are four categories of success factors for the front end of the NPD process: product strategy, product definition, project definition, and organizational roles, all of which focus on the identification of potential problems (Khurana and Rosenthal, 1998). In addition, strong functional interdependencies during the early phase of NPD, between marketing and research and development, are more likely to result in more successful project performance (Kohn, 2006). Consumer research during the opportunity identification stage provides key insights into how consumers' needs are shaped and how perceptions of product concepts are formed (van Kleef et al., 2005). In market-oriented organizations, consumers are viewed as codesigners in the NPD process as they contribute to new food product design and can be integrated into the NPD process at the predevelopment stages of concept ideation, concept screening, and optimization.

In the case study in the next section a market-oriented technique, UCD, is used to identify new functional food concepts with aging consumers at the early stages of the NPD process. The technique involved both marketing and technical personnel in a multifunctional NPD team and identified concepts that could then be further developed in targeted focus groups. Following the focus groups, the information generated can be used in conjoint analysis studies. Conjoint analysis is a multivariate technique that models purchase decision-making processes through an analysis of consumer trade-offs among hypothetical multiattribute products (Green and Srinivasan, 1978). Semiotics can then be used to develop the design elements of the packaging, such as colors, logos, and health messaging for a new functional food concept.

2.3.3 DEVELOPING FOODS FOR AGING CONSUMERS

Functional foods and beverages are becoming increasingly important to the aging cohort as they seek foods that may benefit their health long-term and prevent the onset of noncommunicable diseases (Betoret et al., 2011). However, developing food and beverages for aging consumers is a major challenge for food firms from a marketing and technical perspective. Food firms have been slow to develop new foods for aging consumers, using existing brands, so as not to "age" their brands and also as there is a perception that aging consumers are brand loyal. From a technical perspective, functional decline with age alters aging consumers' perceptions of food and can lead to sensory changes in taste, smell, and olfactory senses. Equally declining oral health and changes in gastrointestinal functions can limit aging consumers' food intake (Mak and Caldeira, 2014; Sura et al., 2012).

Marketing strategists suggest that firms targeting aging consumers should focus on the health benefits and not on the "aging" benefits of the products (Collins and Bogue, 2015a). In addition, marketing activities targeted directly at the aging segment need to address age-related changes. The age-related changes affecting older people are primarily concerned with physical, social, cultural, economic, and demographic changes. The design and development of marketing messages which focus on age-related changes, without conveying a sense of "old" age or deterioration, are more likely to increase a product's likelihood of success (Collins and Bogue, 2015b).

2.4 CONCEPTUALIZATION OF NEW PRODUCTS: CASE STUDY ON CONCEPT DEVELOPMENT OF FUNCTIONAL FOODS FOR AGING CONSUMERS

2.4.1 INTRODUCTION TO THE CASE STUDY

This case study illustrates the use of a market-oriented NPD tool, a UCD interaction, for the conceptualization of new functional food concepts for the aging consumer segment. The objective of the UCD process was to identify new functional food concepts, using aging consumers as codesigners, during the early stages of the NPD process. These concepts could then be further developed by the NPD team. This UCD method proved most suited to aging participants, as previous design process studies found that aging consumers need to be seen as part of the development team, as codesigners, rather than just subjects of experimentation (Newell, 2011; Newell et al., 2007). In order to facilitate a seminonverbal approach and capture consumer tacit knowledge, this study used a drag and drop computer design interface, with moderated group discussions, throughout the idea generation phase of the NPD process.

2.4.2 USER-CENTERED DESIGN METHODOLOGY

A group setting was chosen for the UCD interaction as this provided rational decision-making elements to the product design process, mainly due to the need for explanation and negotiation of product attributes, and the overall design of the final product concept with other group participants. The UCD interactions provided the participants with user toolkits for innovation in the form of a drag and drop product design program (von Hippel, 2001). A sample of 32 Irish consumers, between the ages of 50 and 75 years, was chosen using a purposive sampling technique (Polit and Beck, 2004). At the beginning of the interactions, participants were asked to fill out a brief demographic questionnaire. Each of the four interactions consisted of eight participants and lasted for 90 min and was supported by the moderator. All interactions were audio-recorded to gather relevant data by the marketing and technical team.

Each interaction was divided into two stages. The first stage represented the product design phase of the NPD process. During this first stage, each interaction broke into two subgroups of four. The groups remained separate for the duration of this stage and were unaware of each other's choices. The groups were supplied with a laptop which included a preloaded drag and drop software program. The software enabled participants to drag and drop their chosen product attribute levels onto the screen which provided a visual representation of their finalized product concept. The number of times the attribute levels could be dragged and dropped was limitless which allowed for participants to view hypothetical product concepts and also to facilitate change of mind. The product attributes used in the UCD interactions were identified from an attribute elicitation process which was completed during prior in-depth interviews with key stakeholders (Bech-Larsen and Nielsen, 1999). The corresponding attribute levels were then generated from the marketing and technical NPD team and from relevant literature (Table 2.1).

Table 2.1 Attributes and Attribute Levels

Attributes	Attribute Levels
Health concerns	Bone and muscle
	Brain and cognitive
	Heart and cardiovascular
	Gut and digestive
Food and beverage carriers	Juice
	Yogurt
	Soup
	Home meal replacement
Food and beverage packaging	Tetrapak cartons
	Plastic bottle
	Glass pot
	Foil tray
Marketing messages	Active lifestyle
	Ingredient information
	Medical endorsements
	Food imagery
Health claims	Fortified with calcium, vitamin D, and protein for: *"stronger and improved bone and muscle maintenance"*
	Fortified with whey proteins for: *"improved maintenance of lean muscle mass"*
	Fortified with probiotics and prebiotics for: *"improved digestive health"*
	Fortified with omega 3 and 6 for: *"maintenance of normal brain function"*
	Fortified with plant sterols and stanols for: *"maintenance of cardiovascular health"*

The participants were asked to create a functional food concept using the product attribute levels provided. The only request was that the groups were to assemble the concept in the following order: health concern, food and beverage carrier, packaging, marketing message, and health claim. Each group of four was asked to choose only one attribute level from each of the attributes and choose design concepts that were both appealing and suitable for their lifestyles. A short product presentation by a participant from the interaction on the finalized product concept allowed for verbal feedback on the group's decision-making process.

The second stage of the process included a discussion between all eight participants of each group. The discussion focused around the two different concepts developed in each interaction and the attribute levels selected by the participants. This discussion revealed valuable information related to the motives behind the selection of particular product attribute levels in addition to feedback on why some were more appealing to particular participants. During all stages of the process, members of the multifunctional design team observed the participant interactions to gather "voice of the consumer" information and really understand what they meant when discussing product attributes and the trade-offs.

2.4.3 DATA ANALYSIS

The qualitative data generated from the UCD interactions was transcribed verbatim from the audio recordings and analyzed using thematic analysis using the software package NVivo 10 to identify common themes across the data, rather than solely focusing on individual group transcripts (Braun and Clarke, 2006). The NVivo 10 software package supported the organization of the qualitative data (QSR International, 2013).

2.4.4 CONCEPTUALIZATION THROUGH USER-CENTERED DESIGN

In this section, the results are presented from one UCD interaction and its resultant concept and provide an insight into the UCD process. Overall, the entire UCD study produced a total of eight product concepts designed by eight groups of aging consumers. The results below only relate to part of the overall interactions and represent specifically Product Concept 1, which was designed by Group 1, which consisted of two female and two male participants aged between 55 and 72 years: two participants were employed full-time, one was retired, and one was a student. In addition, in terms of their health status one participant felt they had a very good health status; two believed their health status to be good; and one believed their health status to be moderate. The group's decision-making process is described throughout the subsequent sections and the composite Product Concept 1 is presented in the final section.

2.4.4.1 Attribute 1: product concept health concerns

The health concern chosen by Group 1 was brain and cognitive health. Many participants were interested in brain and cognitive health. From the eight overall UCD interactions, brain and cognitive health emerged as the most selected health concern with heart and cardiovascular health the second most important health concern. From an NPD perspective, the interaction illustrated the potential influence of gender on product design decisions. In particular, when analyzing the discussion by gender, female participants appeared to be concerned about bone, muscle, and joint health and provided strong arguments for those health concerns. However, male participants were more dismissive of those health concerns and felt that they were more relevant to females. Analysis of the eight interactions identified heart and cardiovascular health as a more male-oriented health concern. A large proportion of male participants linked heart and cardiovascular concerns to medical family histories. From an NPD perspective, this raises the issue for food firms of developing functional products targeted at specific genders. The group discussion also revealed that aging participants were very aware of risk factors linked to noncommunicable diseases.

The health function of a potential functional food proved appealing only when it was relevant to the health needs of a particular individual. For example, one female alluded to the menopause and how certain life changes could impact upon food choices. Participants actively chose a health concern based on their own health needs and preexisting medical conditions. Overall, females were more forthcoming with personal information regarding their health status. This was an important observation as high levels of consumer involvement at the early stages of the NPD process can lead to the development of more market-oriented products that ultimately gain consumer acceptance. Female participants tended to think of an overall wellbeing feeling, rather than solely focusing on one health concern:

As females, after the menopause we need to think about osteoporosis so I believe that bone, muscle, and joint health is more relevant especially for us females. (G1: F: 55–72).
Certainly the other concerns, like muscle and bone, are very important, but we felt if you don't have your brain working properly then the rest of your body is in trouble. (G1: F: 55–72).
Anything that would contribute to the brain we think is very important. (G1: M: 55–72).

From a group dynamic perspective, males tended to be more reserved and gave precise and often less detailed responses. This was an interesting observation for the effectiveness of consumer-oriented studies and the importance of getting a gender balance across the interactions to drive the NPD process.

Heart and cardiovascular health featured heavily in the consumer decision-making process. However, for both genders, heart and cardiovascular health remained a close second to brain and cognitive health. One male participant interestingly described the heart as the "motor" of the body and if this wasn't working then "you're not alive." A female participant also noted the importance of heart health:

I felt heart health was very important, there are good sciences there now to produce good food products. It is important to ensure the products we eat would help to keep our heart healthy, especially when we are in retirement and want to remain active. (G1: F: 55–72).

The group felt that there was an abundance of heart health products in the marketplace but there was a lack of products using a brain health platform. There was a clear demand for a future NPD focus from food firms on developing products that target brain and cognitive health:

We went with cognitive health because there are many products out there at the moment that pertain to affect the heart, so we said cognitive health is a little different and there isn't as much on cognitive health as heart health. (G1: M: 55–72).

2.4.4.2 Attribute 2: food and beverage carriers

The food and beverage carrier chosen by Group 1 was a juice-based beverage. Declining oral health and changes in gastrointestinal functions can often limit aging consumers' food intake. In this interaction there was a preference for liquid-based food and beverage carriers. Drinkable dairy yogurts and soups were frequently chosen throughout the eight interactions. The attributes driving the liquid-based preferences were convenience and ease of consumption. The convenience aspect of the juice carrier tended to appeal more to male participants where one described the juice carrier as: "a drink so you can put it in your pocket and drink it anywhere." Interaction results revealed that the integration of convenience design elements into foods for aging consumers may lead to higher levels of acceptance. Overall, the juice carrier was chosen mainly for reasons of ease of consumption:

> *We decided we would go for a fruit juice as we think it would be very easy to consume. (G1: F: 55–72).*

These ease of consumption elements were also linked to drinkable dairy beverages throughout other interactions. Juice was described as being an "easy" product that could be consumed at any time throughout the year. Soup was also considered to be an "easy" product but soup was seen as more seasonal. Importantly, juice was viewed as a more versatile carrier than soup:

> *Juice could be taken all year, but soups really are seasonal. The juice product can be taken at any time of the day and all throughout the year. It fits any season. (G1: F: 55–72).*

Food storage and food preparation requirements emerged as key design considerations. Juice was considered to be a convenient product which was easy to store in the fridge and did not require further preparation or cooking. Additionally, participants described juice as being an easy to digest beverage. Similarly, both soup and dairy yogurt carriers were described as easy to digest foods throughout the eight interactions. In terms of the juice carrier, one female participant expressed her concerns regarding the acidity levels of fruit juices. The discussions revealed how participants linked varied acidity levels to a variety of different fruit juice brands. The fruit juice brand had a strong influence on the purchase decision across all UCD interactions. Participants associated certain brands with perceived higher quality cues. Brand proved to be a key purchase cue and participants tended to be brand loyal, which may impact on the commercial success of new brands targeted at this segment.

2.4.4.3 Attribute 3: product concept packaging

For Group 1, tetrapak cartons were chosen as the packaging format. Chronic aging conditions like arthritis or reduced motor skills require the integration of compensatory strategies into the product design process. Product developers targeting aging consumers need to take into consideration the physical capabilities of this group when making packaging design decisions. The integration of user-friendly design elements ensures food products can be easily handled, prepared, and consumed and may lead to great consumer acceptance of new functional foods for aging consumers.

Tetrapak cartons were chosen for the juice packaging format mainly for the "ease of opening" and "resealable" attributes. Overall product acceptability levels were strongly influenced by packaging formats. The tetrapak cartons were equally preferred by all group members and also proved popular as the pack carrier throughout other interactions. The twist-off cap appealed to the group because they were already familiar with it and also liked the option to reseal it and store it for another consumption occasion. Importantly, packaging familiarity proved to be an essential design feature for products targeted at aging consumers. During the interactions, a food packaging design expert was invited to observe the participant discussions. The expert observed how the physical capabilities of aging consumers influenced their choice of packaging. To demonstrate, one participant described experiences with alternative packaging:

> *The reason I went for the tetrapak cartons is because I find those push up bottles a bit fiddly and often difficult to open. (G1: F: 55–72).*
> *We are used to the tetrapak carton, plus if you don't use it all, you can reseal it and put it back in the fridge. (G1: F: 55–72).*

Female participants most welcomed the resealability option as they were more aware of portion control. From experience, the food packaging expert noted the difficulty associated with the development of optimal portion sizes for aging consumers. As appetite levels decrease with age, resealable packaging is an important design feature. In addition, the tetrapak caps facilitated ease of pouring and the recyclable feature of the packaging was appealing for environmental reasons. Male and female participants were aware of environmental issues and preferred recyclable packaging:

> *These caps enable you to pour, and people can pour easily and the packaging was recyclable as well, which is important. (G1: M: 55–72).*

2.4.4.4 Attribute 4: product concept marketing messages

This Group chose food imagery as the marketing message for the functional food concept and it proved to be the most contentious part of the interaction. Group participants were generally skeptical about marketing and the marketing and health messages on foods and beverages, particularly those in the health and wellness category. A female participant had "bought into marketing and advertising when she was younger" but as she became older she "wised up." Medically endorsed messages were deemed to be "pretty lightweight and poorly referenced." The food imagery message was chosen by participants as it was "attractive" and appeared "inviting." In addition, the vibrant and bright colors on the food imagery grabbed the attention of participants and were deemed to be self-explanatory and easy to see:

> *The image looked very inviting and the colours were bright and vibrant, this image would definitely draw your attention and was attractive for us. (G1: F: 55–72).*

The Group also liked how the message communicated the notion of foods available for maintaining "normal brain function." Other marketing messages were considered, but overall, the food imagery message was chosen as the group deemed it to be a very "precise and clear" message. An important point raised was the need for food firms to use a bigger font for ingredient declarations and marketing messages for aging consumers. Even though the group was required to choose a marketing message, they still remained skeptical about marketing, and took marketing messages and promotional materials "with a grain of salt." From a strategic marketing perspective, messages targeting health concerns that convey simplistic messages may increase levels of first purchase or repeat purchase:

> *For the marketing message, there were two we were interested in, but we went for this one because it came across very clear and precise in a message and more importantly we didn't need our glasses to read it. (G1: F: 55–72).*

2.4.4.5 Attribute 5: product concept health claims

Fortified with omega 3 and 6 for: "Maintenance of Normal Brain Function" was the chosen health claim. Although health claims should add value to a product offering, the participants had negative perceptions of health claims. The chosen health claim corresponded with the group's initial brain and cognitive health concern and the group fully understood the claim and the language used. One male participant stated that where he could not understand scientific information on food packaging he would "simply reject the product and put it back on the shelf."

> *The health claim ties in with maintenance of brain function because without your brain function, how are you going to manage, as everything depends on your brain. (G2: F: 55–72).*

The participants were already familiar with the link between omega 3 and cognitive functioning and noted that "you hear about omega 3 all the time." The awareness was attributed to "education" and an abundance of literature on omega 3 health benefits. Similar to the marketing messages, the group were skeptical about health claims. Certain health claims were considered to be false or misleading and participants felt that healthy aging foods should include more simplistic health claims using consumer-friendly terminology but underpinned with evidence-based science. The participants in this UCD interaction all actively researched new health and wellness products and claims online and had become familiar with health claims and the terminology used. Using terminology consumers understand and are familiar with was deemed to be a prerequisite for product acceptance:

> *You read about the benefits of omega 3 all the time and that would assist me in the decision-making processes. (G1: F: 55–72).*

2.4.4.6 Group 1 composite product concept

The functional product concept identified from this UCD was as follows: a juice fortified with omega 3 and 6 for the maintenance of normal brain function. The juice would be packaged in a recyclable tetrapak carton. The marketing messages for this product concept would focus on food imagery. The health claim for this product concept would focus on maintenance of normal brain function (Table 2.2).

Table 2.2 Group 1 Composite Product

Attributes	Attribute Levels
Health concern	Brain and cognitive health
Food and beverage carrier	Juice
Food and beverage packaging	Tetrapak cartons
Marketing messages	Food imagery
Health claim	Fortified with omega 3 and 6 for: "maintenance of normal brain function"

2.5 CONCLUSIONS

The UCD conceptualization method used in this study allowed for a multidisciplinary process that facilitated the identification of new product concepts that can ultimately be developed into new products targeted at the aging consumer segment. The most important attribute levels driving consumer acceptance in the case study interaction were: products targeted at brain and cognitive health; liquid-based beverages as carriers; easy to use packaging with environmentally friendly elements; simplistic marketing messages with a clear focus on health aspects rather than age-related issues; and realistic health claims that contain consumer-friendly terminology. The interaction acted as an innovation tool which identified consumer needs and requirements and produced high-potential concepts that could then be refined using other conceptualization techniques such as focus groups, conjoint analysis, and semiotics. Integrating aging consumers' opinions and ideas into the early product design stages facilitated the identification of optimal product design attributes.

This approach considers consumers as active participants in the value creation process. Throughout the UCD interaction the aging consumers were considered to be active participants whereby they used their own personal experiences to piece together a composite new functional concept that may provide new opportunities for global food and beverage firms to create market offerings that create superior value for consumers. Generating consumer insights during the early stages of the NPD process using the UCD technique identified potentially innovative product concepts. It also provided a systematic framework for managing the gathering of consumer knowledge during the new food product development process.

SUMMARY

The development of new functional foods provides many market and technological challenges to new product developers. Firms that adopt a market-oriented approach to NPD, through the use of advanced conceptualization research techniques, benefit from a deeper understanding of consumers' value systems. This in turn can assist firms to identify key market segments and more accurately make strategic marketing decisions for new food products. Concept optimization research techniques, such as UCD, promote a multidisciplinary approach to NPD, which can assist firms to manage knowledge more effectively and efficiently and develop more market-oriented food products that gain consumer acceptance.

REFERENCES

Agriculture and Agri-food Canada, 2014. Fortified/Functional Foods in China. Market Access Secretariat, Global Analysis Report, Agriculture and Agri-food Ottawa, Canada, April.

Akbar, H., Tzokas, N., 2013. An exploration of new product development's front-end knowledge conceptualization process in discontinuous innovations. Brit. J. Manage. 24, 245–263.

Bech-Larsen, T., Nielsen, N.A., 1999. A comparison of five elicitation techniques for elicitation of attributes of low involvement products. J. Econ. Psychol. 20 (3), 315–341.

Betoret, E., Betoret, N., Vidal, Fito, P., 2011. Functional foods development: trends and technologies. Trends Food Sci. Technol. 22 (9), 498–508.

Bigliardi, B., Galati, F., 2013. Innovation trends in the food industry: the case of functional foods. Trends Food Sci. Technol. 31, 118–129.

Bleiel, J., 2010. Functional foods from the perspective of the consumer: how to make it a success. Int. Dairy J. 20 (4), 303–306.

Bogue, J., Coleman, T., Sorenson, D., 2005. Determinants of consumers' dietary behaviour for health enhancing foods. Brit. Food J. 107 (1), 4–16.

Braun, V., Clarke, V., 2006. Using thematic analysis in psychology. Qual. Res. Psychol. 3 (2), 77–101.

Chandrasegaran, S.K., Ramani, K., Sriram, R.D., Horvath, I., Bernard, A., Harik, R.F., Gao, W., 2013. The evolution, challenges, and future of knowledge representation in product design systems. Comput. Aided Des. 45 (2), 204–228.

Collins, O., Bogue, J., 2015a. Designing health promoting foods for the ageing population: a qualitative approach. Brit. Food J. 117 (12), 3003–3023.

Collins, O., Bogue, J., 2015b. The role of food in supporting healthy aging: a stakeholder perspective. Int. J. Aging Soc. 4 (3–4), 12–13.

Costa, A.I.A., Jongen, W.M.F., 2006. New insights into consumer-led foods product development. Trends Food Sci. Technol. 17, 457–465.

D'Costa, N., 2014. APAC Functional Beverage Industry: Exploring Trends, Challenges and Development. Frost & Sullivan. Vitafoods Asia Conference, 2–3 September, Asia World-Expo, Hong Kong.

Danone, 2008. Health benefits of yogurts. Danone. http://www.danone.co.uk/benefits-of-yogurt.htm

Datamonitor, 2011. Worldwide Breakfast Manufacturing Report. IFIS Publishing, London.

Datamonitor, 2014. Functional and Drink Trends in Europe. IFIS Publishing, London.

Euromonitor, 2014. Health and Wellness Performance Overview 2013. Euromonitor International, London, UK. www.euromonitor.com

Euromonitor, 2015. Health and Wellness in Brazil. Euromonitor International, London, UK. www.euromonitor.com

Green, P.E., Srinivasan, V., 1978. Conjoint analysis in consumer research: issues and outlook. J. Consum. Res. 5 (2), 103–123.

Hardy, N., 2010. Future Innovation in Food and Drinks to 2015. NPD, Trend Convergence and Emerging Growth Opportunities. Business Insights, London.

Hartman Group, 2013. Health + Wellness: A Culture of Wellness. Winter, Hartman Group, Bellevue, WA.

IFIC, 2013a. Food & Health Survey. International Food Information Council Foundation, Washington, DC.

IFIC, 2013b. 2013 Functional Foods Consumer Survey. International Food Information Council Foundation, Washington, DC.

Industry Arc, 2014. Functional Food and Nutraceuticals Market—by Type (Foods, Beverages, Supplements) Forecast (2015–2020). Industry Arc, United States.

Jakobsen, L.F., Grunert, K.G., Sondergaard, H.A., Steenbekkers, B., Dekker, M., Lahteenmaki, L., 2014. Improving internal communication between marketing and technology functions for successful new food product development. Trends Food Sci. Technol. 37 (2), 106–114.

Khan, R.S., Grigor, J.V., Win, A.G., Boland, M., 2014. Differentiating aspects of product innovation processes in the food industry. An exploratory study on New Zealand. Brit. Food J. 116 (8), 1346–1368.

Khurana, A., Rosenthal, S.R., 1998. Towards holistic 'front ends' in new product development. J. Prod. Innovat. Manag. 15, 57–74.

Kohn, K., 2006. Managing the balance of perspectives in the early phase of NPD. Eur. J. Innovat. Manag. 9 (1), 44–60.

Leatherhead Food Research, 2014. Future Directions for the Global Functional Foods Market. Leatherhead Food Research, Leatherhead, UK.

Mak, T.N., Caldeira, S. 2014. The Role of Nutrition in Active and Healthy Ageing. Joint Research Centre (JRC) Science and Policy Reports. Publications Office of the European Union, Luxembourg.

Malla, S., Hobbs, J.E., Perger, O., 2007. Valuing the health benefits of a novel functional food. Can. J. Agric. Econ. 55, 115–136.

Martins, R.B., 2015. The expanding functional foods market in Brazil. Natural Products Insider, 13 February 2015.

Mellentin, J., 2009. Failures in Functional Foods and Beverages and What They Reveal about Success, second ed. New Nutrition Business, Hammersmith, UK.

MSI, 2012. The 2012 Gallup Study of Nutrient Knowledge & Consumption. Multi-Sponsor Surveys, Princeton, NJ. www.multi-sponsor surveys

Newell, A., Arnott, J., Carmichael, A., Morgan, M., 2007. Methodologies for involving older adults in the design process. In: Stephanidis, C. (Ed.), Universal Access in Human Computer Interaction. Coping with Diversity. Berlin, Springer.

Newell, A.F., 2011. Design and the digital divide: insights from 40 years in computer support for older and disabled people. Baecker, R. (Ed.), Synthesis Lectures on Assistive, Rehabilitative, and Health-Preserving TechnologiesMorgan and Claypool, San Rafael, CA.

Nunes, K., 2014. United States poised to lead functional food market. Food Business News. http://www.foodbusinessnews.net/articles/news_home/Business_News/2014/12/United_States_poised_to_lead_f.aspx?ID=%7B3E104CB5-F892-46A3-B989-4A798CBB1B66%7D

Nutrition Business Journal, 2013. Supplement market grows to $32.5 billion. Nutrition Business Journal, XVIII (6/7), 1–7. Available from: http://newhope360.com/research-and-insights/supplement-market-grows-325-billion

Polit, D.F., Beck, C.T., 2004. Nursing Research. Principles and Methods, seventh ed. Lippincott Williams & Wilkins, Philadelphia.

Prahalad, C.K., Ramaswamy, V., 2004. Co-creation experiences: the next practice in value creation. J. Interact. Mark. 18, 5–14.
QSR International, 2013. NVivo 10. NVivo qualitative data analysis software, QSR International Pty Ltd. Version 10.
Slater, S.F., Narver, J.C., 2000. Intelligence generation and superior customer value. J. Acad. Market. Sci. 28 (1), 120–127.
Sura, L., Madhavan, A., Carnaby, G., Crary, A.M., 2012. Dysphagia in the elderly: management and nutritional considerations. Clin. Interv. Aging 7, 287–298.
Urban, G.L., Hauser, J.R., 1993. Design and Marketing of New Products. Ridge Press, New York.
van Kleef, E., van Trijp, H.C.M., Luning, P., Jongen, W.M.F., 2002. Consumer-oriented functional food development: how well do functional disciplines reflect the 'voice of the consumer'? Trends Food Sci. Tech. 13 (3), 93–101.
van Kleef, E., van Trijp, H.C.M., Luning, P., 2005. Consumer research in the early stages of new product development: a critical review of methods and techniques. Food Qual. Prefer. 16 (3), 181–201.
Vella, M.N., Stratton, L.M., Sheeshka, J., Duncan, A.M., 2014. Functional food awareness and perceptions in relation to information sources in older adults. Nutr. J. 13, 44, Available from: http://www.nutritionj.com/content/pdf/1475-2891-13-44.pdf
von Hippel, E., 2001. User toolkits for innovation. J. Prod. Innovat. Manag. 18, 247–257.

CHAPTER

Marketing challenges and strategies

3

Sourya Datta
Supply Chain Manager at eBay, San Jose, CA, United States

The aim of marketing is to know and understand the customer so well the product or service fits him and sells itself.
Peter Drucker (BrainyQuote, 2016)

3.1 INTRODUCTION

This chapter highlights the key marketing concepts, features, and requirements for successful product launching and marketing of nutraceuticals and functional foods.

In order to come up with an executable marketing strategy, we would need to understand what is our product and who is our customer. This might seem to be a very simple question but in reality it is not. It is extremely difficult to understand the customer requirements and in order to provide an excellent experience for the customers, customer segmentation is key. Correctly segmenting the customers and then creating a strategy to satisfy each segment is key for success. On the other hand it is also very important to understand product categorization and opportunity of various products in the market.

In terms of product categorization, soft drinks and dairy products are the leading categories out of all food items. This immediately shows that there could be greater opportunity in the beverage segment. If we take a deeper dive into the health benefit market, energy drinks are the top category in the segment.

Details of functional food items and the importance of various types of products are captured in the study completed by PricewaterhouseCoopers (PWC) and it clearly shows that energy products occupy the largest segment in health benefit products (PricewaterhouseCoopers, 2015).

3.2 EARLY DEVELOPMENT OF FUNCTIONAL FOOD AND CURRENT STATUS

Development of functional foods requires a lot of details, the process is costly, and there are risks associated with customer acceptance. Over the years, food has undergone a number of transformations: from just satisfying hunger to preventing health

Developing New Functional Food and Nutraceutical Products. http://dx.doi.org/10.1016/B978-0-12-802780-6.00003-1

issues and improving humans both physically and psychologically. The physiological strength comes from the physical satisfaction of well-being and that in turn improves mental health. The idea of functional foods was developed in Japan when it became clear that it was very expensive to provide medical benefits to elderly people. However, on the other hand, countries and companies realized the opportunity of commercial success.

The concept of functional food is different in the United States and Japan though—in the United States functional means "something is added to existing food" while in Japan functional foods is a "different class" completely. The whole idea of functional food in the United States is the fact that it is a form of regular food and not necessary pills, capsules, and tablets. Functional food means regular food, which has the capacity of providing something extra. A literature survey on nutraceuticals and functional foods reveals that the global market lies between \$30B and \$50B (U.S. Department of Health and Human Services, 2015) and the United States holds a majority of that market. Even though Japan, France, Germany and the United Kingdom are the other big players in the functional food industry, studies have shown that the Netherlands is in close proximity to other European giants in terms of functional food development and acceptance.

The challenge with the European market is its heterogeneity, which might be a little easier to deal with in the United States. Generally, for a US company dealing with functional food, the costs of developing functional foods and marketing the products are extremely high. Proof of efficacy, filing required patents, and research and development are extremely costly for various companies involved in the development of functional foods. Another level of complexity has been added by major pharmaceutical companies like GSK and Johnson & Johnson (J&J). These companies have hurried to get commercial success although these companies are not typical food companies but rather pharmaceutical companies. Apart from these giants, the playing field is captured by a number of small and medium sized companies which may not have the power of money behind them but the advantage these companies hold is the ability of traversing from production cycle to product launch in a short span of time. From concept to design is extremely fast and the small companies take much less time for entry into the market than a giant like J&J. "*… can do with a new product. Often times companies launch their product in the market between one and three years, recuperate all the costs, make profit, and jump to the next product while companies like J&J seek long term success and it takes them a longer time to develop the product* (Siro et al., 2008)."

The development of functional food in the United States has been fascinating; initially the whole idea was to have foods with added minerals and vitamins, then there was a shift and food was developed keeping micronutrients in mind, and now the goal is to provide more than one benefit (at least two or three or more) from a single product. It is fascinating how functional food started with dairy, then moved to different forms of juice for a while and now it is in the form of health drinks. Functional food eventually saw two different types of market: one for the "energy enhancing" segment and the other for the "daily lifestyle management & improvement" segment.

3.3 OBJECTIVE OF FUNCTIONAL FOOD

The keys to successful nutraceutical and functional food are mainly the following:

1. Efficacy
2. Price
3. Ease of intake
4. Consistent supply

If any of the above is not completely satisfactory, then the product might not be a success. A very strong marketing strategy needs to go hand in hand with the production of nutraceutical products for any company to be successful in the United States and European markets. Also the marketing plan needs to be built in advance since the product life cycle is long. However, before the actual launch of the product, it is very important that necessary patents are filed and approved in order to increase the value of the product. The product must be accompanied by approvals, certifications, and results from well-respected laboratories both within and outside the country. Well-accepted testimonies could increase the demand of the product strongly and it is important to provide toxicity certificates. Certification from reputed laboratories after testing the product is essential to support the quality, efficacy, and effectiveness of the product and also denounce any negative symptoms and chemical effects from use of the product. The importance of having efficacy tests is vital and functional products could be used in one of the two formats described next:

1. Raw materials used in formulation
2. Finished product

3.4 UPSTREAM MARKETING

3.4.1 PRECONDITIONS FOR SALE OF NUTRACEUTICALS AND FUNCTIONAL FOODS

3.4.1.1 Copyright, patents, and trademarks

Obtaining copyrights of products and necessary patents is not only the key for success of products but it is also important to protect the products in the international market.

There is enormous potential of these nutraceuticals and functional food items manufactured by the manufacturer which act as ingredients for different health drinks. Since the product carries unique features and has its own identity and in a number of cases substitutes aren't available either in the country or outside, it is very important to have the products registered with the appropriate authority/authorities for patents and copyrights. This will help the company to earn benefits on account of originality as well as provide the opportunity for earning royalties. If required, the same formula, process, or technology is transferred to another company for manufacturing the same product either in the same country or outside the country.

3.4.1.2 Brand

If the product has unique features and has no available substitutes, it is very important to register the product under some brand name that is registered and owned by the manufacturing company. The brand name is the exclusive right of the company for use and no other company or corporation can use the same brand in the market. Branding is a very important factor and since the product could be sold as the finished product in retail outlets, it is important not to lose focus on the brand value. If branding is not the core strength of the manufacturing company, then an agency could be engaged for creation of a brand together with a detailed product design and logo.

3.4.1.3 Global competition

Avoiding or reducing global completion for a product could be one of the key successes for the product. In order to minimize global competition, it is important to register the same product in other countries as per the laws of other countries. Since the product was developed and manufactured by the company in a particular country with adequate approvals and clearances, time needs to be allocated for getting all the approvals validated by another country since this could be a lengthy process. The product can be registered in other countries under intellectual property rights (IPR) to enjoy the rights of original invention and thereby there is a high possibility of earning a "know-how fee" if other interested companies are willing to use this brand item/product in their products.

3.4.1.4 Awareness program

Since the product has been invented/developed, it is very important to inform the end user about this product along with creating product literatures with details about merits, demerits, benefits, shortcomings, and side effects along with other risk factors such as toxicity limits. This awareness program is very important and for this purpose organizing various seminars, symposiums, and meetings with the technical team apart from personal visits to the "end user" are equally important to develop an awareness program about the product details before the product is ready for launch. The sales team must be thoroughly knowledgeable and conversant with the product and the sales group should possess a good technical background in order to provide the detailed functional and technical aspects of the products. Before the entry of a particular product into the market, product approvals from the research and development team or technical department clearances from health drink manufacturing companies are required.

Marketing material needs to be made ready before product launch; it's key to capture the initial awareness of the general public. For this purpose, various electronic media, print media, digital media, signboard, and social media marketing are very important. The retail market is large in terms of both size and volume, and thus the marketing depends entirely on budgetary provisions.

3.4.1.5 Market survey

Before launching the product in the market, the company should engage agencies to conduct a proper market survey to understand the use and potential demand of this product in the market. This information will help the company to make plans to decide the quantity of demand along with the size of the market. This will help the company to create a future investment decision plan. Also a smaller subset of the market could potentially provide important details that might be very important for the success of the product in the future.

3.4.1.6 Ascertainment of toxic limit

Since the product could have chemical ingredients that could have an impact on lives, it is very important to get all levels of clearance for toxicity. The necessary permission must be obtained before product launch from the appropriate authorities so that all clearances regarding toxicity are obtained before the product is launched in the market.

3.4.1.7 Permission, clearance, and regulatory approvals from various authorities

Since the product is for the public and the majority of the products will be in the form of either food or drink, it is very important to obtain the necessary clearances from the following authorities in the United States (the same procedure needs to be followed with authorities of other countries). If the products are sold outside the United States, the necessary permission is required from various authorities applicable to the respective country's rules and regulations.

In the United States clearance would be needed from the following:

1. Pollution clearances from the EPA (Environmental Protection Agency)
2. License from the DSHEA (Dietary Supplement Health and Education Act)
3. For good manufacturing practices permission would be required from the USFDA (United States Food and Drug Administration)
4. For better acceptability of the product in the market, permission would be required from the National Supplemental Foundation to use their logo as "NSF"

3.4.1.8 Budget

After getting the clearance from the authorities and based on the market survey, the company now needs to prepare the budget for selling the product to the other health drink manufacturing companies or go solo and simultaneously needs to ascertain the demand in the retail side of the business.

For manufacturing, the company needs to do a thorough estimation of both capital and revenue budgeting. The manufacturing process needs to be examined carefully and thereafter a detailed guide is needed to understand whether any additional equipment/machinery is required. Space is a critical resource and there would be a need to analyze the available capacity and also space availabilities. There should not be space constraints that might create any hindrance for production as per demand in the market. By estimating the quantities that would be sold, the revenue budget is also subsequently prepared.

3.4.1.9 Distribution network

Once the product is ready for the market, the necessary arrangement for creating a distribution network for selling the product needs to be established. If the company has an existing distribution channel, then the same distribution channel could be used or if there is a distribution partner available, the company can dispatch the products to various customers either using its own distribution channel or using the partner's channel. Otherwise, the company needs to examine the process of distribution so that products reach the customer safely, take the minimum time, and have the best economical distribution cost.

In the case of e-retail channel distribution, the company can form a partnership in advance with various distribution companies, and can act in advance to make sure timely delivery is completed against e-orders.

3.4.1.10 Packaging

Packaging is very important and vital. Often companies lose a lot of money due to faulty packaging due to breakage and spills. It is necessary to take all the necessary precautions so that no damages occur during transportation as well as during storage of the same at the customer's location (Gilbert, 2000).

3.4.2 STEPS TO BE FOLLOWED AFTER CLEARANCE OF ALL COMPLIANCES

Before going to downstream marketing, there are still a number of steps that need to be followed. For sales and marketing of these items the following steps need to be followed:

1. Identify the geographical areas that the product will cover both internally within the country and outside the company's own country. It's important to have minute details (state specific, location specific) of where the product will be launched.
2. Identify the lead technical sales team, who will take the initiative to explain the product details to prospective customers. Technical sales representatives have a huge role to play and it is important to hire people who are technically adept in delivering the product features as well as have proven sales background in the food and beverage industry.
3. Ensure that all products come with correct product literatures that contain details of product nomenclature along with all necessary clearances.
4. Organize seminars, symposia, and exhibitions for the product and also lectures by professors and doctors whose words will have value. All of them are vital and key in order to build product awareness campaigns/programs.
5. Identify the prospective customer base that is in close proximity to the final products produced by the company. Close proximity to end customers would help in determining customer reaction to the final product; also close proximity to the final company selling the end products would be helpful in determining changes that might improve the products' quality and effectiveness and may be helpful in reducing the cost as well.

6. State/location-specific conferences should be organized with the prospective customers in various states where health drink manufacturing companies have a presence.
7. Establish relationships or arrangements with various retail chain outlets.
8. E-commerce order handling would be key. A separate team who will handle e-commerce orders needs to be set up.
9. A team needs to put in place all technical requirements which will be used to actively monitor social networking websites and the latest trends. Customer support through social media could be a key factor. A customer support team actively monitoring networking sites and providing guidance and help to questions through the social media platforms could be key to differentiate the product. Millennials often use social media platforms and it's important to answer questions from the prospective customers and provide guidance for product knowledge/development activities for every form of media.

3.5 DOWNSTREAM MARKETING

Downstream marketing is key for the success of the product. Downstream marketing comes into the picture once the product is ready and all clearances have been obtained.

3.5.1 KEY FACTORS FOR SUCCESSFUL PRODUCT MARKETING

While launching a product to the market, several key factors play pivotal roles and it is important to discuss all the factors that will determine the future success of a product.

1. Convenience of taking supplemental food
   The process of intake of functional foods needs to be made very easy. For example, taking pills is more difficult and takes more convincing than providing the same function with regular food and drink.
2. The ease of shopping for functional products
   Three factors determine the ease of obtaining functional foods:
   a. Product availability in retail stores
   b. Ease of shopping via e-commerce/online
   c. Product transparency and whether the product has specified all ingredients and percentages clearly
3. Product integration with technology
   Customers now want to check every important detail through mobile apps. Use of technology has become extremely important. Products like the Apple Watch are used to measure heart rate and also other products like Fitbit or Jawbone help to track daily activities including sleep patterns. Future technology integration will only increase and manufacturing companies need to give customers the choice to make their own food plan.

4. Provide customer support and make the customer an integral part for product success:
   a. Word of mouth marketing becomes key for the success of the product, specially functional foods
   b. If the product can capture the minds of a few people and prove its efficacy, it goes a long way in product success
   c. It is important to keep the "taste factor" while developing the product
5. The customer is the king
   Provide authority and power to the customer to choose products but make sure customers' minds could be influenced.

   Customers themselves are now deciding which food to have whereas the doctors decided this 5 or 10 years back and this is an important factor since products have the capability to influence the customers directly now.
6. Partnership with local retail giants and online giants is one of the keys to success
   Provide the value of the product to local organic food stores like whole foods, sprouts, and Trader Joe's. At the same time companies should form relationships with volume stores like Walmart and Target. An excellent opportunity to get Costco and Sam's Club partnership is also possible and what better way to capture customer demand than actually providing food samples to customers who are coming to the stores for groceries.
7. Provide multivarious benefits
   Just providing one result with the product is not good enough anymore. Customers are looking for a single medicine to take care of multiple needs.

   Providing two results with one product or even multiple results with one product is the key to success.

   For example, an active middle-aged customer is not just looking for a product that will give them energy but he/she is looking for a product that will give them energy and also boost their performance. Even more, the current customer is looking for a product that will give them energy, increase their performance, and alleviate their cardiovascular performance.

   A different set of customers who previously just looked for weight control and weight management is now looking for weight control and also wants to improve their cardiovascular performance.
8. Product differentiation is the key for success
   It is very important to differentiate the product from all other products available in the market. An important step in this regard is proving guidance and a set of instructions and taking customer feedback to make the product even better.
9. Customer segmentation is extremely important
   It is extremely important to have customer segmentation to be successful in the market. One product can't satisfy all types of customers. The needs from various segments need to be captured and products targeting segmented customers are key for success. It's important to provide the right education to both intermediate customers like the "Walmarts" of the world as well as the final set of customers who are buying the product.

A typical segmentation for a functional product could be as shown below:

a. Driven by performance: Millennials who are willing to work hard
b. Give extra energy and stamina: Baby Boomers who are looking to maintain their body
c. Improve cardiovascular joint strength: Swing generation whose bones are becoming brittle but they want to be active
d. Losing weight is the goal: There are a segment of customers who are trying to just lose weight and that is their main goal

Every segment is unique and conditions could be also different. However, at the same time, products that are able to satisfy a broader range of customers could be key to success.

10. Customer education

It is not the responsibility of customers to know everything and be aware of all minute details but it is the responsibility of the manufacturers of the functional foods to make customers aware about the efficacy and ingredients of the product. Not always is the customer aware of all the details. It is the responsibility of manufacturers to educate customers not only about the efficacy but also about the detailed ingredients of the product.

11. Perception of the product in customers' eyes

Having the right perception in customers' eyes will help in the success of the product tremendously. The following factors are very important to improve customer satisfaction:

a. Product effectiveness
b. How healthy is the product: In the customer's eyes, healthier products are perceived to produce better results
c. Speed of action: How fast can the product take effect?
d. Source of product: A natural source improves product perception
e. Taste: However effective the product is, taste is critical for the success of the product. Taste goes hand in hand with effectiveness. It has been proved that the level of consumers' liking decreased with the increase of a defect in flavor

12. Sustainable source

Three important factors for product success are: (1) continuous sourcing, (2) supply chain, and (3) packaging of products.

Apart from the aforementioned three factors, sustainability plays a huge role is the success of the product. When there are two products with all other factors similar, sustainability and green become major factors in customer decision.

13. Meeting demand supply

For a product to be successful, it is important to provide the right product to the right customer at the right time. Creating a demand plan and being able to meet the demand with constant supply is key for product success.

14. Innovative ways to attract customers

There are multiple ways to attract customers. When the customers have become more technology savvy and busy, distraction is greater and it's important to influence the customers in the right way. Messaging and advertising should be provided in various formats viz. audio, visual, and audio and visual.

15. Importance of having the right location to capture future clients
Good locations to capture customers for functional foods are the gym, doctor's office, supermarkets, yoga locations, sports practice, and sports practice sessions where moms of school kids are present.
Not only are the locations important but also the ways to tap the locations are very important. It is important to understand the techniques to capture the right customer from various locations. For example, a doctor's office will not entertain a typical salesperson but if a product has been proved to be effective, the television screen or audio could be great ways to advertise the product when customers are waiting in the doctor's office. To capture moms, the best influence could be another fellow mom who loves the product and can vouch for the benefits of the product.
16. It is very important to be honest about the product in terms of efficacy, ingredients, probable success metrics, and side effects of the product
It's important to be successful as soon as possible and first-time success is the best success and opens the product to future successes. For that reason, it is important to be honest and give a true picture of the product.
17. Understand short-term success versus long-term gains
This is an extremely important factor and companies need to decide which success criteria are important to them. Often times smaller companies try to look for short-term success and they have a high burn rate. They come with products very quickly, try to be the dominant player in the market for a couple of years, fade away, and then come back with another product. On the other hand, more established companies spend more years in research and development and they have a longer-term vision for the same product category. They try to be in the market with the same product for a longer time with minor developments and enhancements to the same product (CBS Moneywatch, 2016).
18. Effective use of social media
Use of effective social media is much more complex than influencing people to post on Facebook. The greatest success would be to use technology to influence other customers through an existing customer. Use of apps for checking and managing one's own vitals, customized apps to influence people by providing rewards for maintaining a better life, and providing rewards for surveys are great ways to influence customers. Using crowdsourcing techniques could be vital for success of products.
19. Product forms and catering to the right customer with the right form of product
Having the right forms of product is a very important success factor. Based on age, most companies try to follow the product forms as follows:
    a. Young customers/millennials: Pills are okay
    b. Middle-aged: Pills are not the best option, chewable forms are okay
    c. Older generation: Powder form works but liquid form is a hit

20. Connecting with customers at an emotional level
    This is a very important step to be successful for both the product and the company. If the company is able to connect at the emotional level, often times they get a set of loyal customers who are willing to take that extra step with the product. They are ready to change their habits according to the specification and they become a lifelong friend for the company.

3.6 CREATING BRAND VALUE

Product branding is very important for the success of any product, more so for food products, and even more for functional products. It's very important to provide the right benefit since the chances of going wrong would have a tremendous detrimental effect both for the customer as well as for the manufacturer. Typically there are two types of branding:

1. Individual branding
   This is the step where the supplier and the manufacturer perform the branding separately and the supplier typically uses branding to sell to the manufacturer and the manufacturer performs a number of steps (research and development, improvement of the product, packaging, marketing, approvals), creates the final product, and brands the product separately. This happens in business to business settings mostly (B2B). Also, this happens with big manufacturing players of the world like J&J even though in the last few years they have started cobranding their products and working closely within the ecosystem.
2. Cobranding
   In order to get the best value for both the supplier and the manufacturer, the suppliers and the manufacturers join hands and cobrand their product. The advantages with this way of conquering the market are the following:
   a. Greater pool of resources since supplier and manufacturer are utilizing each others resources
   b. Reducing the competition from other players of the market
   c. Strong relationship between both parties
   d. Both manufacturer and supplier see end to end processes; supply and demand are captured by both parties and the supply chain system improves
   e. Cost-effective for both parties (shared pool of money)
   f. Quality of research and development increases tremendously since two parties are joining hands and they have knowledge in two different sectors
   g. Win for all parties: supplier, manufacturer, and the customer. The manufacturer helps the supplier to be more involved with customers and the supplier involves the manufacturer to help in product and raw material sourcing and determination. The customer gets a better product and customer feedback is incorporated much faster in the system that helps to come up with an improved product

3.7 EFFECTIVE ADVERTISING

Effective advertising is key for successful marketing. Advertising is a key success factor and it could take various forms. Not only does advertising take different forms but also in order to be effective, it needs to be implemented in various formats for different sets of people. For example, advertising for millennials will be different from advertising for baby boomers.

3.7.1 FORMATS OF ADVERTISING

3.7.1.1 Ads based on age

For millennials, social media ads could be more effective while for baby boomers, ads through television could have a higher impact. In person ads are becoming a rarity but could be used in front of grocery stores or inside grocery stores.

3.7.1.2 Ads based on similar habits

Ads need to be different for sets/groups of people with similar tastes. The type of food sample that will be provided in the gym will be different from food samples provided in the physician's office.

3.7.1.3 Ads based on location

Ads need to be different based on countries but even within a country, there is a need for minor variations. For example, the east coast of the United States has different taste than southern states and ads need to be very specific based on location.

3.7.2 FORMS OF ADVERTISING

Advertising forms also need to be different in order to be successful. They should have various forms depending on the situation. Let's point out the various forms:

1. Type of ads:
   a. Serious ads
   b. Funny ads
2. Ads involving celebrities
3. Ads providing funds for a good cause
4. Ads through scientific conferences
5. Trade shows provide advertisements of the company
6. Working with authors where authors are indirectly providing advertisement for the product
7. Availability of same product in different forms: pill, powder, food, and beverage and numerical proof of efficacy for all forms
8. Use of different media for advertisement:
   a. Television
   b. Radio
   c. Google ads

d. Strong partnership with retail partners like Whole Foods and Costco, online partners like eBay and Amazon, and vitamin shops like GNC (CBS Moneywatch, 2016)
e. Free samples and pamphlets at doctors' offices
f. Advertising products in schools through sponsoring different events like charity events which schools take part in
g. Sponsor events like marathons and even open stalls to distribute products at the finish line
h. Endorsements through programs like Weight Watchers

3.8 CHALLENGES WITH ADVERTISING FUNCTIONAL FOODS

There are however numerous challenges and careful analysis that needs to be done before going for full-blown advertising. Analysis of regulatory aspects and review with a legal team need to be performed before advertising for a product. The challenges of advertising are captured in the following:

1. Drain on cost: Advertising takes a lot of money and resources. It needs to be thought out and planned in terms of both resources and cost.
2. Timing of advertising: Timing of advertising is very critical because it needs to match with the launch of the product. Advertising too far ahead or after a long time when the product is already launched may have only limited impact on the success of the product.
3. Knowledge on competitors' products and their timing of launch: Product knowledge of the competitors and also the timing of competitors' entry into the market are key for the success of the product. Often times strong advertising showing the advantage of the product in comparison to competitors' products has a great impact on the success. Timing of product launch is also key for the success of a product.
4. Match regulatory filing with labeling and marketing: All the claims made during the marketing campaign have to match the FDA approvals in the United States and often times marketing has to be regulated so that it matches product labeling in order to avoid FDA recalls. In the United States, product claims can be done only after thorough research is complete, all data is published in peer-reviewed journals, seminars about the product have been provided, and necessary approvals are obtained from the FDA, FTC, and DSHEA. The same process needs to be followed for other countries and necessary approvals need to be taken from regulatory boards.

3.9 ANALYSIS OF MINDS OF CUSTOMERS

It's been proved that there is a close relationship between functional food and health. Another important factor for marketing is the fact that it should not only help with physical well-being but also improve the physiological or mental aspect of well-being. Product development has various stages:

1. Raw material gathering
2. Conversion of raw material to semifinished goods
3. Final production and preparation
4. Clinical trials
5. Regulatory signature
6. Customer education
7. Taking customer points and using data to enhance the products
8. Technological advancement

The commercial success of functional food depends on how familiar people are with it and their willingness to accept the functional food.

People are looking for quality, quantity, and a great product at an affordable price, with good taste and great nutritional value all in one single product. Marketing might not be able to provide all these features in one product but marketing can definitely help influence customers' minds and improve customer perception greatly.

3.10 RISK FACTORS ANALYSIS

With every new development there are associated risks. Let's look in detail at the risk factors and probable mitigation steps:

1. Analyzing and understanding the demand and supply criteria. The products have a typical shelf life and in order to ensure products are sold in time, an exhaustive demand–supply curve is necessary due to two factors:
   a. Shelf life expiry may force the company to withdraw the product from the market
   b. Disposal of shelf life product affects companies in three forms:
      - Loss in revenue
      - Cost overrun
      - Degradation of brand value in the market

   The only option in this case is to find an alternative for reuse of the product.
2. For a new product, budget estimation is very difficult due to a number of uncertainties. Budget overrun is often a problem in new product development.
3. Validity of the product and clearances from various departments are key to the success of the product. The validity of clearance is directly related to the time the product could be sold.
4. Heavily dependent on both one's own and a foreign country's import and export policy. The finished items/products are dependent on the import of various ingredients and therefore manufacturing of the products is also dependent on a country's import and export policy.
5. The insurance cost of the product might be very high since the product will be directly consumed.
6. Food items are always at very high risks of obsolescence. Often times new products and new items are developed and they flood the market and the cycle time of products is often small.

7. Registering for patent/trademark takes a long time.
8. Processing time for entering other countries might take a long time.

Most of the risks mentioned previously could be mitigated when the product launch is planned well ahead of time and all necessary steps are taken ahead of the launch.

3.11 A FEW EXAMPLES

3.11.1 WEIGHT LOSS

Weight loss is a feature for the product to be successful in the market. Safety plays the most important role but at the same time dosage is critical too. Convenience of dosage (how many times the product needs to be taken), after effects, appetite suppression, and increase in energy could have very high impact on the success of the product. If the product helps in fat oxidation where urinary excretion of fat metabolites occurs, it bolsters the value of the product in the market.

3.11.2 PREVENTION OF SKIN WRINKLING

Skin wrinkles with age, improper nutrition, and stress. If a novel, safe, and cost-effective product can be developed which can minimize skin wrinkling and affirm skin firmness, then the product has the potential to attract a segment of customers and be successful.

3.12 CONCLUSIONS

Even though marketing can help companies in a number of ways, ultimately a lot depends on the quality of a product. Marketing will be successful in the short run and products will be successful in the long run only if the quality of the product is better than an existing product in the market. It's important to launch a product that has cleared the laboratory tests, proved efficacy with a smaller set of customers, and has the necessary patents. The source of the product is also very important. In order to be effective, the source needs to be continuous; there shouldn't be any stoppage or disturbance in the continuity of the source. Another important factor in sales of the product is authenticity. Three factors come into play to enhance the product's authenticity: the science behind the product, publication in journals and other references, and definitely the backing of all claims with a strong patent. In order to market a product, pricing plays a very important role as well. If the price of two products produced by different companies is comparable then other factors come into play. The source of the product plays a critical factor since ideally it should be the cheapest but there can't be any compromise on the quality. Sometimes there are problems with natural products: the chemical composition might not be suitable or ideal, the product will vary from location to location, and there could be minute or drastic changes based on location. The product might be of great quality but sometimes the location might limit the quality of the product and its chemical composition might not match. Product positioning is also key and a product will stand out based on the following factors:

1. Price
2. Quality
3. Supported by peer-reviewed publications including clinical studies
4. Intellectual property and patents
5. Safety affirmed

REFERENCES

BrainyQuote, 2016. Interesting marketing quotes. Peter Drucker. (n.d.). BrainyQuote.com. http://www.brainyquote.com/quotes/quotes/p/peterdruck154444.html

CBS Moneywatch, 2016. Companies who are successful in marketing. http://www.cbsnews.com/news/10-companies-with-insanely-great-marketing/

Gilbert, L., 2000. How to reach target audience. AgBioForum 3, 20–38. http://www.agbioforum.org/v3n1/v3n1a05-gilbert.pdf

PricewaterhouseCoopers, 2015. Functional foods industry snapshot. https://www.pwc.com/us/en/transaction-services/publications/assets/functional-foods.pdf

Siro, I., Kapolna, E., Kapolna, B., Lugasi, A., 2008. Functional food. Product development, marketing and consumer acceptance—a review. Appetite 51, 456–467.

U.S. Department of Health and Human Services, 2015. Overweight and obesity statistics and information. http://www.niddk.nih.gov/health-information/health-statistics/Pages/overweight-obesity-statistics.aspx

CHAPTER

4 The strategic viewpoints of innovation and marketing teams on the development of novel functional foods

Karin Tollin, Antonia Erz, Jesper Vej
Department of Marketing, Copenhagen Business School, Copenhagen, Denmark

4.1 INTRODUCTION

Among researchers and managers in the field of functional food new product development (FFNPD), there seems to be a high level of consensus about the following issues: developing functional food products is a complex process and resembles R&D in the pharma industry; the marketing of functional food products is about fast-moving consumer goods (FMCG) marketing; the market will continue to grow; and no clear definition of functional foods exists. Another recurrent conception is that FFNPD is one of the most interesting areas of research and innovation in the food industry (Bigliardi and Galati, 2013; Siro et al., 2008). Considering the many perspectives and fields of competences involved in food innovation in general and functional food in particular, the belief that FFNPD is a complex process appears appropriate, as does the notion that FFNPD relies upon the involvement of a food technologist, a nutritionist, and a specialist in health benefits, physiologic markers, etc. (Bigliardi and Galati, 2013). However, our thesis is that holding on to these notions too stringently may have a negative effect on the wave of success that functional foods have had until now. We regard the present book as a fruitful initiative toward widening the field. Accordingly, our goal with this chapter is to present an outline of issues and viewpoints that have not so far been discussed in the functional food literature. Inspiration and support in preparing this outline have been obtained from interviews with representatives from the aforementioned fields and those of general management and marketing. In addition to these interviews, reviews of the FFNPD literature, and of the contemporary literature dealing with recent development in innovation and marketing management, have reinforced us in suggesting ideas and paths for the functional food society (research and industry) to embark on. Hence, during our inquiry we have found support in acknowledging divergent views both on FFNPD and on what are essential knowledge, principles, and processes in the marketing of functional food products. Furthermore, we have discovered arguments for

Developing New Functional Food and Nutraceutical Products. http://dx.doi.org/10.1016/B978-0-12-802780-6.00004-3

reflecting critically on the sustainable growth rate of the functional food market and on the value of retaining the term "functional food." This leads us to the following structure of the present chapter: Sections 4.2–4.4.

4.2 INNOVATION OF FUNCTIONAL FOOD PRODUCTS

When scholars discuss what characterizes FFNPD, it is apparent that the market structure for functional food products is setting the scene. It is also evident that the domineering role of probiotics up until now has impacted on the understanding of FFNPD (Bigliardi and Galati, 2013; Jones and Jew, 2007). Another aspect that has had a bearing on how FFNPD is characterized concerns the involved actors and how they understand innovation. Within the probiotic field, a handful of suppliers of ingredients to the food industry seem to have had a central and defining role, due to successful inventions having been made early in the history of functional foods, alongside a continuous and high emphasis on research and development (Khan et al., 2013). From this observation follows the question: what is innovation all about from the perspective of an ingredient supplier in the probiotic field? In order to find an answer to this question we have analyzed the literature dealing with the topic and interviewed managers involved in research and new product development in two prominent ingredient supplier companies in the probiotic area. The interviews followed a framework which is presented later under the heading: Understanding innovation. On the basis of these interviews and an analysis of the literature, we present a framework illustrating a supplier view. As already introduced, it is our notion that this framework or perspective on FFNPD is far too restricted. Hence, we suggest the integration of other perspectives on FFNPD. The core argument is that the performance of investments made in FFNPD by suppliers of probiotics "at the end of the day" relies upon how new concepts are understood and responded to by food companies, by various institutional bodies, the media, and other opinion-leaders, and not least by consumers. This leads us to propose an extension of the supplier view on FFNPD. We call this "an integrated view." Further arguments for our proposal are treated in the section dealing with functional food marketing, and explicitly when we discuss consumer issues being explored and suggested to be explored in relation to the development of novel functional food products. Finally, in our discussion about drivers and barriers to industry development, we outline a number of further arguments for an integrated view.

4.2.1 UNDERSTANDING INNOVATION

From proposing that consumers' engagement and interest in "health and wellness" is an emerging megatrend that presumably will have the same impact on the ways food and retailing companies compete and create value as other previous and concurrent megatrends (e.g., TQM, IT, and corporate sustainability), follows a number of questions about the resulting impact on company practices. One such question is: what impact will the "health and wellness" trend have on the understanding and practice of developing new food products? In our attempt to answer this question, we approached suppliers of probiotics, because they undoubtedly have been a central actor in the evolution of knowledge and innovations in the functional food field.

From perceiving FFNPD as representing a highly knowledge-intensive and science-dependent area, it follows that addressing innovation from a learning perspective appears a relevant path in attempting to answer the aforementioned question. Further support for a learning perspective relates to the notion that the functional food trend presumably will give rise to some changes in the practice of innovation. In our approach to uncover this, we have been inspired by the conceptual model developed by Bell et al. (2002) comprising four schools on organizational learning (i.e., the managerial, the process, the developmental, and the economic school). With inspiration from this model and how it was applied by Tollin and Vej (2012) in identifying the meaning of the corporate sustainability construct and trend in the business world, we developed an interview guide for acquiring an understanding of innovation in the probiotic field. In Tollin and Vej's application and also in our analysis (Table 4.1), the four schools are represented by the following constructs: scope, processes, resources, and outcome. A central idea of the managerial school

Table 4.1 Understanding Innovation

Scope	Process(es)	Resources	Outcome
How do you define innovation? What has been the innovation path so far, and why (type of innovations being focused on)? What do you regard as being the central innovation issue(s) and area for innovation(s) in the future?	What would you say are the distinguishing features of the process of innovation? Which are in your view the important processes and why, i.e. tasks in innovation processes? Which are the important learning partners involved and when (idea, concept test etc.), and how are they involved?	Relationships—besides/in addition to the partners you work with intensively during innovation processes, are there any other relationships of a significant importance in keeping up-to-date on knowledge creation in the field? Which are the important capabilities in order to be at the forefront in innovation? Capabilities to sense new knowledge, to do experiments, to build relationships, to set up a team to implement open innovation, etc.!	How do you measure the performance of innovation—on a project level and at the organizational level (the function, the company)? How are these performance measures communicated (within your team and within the company)? Which are the important performance measurements for driving innovation in your company and for driving the functional food and probiotic field and industry in general?

is that a company's top managers play a decisive central role in addressing evolving issues and developments, developing innovation strategy, and driving the direction and potential of innovation. In our analysis, "scope" expresses a significant element of these initiatives, namely, a definition or understanding of what innovation is; what areas of knowledge and development innovation embrace; and how learning and innovation are attained and secured. In the process school the focus is on managing learning processes, and a central presumption is that what is discovered and developed is determined by: "what lies between rather than within participating individuals" (Bell et al., 2002, p. 79). Hence, the following issue becomes central: what are the important learning processes and relationships (internal and external individuals, networks, and bodies) in developing probiotics or new functional food products?

In the development school, learning is seen as an evolving process, and consequently experiences, capabilities, and other resources gained in previous periods will influence the present nature of and performance outcomes from learning processes in the present. Accordingly, an important question appears to be: what capabilities are important in order to allow an organization to be able to continuously and successfully upgrade and reconstruct core capabilities related to innovation of functional food products or probiotics? The last dimension, as signified by the term "outcome," originates from the economic school of learning following the conceptual model by Bell et al. (2002). Considering the emphasis given to accountability in companies with heavy investments in development and marketing, combined with the negative impact that accountability may have on maintaining a balance between incremental and radical innovation and on keeping innovation at the forefront, the following type of question becomes important: what are the important performance measurements/metrics for gaining and securing a central position in the fields of probiotics, functional food products, as well as the field of "health and wellness?"

4.2.2 A SUPPLIER VIEW

> *We are research-based, so we are mostly concerned with what is happening in research, which are the new insights—which are the health effects to be expected from a new bacterial strain. We got to deliver something new that builds upon new research and which provides health benefits to the end consumer. What we do in research and development is totally product neutral. Development is one thing, production of strains another, and application—to infuse the strains in different products, yet another area and process. Our goal is to detect a strain or a combination of strains that gives rise to a positive health effect. We would never start up something because marketing says customers ask: Don't you have anything for constipation? We say, no we don't.*

The aforementioned quotation originates from one of our interviews with managers of R&D. As stated in the quotation, the scope of innovation consists of bacterial strains that demonstrate the potential to be associated with certain well-defined and important health benefits, beyond basic nutrition. Thus, technological or marketing aspects of a potential carrier, an existing or a new food product, are not embraced by the scope. The issue whether a new strain or a combination of strains will become a component in a food product or in a tablet or capsule is not dealt with in the idea generation or concept development phases. A strong emphasis on being research-based is based in a shared perception in the probiotic industry, namely that there is a huge potential for inventions in the field of live microorganisms. Accordingly, the notion is that success or even survival depends on being aware of the fact that the key driver to industry growth lies in paying attention to research progress underway and remaining ahead.

The notion in the literature of FFNPD being product-oriented (Khan et al., 2013) does not correspond with our understanding of the supplier view. We propose that the supplier view on innovation is neither market nor product-driven, but primarily driven by inventions made in the basic scientific research that deals with the association between human health and various bacterial strains in the human body. Hence, in our opinion, characterizing the supplier view as being product or technology driven results in the loss of an essential dimension and quality of the supplier view, namely its persistent focus on health effects from the point of view of the individual consumer. This was taken up by the interviewed managers in relation to the different issues being discussed. When the managers were asked to describe "what innovation is all about," they gave the following answers: "Innovation is to use new knowledge, new insights from research, in a way that will provide value for the consumer—a health effect"; "Innovation is to come up with clinically well-documented bacterial strains that can be put into food products so food becomes a vehicle for the consumer."

From the focus on detecting in what directions science is developing in the field of bacterial genomes, or how it may evolve in the future, it follows that important sources of knowledge and learning relationships to a large extent reside in academic research institutions. In accordance with what is being treated as distinguishing features of FFNDP (Jones and Jew, 2007; Siedlok et al., 2010; Khan et al., 2013), the interviewed managers stressed the importance of establishing and nurturing collaborative relationships in the research society of biotechnology and preventive medicine. Our inquiry does not allow us to formulate statements about the nature of these relationships with respect to some general dimensions, such as, collaboration breadth and depth (as outlined by Martinez et al., 2014, in their discussion of the open innovation construct). However, our inquiry makes us suggest that in development, the collaborative relationships do to a considerable extent consist of relationships with individual specialists (i.e., academic researchers), and to a lesser degree of relationships with groups or organizations, as described in the previous quotation by one of the interviewed managers. Furthermore, that this feature also relates to the more or less continuous scientific advisory board affiliated with development, consisting of a small and carefully selected group of professors from academic research institutions and hospitals.

> *We have a scientific advisory board consisting of very well-known professors. We have formal meetings two times per year, but are in contact rather continuously. The board provides us with an evaluation of our ideas and projects from a purely scientific perspective. Of course, we have to be out there and be distinguished. Our academic profile gives us credibility. We take part in some larger networks, within the framework of some EU projects, mainly because it gives us good contacts. However, for us it is not about establishing relationships with companies or institutions. Our primary relations are with academic partners at an individual or group level. We work with scientists, not with research institutions.*

Our inquiry does not concur fully with the model by Bigliardi and Galati (2013) of the three main actors in the development of functional food (i.e., specialist, nutritionist, and food technologist). In accordance with their view, we suggest that nutritionists and specialists involved in research in the areas of health benefits, physiologic markers, or pathogenic bacteria are central actors in the supplier view. However, according to our study food technologists represent a main actor in "application," but not in development. Hence, even though "application" (i.e., the integration of a strain of bacteria into a food product) is related to several challenges (due to the sensitive quality of probiotics and the impact that living compounds in general have on the taste and shelf life of a food product, among other things), development and application are regarded as two different and separated processes and fields of knowledge. However, based on our interviews we conclude that the conception of the separation and its value for innovation may vary. In the first quotation, the proclaimed approach is to keep development totally "product neutral." Quite a different approach was outlined as follows by a manager in another company: "we use a stage-gate model, and the ideas we work with come from both technology (push) and the market (pull)." However, when the same manager was asked to describe what represents essential knowledge for the development of probiotics, the answer provided represented the opposite view:

> *It's definitely knowledge about the impact of microorganisms on health and on various disorders and sicknesses. We have medical doctors in our organization and we cooperate with the university hospital. Our technological platform is to produce microorganisms.*

The interface between development and application is a critical strategic issue for suppliers in the probiotic field. On the one hand, the two processes have to be kept apart in order for a supplier to attain or keep a leading position in the scientific community, that is, to keep the focus on developing new knowledge in the field of probiotics. On the other hand, the two processes have to be related though the market, and business growth relies upon not only the development of new probiotics, but also the development of new food products. In our understanding, the drivers for the latter are not primarily to be found in the geographical markets that presently, according to our interviews, demonstrate the strongest interest in probiotics.

Our sales, and our innovation, are driven by some customers and markets that show huge economic growth. Due to that, a large number of people in Asia have gained an increased standard of living. The markets for fresh dairy products such as yogurt are growing considerably. They can now afford to buy these products, and they have become aware of what creates health and wellbeing. The first step in these markets is traditional milk. The next stage is refined milk. That is, yoghurt. We try to target our products to the pains and needs of the populations in Japan, China and South Korea. Our strategy is very country oriented. We analyze the market with respect to, among other things, people's attitude towards vitamins.

4.2.3 AN INTEGRATED VIEW

We agree with Khan et al. (2013) that the conception of FFNPD needs to be radically changed. They state: "The development of innovative bioactive functional food products may require a paradigm shift in the process of food product development" (p. 35). The essence of the suggested change is the establishment of a diverse network of external partners connected to and engaged in the process. Hence, it is about adopting and enacting an integrated view on innovation. We recognize this as representing an essential, strategic viewpoint on the development of novel functional food products. As discussed in the previous section, our study into how FFNPD is understood in the literature and in practice from an ingredient supplier view demonstrates a considerable correspondence: high preoccupation on capturing novel scientific knowledge about the effects of a functional food ingredient on physiological functions in the human body on the one hand, and on the other, low preoccupation with monitoring of issues and developments related to food consumption and to food innovation in general. Our key argument for suggesting an integrated view is that the supplier view provides a partial understanding of FFNPD, as well as of the potential developments and growth possibilities that lie in the "Health and Wellness" market segment. As expected, we have found substantial support for our suggestion in the innovation management and strategy literature. The prevalent conception there is that innovation only occurs when the new knowledge being created has been put "into use" and thereby enhances value and newness for the company, its suppliers, its customers, and end users (Amabile, 1988; Teece, 2010; Tidd and Bessant, 2014).

In response to our proposition that the supplier view restricts the understanding of FFNPD, it may be claimed that the insights and experiences required to convert an invention of a new ingredient to a novel functional food product are supplied by the food manufacturing companies. However, according to the literature, the majority of food companies are still oriented toward continuously making small and incremental developments of existing products, and toward further developing their knowledge about how to become even more effective and efficient in the production and distribution of food products (Khan et al., 2013; Martinez et al., 2014; Pellegrini et al., 2014). In order to realize a shift from incremental innovation toward development of novel food products, the suggestion, or strategic viewpoint, offered in the literature is for the food industry to "open up" to external partners in knowledge creation processes related to innovation (Martinez et al., 2014; Pellegrini et al., 2014). The term "open innovation" has gained considerable recognition in the innovation management literature (Chesbrough, 2003; Dahlander and Gann, 2010; Huizingh, 2011; Tidd and Bessant, 2014). The core essence and principle of "open innovation" may be captured as follows: "Put simply, open innovation involves the recognition that 'not all the smart guys work for us' (Tidd and Bessant, 2014, p. 133). Another way to highlight its essence lies in recognizing that collaborating through the same communication channel continuously and over time (with suppliers, resellers, or business customers) "can blind the firm to other technological possibilities from other channels" (Martinez et al., 2014, p. 216).

The thesis that the existing nature of most companies' environments presupposes an extensive monitoring of knowledge (related to product and process technology, trends and developments on a societal level, and among consumers, etc.) cannot be repudiated. In order to proceed with the idea to "open up," an important strategic issue to be settled is: who might be the important cocreators of novel food products? Inspiration to answering this question is provided in the literature that takes on a business model perspective in dealing with innovation. The object of focus in this literature may appear not to be innovation of novel products or services, but rather design of business models. However, from agreeing that "a business model defines how the enterprise creates and delivers value to customers, and then converts payments received to profits" (Teece, 2010, p. 173), and that "every new product development effort should be coupled with the development of a business model which defines its 'go to the market' and 'capturing value' strategies" (p. 183), it follows that innovation and business model design are, or ought to be, two integrated fields of consideration and processes. However, as argued by Teece (2010), the integration is often overlooked by managers in charge of innovation.

As argued by DaSilva and Trkman, although the business model (BM) construct was introduced in the academic literature more than 50 years ago, its rate of diffusion in the business as well as in the academic world did not accelerate until 10 years ago. The explanation for this relates to what Teece (2010) discusses, that is, that a careful analysis and the design of a business model have not been central issues in the mindset of managers when considering innovation. However, it is apparent that a change is underway. One reason is that the BM construct opens up the scope for innovation and that this has been found to be related to the success of new product development (NPD). That is, BM makes managers associate NPD with other and related innovation areas, such as innovation of supply-chain processes and practices, innovation of communication processes to end users, innovation of a company's dominant logic, or core values, etc. (Tidd and Bessant, 2014). A second reason, which is important for our investigation, is that the BM construct encourages learning and continuous innovation by representing a "specific combination of resources which through transactions generate value for both customers and the organization" (DaSilva and Trkman, 2014). Hence, from linking the BM construct to NPD it follows that a strategic perspective on innovation (i.e., considering dynamic capabilities and other organizational resources) becomes as important as reflection on the scope, the processes, and the outcome of innovation. Consequently, the understanding of innovation becomes complete in the sense that it embraces more than one perspective on, or school of, learning (see the previous section). Finally, a third argument to consider the BM construct in relation to NPD is that the manifold set of potential partners for knowledge and value-creation becomes apparent. Accordingly, the BM construct demonstrates that there are many partners involved not just in delivering a certain value proposition, but also in securing its viable future existence.

Chesbrough (2007), who is one leading academic proponent of "open innovation," presents a framework that highlights the association between the nature of a company's BM and innovation strategy. The aim of the framework is to start up a change process by encouraging managers to determine at what stage their company is and to discuss why and how to progress along the path consisting of six stages, or company archetypes. The difference between the first and the sixth archetypes is significant. According to Chesbrough (2007), the vast majority of companies are at the first stage. They operate without having a distinct BM or an integrated process for managing it. These companies sell commodities, and their managers focus on pricing strategies and tactics, and how to increase the availability of their companies' market offerings. At the other end of the continuum are companies that are open to external ideas and technologies in ongoing processes to develop the company and its value proposition. However, these companies may differ with respect to the integrative role given to their BM, and the access that suppliers and customers have to innovation processes. In companies operating at the sixth stage, "key suppliers and customers become business partners, entering into relationships in which both technical and business risk may be shared" (Chesbrough, 2007, p. 15).

In order to develop the understanding of FFNPD in the literature and to proceed in practice with exploring the "health and wellness" segment, we suggest the introduction of the BM construct into the mindset of scholars and managers involved in innovation. Hence, following the argument by Teece (2010), we propose that the performance of investments made in FFNPD from the point of view of a supplier or a food manufacturer does not depend solely upon the invention of a new ingredient; it also depends on taking an integrated view on innovation, and implicitly on imagining a BM for functional foods. Another thesis, or strategic viewpoint, that we have detected on innovation of novel functional food products is the necessity to start up a debate about the nature of those business models that deliver value propositions to the "health and wellness" market segment or to a particular subsegment of it, such as probiotics. Hence, the following questions need to be addressed: what should the value proposition be? What are the essential processes and resources? Who are the important partners and relationships? In our study about how innovation and marketing are understood and implemented in the functional food area, these questions are most commonly answered by reflecting on what characterizes a business model in the pharmaceutical industry as well as in the FMCG industry. Supported by the references provided previously, we believe that a hidden potential for further development and growth of novel functional food products can be found by abandoning these business cases (i.e., business models in the pharma and the food industries).

How can this be achieved? How can a shift in the understanding of innovation be realized? How can new answers to the aforementioned questions be found? Our suggestion is very much influenced by our findings about the "health and wellness" market and the functional food sector, within and outside ingredient supplier and food companies, namely the presence of many "innovation champions," that is, people who are not only interested in what is happening in the "health and wellness" sector, but also have ideas about what are the essential issues and tasks to be performed by various partners (suppliers, manufacturers, resellers, and consumers) in order to drive and support the field. In response to this, research in the field of innovation and knowledge management keeps supporting the notion that successful new products originate from combining ideas, people, and objects in new and creative ways. This leads us to a final issue or strategic viewpoint, namely to connect people who do not work in the same department or organization, but who share a preoccupation with topics that deal with "health and wellness," food, and food innovation. In the literature, the construct of communities of innovations (CoInvs) has been introduced in an attempt to clarify this phenomenon. According to Coakes and Smith (2007), CoInvs represent a form of communities of practice. That is, a group of people "who share a concern, a set of problems, or a passion about a topic, and who deepen their knowledge and expertise in this area by interacting on an ongoing basis" (Wenger et al. 2002, p. 4). What distinguishes CoInvs, as indicated by the term, is that they are "very specifically dedicated to the support of innovation, and [that] their formation and sustainability are the responsibility of those individuals charged with organizational entrepreneurship" (Coakes and Smith, 2007, p. 77). In our understanding of FFNDP, in the literature and in practice, the long-existing and nourished CoInvs consists of an all too small and restricted community of specialists in food technology, nutrition, and medicine. Hence, our last viewpoint concerns recognizing "health and wellness" as being not only a megatrend to respond to, but also a megatrend to connect with.

4.3 MARKETING OF FUNCTIONAL FOOD PRODUCTS: UNDERSTANDING THE CONSUMER

Considering the high failure rates of innovations in general and functional food innovations in particular (Menrad, 2003), it becomes pivotal to gain a better understanding of the consumer. In recent years, scholarly work on consumers of functional foods has gained increasing interest. Thereby, extant research is characterized by a focus on consumer acceptance of functional foods. Although inherently closely connected and often investigated in one study, the body of work on consumer acceptance can be broadly divided into three areas of investigation: first, product-oriented research investigates differences in products and their effect on consumer acceptance with a focus on the carrier product; second, communication-oriented research focuses primarily on health claims; and, third, research with a focus on consumer characteristics is mostly concerned with market segmenting. Methodologically, all three areas are dominated by correlational research, exploratory experimental research, and conjoint analyses. In the following section, we provide an overview of extant findings in these three broad areas. Against this line of work, we then outline a new view on the consumer, based on the premise of the consumer as a cocreator of value.

4.3.1 EXTANT RESEARCH: THE PRODUCT AT CENTER-STAGE

Product-oriented market research is dominated by investigations of questions surrounding the role of the carrier product and is inherently concerned with outcomes such as consumers' purchase intentions, willingness to buy, or general preferences. Researchers are in general agreement that the carrier product plays a crucial role in consumer preferences, a factor that may even outperform any socio-demographic or other drivers of acceptance (Ares and Gámbaro, 2007). Whereas prior research has found that consumers may find enrichments in nonhealthy food more reasonable than in food that is already healthy (Krutulyte et al., 2011), other work has claimed that inherently healthy food products, such as yogurt, are more credible carriers of functional benefits (Siegrist et al., 2008). Ares et al. (2008) find that, although inherently healthy food products may be credible carriers of enrichments, these kinds of food may not benefit from enrichments in their healthiness perceptions, which, in turn, is a pivotal driver of consumers' willingness to try new products. The authors suggest that products of an intermediate healthiness level would potentially benefit most from enrichments. Krutulyte et al. (2011) extend this research and show that it is not the carrier product alone that determines consumers' acceptance, but rather that it is the perceived fit between carrier and ingredient which predicts the intention to purchase functional food more than health concerns or consumers' general attitude toward functional food. Lu (2015) builds on this finding and demonstrates that purchase intention depends on the perceived fit between carrier and ingredient, which in turn is found to depend on the individual's knowledge about nutrition. More particularly, expert consumers are not affected as much by a carrier–ingredient fit as are novice consumers. Moreover, the results suggest that the weaker the carrier–ingredient fit, the more important the health claim becomes.

The importance of health claims in consumer acceptance of functional foods is investigated in communication-oriented research. In general, research has established that physiological and prevention health claims affect purchase intentions more positively than do psychological or general health claims (Annunziata and Vecchio, 2013; Siegrist et al., 2008). However, the effects of health claims are not independent of other factors, such as actual sensory and nonsensory product attributes. For example, Siegrist et al. (2008) find that consumers do not perceive health claims independently from the carrier. Extending this finding, Ares et al. (2010) find that the influence of the carrier product (here, yogurt) and of the brand outperforms that of price and health claim. It may also depend on consumers' evaluation and their perceived relevance of healthiness whether a health claim affects consumers' purchasing behavior, a topic that has been taken up in research investigating consumer characteristics.

This line of research is mostly concerned with understanding general trends and specific consumer characteristics in order to be able to describe and segment the target market. On a more general level, scholars have recognized that consumers are developing an increasing interest in and understanding of the relation between health and nutrition, with consumers becoming increasingly reflective and willing to adopt health-oriented changes in their eating habits (for an extensive review, see Siro et al., 2008). Whereas the common, lay belief of "unhealthy = tasty" seems to be fading, that is, good taste and healthiness are no longer necessarily perceived as trade-offs (Verbeke, 2006), it appears that consumers' willingness to compromise on taste for health benefits depends on several factors. Verbeke (2006), for example, suggests that the product category and the strength of the health claim may play important roles when consumers consider compromising. However, the author also raises concerns as to whether consumers actually consider functional foods as part of a healthy and tasteful diet, and finds a decreasing trend. Indeed, consumers' belief in the health benefits of functional foods appears to be a crucial factor in functional food acceptance, including when due to the presence of an ill family member, even outweighing socio-demographics as potential determinants (Verbeke, 2005). Knowledge and awareness of the concept, however, appear to have a contrary effect on functional food acceptance, at least at a younger age. In a similar vein, investigating the role of general nutritional knowledge on consumers' health perceptions and willingness to try functional foods, Ares et al. (2008) find that nutritional knowledge affects healthiness perceptions of functional foods positively, but does not necessarily increase preference of functional foods for all consumer groups.

General nutritional knowledge or awareness of the functional food concept in particular may interact with other consumer characteristics on the individual level, such as age, gender, and culture. For example, age is only partly found to affect willingness to buy (Siegrist et al., 2008). Older consumers appear to be more willing to compromise sensory characteristics for health (Ares et al., 2010; Verbeke, 2006). Although women are primarily responsible for food purchasing and appear to be more reflective of the connection between health and nutrition, gender has not been reliably found as a predictor for preference (Siro et al., 2008). Instead, other factors might weigh heavier, such as the carrier product or health claims. For example, Ares and Gámbaro (2007) find different preference patterns for different types of consumer groups in Uruguay, characterized by age, gender, and motives. That is, some consume certain types of functional foods for health reasons (e.g., nutritional factors, weight loss, high in protein or fiber) and others for reasons of sensory appeal (e.g., taste, texture, looks). However, the strongest predictor for preference remains the carrier. The authors conclude that there is no unconditional acceptance of functional food per se, but rather that the product itself determines acceptance or rejection. Eventually, all product- and communication-oriented factors may speak to different target groups. Annunziata and Vecchio (2013), for example, distinguish between three different consumer groups. The first one is found to be familiar with functional foods, exhibiting higher consumption rates of functional foods and considering the carrier and health claim as the most important attributes. The second group is characterized by a lower consumption of functional foods, being driven by arguments such as brand and price. The third group has less trust in the effectiveness of functional foods, being mostly affected by the carrier brand and price.

Acknowledging differences on the regional and cultural levels, scholarly work has generally suggested that Americans appear to be more accepting of the functional foods concept, although healthy food consumption is still low. Europeans are considered more skeptical, and food consumption is more deliberate, and especially Southern European countries stick to a natural diet (for a summary, see Siro et al., 2008). Investigating consumers in three different countries, Bech-Larsen and Grunert (2003) find only small differences in healthiness perceptions between Finns, Danes, and Americans, with Danes exhibiting an increased rejection of functional foods. Again, the role of the carrier product appears to be pivotal, also in intercountry comparisons. Bech-Larsen and Grunert (2003) find that health claims can enhance consumers' perception of the healthiness of functional foods, but that this perception is dependent on perceived nutritional qualities of the base-product. "Therefore, the results imply that food producers considering marketing a functionally enriched alternative should be very particular in their research of consumer attitudes to the particular base-product and enrichment involved" (Bech-Larsen and Grunert, 2003, p. 13). On a more general level, recent studies have also emphasized the important function of trust in the food industry (Siegrist et al., 2008). Whereas food-neophobia, that is, the reluctance toward novel food concepts, appears to play a role only for some product categories (Siegrist et al., 2008; Urala and Lähteenmäki, 2007), culturally rooted distrust in the industry may account for the rejection of novel concepts more generally. For example, investigating German and Russian consumers, Dolgopolova et al. (2015) suggest that distrust in the food industry can lead to rejection of functional foods in both cultures. However, since distrust has different cultural roots, it may be harder to overcome in Russia where distrust in information systems is more culturally rooted. The authors propose different educational strategies: whereas informal self-developed networks are at work in Russia, German consumers are characterized by a strong belief in scientific results.

4.3.2 AN INTEGRATED VIEW: THE CUSTOMER AS A COCREATOR/"BUSINESS PARTNER"

Whereas extant research on consumer acceptance of functional foods greatly enhances our knowledge in the areas of market segmentation and targeting, a more comprehensive understanding of the consumer is lacking (Siro et al., 2008; Verbeke, 2005). We argue that marketing and innovation knowledge in the area of functional foods would benefit from a more holistic view on the consumer. More specifically, as proposed by fundamental work based on the service-dominant logic (Vargo and Lusch, 2004, 2008), contemporary marketing has acknowledged the consumer beyond acts of pure consumption. The point of departure is not the product as the "organizer" of opportunities for the firm (Payne et al., 2008). Instead, the consumer is described as a cocreator of value, and "relevant meanings are created by the experiences a customer has over time" in a constant exchange with firms (Payne et al., 2008, p. 86). As a cocreator of value, the consumer is not simply regarded as a rational decision-maker who makes decisions about accepting or not accepting firm-generated products and their communicated benefits. Instead, understanding consumers' experiences with a firm, which are a culmination of their cognitions, emotions, and behaviors, becomes pivotal. This view on cocreation entails, but goes beyond, the role of consumers as idea generators or cocreators at the product development stage. Value is cocreated at all stages, or more specifically, in all potential encounters with the firm. The notion of value cocreation by the consumer therefore does not only concern activities in product development, but profoundly changes the way we look at all marketing efforts, including product innovation. This view, therefore, is helpful in opening new perspectives that go beyond understanding the consumer as an adopter and recipient of firm-generated products and messages. The following section outlines new opportunities in the light of the notion of the consumer as the cocreator of value against the previously described consumer acceptance research and sketches new paths for a consumer understanding that changes the way firms look at product development and marketing communications.

The notion of cocreation is not completely new to functional food research. A few scholars have stipulated a better integration of consumers' demands in the early phases of the product development process (Van Kleef et al., 2002). It is argued that consumer input is pivotal in the development process, as it "can help to better understand these issues so that product developers can overcome confusing or fear-raising aspects, for example, in their marketing and communication activities" (Van Kleef et al., 2002, pp. 99f). Researchers are very well aware of the potential obstacles and concerns when involving consumers in the product development process that have been raised by practitioners and researchers alike, such as consumers demanding new functions and features that are neither innovative, and as such not competitive, nor technologically feasible (Hoyer et al., 2010). Van Kleef et al. (2002), for example, compare ideas of internal experts to those of consumers and suggest that, despite several challenges, consumers need to be better integrated into the development process. Siro et al. (2008) suggest in particular that market research should better understand what diseases consumers are concerned with, and then identify those diseases that consumers would prevent by a change in diet. Scholars also point to the complexity of the stakeholder system, and, in order to meet these challenges, they suggest that consumer desire, legislative boundaries, and technical feasibility should help define the priorities of R&D programs (Siro et al., 2008; Van Kleef et al., 2002). More research is needed in this area.

Value cocreation, as defined by Payne et al. (2008), goes beyond the idea of integrating consumers' demands in product development processes. It understands the customer as a cocreator of value when getting in contact with the firm and its products, services, and messages. More specifically, this understanding takes its point of departure in the customers who engage in a series of cocreational activities to reach a particular goal by using their knowledge, skills, information, and other resources. If firms want to improve their competitiveness, they have to either provide the customer with some sort of resource or influence the customer's process to reach goals by making it more efficient and effective (Payne et al., 2008). It becomes a requirement for firms to understand the customer encounter processes, that is the "touchpoints," which can be firm- or customer-initiated or both, such as direct mailings, trade shows, or the actual consumption of the product. For example, when consuming food in general, a consumer's goal may go beyond merely satisfying hunger. It may involve preparation of the food that consumers derive value from, the social aspect of consuming food together, and the actual taste of the food. Different food products may serve all goals in different ways. Some foods may ease the preparation of food or make it more enjoyable, other foods are particularly designed for being enjoyed in the company of others, and yet other foods satisfy sensory goals. In all cases, a different value is created by the consumer by consuming the product or even taking part in its production, as in the case of preparing food, termed "prosumption" (Xie et al., 2008). This cocreational perspective moves the focus away from merely material substances (i.e., the food product and its attributes) to processes, ultimate goals, and values, and, in doing so, "detangibilizes" the traditional concept of a product (Xie et al., 2008). This has repercussions for all marketing and product development efforts.

As discussed previously, consumer acceptance research to date has been driven by an understanding that puts the product and its attributes at center-stage where the consumer is understood as a passive buyer of what firms produce, involving only single acts, such as the purchase (Xie et al., 2008). Only a few researchers have tapped into consumers' goals of consuming functional foods that go beyond the mere health-beneficial function. For example, Urala and Lähteenmäki (2007) find that consumers derive pleasure not only from improved health per se, but also from an improved mood and performance together with the possibility of preventing disease and gaining the healthy lifestyle that these foods enable. This finding corroborates the cocreational perspective and, at the same time, calls for further investigation of consumers' goals of consuming functional foods on the individual and collective levels. For example, whereas consumers' search for and experimentation with individual and social identities has been acknowledged in many other consumption research areas, including general food consumption (Bisogni et al., 2002), it has not been addressed in functional foods research. This is rather surprising, since today's consumers, especially the younger generation, who are particularly interested in health, use products and brands for goals of identity enhancement, especially in social media (for an overview, see Bolton et al., 2013). Considering research results that suggest an increasing interest in functional foods with increasing age, it might be worthwhile to investigate whether these results come about due to a restricted view on functional foods consumers. Taken together, understanding cocreational processes can provide firms with a deeper understanding of customers and their experiences if they accept that customers' goals may go beyond a pure health function when consuming functional food. Customer insights from monitoring and understanding encounter processes and the experiences customers make in these could also feed back into the product development process. More specifically, if firms acknowledge marketing's role in understanding the different values consumers derive from functional food based on individual and collective goals, it will be able to uncover consumers' latent and future needs and wants.

Taking a cocreational perspective will also change the role of marketers "from being merely mass-communicators to becoming instructors, coaches, and advisers" (Xie et al., 2008, p. 119). A more fundamental challenge is the often-discussed perception of functional foods as a distinct product category. Several authors have pointed out consumers' confusion about functional food. Consumers seem to perceive functional food not as a product category in itself (Annunziata and Vecchio, 2013; Urala and Lähteenmäki, 2003; Urala and Lähteenmäki, 2007). Instead, they appear to evaluate "functional foods first and foremost as foods. Functional benefits may provide added value to consumers but cannot outweigh the sensory properties of foods" (Siro et al., 2008, p. 465). This perception may at least partly explain the different acceptance levels of different functional food products. More specifically, if consumers lack an understanding of and a reference point to functional food, they will turn to other characteristics of the product, such as the carrier product or the brand, and acceptance will depend on those. As a consequence, an acquired familiarity with carrier products and their health benefits, as in the case of yogurt (Siegrist et al., 2008), and with established combinations of carrier and enrichment becomes a pivotal driver of actual acceptance (Krutulyte et al., 2011). We argue that innovation and marketing efforts that ignore this circle are short-sighted: if research keeps defining the consumer as a passive "acceptor" of the product, it obstructs new ways of innovating and positioning functional food products. That is, it will continuously arrive at the conclusion that only a few carrier products or combinations are "acceptable," ignoring the fact that different factors, such as familiarity, may be at play. Familiarity, however, is a perceptual, subjective measure and can be established through various marketing efforts, including the product itself, package designs, and positioning efforts (Pine and Gilmore, 2011).

In a similar vein, extant research that has focused on the communication of a health claim may be enhanced by considering different angles to message framing. For example, scholarly work has largely ignored insights into why consumers reject functional foods, understanding rejection not as "nonaccepting," but as a distinct process (Kleijnen et al., 2009). If perceived risk and a perceived lack of compatibility with consumers' values and goals lead to rejection and therefore prevent adoption, communication strategies may have to consider other than merely health-claiming messages. Further, understanding the different kinds of values consumers can derive from consuming functional foods, such as identity-enhancing or experiential values (for a framework, see Smith and Colgate, 2007), will provide new paths for marketing communications that acknowledge the consumer beyond single acts of passive consumption.

In summary, communication research and practice in the area of functional foods have to enter a new debate and discuss whether it is actually important for consumers to perceive functional foods as firstly a product category in itself, and secondly as (purely) functional. It is questionable whether the term "functional" meets consumers' modern lifestyles, which revolve around experiences rather than functions, and whether the conventional technical focus resonates with other trends, such as organic and natural food. Gaining consumer insights based on a cocreational understanding of the consumer appears to be a fruitful path to new innovation and marketing strategies.

4.4 DRIVERS AND BARRIERS TO INDUSTRY DEVELOPMENT

As discussed initially, in the course of our study we have found a seemingly high degree of consensus in the literature and also in practice about a number of overall characteristics of the FFNPD field, regarding its distinguishing characteristics and challenges. Being now in the phase of summing up strategic viewpoints of innovation and marketing, we see the "point" in returning to this quality and in reflecting upon two drivers of industry development about which the consensus also appears to be high. "To reflect" may have different meanings. Here, we delimit our reflection to strategic viewpoints on innovation and marketing that are presently strongly emphasized in the general literature, but less emphasized in the literature dealing with innovation and marketing in the functional food field. The most important driver mentioned by the interviewed managers, irrespective of their field of practice—research, development, marketing, or general management—concerns the potential for new research insights and inventions in the field of bacterial genomes. Hence, a recurrent statement made by the managers we interviewed was the following: "Research is the most important driver. It's a huge focus on the impact of bacterial strains on health. That's the big research area. That's where the opportunities are." From a scientific point of view in the fields of biotechnology or preventive health, this proposition or scope in innovation cannot be denied.

Without denying the crucial role of biotechnical or preventive health research in FFNPD, it is limited to a focus on health and the functions that a product can fulfill for the end consumer's body. However, acknowledging the variety of motivations and the different individual and social situations of food consumption, the relevance for FFNPD of understanding consumers' minds cannot be ignored. As presented previously, marketing theory has moved away from a product-centered perspective—and with it, from a purely functional view on the product—to a service-dominant logic. This logic challenges firms to understand consumers as cocreators of value, acknowledging that acts of consumption are experienced subjectively and from these experiences value can be derived beyond purely functional product attributes. More generally, consumers follow different goals in life on individual and social levels, and food, including functional food, may play major or minor roles in their lives. On a more specific level, different situations and contexts may determine what kinds of value consumers cocreate in the consumption process: whether it has been consumers' striving for general health and well-being that brought them to functional food; whether life events, such as the death of a relative, changed their eating habits, and functional food serves as a prevention measure; whether purchasing and consuming functional food is seen as part of a lifestyle and identity; or whether someone wants to do something good to others, and functional food helps fulfill a social role. Along the same lines, consumers' motivations for more or less explicitly rejecting functional food may be an equal source for product development, including the quest for new carrier products and different combinations of bacterial strings and carrier products. Ignoring consumers' roles as cocreators of value, thus ignoring that consumers derive different kinds of values from subjective consumption experiences, may mean ignoring opportunities and building a barrier to innovation, in both product development and marketing. A more comprehensive consumer understanding—acknowledging the needs of consumers' minds and bodies—is consequently not understood as a substitute for biotechnological research or as the most pivotal driver of new product development, but rather as an equally important source for opening up and tapping into unexplored fields in FFNPD.

Presently, more than 6 billion people drink milk and various fertilized milk products regularly, but still the majority of milk consumers live in developing countries. However, the gap in per capita consumption between European countries and many developing countries in East and Southeast Asia, particularly in China, Indonesia, and Viet Nam, is decreasing. In accordance with what international bodies in the food consumption field conclude (www.fao.org/agriculture/), the interviewed managers emphasized this evolving and new market demand for milk products as representing an essential business opportunity and an important future growth driver for the dairy sector as well for suppliers of ingredients in the functional food field. As noted in the section dealing with the supplier perspective, implanting probiotics into dairy and other food products (i.e., 'application') is perceived as representing a particular field of knowledge and competence. The reason is that application can presuppose and give rise to different degrees of novelty, proceeding from minor, that is, incremental, changes of the formula of an existing fruit yogurt product to more or less radical changes in the way a dairy produces, packages, and markets a fertilized milk product as yogurt, and eventually also in the way consumers will consume it. Our reflection on the aforementioned driver to industry development concerns the separation of the two innovation processes: "development" and "application" in the supplier view. In the innovation management literature, the issue of separating different innovation processes and securing a balance between exploitation and exploration has for a long time been regarded as a key challenge in businesses and as a core topic in the field of management and strategy (Lavie et al., 2010). The challenge stems from the view that exploitation "represents a capability to refine existing competencies and resources to improve operational efficiency" (Menguc and Auh, 2008, p. 457) whereas exploration represents "the capability of an organization to learn through questioning, pushing the knowledge frontier, and engaging in proactivity and risk taking" (p. 457). The ability of organizations to be oriented toward both exploitation- and exploration-oriented continuous learning processes is being captured by the term ambidexterity in the literature (Lavie et al., 2010; Raisch and Birkinshaw, 2008). As a result, according to the literature, the separation in the supplier view of "development" and "application" appears as a productive path toward securing that "development" keeps its focus on science-based radical innovation of food ingredients. However, considering the scope in "development" (i.e., excluding food technology and market issues) together with the emphasis given to the increased demand for functional dairy products on a world level, we suggest that ambidexterity become a core construct and path for "application."

The main argument is that a future key driver for the probiotic industry presumably will revolve around supporting the food industry in becoming more focused on developing new functional food products (exploration) and less oriented toward adding a new or an already well-established ingredient to an existing and mature product (exploitation). In enacting this role, we suggest that suppliers in the probiotic field and their business partners (the dairy and the food industries) take an integrative view on innovation and marketing, as it has been outlined in the previous sections. We are aware that much is left to be discussed and carefully explored with respect to the integrative views, from a theoretical perspective and in relation to development of new functional food products. Furthermore, we are aware that some of the strategic viewpoints presented are already in practice in a few food companies and their suppliers. However, from a research point of view in innovation management and marketing, many aspects of the views remain to be studied and communicated to various stakeholders in the functional food community.

REFERENCES

Amabile, T.M., 1988. A model of creativity and innovation in organizations. Straw, B.M., Cummings, L.L. (Eds.), Research in Organizational Behavior, vol. 10, JAI Press, Greenwich, CT, pp. 123–167.

Annunziata, A., Vecchio, R., 2013. Consumer perception of functional foods: a conjoint analysis with probiotics. Food Qual. Prefer. 28, 348–355.

Ares, G., Gámbaro, A., 2007. Influence of gender, age and motives underlying food choice on perceived healthiness and willingness to try functional foods. Appetite 49, 148–158.

Ares, G., Giménez, A., Gámbaro, A., 2008. Influence of nutritional knowledge on perceived healthiness and willingness to try functional foods. Appetite 51, 663–668.

Ares, G., Giménez, A., Deliza, R., 2010. Influence of three non-sensory factors on consumer choice of functional yogurts over regular ones. Food Qual. Prefer. 21, 361–367.

Bech-Larsen, T., Grunert, K.G., 2003. The perceived healthiness of functional foods: a conjoint study of Danish, Finnish and American consumers' perception of functional foods. Appetite 40, 9–14.

Bell, S.J., Whitwell, G.J., Lukas, B.A., 2002. Schools of thought in organizational learning. J. Acad. Market. Sci. 30, 70–86.

Bigliardi, B., Galati, F., 2013. Innovation trends in the food industry: the case of functional foods. Trends Food Sci. Technol. 31, 118–129.

Bisogni, C.A., Connors, M., Devine, C.M., Sobal, J., 2002. Who we are and how we eat: a qualitative study of identities in food choice. J. Nutr. Educ. Behav. 34, 128–139.

Bolton, R.N., Parasuraman, A., Hoefnagels, A., Migchels, N., Kabadayi, S., Gruber, T., Komarova Loureiro, Y., Solnet, D., 2013. Understanding Generation Y and their use of social media: a review and research agenda. J. Service Manag. 24, 245–267.

Chesbrough, H., 2003. Open Innovation: The New Imperative for Creating and Profiting from Technology. Harvard Business School Press, Boston, MA.
Chesbrough, H., 2007. Business model innovation: it's not just about technology anymore. Strat. Leadership 35, 12–17.
Coakes, E., Smith, P., 2007. Developing communities of innovation by identifying innovation champions. Learn. Organ. 14, 74–85.
Dahlander, L., Gann, D., 2010. How open is innovation? Res. Policy 39, 699–709.
DaSilva, C.M., Trkman, P., 2014. Business model: what it is and what it is not. Long Range Plann. 47, 379–389.
Dolgopolova, I., Teuber, R., Bruschi, V., 2015. Consumers' perceptions of functional foods: trust and food-neophobia in a cross-cultural context. Int. J. Consum. Stud. 39, 708–715.
Hoyer, W.D., Chandy, R., Dorotic, M., Krafft, M., Singh, S.S., 2010. Consumer cocreation in new product development. J. Serv. Res. 13, 283–296.
Huizingh, E.K., 2011. Open innovation state of the art and future perspectives. Technovation 31, 2–9.
Jones, P.J., Jew, S., 2007. Functional food development: concept to reality? Trends Food Sci. Technol. 18, 387–390.
Khan, R.S., Grigor, J., Winger, R., Wind, A., 2013. Functional food product development—opportunities and challenges for food manufacturers. Trends Food Sci. Technol. 30, 27–37.
Kleijnen, M., Lee, N., Wetzels, M., 2009. An exploration of consumer resistance to innovation and its antecedents. J. Econ. Psychol. 30, 344–357.
Krutulyte, R., Grunert, K.G., Scholderer, J., Lähteenmäki, L., Hagemann, K.S., Elgaard, P., Nielsen, B., Graverholt, J.P., 2011. Perceived fit of different combinations of carriers and functional ingredients and its effect on purchase intention. Food Qual. Prefer. 22, 11–16.
Lavie, D., Stettner, U., Tushman, M.L., 2010. Exploration and exploitation within and across organizations. Acad. Manag. Ann. 4, 109–155.
Lu, J., 2015. The effect of perceived carrier-ingredient fit on purchase intention of functional food moderated by nutrition knowledge and health claim. Brit. Food J. 117, 1872–1885.
Martinez, M.G., Lazzarotti, V., Manzini, R., Garcia, M.S., 2014. Open innovation strategies in the food and drink industry: determinants and impact on innovation performance. Int. J. Technol. Manag. 66, 212–241.
Menguc, B., Auh, S., 2008. The asymmetric moderating role of market orientation on the ambidexterity–firm performance relationship for prospectors and defenders. Ind. Market. Manag. 37, 455–470.
Menrad, K., 2003. Market and marketing of functional food in Europe. J. Food Eng. 56, 181–188.
Payne, A.F., Storbacka, K., Frow, P., 2008. Managing the co-creation of value. J. Acad. Market. Sci. 36, 83–96.
Pellegrini, L., Lazzarotti, V., Manzini, R., 2014. Open innovation in the food and drink industry. J. Agric. Food Ind. Org. 12, 75–94.
Pine, B.J., Gilmore, J.H., 2011. The Experience Economy. Harvard Business Press, Boston, MA.
Raisch, S., Birkinshaw, J., 2008. Organizational ambidexterity: antecedents, outcomes, and moderators. J. Manag. 34, 375–409.
Siedlok, F., Smart, P., Gupta, A., 2010. Convergence and reorientation via open innovation: the emergence of nutraceuticals. Technol. Anal. Strat. Manag. 22, 571–592.
Siegrist, M., Stampfli, N., Kastenholz, H., 2008. Consumers' willingness to buy functional foods. The influence of carrier, benefit and trust. Appetite 51, 526–529.
Siro, I., Kapolna, E., Kapolna, B., Lugasi, A., 2008. Functional food. Product development, marketing and consumer acceptance—a review. Appetite 51, 456–467.

Smith, J.B., Colgate, M., 2007. Customer value creation: a practical framework. J. Market. Theor. Prac. 15, 7–23.

Teece, D.J., 2010. Business models, business strategy and innovation. Long Range Plann. 43, 172–194.

Tidd, J., Bessant, J., 2014. Strategic Innovation Management. John Wiley & Sons Ltd, Chichester, UK.

Tollin, K., Vej, J., 2012. Sustainability in business: understanding meanings, triggers and enablers. J. Strat. Market. 20, 625–641.

Urala, N., Lähteenmäki, L., 2003. Reasons behind consumers' functional food choices. Nutr. Food Sci. 33, 148–158.

Urala, N., Lähteenmäki, L., 2007. Consumers' changing attitudes towards functional foods. Food Qual. Prefer. 18, 1–12.

Van Kleef, E., Van Trijp, H.C., Luning, P., Jongen, W.M., 2002. Consumer-oriented functional food development: how well do functional disciplines reflect the 'voice of the consumer'? Trends Food Sci. Technol. 13, 93–101.

Vargo, S.L., Lusch, R.F., 2004. Evolving to a new dominant logic for marketing. J. Market. 68, 1–17.

Vargo, S.L., Lusch, R.F., 2008. Service-dominant logic: continuing the evolution. J. Acad. Market. Sci. 36, 1–10.

Verbeke, W., 2005. Consumer acceptance of functional foods: socio-demographic, cognitive and attitudinal determinants. Food Qual. Prefer. 16, 45–57.

Verbeke, W., 2006. Functional foods: consumer willingness to compromise on taste for health? Food Qual. Prefer. 17, 126–131.

Wenger, E., McDermott, P., Snyder, W.M., 2002. Cultivating Communities of Practice. Harvard Business School Press, Boston, MA.

Xie, C., Bagozzi, R.P., Troye, S.V., 2008. Trying to prosume: toward a theory of consumers as co-creators of value. J. Acad. Market. Sci. 36, 109–122.

CHAPTER

5 Food safety: marketing trends intended for consumer awareness and partnership

Palma Ann Marone
Toxicology and Pathology Associates, Department of Pharmacology and Toxicology, Medical College of Virginia, Virginia Commonwealth University, Richmond, VA, United States

5.1 INTRODUCTION

Like two galaxies moving past one another, the forces of food availability, sufficiency, and sustainability for a growing population are colliding with the increasing demand for healthy food choices, the functional foods market. Add to that the interdependence of nations and the expanding burden on natural resources, now considered approaching the earth's quantitative limits, and good food choice based on education becomes a critical partnership between manufacturers and consumers alike for assurance of a safe food supply. The growing trend toward health awareness among an increasingly educated world population is exerting an increasingly greater influence over food choices, particularly in developed countries with affluent consumers. Biologically, environmental conditions including dietary factors can induce reversible heritable traits through epigenetic mechanisms, defined as somatically heritable states of functional gene expression resulting from changes in chromatin structure without alterations in the primary DNA sequence (change in phenotype without change in genotype), thereby affecting gene expression in a transgenerational manner (Choi and Frisco, 2010). The importance of identifying positive epigenomic markers through diet (nutriepigenomics) and nutritional supplementation coupled with other lifestyle changes has the capability to turn medical science from disease treatment to disease prevention and with it, the saving of countless human lives and medical resources. One challenge for nutrition experts is to identify the precise epigenomic signatures that correlate with improved overall health resulting from positive changes in diet or lifestyle behaviors. Presently, the search for positive dietary influences has addressed a number of common conditions, including aging, in the form of cosmeceuticals and antioxidants; inflammation for free radical control; mental acuity and alleviation of Alzheimer's-like symptoms; metabolic disease, diabetes, and obesity; joint and bone health; and cancer and chronic disease, as examples. Conversely,

Developing New Functional Food and Nutraceutical Products. http://dx.doi.org/10.1016/B978-0-12-802780-6.00005-5

it is well known that presumably negative epigenetic changes relating to poor diet are more difficult to reverse as developmental delays experienced during gestation and early childhood, particularly, tend to become long lasting. Although these effects may vary in their severity, those relating to critical central nervous system function or predisposing individuals to chronic conditions such as obesity, hypertension, and cancer exact the greatest costs in both human and economic terms (Gallou-Kabani et al., 2007; Yadav and Maurya, 2014; Donkin et al., 2016).

5.2 FACTORS INFLUENCING FOOD CHOICES

The consequences of individual food choices become cumulative over time. In order to effect a positive impact on health and lifestyle overall, a greater understanding of the determinants of food choice is required. The European Food Council (EUFIC) offers an excellent review and categorizes the major determinants of food choices into biological, economic/physical, social, psychological, and attitudinal factors, all of which vary independently over time under changing conditions and between populations, communities, and individuals and are integrated to influence collective food choice behaviors (EUFIC, 2005).

5.2.1 BIOLOGICAL

Physiologic need is considered the most basic determinant of food choice. The ability to satisfy fundamental energy needs for health and maintenance of the body is largely driven by regulation of appetite through the central nervous system (Morton et al., 2014). The complexity of signals directing behaviors of hunger and satiety is thought to originate in the hypothalamus, requiring specialized cells with the ability to assess peripheral metabolic status and to process this information in a coordinated response for balance of energy intake and expenditure in both short and long terms (Stubbs et al., 1996; Smith and Ferguson, 2008; Rui, 2013). The need to alleviate sensations of hunger is a strong motivator that will influence both food choice and portion sizing according to these integrated neurologic cues. When food is readily available, the ability to make choices based upon good nutrition and high satiety indices is preferable for weight control and general health. Optimized choices for general overall health comprise low-energy density foods high in protein and fiber. Higher energy density carbohydrate-rich foods for which the glycemic index provides an indicator of blood glucose elevation, followed by foods high in fat are considered less satiating and contribute toward overconsumption and obesity. When food is not readily available, the drive to alleviate hunger oftentimes results in nutritionally poor food choices based simply on what is accessible. This situation is more common in developing countries and in vulnerable populations (children, the elderly) who generally rely on selection assistance.

Most people generally claim taste as the most important factor in choosing what to eat. The sensation of "taste" is actually a combination of taste, appearance, smell, and texture, all sensations of fundamental biologic origin, specifically, under the control of the gustatory cortex of the brain (Drewnowski, 1997). Perceptions of taste may be acquired but its overall broad organoleptic appeal is a universally acknowledged strong component of food selection, contributing to enhanced awareness of palatability for the pleasure associated with eating. Numerous clinical studies and population surveys have found that, given the choice, combinations of sugar and fat are generally most appealing, with sweetness preferable and bitterness avoided. So universal are these cross-cultural selections that they are considered innate human traits present from birth (Steiner, 1977).

Other biologic components of food selection may be related to specific medical conditions. The need to limit or avoid some foods from the normal diet due to organic conditions associated with diabetes or food allergies/intolerances will restrict food choice. Depending on the health condition, the effects of this type of food avoidance may be profound, as in the case of carbohydrate avoidance among diabetics or of simple elimination of the singular offending food, such as dairy products in lactose intolerance or a specific allergy. When these conditions occur in children or the elderly, food choice must be carefully monitored through medical consultation to assure balanced nutrition when an essential nutrient is eliminated.

Finally, foods may be consciously selected based on the substantive evidence of the health benefits of ingredients. It is well known that diets higher in fruits, vegetables, and whole grains have been associated with reduced risks of atherosclerosis, type 2 diabetes, heart disease, certain cancers, obesity, eye diseases, dementia, osteoporosis, and overall mortality. Similarly, excess sodium intake has been associated with hypertension, stroke, kidney disease, and coronary heart disease, while diets high in certain processed meats may increase the risk of breast, colon, and prostate cancers (Bingham, 1999; World Cancer Research Fund, 2007; Morrison and Ness, 2011; Boeing et al., 2012).

5.2.2 ECONOMIC AND PHYSICAL

The cost of food is an important driver in food choice influenced on an individual level by income and socioeconomic status, and further complicated by gender, age, culture, and environment. On the community level, economic marketing trends will determine available food offerings and possible food quality, particularly as growth in local ethnic populations influences food availabilities and economies. Oftentimes, the most healthful of choices are also the least affordable. High-fat and high-calorie sugar foods tend to cost less than more healthful alternatives such as fruits and vegetables, making it difficult for low-income consumers to afford options needed for optimal nutrition (De Irala-Estévez et al., 2000; Darmon and Drewnowski, 2008). Poor choices result in undernutrition (lack of vital nutrients) despite overconsumption that predisposes toward obesity. Food insecurity resulting from poverty is exacerbated when areas become logistically isolated, limiting grocery store options (inventories) and transportation access for better alternatives. The cities and small towns where poverty most often exists are often more expensive with fewer consumer options than are available in the supermarkets of the more wealthy suburban areas.

Access to education and nutrition knowledge also directly affect food selection. Oftentimes, individuals in lower socioeconomic groups lack food education or receive conflicting information, making it even more difficult to break the poor nutrition cycle for themselves and their families. Limited access to reliable media-disseminated information or professional personnel reduces comprehension and trust, consequently leading to conflicting or poor food behaviors. Limitations in income (occupation) and education become the barriers against good food choice and the eventual cause of diet-related diseases such as diabetes, obesity, cardiovascular dysfunction, and cancer, often with premature onset. Governmental subsidies such as school lunch programs target schoolchildren among such populations to provide intervention, assistance, and nutrition education. For all ages, access to accurate, reliable information through qualified health personnel or reliable media outlets stands as a critical component for the healthy food choices necessary for positive health outcomes. Extending from greater awareness of food choice is the optimization of food nutrient availability through proper processing and cooking techniques. Selecting fresh foods over frozen or highly processed foods and avoiding frying and grilling preparation methods are among a few of the basic recommendations of food education, understandable at all ages in all socioeconomic levels.

5.2.3 SOCIAL

Eating habits, and therefore good nutrition, are formed from the strong influence of social setting. The people with whom we live, work, and associate, and the occasions and circumstances under which those associations exist, directly affect our eating behaviors as we tend to mimic those of families, friends, and coworkers. The EUFIC distinguishes these determinants by culture, social class, context, and setting. Culturally, habitual selection of the same foods, prepared in a traditional way, may tend to limit selection and the possible exposure to a diversity of nutrients. Awareness of the possible ramifications of such diets requires knowledgeable personnel with the ability and sensitivity to interact within these populations. Similarly, cultural attitudes adapt to the environmental setting when cultural groups move to new geographic areas. As mentioned previously, higher social class and economic stability tend to promote healthier diets and economic issues pertaining to food choice remain problematic. The economic and social determinants of food selection are more difficult to control since it is often individual motivation alone, apart from community support, that provides the impetus toward positive change. Various aspects of social interaction with peers, including the time, place, and social context of those interactions, may profoundly affect what we eat both consciously and unconsciously. While studies show that the family is the primary motivator in food selection, social settings influence food options in school or work environments where most individuals spend a good part of their time. As an example, children first learn and adopt the practices of their parents when forming food preferences, demonstrating the importance of "leading by example" (Johns Hopkins Center for a Livable Future, 2011). Teenagers and millennials, defined generally as those persons born between the 1980s and early

2000s, are favorite targets of advertisers due to their susceptibility to peer influence and tendency toward change or novelty. Traditions and attitudes also extend to non-conventional meal patterns where availability of convenience foods that conserve on preparation time offers even more food alternatives.

Finally, social support and a sense of belonging can reinforce positive food selection behaviors through confidence and self-sufficiency (Berkman, 1995). Food-as-status symbol may be an outward manifestation of upward mobility. Look no further than the popular upscale trends of sushi and high-end restaurant dining to provide a sense of importance and success to observe evidence of this phenomenon.

5.2.4 PSYCHOLOGICAL

Psychological factors resulting from random transient events related to everyday life and the response to these events will influence food behaviors on a temporary, day-to-day basis. However, with repetition or increased regularity of occurrence, these random events may induce permanent alterations in consumption behavior that may result in chronic poor eating habits. While still in its infancy in our understanding, the complexities of temperament and mood changes from physiological states of stress, guilt, or other internalized emotions may cause changes in food selection. These may take the form of overeating or undereating, or hedonic cravings for certain nonnutritious foods, all of which may result in adverse dietary selections and health risk if prolonged. In addition to the vagaries of everyday life, the existence of cognitive-controlled eating disorders in the form of unhealthy obsessions such as bulimia or anorexia nervosa may require medical intervention for severe nutritional imbalance.

5.2.5 ATTITUDINAL

Consumers who prioritize nutrition will make food selection choices that differ from those of individuals who value convenience. This simple, true statement demonstrates how consumer food choices are directly influenced by attitudes and values that may be individual or communal. The impactful result is purposeful marketing where, for example, millennials might purchase snack bars for energy before exercise and as a meal replacement, while Baby Boomers do so as part of a weight loss program—underscoring the importance of intergenerational age and social factors toward the complexity of food attitudes. Here, traditional beliefs may result from or give rise to brand loyalty and the targeting of specific groups by age, gender, and culture for the comfort and expected quality the manufacturer affords. Garnering this support is economically advantageous to the manufacturer as most consumers choose and stay with brands recommended or used by friends or family members. Brand loyalty over the lifetime of the consumer is a powerful economic incentive and the ultimate goal of the responsible manufacturer. Similarly, individuals or groups concerned with food and agricultural policy, environmental, or regulatory issues will gravitate toward those manufacturers and brands that most reflect their values

(Johns Hopkins Center for a Livable Future, 2011). These determined groups can be economically powerful, not only directing what people eat, but also how a food is produced and the quantity of its production. Governmental regulation in the form of farm subsidies and genetically modified organism (GMO) initiatives are examples of interest group involvement into food selection and availability. Also influential is the public support for government nutritional assistance programs such as the Supplemental Nutrition Food Assistance (SNAP) program, food stamps, Women, Infants, and Children (WIC), and Head Start, for those of special need. All of these become interconnected in food production as the food chain supply becomes increasingly complex. Demand by consumers for social responsibility on the part of manufacturers in the form of environmental protection and fair labor practices for their workers is partly driven by a global media presence and generates substantial incentive toward manufacturers ensuring a safe food supply for their customers (Fontanazza, 2015).

5.3 MEGATRENDS IN MARKETING AWARENESS

The food industry is undergoing important and broad-ranging modifications designed to enhance overall food safety both during food production, processing, and distribution by the manufacturer, and for the consumer as the end user. The result is expected to make both stakeholders responsible partners in food education/awareness as a prelude to healthy food choice. As the regulatory body ultimately responsible for food safety within the United States, it is the Food and Drug Administration (FDA) that is the arbiter of enforcing policy. Recent developments in (1) policy, in the form of the final rules for preventive controls as part of the Food Safety Modernization Act (FSMA), and proposed food label changes; (2) technologies to enhance food production, freshness, and convenience; and (3) wellness and lifestyle diets tailored to body maintenance and benefit, particularly if coupled with (4) convenience, are considered among the prominent megatrends most poised toward influencing food marketing and consumer awareness.

5.3.1 RECENT POLICY INITIATIVES DIRECTING FOOD SAFETY AWARENESS

Since the passage of the 1906 Pure Food and Drugs Act, the regulatory functions of the FDA (1930) through the Federal Food, Drug, and Cosmetic Act (1938, FFDCA) have undergone transformation and enhancement in the directive to improve food safety. These changes culminated in January 2011 in the Food Safety Modernization Act that provided the FDA with a sweeping mandate over food manufacturers, foreign and domestic, from adverse event response to prevention. The most recent directive within this law (September 2015) is the final rule for current good manufacturing practice, hazard analysis, and risk-based preventive controls for human and animal foods (http://www.fda.gov/Food/GuidanceRegulation/FSMA/ucm334115.htm). In response to calls from consumer advocacy groups, local regulatory bodies, and industry, the FDA has mandated preventative critical control points for minimum standards for both the safe production and harvesting of produce, and across the food manufacturing and processing supply chain. With the final rule recently passed (Aug. 2015) and compliance dates expected to begin in Sep. 2016, manufacturers, through comprehensive documentation of process, become the initial and vital responsible party in the supply of safe food.

In an effort to better assist consumers toward making and maintaining healthy food choices and to avoid those ingredients with potentially adverse effects in allergic individuals, the FDA has proposed to update the 20-year-old Nutrition Facts label found on food packaging (http://www.fda.gov/Food/GuidanceRegulation/GuidanceDocumentsRegulatoryInformation/LabelingNutrition/ucm385663.htm). A comprehensive review of the proposed changes can be found in Marone and Birkenbach (2016). To summarize here briefly, the changes include updates for serving size requirements to reflect the larger portion sizes people eat today, prominently displaying serving sizes and calories, adding a declaration for "added sugars," and indication for the *type* of fat (saturated, trans) over the amount. Public comment on the proposed changes has just concluded with formal label changes expected within 2 years of ratification. In the most recent amendment to strengthen the Food Labeling Modernization Act (FLMA, 2013, https://www.opencongress.org/bill/hr3147-113/text), a new provision adds sesame to the list of the major food allergens and the requirements for submission of food label information with imported foods. The food ingredient label is meant to empower the consumer with the basic information to formulate his/her own position in food selection with as little influence from the manufacturer as possible. In a recent Supreme Court decision (POM Wonderful LLC v. FTC, 13-1060, U.S. Court of Appeals, District of Columbia (Washington)) ruling against false advertising, the original FFDCA mislabeling law was upheld to ensure consumer protection against foods or beverages having been adulterated, knowingly contaminated, misbranded, or carrying false claims regarding health benefits (Zajac, 2015). This ruling demonstrates the seriousness with which regulatory authorities regard the need for accuracy in advertising in order to prevent dissemination of objectively false information to consumers.

5.3.2 AT THE FOREFRONT OF FOOD TECHNOLOGY

In an effort to satisfy the nutritional needs of a growing world population, numerous advances in agricultural and food technologies are continually being developed. With the understanding that food policy, technology, economics, and consumer choice are interrelated, three technology areas, particularly, have the capability to strongly impact food awareness and, in so doing, shape future directions.

Agrotechnology is poised to become today what the computer technology industry was 20–30 years ago. Centered in California but advancing into areas needing to maximize land use efficiency, precision agriculture aims to improve farm production and efficiency while reducing environmental impact. The use of robotics for harvesting produce; implementation of software to calculate grass usage per head of cattle;

ability to monitor real-time soil conditions for nitrogen and moisture content; assistance of Global Positioning System satellites to target individual crop treatments to the smallest plots of soil; identification of green substitutes for hormones, pesticides, and meat and poultry; climate control; and nutritional quality of feed for livestock, are just some of the farm innovations directed toward targeted land use. These practices, along with advances in food and agricultural biotechnology, particularly the use of genetic methods of selection, continue to enhance beneficial traits for plants and animals (Reed, 2015). Biotechnology-derived plants and animals serve to protect the environment and enhance nutrition through the reduced use of herbicides and pesticides, allowing addition of more healthy food components (vitamins/minerals/antioxidants), and reducing loss through plant and livestock disease, as just some examples.

Commercial sale of genetically modified crops began in 1994, when Calgene first marketed its Flavr Savr delayed ripening tomato. Today, it is estimated that over 88% of grains such as soybean, corn, canola, and cotton seed oil are GMO-derived, generally defined as <0.9% genetically modified content by the US non-GMO Project (officially, there are no established tolerance limits by US regulatory agencies) and the European Union (EU). These have been engineered for enhanced nutrient profiles and for resistance to pathogens and herbicides. In the United States a debate continues as to whether genetic modifications of food should require mandatory labeling at the federal level. Vermont has already required mandatory labeling and similar measures have either passed or remain on the ballot in several other states. Most recently however, legislation in the form of the "Safe and Accurate Food Labeling Act (SAFE)" was passed by the House of Representatives (July 2015) blocking states from *requiring* identification of food made with genetically modified ingredients but substantiating their inherent safety through the establishment of a *voluntary*, United States Department of Agriculture (USDA)-led, nationwide nonGMO certification (Dokoupil, 2015; Grebow, 2015). The Senate has yet to rule as of this writing. While consumer advocacy groups, small-scale food producers, and organic farmers are promoters of mandatory labeling, Congress has generally opposed such measures, typically with support from large food and chemical corporations. To date, there is no indication that GMO foods are a health risk, including any tendency toward allergenicity. Toxicologically, safety guidelines are the same for GMO foods as those of conventional foods. The FDA considers foods developed by GMOs "substantially equivalent" to conventional foods including the Generally Recognized as Safe determination (SOT, 2003; http://www.fda.gov/Food/GuidanceRegulation/GuidanceDocuments-RegulatoryInformation/Biotechnology/ucm096095.htm). In the EU, determination of safety is performed on a case-by-case basis. For now, the National Organic Program (NOP) under the jurisdiction of the USDA prohibits the use of GMO through the use of the "organic" or "natural" label (Section 5.3.3). The FDA's final guidance on GMO labeling of foods derived from plants affirms GMO labeling as voluntary and extends its recommendations to terminology (http://www.fda.gov/Food/GuidanceRegulation/GuidanceDocumentsRegulatoryInformation/ucm059098.htm). At the same time, particularly noteworthy is the FDA's new, first-of-its-kind approval for a genetically engineered animal intended for food, AquAdvantage Atlantic Salmon (http://www.fda.gov/Food/GuidanceRegulation/GuidanceDocumentsRegulatoryInformation/ucm469802.htm). The salmon, engineered such that it attains market size faster than nonengineered product, has been under debate since its development and approval request by AquaBounty Technologies 20 years ago (Pollack, 2015). The contested deliberation between the company, members of Congress, scientists, environmental and consumer groups, and store retailers is sure to continue while the processes and regulations necessary to insure safety are securely in place. For now, use of the terms "GMO-free" and "nonGMO" is discouraged as being too difficult to substantiate; to the extent that many foods contain minor genetic modifications through historic hybridization and breeding techniques, companies should instead use wording such as "not genetically engineered" or "not genetically modified through the use of *modern* biotechnology" (Grebow, 2015). Nonetheless, the FDA affirms its position that all claims, be they with or without GMO-containing ingredients, must be truthful, not misleading, and be able to be substantiated through records from farmers, distributors, manufacturers, or USDA NOP certification. With the understanding that GMO labeling is voluntary, and the use of the term "organic" while not a guarantee of nonGMO conformity still carries with it certified standards, consumers are compelled to seek their own information and draw their own conclusions regarding GMO-containing foods, as with all other foods.

As in the case of genetically modified foods, applications are rapidly developing for use of nanoenabled technology in three areas, specifically: (1) in food, for development of improved tastes (flavorants), color, flavor, texture, and consistency of foodstuffs ("mouth feel") (Kwak, 2014); (2) to moderate metabolism for increased absorption and bio-availability of nutrients and health supplements, incorporating such forms as microemulsions, liposomes, solid lipid nanoparticles, and nanofibers (Oehlke et al., 2014); and (3) in food packaging materials for improved mechanical barriers and antimicrobial properties, and for nanosensors to enhance traceability and monitoring the condition of food during transport and storage (Duncan, 2011). From a consumer standpoint, the greatest impact would be on palatability and efficient nutrient delivery (assuming that various preparations will have different surface qualities). It is in the area of food sensation enhancement that exciting and rapid developments are being made, garnering the attention of food manufacturers and consumers alike. Food formulators have captured the attention of aficionados with novel organoleptic sensations in appearance (color), aroma, flavor, and even sound through textural "crunch/crispy" capacity. The newest sensation of fat (oleogustus) as the sixth sense, along with sweet, salty, bitter, sour, and umami, lends an added hedonic quality to food.

With the unprecedented, insatiable craving of consumers for new and unique tastes and preparations, manufacturers are recognizing the potential economic impact of being first to market in staking out a position for loyal followers. These efforts have resulted in ever-increasing resources being committed to developing products as programs involving rigorous taste-test sampling and consumer surveys have come to precede market presentation.

5.3.3 WELLNESS AND LIFESTYLE

Only since the historically recent awareness that certain (often isolated) societies exhibit unusually long lifespans has a growing body of evidence suggested that diet is an important contributor to longevity and healthy living. With a greater global awareness stemming from media connectivity, understanding of the direct influence of food selection over general health and well-being has improved and has commensurately assumed a greater role in determining food choice for recent generations. From a 2014 Shopping for Health Report from the Food Marketing Institute, "80% of shoppers make some effort to eat a healthy diet – while nearly as many (67%) believe that what they consume affects their physical and mental wellness" (http://www.factsfiguresfuture.com/issues/july-2014/how-consumers-eat-healthier-fmi,-rodale.html).

Presently, with consumers and particularly younger shoppers as key drivers in food for health and functional food choice, the demand for freshness has led a commanding "think globally, act locally" trend that continues to grow. At the heart of these trends toward authentic, more natural fare are whole grains and fresh fruits and vegetables, often locally grown without the aid of chemical pesticides and herbicides, including foods that are minimally processed with no preservatives, nor unrecognizable or frozen ingredients. It is well known that fresh produce has the highest concentration of nutrients, and the longer the time from harvest to consumption, the greater the degradation of those nutrients (Barrett, 2007; Rickman et al., 2007). Sales of fresh ingredients, distinguished by specific designations including "natural," "organic," "free-from," or "clean label" among others, now constitute 44% of all food sales, in contrast to 34% as recently as 2 years ago (highest growth market), despite the fact that consumers are likely to pay more for a food described as "fresh." Fresh-format stores are the most successful supermarket stores with local/seasonal ingredients of "grass-fed," "free-range," "farm-raised," "homemade," and "made-from-scratch" descriptors as the most strongly tied in consumers' minds to the notion of freshness. Fresh ingredients lend themselves to cooking from scratch and the development of new homemade food combinations with specialty regimens and individualized exclusion diets. Diets consisting of individualized taste and nutrient preferences are particularly beneficial for feeding sensitive populations (children, chronic disease conditions, the elderly), those households with multiple cooks, for men buying and helping to prepare meals, and households of single persons, all of which are becoming more commonplace (Produce for Better Health Foundation, 2015). Curiously, the freshness movement has sparked a small business, cottage industry, as individual entrepreneurs sell their favorite food compositions in an attempt to fill market gaps. This is especially true in the baby food market as young mothers prepare and market specialized mixtures.

Restaurant fare also increasingly reflects the trend toward healthful eating. Sandwiches featuring fresh vegetables, and fresh-squeezed, cold-pressed juices from Subway, Schlotzsky's, and Zoës Kitchen, among other culinary establishments, have become increasingly popular. Although the traditional use of breakfast cereals—the ones Baby Boomers were raised on (sentimentality)—is generally on the decline (7–30% reduction), breakfast is being eaten by more people, as high-carbohydrate cereals are replaced with high-protein yogurts, breakfast drinks, some with ethnic flavors (chutney, chimichurri, chipotle), breakfast sandwiches, and more healthy children's options to start the day. Generally, both home and away, consumer preferences are already trending against added no-calorie sweeteners, believing "natural cane sugar" to be the healthier option (Mellentin, 2015).

While use of the words "natural" and "organic" may be prevalent for promotion purposes and to augment consumer demand, the Natural Marketing Institute (NMI, 2015) found that 36% of consumers believe there is no significant difference between items designated with these terms. Moreover, 46% of consumers believe that the US government regulates foods labeled as "natural" and 61% believe the government regulates food designated as "organic" (Crane, 2015a). Notably, organic foods may be indeed certified on the basis of containing no antibiotics or synthetic hormones, pesticides, or fertilizer, and made without artificial colors, dyes, or preservatives, and may be regulated under the NOP. The latter is a regulatory agency housed within the USDA Agricultural Marketing Service and authorized by the Organic Foods Production Act (OFPA; Title 21 of Food, Agriculture, Conservation, and Trade Act of 1990, codified at 7 U.S.C. ch. 94) through the National Organic Standards Board to develop standards for definition, applicability, labeling, production handling, marketing, and accreditation of organic products as mentioned previously (http://www.fda.gov/AboutFDA/Transparency/Basics/ucm214871.htm; eCFR Part 205, November 12, 2015, http://www.ecfr.gov/cgi-bin/text-idx?tpl=/ecfrbrowse/Title07/7cfr205_main_02.tpl). In contrast, the FDA has not yet developed a definition for the term "natural" or its derivatives, leaving manufacturers to decide what they want it to mean. However, in a move that may signal that the agency is considering a formal definition for the term, the FDA is currently requesting comments on the "natural" label. This action may further attempt to address questions regarding previously unconsidered GMO-derived products, those resulting from other newer processing/manufacturing techniques, and/or foods of potential health benefit (Crane, 2015b; http://www.fda.gov/Food/GuidanceRegulation/GuidanceDocumentsRegulatoryInformation/LabelingNutrition/ucm456090.htm).

In addition to the growing demand for natural basic ingredients in foods, a growing trend is the use of novel ingredient combinations specifically designed for health benefit. Some examples include basic flavors of food and drink manipulated by chefs and manufacturers in ingredient mashups, making "ethnic authenticity" in food, apart from ingredients, no longer relevant. The trends are toward *looking beyond* the traditional, incorporating such novelties as seaweed for sushi, cocktails with beer, creative flavors for high proteins for hummus and select insect powders/flours; the rise of cultural foods and "Flavor Tourism"—hybrid ethnic cuisines and spicy flavors; and a focus on food discovery for new must-have food attributes and more sophisticated healthy food options, such as gourmet Asian and Middle Eastern products, ethnic fusion, regional cuisines, unique pastas, and whole food nutrition in fiber/grains, dark greens (broccoli, kale, Brussels sprouts) and purple produce (purple apples, carrots, beets), and vegetable juices containing antioxidant anthocyanins (Sloane, 2015; http://www.eufic.org/article/en/show/spotlight/rid/colours-of-fruits-and-vegetables-and-health/).

Functional foods and nutraceuticals are generally defined as foods and beverages purportedly possessing health benefits beyond their nutritional purpose. These often focus on macro ingredients and generally fall into one of three categories: (1) natural or fortified foods intended to treat deficiency-based disease; (2) foods containing live microorganisms intended to promote healthy gut flora; and (3) foods that support normal physiologic processes of metabolism (Williams, 2008). Examples include antioxidants, in the form of vitamins A, C, and E, selenium, CoQ10, and pycnogenol, found in fruits, vegetables, teas, and coffees to facilitate free radical scavenger activity for skin and antiaging; glucosamine and chondroitins and collagens for bone, joint, and skin elasticity; hemp fibers for energy/mood enhancement; whole fibers and grains for weight management and heart/cardiovascular health; proteins for weight management and body fitness; and probiotics for gastrointestinal and immunologic health. Many of these have validated health claims (Marone and Birkenbach, 2016). These functional foods are expected to continue their market growth well into the future as more consumers, through global market availability and their own advancing economic means, become aware of the potential impact of food for enhanced wellness. The ability to select foods with the intention of acting preventively rather than reactively on health issues allows the consumer control and partnership in his/her own well-being. By extension, society benefits as a whole as this knowledge and empowerment spread through the family and community (Leatherhead Food Research, 2014).

In addition to orally ingested functional foods and nutraceuticals, inner beauty, cosmeceuticals as topical agents of biologically active components provide dual outer beauty health. These products often comprise botanically derived antioxidants and antiinflammatories in the form of lotions and creams that are applied to the epidermal layers of the skin and penetrate to the deeper dermis to deliver nutrients that have the capability to restore a youthful appearance to skin. Examples include extracts of fruits and vegetables containing phytosterols and vitamins C and E that support and promote synthesis of collagen in the deeper dermal connective tissue to improve skin elasticity, resulting in the subsequent reduction of fine lines and wrinkles (Marone, 2016). From a marketing perspective, the consumer who is educated as to the safety and health benefits of food is also the one most likely to seek out natural and safe alternatives for cosmetic and general lifestyle improvement.

The Council for Responsible Nutrition consumer survey on dietary supplements indicates that 68% of adults in the US take some form of dietary supplements and 83% express overall confidence in the safety, quality, and effectiveness of those products (Council for Responsible Nutrition, 2014). The downside of this health awareness is the potential for exploitation by manufacturers, marketers, and suppliers who adulterate or misbrand popular foods and ingredients for the purpose of economic gain, resulting in possible widespread health hazards. This is particularly true in the seafood, sports nutrition (energy drink), herbal and botanical, weight management, and dietary supplement industries (see Johnson, 2014 for a comprehensive review of food and ingredient adulteration; Rocha et al., 2015). In the most recent, comprehensive, and severe civil actions to date, federal agencies of the FDA, Federal Trade Commission, Internal Revenue Service, and Pentagon, through the Department of Justice, brought criminal charges against more than 100 manufacturers and marketers of dietary supplements and related items for selling products laced with "amphetamine-like" substances, some of which incorporated Chinese-made chemicals in place of natural ingredients (Flynn, 2015). The FDA continues to carefully monitor caffeine-containing products and has issued warning letters to manufacturers of pure, bulk caffeine powders available to consumers. Since it is the responsibility of the manufacturer to ensure safety, they are also liable for the traceability and proper labeling of the product according to the Dietary Supplement Health and Education Act of 1994 (DSHEA) and FDA regulations (http://www.fda.gov/Food/DietarySupplements/). In a ripple effect, some prominent retailers have seen their stock values plummet—a consequence of lack of vigilance for their inventory and a warning for those who esteem profit over consumer advocacy and trust. While adulteration is considered by regulatory authorities to be a more serious offense than the misleading advertising described earlier (see Section 5.3.1), both actions are unlawful and carry legal liability. As a note of caution: Regulatory authorities will hold the purveyor responsible by all means, whether consciously aware or not of the composition of their market fare if consumer safety is at all in question.

Generally, awareness of functional food health benefits has put the onus on the consumer to separate useful information from fads in the industry. A notable recent example is found in the marketing of "gluten-free" products manufactured specifically to exclude the protein gluten found in grains such as wheat, barley, rye, and others. Where once consumption of gluten-free food was limited to effective medical management for persons with gluten sensitivity and celiac disease (1% of the US population), now numerous food items are being formulated, reformulated, and marketed as such for purported health benefits for the general population. As a result, one in five consumers are now exploring gluten-free lifestyles without medical necessity, as part of a distinctly growing trend, citing "no reason at all" as the basis for their purchase decision (http://www.nutraceuticalsworld.com/contents/view_infographics/2015-09-08/gluten-free-trends).[a] The expanding gluten-free trend is one example of consumer demand for the "free-from …" market extending into mainstream food options worldwide (Mintel Group Ltd., 2014).

[a]In 2013 the FDA issued its ruling on the labeling of "gluten-free" foods as those foods that are inherently gluten-free or not containing an ingredient that is: (1) a gluten-containing grain, (2) an unprocessed gluten-containing grain, or (3) derived from a processed gluten-containing grain resulting in the presence of more than 20 ppm gluten in the food (http://www.fda.gov/food/guidanceregulation/guidancedocumentsregulatoryinformation/allergens/ucm362510.htm).

5.3.4 CONVENIENCE

With the world becoming increasingly connected through technological advancement and ever faster communication channels, is it any wonder we are all busier than ever just trying to keep up? Food choice has reflected this hurried (harried?) lifestyle trend, leading, in turn, to convenience packaging, preparation, consumption, and service. In addition, the simpler, scaled-down needs of a generally aging world population, as well as singular household occupants, have heightened the demand for convenience foods. Consumers are attracted to these products by their relatively low cost and complexity, ease of use and reduced preparation effort, and frequent single-serving packaging permitting consumption quickly and in one sitting. Additional appeal derives from the variety and availability of these foods despite the season, as well as their perceived nutritional value and safety. In turn, the market (manufacturers) must meet the demand for taste and high nutritional quality while maintaining product appeal. While price remains the most significant fresh food selection factor, nutritional content and preparation time and difficulty are increasingly important, particularly for millennials and Baby Boomers who have greater financial resources. Aforementioned all is the consideration to harmonize healthy, safe food choices with lifestyle without the need to compromise on taste preferences. Some examples of convenience foods include slow-cooker/crockpot meals, fresh protein snacks, drinkable meals (nutritional smoothies in greater varieties and combinations), meals-on-the-go (food trucks), and yogurts, to name just a few. Generally, resolution of the often conflicting needs for timesaving and reduced effort while maintaining wholesomeness is best observed through three prevailing trends: (1) an emphasis on sensible snacking; (2) flexible food packaging and meal preparation; and (3) greater use of health beverages.

5.3.4.1 Sensible snacking

The trend toward more mini- and "on-the-go" meals and away from longer, formal dinners is evidenced in the burgeoning food culture of *nutritious* snacking. The continued growth and popularity of this trend has caused a generalized blurring of the lines between meals, giving rise to a competitive industry driven primarily by taste and health demand. According to the Mintel Group, 95% of Americans snack daily with 62% reporting the need to satisfy a craving as the principal motivating factor (Mintel Market Research, 2015). Moreover, the number of between-meal snacks has increased from 1.9 per day per person in 2010 to 2.8 per day per person in 2014. However, one-third of consumers polled acknowledge making a conscious effort to snack more healthfully than in the previous year, and nearly half of consumers (45%) look for snacks that go beyond basic nutrition, focusing on fruits and vegetables in novel forms in order to replace high-salt, high-sugar, or calorie-dense snacks, the consumption of which has declined over the last 5 years (Wyatt, 2014). Foods high in protein, fiber, vitamins and minerals, whole grains, and probiotics, particularly those derived from fresh fruits/vegetables, are the most favored choices among the health-conscious snacking electorate. Chocolate and peanut butter are listed as the most universally preferred flavors as millennials gravitate to trendy hazelnut spreads (Nutella), coffees, chia derivatives, and Greek yogurts, while "chocolate, fresh fruit, chips, bread/sandwich (outside a meal), cheese, ice cream, vegetables, and cereal are the snacks Americans would choose above all others" (Nielsen, 2014; Sloane, 2015). Particularly trendy are concentrated foods such as nutrition-packed powders, freeze-dried preparations in the form of fruit chips and pulverized fruits and vegetables, and various items derived from marine sources high in choline, potassium, and magnesium. With market growth of 30% since 2014, seaweed products are now becoming very popular as snacks due to their low-calorie, mineral-rich nutrients and sustainable character. In addition to fruits and vegetables, low-calorie, organic meat snacks that incorporate premium lean meat with seeds, grains, herbs, spices, and superfruit combinations are new to the market. Despite the increasing number of snacking options available, 60% of consumers would prefer more choices, particularly for children (Mintel Market Research, 2015). This translates to a continued advancing market opportunity in this space.

Coincident with these snacking trends, heavily fortified, high-performing, "complete" foods containing all the essential amino and fatty acids in optimized ratios to meet human nutritional needs, are most sought and satisfying, particularly among the 53% of health-conscious adults. Coordinate with this focused, high-impact, nutrient-rich food trend is the recent appearance of even more highly concentrated (ultra) supplements that deliver greater micronutrient quantities in smaller pill sizes (Kingsley, 2015). Prevalent among these are vitamin and whey protein supplements and products enriched in omega-3 polyunsaturated fatty acids, such as docosahexaenoic acid (DHA) and eicosapentaenoic acid (EPA). As competition drives the industry toward ever-increasing nutrient concentrations, the distinction between foods and pharmaceuticals becomes obscured. The result of these trends will be closer scrutiny from regulatory authorities for inherent safety concerns.

5.3.4.2 Flexible food packaging, meal preparation, and health beverages

Coordinate with the increased convenience of high nutrient snacking, "on-the-go" packaging requirements are a critical consideration for manufacturers and consumers alike. Food packaging is increasingly designed for immediate, single-serving consumption. Recent data indicate that 15% of all eating occasions involve foods and beverages consumed within an hour of purchase, and 47% of all eating occasions are solo experiences (Hartman Group, 2013; Sloane, 2015). Simultaneously, it is curious to note a distinct movement away from *frozen* packaged foods. Consumers are demanding freshness with convenience and are not inclined to compromise on this point. Although fresh fruits and vegetables are theoretically stable from a nutritional standpoint from the time they are flash-frozen, this stability is somewhat variable with increased time and distance to market. These factors have caused many food manufacturers to reposition themselves in the market for better appeal. A good example of this is the newer "freshly-made, simply frozen" frozen food marketing campaign by Nestlé as an attempt to improve the flagging frozen food industry (Watrous, 2015).

For the fresh food sector, there is always a "race against the clock" to ensure delivery of the most wholesome food. Continued improvements in packaging offer manufacturers a competitive edge in the freshness market. This coordinates well with sustainable, biodegradable food packaging trends where new biologically flexible, resealable materials provide greater consumer-friendly food freshness protection for their functional toughness, sealant, and barrier capabilities. Newer modified atmosphere packaging (MAP) involves use of high-tech plastic films that offer flexibility to modulate the level of oxygen permeability based on temperature. These, coupled with minimal/gentle processing techniques, slow the food deterioration process related to oxygenation in order to increase product consistency and shelf life (Kuhn, 2015). Simple consumption of 4–5 mini-meals/nutritious snacks in place of long, preparatory sit-down repasts allows shorter time and higher turnover from food storage both from the manufacturer and in the consumer refrigerator, resulting in greater preservation of nutrients.

Nowhere is the trend to convenience more evident than in the food preparation area, both home and away. The increased time constraints of busy lifestyles also put increasing pressure on manufacturers to develop and supply products that minimize preparation time to less than the favored 30 min per meal. Coincident with this demand is the popularity of countless, nutritious beverage (smoothie) combinations largely incorporating fruits, vegetables, nuts, and even homemade alcoholic beverages. As kitchen appliances go, the blender and juicer have never been so valuable, particularly for the younger gastronomes. To this end, the recent uptrend in consumption of tea (Teavana) and coconut water are specific examples of increased culinary beverage interest for enhanced health benefit. Drinks such as these make the perfect accompaniment to a wholesome, easy-to-prepare, on-the-go snack or meal, maximizing the health potential of all food components. Another new appliance that is expected to populate the store shelves within the next year is the 3D food printer. With orders being taken now for the "Foodini" and similar devices, printing food from an unlimited combination of nutritious ingredients in any format imaginable can provide fast delivery of visually appealing foods unique to the preparer, in "gee-whiz" wonder. This capacity allows the consumer even greater flexibility and variety in health food choice. The ability of the consumer to combine any one of countless food ingredients into any form also demands vigilant knowledge and trust for the correct, healthful raw material choice, to say nothing of the potential regulatory control over such an accessible convenience.

Expedient, healthy eating experiences away from home are also adapting to become the norm. Where once the traditional convenience store offered only high fat, salt, and sugar snacks for incidental, compulsive purchase by customers primarily intending to purchase gasoline or other necessities, savvy retailers are now stocking combinations of fresh fruits, grains, nuts, and vegetables, including wholesome, high-protein snack bars and salads (Kuhn, 2015). The change is paying off for retailers as fresh produce accounts for an increasing proportion of convenience food sales, and convenience store sales constitute a greater percentage of the produce market in general (National Association of Convenience Stores, 2015). Noteworthy in the transition toward nutritious food and beverage offerings is the increasing prevalence of high-visibility, grab-and-go vending machines *with healthy food choices* within stores, workplaces, and schools. PepsiCo is leading the way by revamping an invention long associated with junk food and cigarettes through its rollout of dual-temperature machines that offer food and drink under the new "Hello Goodness" brand.

Finally, convenience dominates in restaurant dining as well. Fast-casual eateries occupy a market niche intermediate between fast-food and traditional sit-down restaurants by offering expeditious, high-quality food that is freshly prepared to order but accompanied by only minimal service. Delivering healthy food options apart from the traditional burger and fries, these establishments constitute the fastest growing segment of the restaurant industry and include Panera Bread, Chipotle, Cozi, Noodles and Company, and Quiznos as examples. This dining modality minimizes time-consuming food preparation and high service labor costs. By using replacement technology, smartphone ordering and payment substitute for a formal server who is only met when the food is delivered. The distinguishing component of these restaurants is their commitment to quality. "Fast casual concepts are typically more committed to wholesome ingredients, professional preparation and aesthetic presentation. When consumers feel better about what they eat, they become loyal to a brand or concept. Food quality is a huge part of what makes fast-casual restaurants successful" (Parpal, 2015).

5.4 BEHAVIORAL INTERVENTIONS

Interventions aimed at promoting healthy food choices are based on increasing consumer awareness of purchasing behavior through education. In conjunction with the manufacturer's responsibility for safety and truth in advertising according to regulatory directives, the onus given the consumer to read and understand food labeling, is a start in this process. Extending the process to define the more important social (who), dietary patterns and practices (what), and environmental (where) determinants of food awareness allows greater consumer control in recognizing the importance of food selection to overall health.

5.4.1 SOCIAL

Some of the solutions proposed by public health advocates include offering discounts on nutrient-rich foods in economically depressed areas, changing social norms to encourage healthy eating habits, reducing marketing of foods that are particularly high in sugar and fat to impressionable children, and making nutrition advice simpler for parents and caregivers to understand and follow.

As a consequence of the understanding that marketing to youth also gives greater opportunity to maintain brand loyalty for a long time, millennials, specifically, are driving the market with their economic independence, gourmet tastes, and discriminating preferences for more choices and combinations. And millennials are only the beginning. The health food trend will be further accelerated by the postmillennial generation, those now under the age of 18, also referred to as Generation Z. Aside from youth, food manufacturers continue to develop more diverse food choices to cater to growing ethnic markets. Whatever the environmental setting, however, social support is an excellent prognosticator of lasting positive behavioral change.

5.4.2 DIETARY PATTERNS AND PRACTICES

Focusing on eating habits for conscious awareness of the nutritional value of foods, that is, knowing how to read a food label and recognizing the health consequences of poor choices (high fat and sugar foods), is just the beginning. Routinely scheduling eating times, taking time to sit down for a meal, and avoiding rushing allow neurologic control mechanisms to regulate more effectively the physiologic sensations of hunger and satiety for homeostatic body maintenance. This is in contrast to the impulsive, psychologically driven eating that is often motivated by transient social or environmental settings, life stresses, and the influence of peers, particularly during adolescence when self-image is developing. Routine self-monitoring and record keeping of intake have been shown to be strong predictors of long-term weight loss success (Nestle et al., 1998), as have diets and lifestyle changes that are sustainable over time.

Finally, general consumer awareness of producer marketing techniques and claims, along with an understanding of new technologies, governmental policies, and cost fluctuations regarding foods, is readily available to any smart phone user through the various and numerous available social media outlets. As an example, the Grocery Manufacturers Association (GMA) has announced the launch of SmartLabel—a new technology for 2016 that will enable consumers to have access to detailed information from the manufacturer on ingredients and other features about thousands of retail products simply by scanning a code on the package, using a web search of the product or company, or through a phone application (http://www.gmaonline.org/news-events/newsroom/new-smartlabel-initiative-gives-consumers-easy-access-to-detailed-ingredien/).

Producer obligations include a remediation of imbalances for healthy food affordability and access to assure maximum freshness and nutritional value, attention to the plight of farm workers and observation of fair trade practices, and overall environmental impact all along the food supply chain, including storage, transportation, and marketing (Garnett, 2014; Johnson et al., 2014). It is estimated that 30–50% of all food is wasted along the supply chain. Wastage adversely influences food security and contributes in part to food insufficiency that presently affects one in nine people worldwide. Compensatory overproduction and misallocation promote natural resource misuse and potential environmental devastation. Aligning these considerations in parallel offers a good opportunity to balance the needs of the food industry and the consumer alike.

5.4.3 ENVIRONMENTAL

For the purpose of promoting consumption of healthful foods, nutrition experts recommend modification of the physical environment in which foods are most often eaten, specifically, within homes, schools, workplaces, and restaurants in descending order of significance. As the first and most powerful influence over personal eating habits, family mealtime provides the greatest opportunity to instill lasting healthy nutrition and lifestyle awareness while simultaneously offering a sense of well-being and building strong interpersonal bonds through communication (Forthun, 2015). The environment provided by the school cafeteria also constitutes a critical determinant in the development of long-term eating habits. At a time when children grow in independence and should be developing good eating choices and habits, they are subjected to the simultaneous influences of peer pressure and marketing strategies directly targeted to youth, both of which may negatively affect this process. Many schools are helping to foster healthy eating habits through school lunch and breakfast programs such as the SNAP program and Head Start mentioned previously, through the elimination of school vending machines with poor food and beverage choices, and by offering nutrition education classes with take-home messages for the entire family (Moore et al., 2013; Richards and Sindelar, 2013). Making quality food available by opening grocery stores with a variety of fresh produce options closer to family homes provides the means for practical application and reinforcement of newly acquired nutrition education, while also building community support (Escaron et al., 2013). Along with reduced caloric intake from fresh food options, other interventions that encourage physical activity by providing access to bicycle paths and gymnasiums for increased energy expenditure promote healthy weight maintenance and lifelong healthy habits.

5.5 GENERATING AWARENESS: THE CONSUMER AS PARTNER IN FOOD SAFETY

Safe, affordable, nutritious food consistently ranks among the top priorities of consumers (Center for Food and Integrity, 2011). Consumers and manufacturers alike have a responsibility to assure healthy food options in the marketplace. For the consumer, smart food choice over a lifetime can indeed influence overall wellness. These changes are positively reinforced through an equally active lifestyle. In order to make good food choices, the consumer is often required to distinguish acceptable or unacceptable risk based upon the identification or recognition of potential hazard and an assessment of the appropriate level of exposure to it. These decisions are predicated on the availability of reliable product information. Here the news media play a powerful role in the dissemination of information. However, that information is often confusing, misleading, or contradictory (recall the coffee as carcinogen debate). As a consequence, it is understandable that consumers become suspicious, frustrated, and angry at so-called "experts" in the food space (Lipp and Chase, 2015). Nonetheless, skepticism is warranted in view of the critical need for reliable, scientific substantiation of facts. When asked whom they trust most to provide the most accurate food safety information, American consumers ranked US government agencies (39%) followed by health care professionals (17%), scientific journals (14%), and health-focused websites (13%) as their most trusted sources (IFIC, 2014). To this end, "effective communication with timely, understandable, accurate messages conveyed by health authorities ... will keep consumers informed of the real threats while teaching them to spot the false alarms that pose no risk" (Lipp and Chase, 2015).

When the information from industry and governmental authorities becomes contradictory, consumer advocacy groups have a certain obligation to serve as intermediaries while maintaining consumer welfare as their highest priority. Specifically, these groups may act to inform consumers by resolving seemingly conflicting data from medical professional, governmental, and food industry sources, to present this information to consumers in an understandable and generally useful form, and to encourage the individual consumer to become involved on his/her own behalf.

In summary, consumer food choices are based upon diverse considerations that include availability, taste, safety and perceived health benefits, convenience, and a variety of cultural, social, and marketing influences. However, as availability of health information continues to expand, consumer education that will promote food choices based upon health benefits should assume an ever increasingly important role in the decision-making process. Manufacturers have a responsibility to maintain food safety and regulatory compliance in the items they market and to provide consumers with accurate and reliable information regarding the derivation and nutritional composition of their products. Working together, informed consumers, effective governmental regulatory agencies, and responsible food industry stakeholders are presented with an opportunity to optimize personal health benefit and food availability in balance with environmental sustainability.

ACKNOWLEDGEMENTS

This manuscript is taken in part from the original presentation, "Marketing Trends: How America's Food Choices are Changing," delivered at Food Week, Food Drug Law Institute, Washington, DC, February 2015.

REFERENCES

Barrett, D.M., 2007. Maximizing the nutritional value of fruits & vegetables. Food Technol. 61 (4), 40–44, Available from: http://www.fruitandvegetable.ucdavis.edu/files/197179.pdf

Berkman, L.F., 1995. The role of social relations in health promotion. Psychosom. Med. 57 (3), 245–254.

Bingham, S.A., 1999. High-meat diets and cancer risk. Proc. Nutr. Soc. 58 (2), 243–248.

Boeing, H., Bechthold, A., Bub, A., Ellinger, S., Haller, D., Kroke, A., Leschik-Bonnet, E., Müller, M.J., Oberritter, H., Schulze, M., Stehle, P., Watzl, B., 2012. Critical review: vegetables and fruit in the prevention of chronic diseases. Eur. J. Nutr. 51 (6), 637–663.

Center for Food Integrity, 2011. 2011 consumer trust research. http://www.motherjones.com/files/cfi_research_booklet_final_web.pdf

Choi, S.W., Frisco, S., 2010. Epigenetics: a new bridge between nutrition and health. Adv. Nutr. 1, 8–16.

Council for Responsible Nutrition, 2014. New survey reveals high percentage of U.S. population take dietary supplements—and with high confidence. October 14, 2014, Washington, DC. http://www.crnusa.org/prpdfs/CRNPR14-CRNCCSurvey103014.pdf

Crane, M., 2015a. What's the difference between natural and organic foods? Don't ask consumers. Nutritional Outlook. November 6, 2015. http://www.nutritionaloutlook.com/trends-business/what%E2%80%99s-difference-between-natural-and-organic-foods-don%E2%80%99t-ask-consumers

Crane, M., 2015b. FDA requesting comments on use of the term "natural". Nutritional Outlook. November 10, 2015. http://www.nutritionaloutlook.com/regulatory/fda-requesting-comments-use-term-natural

Darmon, N., Drewnowski, A., 2008. Does social class predict diet quality? Am. J. Clin. Nutr. 87, 1107–1117.

De Irala-Estévez, J., Groth, M., Johansson, L., Oltersdorf, U., Prattala, R., Martinez-Gonzalez, M.A., 2000. A systematic review of socioeconomic differences in food habits in Europe: consumption of fruit and vegetables. Eur. J. Clin. Nutr. 54, 706–714.

Dokoupil, T., 2015. Why Congress is having a food fight over GMO labeling. MSNBC. http://www.msnbc.com/msnbc/why-congress-having-food-fight-over-gmo-labeling

Donkin, I., Versteyhe, S., Ingerslev, L.R., Qian, K., Mechta, M., Nordkap, L., Mortensen, B., Appel, E.V.R., Jørgensen, N., Kristiansen, V.B., Hansen, T., Workman, C.T., Zierath, J.R., Barrès, R., 2016. Obesity and bariatric surgery drive epigenetic variation of spermatozoa in humans. Cell Metabol. 23, 369–378, http://www.sciencedirect.com/science/article/pii/S1550413115005719

Drewnowski, A., 1997. Taste preferences and food intake. Ann. Rev. Nutr. 17, 237–253.

Duncan, T.V., 2011. Applications of nanotechnology in food packaging and food safety: barrier materials, antimicrobials and sensors. J. Coll. Inter. Sci. 363, 1–24.

Escaron, A.L., Meinen, A.M., Nitzke, S.A., Martinez-Donate, A.P., 2013. Supermarket and grocery store-based interventions to promote healthful food choices and eating practices: a systematic review. Prev. Chronic Dis. 10, 1–20.

European Food Information Council (EUFIC), 2005. EUFIC Review. The determinants of food choice. April, 2005. http://www.eufic.org/article/en/expid/review-food-choice

Flynn, D., 2015. Letter from the Editor: who cares about dangerous supplements? Food Safety News. November 22, 2015. http://www.foodsafetynews.com/2015/11/letter-from-the-editor-who-cares-about-dangerous-supplements/#.VlIX2XarRD9

Fontanazza, M., 2015. How social responsibility affects food safety. Food Manuf. 8, October 8, 2015. https://foodsafetytech.com/news_article/how-social-responsibility-affects-food-safety

Forthun, L., 2015. Family nutrition: the truth about family meals. The Institute of Food and Agricultural Sciences (IFAS), Public No. FCS8871. University of Florida, pp. 1–4. https://edis.ifas.ufl.edu/pdffiles/FY/FY106100.pdf

Gallou-Kabani, C., Vige, A., Gross, M.S., Junien, C., 2007. Nutri-epigenomics: lifelong remodelling of our epigenomes by nutritional and metabolic factors and beyond. Clin. Chem. Lab. Med. 45 (3), 321–327.

Garnett, T., 2014. What is a sustainable healthy diet? A discussion paper. Food Climate Research Network. http://www.fcrn.org.uk/sites/default/files/fcrn_what_is_a_sustainable_healthy_diet_final.pdf

Grebow, J., 2015. FDA issues GMO labeling final guidance for foods derived from plants. Nutritional Outlook. November 19, 2015. http://www.nutritionaloutlook.com/regulatory/fda-issues-gmo-labeling-final-guidance-foods-derived-plants#sthash.JUZqLLPw.dpuf

Hartman Group, 2013. Modern eating: cultural roots, daily behaviors. The Hartman Group, Bellevue, WA. http://store.hartman-group.com/content/modern-eating-2013-overview.pdf
International Food Information Council (IFIC) Foundation, 2014. 2014 Food & Health Survey. The pulse of America's diet: from beliefs to behaviors. Washington, DC, pp. 1–14. http://www.foodinsight.org/sites/default/files/FINAL%202014%20Food%20and%20Health%20Survey%20Executive%20Summary_0.pdf
Johns Hopkins Center for a Livable Future, 2011. Diet and influences on food choice. Teaching the food system. A project of the Johns Hopkins Center for a Livable Future. http://www.jhsph.edu/research/centers-and-institutes/teaching-the-food-system/curriculum/_pdf/Diet_and_Influences-Background.pdf
Johnson, R., 2014. Food fraud and "economically motivated adulteration" of food and food ingredients. Congressional Research Service 7-5700, pp. 1–40. https://www.fas.org/sgp/crs/misc/R43358.pdf
Johnson, J.L., Fanzo, J.C., Cogill, B., 2014. Understanding sustainable diets: a descriptive analysis of the determinants and processes that influence diets and their impact on health, food security, and environmental sustainability. Adv. Nutr. 5, 418–429.
Kingsley, M., 2015. High times for high concentrates. Nutritional Outlook September, 70–76.
Kuhn, M.E., 2015. Health+convenience=market opportunity. Food Technol. November, 32–41.
Kwak, H., 2014. Nano-and Microencapsulation for Foods. Wiley-Blackwell Publishers, Oxford, UK, pp. 1–432.
Leatherhead Food Research, 2014. Future directions for the global functional foods market. Market Report, October 2014. https://www.leatherheadfood.com/sites/default/files/Future-Directions-for-the-Global-Functional-Foods-Market.pdf
Lipp, M., Chase, C.G., 2015. Chemical contaminants in foods: health risks and public perception. Food Technol. November, 43–48.
Marone, P.A., 2016. Nutraceuticals in cosmeceuticals. In: Bagchi, D., Preuss, H.G., Swaroop, A. (Eds.), Nutraceuticals and Functional Foods in Human Health and Disease Prevention. CRC Press, Taylor and Francis, Boca Raton, FL, pp. 593–605.
Marone, P.A., Birkenbach, K.E., 2016. Safety of functional foods, Chapter 2. In: Bagchi, D., Preuss, H.G., Swaroop, A. (Eds.), Nutraceuticals and Functional Foods in Human Health and Disease Prevention. CRC Press, Taylor and Francis, Boca Raton, FL, pp. 9–42.
Mellentin, J., 2015. Key trends in functional foods & beverages for 2016. New Nutrition Business. November 2, 2015. http://www.nutraceuticalsworld.com/issues/2015-11/view_features/key-trends-in-functional-foods-beverages-for-2016
Mintel Group Ltd., 2014. Trends and opportunities in the growing market for gluten-free foods. Free From Food Expo, Barcelona, Spain, June 2015. http://www.freefromfood-expo.com/pdf/2015-conference-mintel.pdf
Mintel Market Research, 2015. Snacking motivations and attitudes. US Mintel Group, Chicago, IL. http://www.mintel.com/press-centre/food-and-drink/a-snacking-nation-94-of-americans-snack-daily
Moore, L., de Silva-Sanigorski, A., Moore, S.N., 2013. A socio-ecological perspective on behavioural interventions to influence food choice in schools: alternative, complementary or synergistic? Public Health Nutr. 16 (6), 1000–1005.
Morrison, A.C., Ness, R.B., 2011. Sodium intake and cardiovascular disease. Ann. Rev. Publ. Health 32, 71–90.
Morton, G.J., Meek, T.H., Schwartz, M.W., 2014. Neurobiology of food intake in health and disease. Nat. Rev. Neurosci. 15, 367–378.

National Association of Convenience Stores (NAC), 2015. Convenience stores grow sales, add more fresh foods in first half of 2015. http://www.nacsonline.com/News/Press_Releases/2015/Pages/PR063015.aspx#.VlN533arRD8

Nestle, M., Wing, R., Birch, L., DiSogra, L., Drewnowski, A., Middleton, S., Sigman-Grant, M.W., Sobal, J., Winston, M., Economos, C., 1998. Behavioral and social influences on food choice. Nutr. Rev. 56 (5), S50–S74.

Nielsen, 2014. Snack attack. What consumers are reaching for around the world. Nielsen Global Snacking Survey. September 2014. The Nielsen Company. http://www.nielsen.com/content/dam/nielsenglobal/kr/docs/global-report/2014/Nielsen%20Global%20Snacking%20Report%20September%202014.pdf

Natural Marketing Institute (NMI), 2015. Consumer insights on organic and natural. Organic & Natural Health Association. A Research Study Prepared by Natural Marketing Institute, February 2015. http://organicandnatural.org/wp-content/uploads/2015/10/Consumer-Insights-on-Organic-and-Natural-A-Research-Study.pdf

Oehlke, K., Adamiuk, M., Behsnilian, D., Graf, V., Mayer-Miebach, E., Walz, E., Greiner, R., 2014. Potential bioavailability enhancement of bioactive compounds using food-grade engineered nanomaterials: a review of the existing evidence. Royal Soc. Chem. Food Funct. 5, 1341–1359.

Parpal, M., 2015. What is a fast casual restaurant? July 16, 2015. http://www.foodservicewarehouse.com/blog/fast-casual-restaurant/

Pollack, A., 2015. Genetically engineered salmon approved for consumption. New York Times. November 19, 2015. http://www.nytimes.com/2015/11/20/business/genetically-engineered-salmon-approved-for-consumption.html?_r=0

Produce for Better Health Foundation, 2015. State of the plate. 2015 study on America's consumption of fruit and vegetables. Produce for Better Health Foundation, Hockessin, DE, pp. 1–59. http://www.pbhfoundation.org/pdfs/about/res/pbh_res/State_of_the_Plate_2015_WEB_Bookmarked.pdf

Reed, K., 2015. The future of food: food production, innovation, and technology. http://www.foodinsight.org/newsletters/future-food-food-production-innovation-and-technology#sthash.D3rHVyMe.dpbs

Richards, M.R., Sindelar, J.L., 2013. Rewarding healthy food choices in SNAP: behavioral economic applications. Milbank Q. 91 (2), 395–412.

Rickman, J.C., Barrett, D.M., Bruhn, C.M., 2007. Review. Nutritional comparison of fresh, frozen and canned fruits and vegetables. Part 1. Vitamins C and B and phenolic compounds. J. Sci. Food. Agric. 87, 930–944.

Rocha, T., Amaral, J.S., Oliveira, M.B.P.P., 2015. Adulteration of dietary supplements by the illegal addition of synthetic drugs: a review. Comp. Rev. Food Sci. Food Safety. Epub. October 19, 2015. Available from: http://onlinelibrary.wiley.com/doi/10.1111/1541-4337.12173/full

Rui, L., 2013. Brain regulation of energy balance and body weight. Rev. End. Met. Dis. 14 (4), 387–407.

Sloane, A.E., 2015. The top ten food trends. Food Technol. 69 (4), 24–43.

Smith, P.M., Ferguson, A.V., 2008. Neurophysiology of hunger and satiety. Dev. Disabil. Res. Rev. 14 (2), 96–104.

Society of Toxicology (SOT), 2003. Position paper. The safety of genetically modified foods produced through biotechnology. Toxicol. Sci. 71, 2–8.

Steiner, J.E., 1977. Facial expressions of the neonate infant indicating the hedonics of food-related chemical stimuli. In: Weiffenbach, J. (Ed.), Taste and Development: The Genesis of Sweet Preference (DHEW Publication No. NIH 77-1068). US Government Printing Office, Washington, DC.

Stubbs, R.J., van Wyk, M.C., Johnstone, A.M., Harbron, C.G., 1996. Breakfasts high in protein, fat or carbohydrate: effect on within-day appetite and energy balance. Eur. J. Clin. Nutr. 50, 409–417.

Watrous, W., 2015. Nestle refreshing frozen portfolio. February 23, 2015. http://www.foodbusinessnews.net/articles/news_home/Financial-Performance/2015/02/Nestle_refreshing_frozen_portf.aspx?ID=%7BBD55D550-2D80-48AC-86A7-09E08ABB5FBA%7D

Williams, J., 2008. Factoring safety into functional foods. Food Safe. Mag., 14(6), 42.

World Cancer Research Fund, American Institute for Cancer Research, 2007. Food, nutrition, physical activity, and the prevention of cancer: a global perspective. AICR, World Cancer Research Fund International, Washington, DC, pp. 1–517.

Wyatt, S.L., 2014. The state of the snack food industry. Presented at SNAXPO Snack Food Assoc. Annual Meeting, Dallas, TX, March 5, 2014. http://www.potatopro.com/news/2014/state-snack-food-industry-revealed-snaxpo

Yadav, L., Maurya, P., 2014. Epigenetics: new relation of health and nutrition. As. J. Biosci. 9, 288–292.

Zajac, A., 2015. Pom Wonderful loses bid to block FTC deceptive-ad claims. Bloomberg Business 30, January 30, 2015. http://www.bloomberg.com/news/articles/2015-01-30/pom-wonderful-loses-bid-to-block-ftc-deceptive-ad-claim

CHAPTER

Intellectual property considerations

6

Brent Batzer, Joseph Cwik, George Carrera, Jr.
Amin Talati & Upadhye, LLC, Chicago, IL, United States

6.1 INTRODUCTION

Intellectual property (or IP), including patents, trademarks, copyright, and trade secrets, is often among the most valuable categories of assets for a company. Your brand, your innovations, your improvements, and your institutional knowledge can all fall under the ambit of intellectual property. In this chapter, we will give an overview of each category, along with key considerations for developing your IP portfolio, protecting your IP assets including enforcement, and defending against potential infringement claims of others.

6.2 PATENTS

Patents are a powerful piece of an IP portfolio. A patent gives the owner the right to exclude others from practicing the invention that it claims. Whether you practice the invention yourself, or license the technology to others, patents can be a competitive advantage and/or source of cash flow. To understand how you can develop a portfolio, as well as enforcement and defense, you first need to understand some patent basics.

Patents are granted by the United States Patent and Trademark Office (USPTO) based on applications filed there and examined by its examining corps. A US patent has several parts. The cover sheet of the patent contains a wealth of information, including the date the patent was granted, the inventors, the filing date of the patent, any other patent applications the patent claims priority to, the title of the patent, and an abstract that is a short summary of the invention. The filing date, including the filing dates of prior US applications from which the patent claims priority, is important because it sets the term of the patent. US patents are valid for 20 years from the earliest filing date of a full US utility patent application in the chain. The filing dates of the applications, including foreign applications, to which the patent claims priority are important because that sets the date that determines the world of prior art against which the patent application is judged. We will discuss this later. The patent may

Developing New Functional Food and Nutraceutical Products. http://dx.doi.org/10.1016/B978-0-12-802780-6.00006-7

contain drawings of the invention. The bulk of the patent document is known as the specification. The specification is a detailed description of the invention and should instruct one of ordinary skill in the art how to practice the invention. Last, and most importantly, are the claims. The claims appear at the end of the specification and are numbered. It is the claims that set the metes and bounds of the rights granted by the patent. The patent owner only has the right to exclude activities that fall within the scope of the claims.

There are two types of utility patent applications in the United States: provisional, and nonprovisional (or "full utility"). The nonprovisional utility application is the traditional patent application that contains claims, is examined by the USPTO, and can eventually become a patent. A provisional application can be thought of as more of a placeholder or preapplication. A provisional application does not have as many formal requirements as a utility application, and is not examined. It is particularly useful for establishing your filing date. The United States, like much of the rest of the world, now grants patents based on a first-inventor-to-file system. This means that generally the first person to file on a given invention will have the right to the patent on that invention, assuming it is patentable and not derived. Formerly, the US operated on a "first-to-invent" system. Because the filing date is now the sole measure of who has priority to an invention, getting that earlier filing date could potentially be of great importance. Once you have filed a provisional application, you have 1 year in which to file a full utility application and claim the filing date of the provisional application. A provisional application allows you to claim that your invention is "patent pending" and also can allow you to continue development work, especially if you are conducting trials where you can expect additional data in that 1- year time frame. That data can be added to the full utility application to bolster the description of your invention, and importantly, support the claims. A provisional application will also allow you to test market your invention during this 1 year before investing in the more expensive full utility application.

What is patentable? In general, anything new and useful under the sun that is created by man can potentially be patentable. This includes compositions, processes, as well as methods such as methods of use of a compound. There are some important caveats here though that has arisen particularly in recent years due to Supreme Court rulings. Importantly, "products of nature" are excluded from being patented. What exactly is a "product of nature" under the law? This question is currently in some state of flux. What is clear is that purely natural products, such as a plant that occurs in nature, an extract of a plant that occurs in nature, or other such naturally occurring products, are *not* patentable. Under US patent law, the products and laws of nature are not considered to be an invention of man, and thus are ineligible for patenting. Thus, you cannot patent human DNA, nor can you patent the result of a naturally occurring process.

On the other hand, the discovery of a new use for a naturally occurring product may be patentable. For example, a method of treatment using a natural substance is potentially patentable. Likewise, when a person combines naturally occurring ingredients to make something new, the new thing created may be patentable.

An example given by the USPTO in guidance (USPTO, In Press-a) to its examiners is gunpowder. Gunpowder is a combination of potassium nitrate, charcoal, and sulfur, all naturally occurring materials. However, when mixed they have a different characteristic (explosiveness). Thus, because the mixture is made by man and has a different characteristic from the components, the combination is patentable subject matter. The law around patentable subject matter and how it is applied by the courts and the USPTO continue to develop, so if there is any question about whether something is patentable subject matter, a patent attorney should be consulted for the latest on how the USPTO is treating natural products and combinations.

Patentable subject matter is a threshold matter that rests on the principle of "utility" or usefulness. If the invention falls within patentable subject matter, it must also be novel and nonobvious as compared to the prior art. Prior art includes essentially everything that was published prior to the earliest effective filing date (known as the priority date) of the patent application. This includes other patents, published patent applications, and other publications such as scholarly journals, newspapers and magazines, handouts or poster presentations from conventions, and any other material that was published and made available to the public, including posted on the Internet. It includes not just material published in the US, but worldwide, even if in another language. The claims of the patent application are compared to the prior art. A claimed invention is "novel" if it is not disclosed in a single prior art reference. However, even if not disclosed in a single reference, a claimed invention can be "obvious" in light of one or more prior art references. This is the primary focus of the patent examination process at the USPTO (In Press-b). An examiner may reject claims based on one or more prior art references, and the applicant (or applicant's attorney, most likely) will respond with arguments as to why the rejection is incorrect, and/or amend the claims to change their scope to not include what the examiner says is within the prior art.

There are other requirements for claims. The claims must be "enabled" and described by the specification. The right to exclude others that comes from owning the patent is given in exchange for the disclosure of the invention to the public, as set out in the patent. The enablement and written description requirements exist to enforce that disclosure requirement. A claim is not "enabled" if the patent document does not allow one of ordinary skill in the art to make and use the claimed invention. Likewise, a claim fails the written description requirement when the specification does not describe the claimed material to one of ordinary skill in the art. By requiring this disclosure, it allows others to build on the discovery, and to use it after the patent expires. Thus, the Constitutional purpose of the patent system, to advance science and the useful arts, is fulfilled.

6.2.1 DEVELOPING A PATENT PORTFOLIO

The first, and best, place to look for potential patentable inventions is within the company. Your internal review should include a basic recording system, including an internal process by which inventions are reported, that will allow you to review any potential inventions as they are developed for possible patent opportunities. One such way of collecting the relevant information is by using a record of invention or an invention disclosure form. This type of form puts the information into one place for the person in the company who is assigned to review any potentially patentable material. It will also save time with your patent attorney if you decide to pursue a patent application, since a good form will have much of the information an attorney will initially need in order to pursue the application. The form should include the names, addresses, residence, and citizenship of the inventors, a summary of the invention, and it is helpful to have such information as what problem the invention solves, why other known solutions do not work or do not work as well, any surprising results, drawings, etc. The more detailed technical information you can include in the disclosure form, the easier it is for your in-house reviewer to understand what the company may have that might be protectable IP, and the easier it is for the patent attorney to draft an application, which should save the company money on attorney fees.

Your patent attorney can likely provide you with a record of invention or disclosure form for your company's use or develop one with you, and help set up a plan to harvest the inventions and ideas from your R&D operations. As discussed previously, time is of the essence in filing for protection, as the first inventor to file a patent application has priority to get the patent on the invention, even when another may have invented the same invention first. Making a record of invention part of your standard process helps under this essentially "first-to-file" regime by speeding up your internal fact-gathering.

Another common way that inventions are recorded is through use of laboratory notebooks. If your company chooses to use lab notebooks, your employees should be instructed to make their entries as detailed as possible. Lab notebooks that keep a contemporaneous record of development activities formerly had additional value under the "first to invent" system. However, under the "first to file" system now in effect, proof of prior invention is not likely to make much of a difference going forward, absent special circumstances. Regardless of the methods you choose to record innovations, it is important to make these records as detailed as possible if only to save time and money later when pursuing a patent. All additional details provided mean less time later spent by your attorney or patent agent in trying to determine all the relevant facts and information that will go into the application. In the final analysis, good recordkeeping remains essential to capturing and protecting your inventions, and fostering your R&D program through good science.

If you have identified a possibly patentable invention, you might consider commissioning a patentability search of the prior art to gain further insight on the competitive landscape and on whether it is worth the time and money to file a patent application. Sometimes, the search results save you from the cost of filing a full application, because it turns out that the invention, or something very close to it, has been previously disclosed. And in a crowded art area, the search results can be used to help focus a patent application on the truly innovative material, ideally streamlining the process during examination. However, pertinent search results will need to be disclosed to the Patent Office.

As part of your patenting strategy, you should also consider national or international patent filings. Patents in many other countries can be applied for using your United States patent application as the starting point, including the priority date for prior art purposes. Your initial decision on foreign filing must be made within a year of your US application filing. This applies whether your application is a provisional application or full utility application. Thus, if your initial application is a provisional, you will need to make a decision on whether to pursue foreign filing at the same time you decide whether to file the full US utility application. The process in many countries, including all of Europe, China, Japan, Korea, Australia, and Canada (WIPO, In Press), can be started using an international application known as a PCT (Patent Cooperation Treaty) application. The PCT allows you to begin the process in one or more of its 149 member countries, known as the "international stage," but you do not have to make your final decision on which specific countries to pursue until about 2.5 years (30 months) after your initial application. These filings in individual countries are known as "national stage" filings. The 2.5 year time frame before national stage filings are due from your application can allow you to obtain market data to develop a business plan as to which countries are most important for your product.

In the national stage, each individual country will typically have periodic or generally yearly fees (known as annuities or maintenance fees) in order to maintain the patent as active. Unsurprisingly, with each country charging such fees, the fees can become substantial if you are filing in many countries. If the cost of pursuing patents in many countries is unmanageable, your international strategy should focus on specific countries that are important markets for the company, or even on specific countries for just that particular product within a broader market. However, keep in mind that you cannot file applications later, so if you decide to not pursue protection in a country, or discontinue paying the fees in a country and allow the patent or application to lapse, you will not be able to have patent protection in that country for that invention.

In addition to developing your own patents, patent rights can also be purchased from others. One common way of obtaining patent rights from third parties is through a license. Patent licenses can be mutually beneficial; a patentee may not be able to commercialize an invention due to lack of funds or other barriers to entry, while a licensee can obtain rights that give it an advantage in the marketplace. The licensee gains especially valuable rights when the license is an exclusive license, which effectively gives you most of the same rights as if you owned the patents yourself. Licenses can be structured in a variety of ways, including flat-fee, royalties based on a percentage of the profits or sales, tiered pricing models, and other creative structures. Patents can also be purchased outright. You can purchase patents directly from a third party who is interested in selling the patent. Patents are also available at patent auctions or bankruptcy asset sales. It is important to note that patent ownership via assignment of rights can be recorded with the USPTO in a public database (USPTO, In Press-c).

Joint ventures, including research agreements with universities, may also generate patentable inventions, or be based on the patents already owned by one party with the goal of improvement in some way, with the improvements owned by both parties. When in a joint venture, it is extremely important that your initial joint venture agreement set forth who owns the IP, including patents. Both ownership of patents that each party already owned, as well as the ownership of any inventions developed as part of the joint venture, must be addressed. If possible, the question of who will own the patents upon termination of the joint venture should also be answered in the initial agreement. If not addressed then and highly valuable patents are developed during the joint venture, a messy fight over the patent rights could well ensue.

6.2.2 ENFORCEMENT CONSIDERATIONS

Protecting your patent portfolio is first and foremost an exercise in monitoring the market. Thus, you should be monitoring for other products that are being sold or made in the US that may infringe one or more claims of your patents. Likewise, if you have foreign patents, you should also be monitoring in those countries. Something as simple as an occasional trip to stores to take a look at the products competing with you can be a useful tool in keeping abreast of the latest product offerings, including ones that might tread on your patent rights, and this can be part of your monitoring process. Industry press releases, product catalogs, and any other sources that help keep you up to date with what others are doing in your space are all useful to review for possible infringers.

If you encounter a potentially infringing product or an infringer, a common first step is to send a letter informing the third party of your patent rights, commonly called a cease-and-desist (C&D) letter. These letters can be particularly important when the patent you believe is being infringed covers a method or a process, as they can "start the clock" from when damages can accrue.

The letter may or may not notify the target of specific products that you believe infringe your patent. There are multiple considerations at play, including the risk of being forced into premature litigation. Certain letters give the recipient the right to immediately seek a declaration that no infringement exists, or that the patent is invalid. These actions are known as declaratory judgment actions. If the target files a declaratory judgment action, they may force you to litigate in a forum that is not convenient for you, or preferred by the target. Furthermore, some states are enacting laws that are intended to target entities known by a number of names, such as "patent assertion entities," "nonpracticing entities," or "patent trolls" but these laws may have an effect on C&D letters sent by anyone. You should strongly consider consulting with an attorney well versed in patent infringement matters before sending any letter to a potential infringer because of these considerations.

Before sending a C&D letter, it is important to "do your homework" on both your own claim as well as the target of the C&D letter. With respect to your claim of infringement, you need to perform an adequate investigation into the infringement in order to have a reasonably good idea that the target is infringing. Sometimes this is easy, but often many of the key pieces of evidence are in the hands of the target, in which case you do the best you can. With respect to the target of the letter, you will want to know things like how aggressive the target has been in the past, including whether they have filed declaratory judgment suits. If the product you believe might be infringing is a key product for the target, they could be more likely to act quickly in an attempt to protect their rights. By filing the declaratory judgment suit, they will get to pick the venue where the suit will be heard. Some courts have specialized patent rules where numerous actions are required in a relatively short time frame, forcing you to spend substantial funds in legal fees at an early stage in the case in order to meet the requirements. If you know your target has filed for declaratory judgment before, you might consider filing your lawsuit but not serving it, in order to file in your preferred venue.

Patents are of course often enforced through the courts via a patent infringement lawsuit. These lawsuits can be very expensive due to the costs of attorneys, discovery, expert witnesses, and other expenses. However, it can be necessary in order to enforce your right to exclude others from practicing your patented invention. A victory in a patent infringement lawsuit often includes an injunction against the infringer that they cease selling the infringing product. Patent infringement suits also carry damage awards, which can be determined by a "reasonable royalty" rate, by lost profits, and by assorted other measures of damages. Furthermore, an infringer who is aware of your patent and willfully infringes the patent anyway may be subject to increased damages that can double or triple the damage award determined by the court or jury.

If the infringer is importing the infringing goods, one possible venue is a proceeding at the US International Trade Commission (ITC), sometimes called a "Section 337 Action" or just "337 Action." The ITC has rules that push a case along in a fairly rapid time frame, which can rack up legal fees quickly but also will likely get you to a result more quickly. Monetary damages are not available in an ITC proceeding; the only remedy available is an import ban on the infringing goods.

Enforcement is not the only aspect of protecting your patent rights. A patent is presumed valid once it is issued. However, given the vast universe of potential prior art, between worldwide patents, published patent applications, and nonpatent publications including journal articles, newspapers, or any other publication available to the public, it is possible that the patent office did not have a potentially important piece of prior art in front of the examiner when it granted the patent. If in the course of your business you come across prior art that was not reviewed by the examiner, it is possible to have your patent reexamined.

There are both disadvantages and advantages to having your patent reexamined in light of the prior art reference you found. The disadvantage is clear: your patent claims could be significantly narrowed, or you could even lose your patent entirely if the prior art reference or references are a close match to your patent. However, there are several advantages to the reexamination as well. First, if you found this prior art, you can be fairly confident that anyone you attempt to enforce the patent on through litigation will probably find it too. In a court case, the court will invalidate entire claims. But in a reexamination, you have the opportunity to amend your claims to get around that prior art reference. Furthermore, in an ex parte reexamination (which is the type of reexamination you would file for in this situation) it is just you and the patent examiner involved; there is no litigation opponent to interject or put forth additional arguments. The end result of a reexamination could be that the prior art reference you found would be removed or at least severely weakened as evidence in future enforcement litigation, making the resultant patent perhaps a more valuable and more certain asset to your company.

6.2.3 RISK MANAGEMENT AND DEFENSE CONSIDERATIONS

Defending against a patent suit is obviously a position most companies would prefer to avoid. As discussed earlier, patent infringement cases are expensive. The cost of defense alone can run into the millions even if you are victorious in defending the claim; lose, and the damages owed can in some cases be massive. And furthermore, unlike when pursuing those infringing your patents, you do not have the possible damage award at the end of the road. There are, however, several actions that a company can take to reduce its likelihood of being sued for patent infringement. Putting one or more of these programs into place, while costing some up-front, can save the company significant money later, and further allows you to avoid shifting time and attention away from your core business to a patent suit.

Perhaps the most familiar of the actions your company can take is to obtain freedom to operate (or FTO) opinions when you release new products. This process involves having a patent search performed to find any patents of which your new product might run afoul. An experienced patent attorney will then review the results and come to an opinion as to your risk of infringement relative to those patents. Knowledge of the patent landscape and your relative risk can help you avoid a later patent infringement suit; indeed, in some cases your company may decide to scrap a project entirely because the risk is too great.

A written opinion of outside counsel may also come in handy later should you go forward with the project and then be sued by the owner of one of those patents. A well-reasoned opinion of independent counsel can be used to show that any infringement was not willful, an important point as willful infringement can put a company on the hook for increased damage awards. Because of the chance that the opinion will be disclosed in litigation at a later time, it is important to consider all communications in relation to the opinion as potentially discoverable, even though they would ordinarily be covered by the attorney–client privilege. Thus, when discussing any opinion-related material, consider doing so by phone.

It is also possible to run FTO analysis using your own company employees. This will not necessarily carry the benefit that the outside counsel opinion has with respect to willfulness. However, it can have the same benefits in the go/stop decision calculus on a project.

Another program you might wish to put into place is the monitoring of competitors' patents and published patent applications, and associated technical or scientific publications. This is somewhat related to FTO searches and analysis in that the analysis of competitor patents as they issue might reveal an issue with a current product or a product still in the development stage. However, monitoring of published patent applications that have not yet matured to granted patents has an additional advantage: the USPTO rules allow third parties to submit prior art to the patent examiner. The time period you have for submitting to the examiner is quite short, so if your monitoring program spots an application that could be a problem for you, but that also does not seem to have the most pertinent prior art in front of the examiner, you need to act quickly. In some cases, the prior art you submit might be the art that prevents that patent from issuing, or causes it to issue with a significantly narrowed claim scope. Thus, competitive intelligence may rest on putting into effect a properly defined, diligently monitored patent/literature alert protocol.

For patents that have already been granted, there are a number of postgrant review procedures currently in place at the USPTO. Reexaminations can be granted by submitting a request that shows a substantial new question of patentability in light of art that was not in front of the examiner during the prosecution of the patent. Ex parte reexaminations allow a party to submit a request for reexamination including a description of how the prior art makes one or more of the claims unpatentable. The reexamination then proceeds only between the patent holder and the examiner. However, patent reforms passed in 2011 include a new *inter partes* review option, or IPR. In an IPR, the requestor plays an active role and can involve a discovery-like process of fact finding. Because an IPR includes this interactive process, they cost substantially more than an ex parte reexamination. However, they allow you to participate all the way through the process. If you wish to consider these reexaminations or reviews, it is highly recommended to consult with an experienced patent attorney, as there are additional pros and cons to each that you will need to fully understand before making a decision.

Risk management is not limited to your own products. In recent years there has been a significant increase in the filing of patent infringement suits against users or purchasers of a product rather than the original manufacturer. This is an especially common tactic of "patent assertion entities" (also called "nonpracticing entities" or even "patent trolls") that as a business model buy or obtain patents solely in order to assert them in litigation or extract a license fee under the threat of litigation, rather than run as a traditional company that sells a product or service of some kind. You certainly aren't going to be running FTO searches or monitoring for products you buy, so what can you do to avoid this kind of lawsuit?

The primary way to protect the company is to negotiate careful contracts when purchasing from others. In any agreement, you will want to have a good indemnity provision where the other party agrees to defend you, including paying for the attorneys and costs, against patent infringement suits based on the product or service they sold you. Likewise, you will want the contract to contain a warranty that the product does not infringe any patents. You will want these types of provisions in contracts for a variety of goods and services, not just components for your products. Indeed, some of the most important areas to have these provisions include your information technology and marketing functions. When getting software licenses from a software publisher or manufacturer, negotiate these terms into the agreement. There is a good chance they will not appear in the standard terms they offer, so you will likely have to ask for them. When working with a marketing firm, negotiate these terms into the agreement for their services. IT and software, including some fairly ubiquitous technology at this point, are regular targets of patent assertion entities at the user or customer level. While your company may not have been hit with a suit from one of these entities yet, there is a reasonable chance it may happen to you in the future. Protecting yourself in as many contracts as possible at least can push the cost onto the party that actually caused the possible infringement in the first place.

Finally, there are insurance policies that can cover patent infringement costs. If you choose to get such insurance, you will of course need to carefully read the policy to determine precisely what is covered and what is excluded, and make an informed choice based on the cost–benefit analysis for your particular situation.

6.3 TRADE SECRETS

Trade secrets cover any proprietary material that can be used for a business advantage. They are governed by common law (e.g., law made by judges through cases that have set precedent over the years) as well as state and federal statutes. They can include proprietary formulae, proprietary processes, as well as business-related information such as customer and leads lists.

For some types of intellectual property, trade secret is your primary method of protection. This would include information like the aforementioned customer lists. Such information is not likely to be protectable under other forms of IP protection. However, in other cases, the trade secret is also something that could be patentable. There are advantages and disadvantages to picking either patent or trade secret. A key advantage to keeping a trade secret is that a trade secret is protected as long as you can maintain the secrecy. Famous trade secrets that have been maintained for periods much longer than a patent could have lasted include the formula for Coca-Cola, or the "secret blend of 11 herbs and spices" of Kentucky Fried Chicken. These secrets have incredible value to those companies.

However, on the flipside a patent gives you a certain term without needing to worry about maintaining the secret. Just as a trade secret can last much longer if the secret is maintained, it can be much shorter if it cannot be maintained. Thus, if a product or process can be reverse-engineered, patent protection is likely a superior option. Furthermore, even if the secrecy can be maintained, it will require substantial vigilance by the company to keep it so. Access restrictions, document policies and restrictions, keeping the information limited strictly to those having a need to know, confidentiality agreements including provisions in your employment agreements with employees and contractors, exit interviews with employees to remind them of their secrecy obligations, all of these are part of a well-considered trade secret plan. A patent, on the other hand, requires no such safeguards once you have filed, as you have already publicly disclosed the information in exchange for the potential patent protection.

Because trade secret laws can vary by state, if you have trade secret information and wish to ensure that your policies are sufficient for your protection, an attorney should be able to help you with developing a plan that works for your business.

6.4 TRADEMARKS

Your branding is also a key piece of your IP portfolio and helps to establish goodwill in the marketplace. In the United States, trademark rights accrue based on the use of a trademark in commerce. You can have common law trademark rights based solely on your use, as well as rights based on registrations through state trademark offices and of course, through the USPTO.

However, while you can have some common law rights based on your use, those rights are limited to the geographical area where you use the mark. Thus, it is strongly recommended that you attempt to register your trademarks with the USPTO. Such a registration gives you national coverage, and includes other rights and remedies, as well as evidentiary advantages should litigation be necessary, that you would not have in relying solely on common law.

It is important to select marks that are distinctive to identify your products. Trademarks that are generic or merely descriptive of a product are not eligible for registration, because they are not identifiers of the source of the product but rather of the product itself. Likewise, it is important to avoid being the same as or very close to another company's marks. Trademarks, both in examination before the USPTO in the registration process, and in the infringement context, are judged based on the likelihood a consumer would be confused by the similarity of the marks. If you pick a mark too close to another, you may not be able to obtain registration, and furthermore, you may open yourself up to an infringement suit from the other party that owns the similar mark.

Thus, it is strongly recommended that you commission trademark searches on your proposed marks before you begin to use them. You do not want to spend the time and energy developing a brand and marketing plan, only to discover that someone else was already using that mark. Furthermore, if you do no search at all, your discovery that someone else had rights in the mark could come when you receive a letter from the trademark holder demanding that you cease use, or worse, when you are served with a lawsuit for trademark infringement. These risks can be limited by having the search done at the outset of the process.

While trademark rights do not vest until you use the mark in the United States, the USPTO does have an option to allow you to in essence reserve the mark via an "intent to use" application. This application is examined much like a regular trademark application, based on whether your proposed mark is confusingly similar to another active mark on the trademark register. If the mark passes examination, rather than becoming a registration, it then waits at the USPTO until you use the mark in commerce. Once you have used the mark, you can file the statement of use with the USPTO, and the registration issues on the Principal Register. You have 6 months from the time the application finishes examination and is approved, with additional extensions available for successive 6-month periods up to a total of 36 months from the allowance of the application.

If your company sells products outside the United States, international trademark protection is available. As is available with patents, an international treaty allows for an international trademark application, referred to as the Madrid Protocol. There are currently 93 member countries, including many of the most commercially relevant nations. Unlike with patents, you do not lose the right to file if you choose not to file initially. However, in order to claim the filing date of your US application, you must file your Madrid Protocol application within 6 months of the filing of your US application.

Most foreign countries base priority for trademarks solely on filing date, rather than use. Because of this, you must take care when dealing with distributors abroad if you have not already filed for your trademark in those countries. It is generally recommended to have your trademark application filed before commencing sales in a new country; however, if this cannot be done, one must take care to ensure that your contracts with distributors or agents in those countries require that they not register the mark nor take any action against your ownership of the mark. Without such protection, it is not uncommon for someone to file a registration for the mark and then attempt to extract a payment from the proper rights holder.

Once you have a registration, it is highly important to monitor the market for any unauthorized use of your mark, particularly in the United States. First, you want to protect your brand from any unauthorized use because use by another can damage your brand should they sell substandard or defective products under your name. Furthermore, because trademark rights are based on use and because of their being an identification of source, if you do not adequately police the use of your mark, it is actually possible to lose your rights in the mark. Thus, you will need to send C&D letters to those using your mark or a confusingly similar mark. In some cases, a lawsuit may be necessary. As policing can be time and labor intensive, there are services available to monitor for usage if you feel your company does not have adequate resources to properly monitor for improper usage.

It is also advisable to monitor for new trademark applications at the USPTO. Applications are published for opposition before registration is granted. If you spot a mark that you believe is too similar to yours, you have 30 days to file an opposition after the mark is published. As with general usage monitoring, there are services that can do this monitoring for you for a fee.

Trademarks can also be acquired in mergers or acquisitions where your company buys the entire business that relates to the trademark. If you acquire a product, you will want to immediately add the new marks to your monitoring program, and if for some reason the product does not currently have a trademark registration, consider filing. Trademarks can be licensed in certain situations, but there are traps which if not avoided present a risk of losing rights in the mark entirely. Thus, it is important to review the ownership and status of any trademarks you acquire in a transaction when acquiring a business.

6.5 COPYRIGHT

Copyright is available for any original works of authorship fixed in a tangible medium. Copyright, like trademarks, arises from common law as well as statute. You have a copyright in your original works by common law, even if you do not file for a copyright registration. However, once again, registration provides several benefits, including additional remedies not available with common law rights alone. Furthermore, copyright registration is fairly inexpensive. Presentations and instruction manuals are two types of material that might be something you would consider for copyright protection. Websites and source codes may be protected in this fashion, keeping in mind the "fixed" nature of the protection, in which dynamic changes are not allowed.

6.6 CONCLUSIONS

An intellectual property portfolio can bring significant value to a company. Developing an IP portfolio to leverage your branding and your technical advances, and using both offensive and defensive IP strategies, will place you in the best position to compete in today's marketplace.

REFERENCES

USPTO, In Press-a. Nature-based products examples from subject-matter eligibility guidance. http://www.uspto.gov/patents/law/exam/mdc_examples_nature-based_products.pdf

USPTO, In Press-b. Third-party preissuance submissions. http://www.uspto.gov/patent/initiatives/third-party-preissuance-submissions

USPTO, In Press-c. Assignments webpage. http://assignment.uspto.gov/

WIPO, In Press. List of PCT contracting states. http://www.wipo.int/pct/en/pct_contracting_states.html

CHAPTER

7 Regulatory constraints on new product development and approval procedures in the United States

Jennifer Shield, Jacqueline Kuler, Abhishek Gurnani
Amin Talati & Upadhye, LLC, Chicago, IL, United States

7.1 INTRODUCTION

Much of the functional food industry believes that onerous regulations hamper innovative efforts by the industry. Others argue that the regulatory framework for bringing a product to market is appropriate and geared toward the adequate protection of public health. In developing a new product it is imperative that companies first understand not only the regulatory hurdles they face in bringing a novel product to market, but also regulations that govern the marketing of the product as well. The following is a summary of key regulatory requirements that affect the functional food industry; requirements that all innovative companies must be mindful of when bringing a product to market and crafting a marketing position. It is one thing to have a highly innovative product that serves a profound consumer need, yet it is another to be able to successfully navigate compliance when selling the innovation. This chapter outlines a number of requirements for conventional foods, medical foods, and dietary supplements, including when an ingredient requires approval, requirements for manufacturing processes and controls, and certain advertising claims that can and cannot be made for a product.

7.2 PRODUCT FORMULATION

7.2.1 CONVENTIONAL FOODS

Under section 201(s) of the Federal Food, Drug, and Cosmetic Act (FD&C Act), any substance intentionally added to food is a "food additive" and requires premarket approval by the Food and Drug Administration (FDA) unless the substance is generally recognized as safe (GRAS) by qualified experts for its intended use in food, or qualifies for an exemption from the "food additive" definition. The term "food additive" is broadly defined to include substances which directly or indirectly

Developing New Functional Food and Nutraceutical Products. http://dx.doi.org/10.1016/B978-0-12-802780-6.00007-9

become a component of food or otherwise affect the characteristics of food, including substances used in manufacturing, packing, processing, treating, transporting, or holding of food. The term "direct food additive" refers to substances that are directly added to food, while the term "indirect food additive" or "food contact substance" refers to substances which become a component of food during the manufacturing, packing, treating, transporting, or holding of food (e.g., substances used on processing equipment).

The following substances are excluded, by statute, from the definition of "food additive": (1) pesticide chemical residues in or on a raw agricultural commodity or processed food; (2) pesticide chemicals; (3) color additives; (4) any "prior sanctioned" substance (i.e., one used in accordance with a sanction or approval granted prior to Sep. 6, 1958 under the FD&C Act, the Poultry Products Inspection Act, or the Meat Inspection Act); (5) new animal drugs; and (6) dietary ingredients in dietary supplements.[a]

Also exempt from the definition of a "food additive" are substances that are generally recognized as having been shown to be safe under the conditions of their intended use. A determination that a particular use of a substance is GRAS requires both technical evidence of safety and a basis to conclude that the technical evidence of safety is generally known and accepted throughout the scientific community knowledgeable about the safety of the substance.[b] A determination of "safety" is the same regardless of whether a substance is GRAS or a food additive subject to premarket approval. "Safe" means that there is a reasonable certainty in the minds of competent scientists that the substance is not harmful under the intended conditions of use.[c]

The use of a substance may be determined GRAS either through scientific procedures or, for substances used in food prior to 1958, through experience based on common food use. A general recognition of safety through scientific procedures requires the same quantity and quality of scientific evidence as is required to obtain approval of the substance as a food additive.[d] A GRAS determination based on use in food prior to 1958 requires a substantial history of consumption of the substance by a significant number of consumers prior to 1958. This method of GRAS determination is seldom used today.

The common knowledge element of a GRAS determination requires that the data and information relied upon to establish the technical evidence of safety be generally available and that there be a basis to conclude that there is consensus (not unanimity) among qualified experts about the safety of the substance for its intended use.[e] GRAS determinations through scientific procedures must ordinarily be based upon published studies, which may be corroborated by unpublished studies and other data and information.

[a]21 U.S.C. § 321(s) (2016).
[b]21 C.F.R. § 170.30 (2016).
[c]21 C.F.R. § 170.3(i) (2016).
[d]21 C.F.R. § 170.30 (2016).
[e]62 Fed. Reg. 18938, April 17, 1997.

While the regulations provide a voluntary administrative mechanism by which companies may petition the FDA to obtain an official affirmation of lawfully made GRAS determinations, this rulemaking process proved to be too resource-intensive for the agency, and in 1997 the FDA issued a proposed rule to replace the GRAS affirmation petition process with a notification procedure. Despite never being finalized, the FDA has been administering GRAS notices under this interim policy ever since.[f]

Under the GRAS notification procedure, companies can submit information regarding substances determined to be GRAS, their intended conditions of use, and details regarding the basis for the GRAS determination to the FDA, and the agency will respond, in writing, by either informing the company that it does not question the basis for the notifier's GRAS determination, or that it concludes that the notice does not provide a sufficient basis for a GRAS determination.

Substances added to food or which may reasonably be expected to become a component of food and that are not excluded from the definition of "food additive" are subject to premarket approval by the FDA. The marketing of a new direct food additive (or a previously approved additive for a new unapproved use) requires the submission of a petition to the FDA. The primary means by which the FDA evaluates the safety of indirect food additives is through the food contact notification (FCN) process.[g] The FCN process is used for authorizing the marketing of food contact substances (FCS) except where the FDA determines that the submission of a food additive petition is necessary to provide adequate assurance of safety, including either of the following situations: (1) When the use of the FCS will increase the cumulative estimated daily intake (CEDI) of the FCS from both food and food-contact uses to a level equal to or greater than 1 part per million (ppm) for a substance that is not a biocide or, in the case of a biocide, to a level equal to or greater than 200 parts per billion (ppb); or (2) when existing data include one or more bioassays on the FCS that the FDA has not reviewed already and such studies are not clearly negative for carcinogenicity.[h]

Because the safety standards are the same for all food additives, the information submitted in an FCN and a food additive petition is comparable. The basic elements of a food additive petition include a description of the identity of the substance, an assessment of probable exposure to the substance, an evaluation of safety of the substance under the intended conditions of use, and consideration of potential limitations of the conditions of use if necessary to ensure safe use of the substance.[i] Once received, the FDA and other governmental and nongovernmental experts (if necessary) evaluate the chemistry, toxicology, and environmental (if applicable) data and information to determine whether a safety conclusion can be made. If a final safety

[f]Idem.
[g]21 U.S.C. § 348 (2016).
[h]21 C.F.R. § 170.100(c) (2016).
[i]See 21 C.F.R. Part 171 (2016).

determination is made, the FDA publishes its findings in the Federal Register. Once a final rule is published, the regulation appears in the Code of Federal Regulations. The FDA estimates that the average time between submission and when a final rule is published for a direct food additive petition is 24 months.[j]

As previously mentioned, dietary ingredients in dietary supplements are expressly exempt from the food additive definition.[k] Under section 402(f) of the FD&C Act, dietary ingredients cannot present a significant or unreasonable risk of illness or injury under conditions of use recommended or suggested in labeling, or if no conditions of use are suggested or recommended in the labeling, under ordinary conditions of use. Also, new dietary ingredients (i.e., dietary ingredients not marketed in the United States prior to Oct. 15, 1994) may be subject to premarket notification requirements discussed later in this chapter. It is important to note, however, that other ingredients intended for use in dietary supplements, such as binders, excipients, fillers, and flavors, are not exempt from the food additive definition and must meet the same requirements as substances added to conventional foods.

7.2.2 MEDICAL FOODS

A medical food is defined as a food which is formulated to be consumed or administered enterally under the supervision of a physician and which is intended for the specific dietary management of a disease or condition for which distinctive nutritional requirements, based on recognized scientific principles, are established by medical evaluation.[l] This particular category of food has gained traction recently and many companies are looking to develop an innovative medical food. However, FDA narrowly defines the conditions under which a medical food may be marketed:

1. It must be a specially formulated and processed product (as opposed to a naturally occurring foodstuff used in its natural state) for the partial or exclusive feeding of a patient by means of oral intake or enteral feeding by tube;
2. It must be intended for the dietary management of a patient who, because of therapeutic or chronic medical needs, has limited or impaired capacity to ingest, digest, absorb, or metabolize ordinary foodstuffs or certain nutrients, or who has other special medically determined nutrient requirements, the dietary management of which cannot be achieved by the modification of the normal diet alone;
3. It must provide nutritional support specifically modified for the management of the unique nutrient needs that result from the specific disease or condition, as determined by medical evaluation;
4. It must be intended to be used under medical supervision; and

[j] Guidance for Industry: Questions and Answers about the Petition Process, April 2011., http://www.fda.gov/food/guidanceregulation/guidancedocumentsregulatoryinformation/ucm253328.htm.
[k] 21 U.S.C. § 321(s) (2016).
[l] 21 U.S.C. § 360(ee)(b)(3) (2016).

5. It must be intended only for a patient receiving active and ongoing medical supervision wherein the patient requires medical care on a recurring basis for, among other things, instructions on the use of the medical food.[m]

In terms of ingredients that may be used in medical foods, the ingredients should be either (1) a food additive used in accordance with FDA's food additive regulations;[n] (2) a color additive used in accordance with the color additive regulations;[o] (3) a substance that is generally recognized, by qualified experts, to be safe under the conditions of its intended use (GRAS);[p] or (4) a substance that is authorized by a prior sanction issued by FDA.[q]

A potential benefit to marketing of a medical food product is that medical foods are exempt from the labeling requirements for health claims and nutrient content claims under the Nutrition Labeling and Education Act of 1990.[r]

While medical foods do not require premarketing approval by FDA, there are a limited number of conditions which FDA believes are appropriate for medical food use. For example, FDA generally considers inborn errors of metabolism to be diseases or conditions that a medical food could be used to manage. Inborn errors of metabolism include inherited biochemical disorders in which a specific enzyme defect interferes with the normal metabolism of protein, fat, or carbohydrate. On the other hand, FDA generally would not consider a product labeled and marketed for diabetes mellitus to meet the regulatory criteria for a medical food as a regular diet can be modified to meet the needs of an individual affected by diabetes mellitus.

In the past, FDA has designated certain conditions that it believes do not have a nutritive need and thus products marketed for such conditions would not meet the established definition of a medical food. FDA has stated in a warning letter to a medical food distributor that there are no distinctive nutritional requirements or unique nutrient needs for individuals with mild to moderate Alzheimer's disease.[s] As such, FDA considered the product to be misbranded within the meaning of section 403(a)(1) of the Act because the product labeling is false and misleading in that the product is labeled and promoted as a medical food, but it does not meet the definition of a medical food. In another warning letter to a prominent medical food distributor, FDA stated that it was not aware of any nutritive requirements for chronic fatigue syndrome, fibromyalgia, leaky gut syndrome, metabolic syndrome, cardiovascular disease, inflammatory bowel conditions, inflammatory bowel disease, or atopic disorders, rendering a number of medical food products marketed for these conditions as misbranded.[t]

---

[m]21 C.F.R. § 101.9(j)(8) (2016).
[n]21 C.F.R. § 172 (2016).
[o]21 C.F.R. § § 73, 74 (2016).
[p]21 C.F.R. § 170.30 (2016); 21 U.S.C § 321 (s) (2016).
[q]21 C.F.R. § 170.3(l) (2016).
[r]21 U.S.C § 343(q)(5)(A)(iv) (2016).
[s]FDA Warning Letter to Accera, Inc., December 26, 2013., http://www.fda.gov/ICECI/EnforcementActions/WarningLetters/2013/ucm381320.htm.
[t]FDA Warning Letter to Metagenics, August 13, 2013., http://www.fda.gov/ICECI/EnforcementActions/WarningLetters/2013/ucm367142.htm.

Thus, in developing an innovative medical food product a company must be aware of FDA's narrow definition and ensure that the various conditions it seeks to target have distinct nutritive needs which would not be addressed through normal diet changes alone. Otherwise, its product could be considered misbranded in violation of the FD&C Act.

Other than industry guidance and a handful of warning letters to only a few of the dozens of medical food companies, FDA has not undertaken any rigorous enforcement action against the medical food industry. With the relatively light regulatory oversight from FDA and minimal barriers to market entry, medical foods prove to be another way to bring an innovative product to market keeping the constraints outlined previously in mind.

7.2.3 DIETARY SUPPLEMENTS

At one time, supplement products were regulated under either food or drug regulations. However, this changed in 1994 when the new FDA-regulated class of "dietary supplements" was created with the enactment of the Dietary Supplement Health and Education Act (DSHEA). DSHEA amended the FD&C Act to define dietary supplements and dietary ingredients, and to set forth other requirements related to labeling, claims, and good manufacturing practices.

Generally, a "dietary supplement" is a product intended to supplement the diet that contains a "dietary ingredient"[u] Under DSHEA, there are six types of "dietary ingredients":[v]

(A) a vitamin,
(B) a mineral,
(C) an herb or other botanical,
(D) an amino acid,
(E) a dietary substance for use by man to supplement the diet by increasing the total dietary intake, or
(F) a concentrate, metabolite, constituent, extract, or combination of any ingredient described in (A) through (E).

However, DSHEA also created two categories of dietary ingredients: (1) grandfathered or "old dietary ingredients" (ODI); or (2) new dietary ingredients (NDI). A "new dietary ingredient" is defined as "a dietary ingredient that was not marketed in the United States before October 15, 1994."[w] While DSHEA did not explicitly define "old dietary ingredients," it can be concluded from the definition of "new dietary ingredient" that an old dietary ingredient is the opposite: an ingredient that

[u]Dietary supplements are products intended for ingestion, are not represented for use as a conventional food or as a sole item of a meal or the diet, and are labeled as dietary supplements. The definition also excludes an article that is approved as a new drug, certified as an antibiotic, or licensed as a biologic, or if substantial clinical investigation has been undertaken and publicized for one of those three approvals.
[v]21 U.S.C. § 321(ff)(1) (2016).
[w]21 U.S.C. § 350b(d) (2016).

was marketed in the United States prior to Oct. 15, 1994. Determining whether a dietary ingredient is ODI or NDI is essential for knowing whether a product is being marketed in compliance with FDA regulations.

7.2.3.1 Old dietary ingredients

For an ingredient to be considered ODI, a company must show that the dietary ingredient was marketed in the United States prior to Oct. 15, 1994. "Marketing of a dietary ingredient" is considered by FDA to mean offering the dietary ingredient for sale as a dietary supplement, a bulk ingredient for use in dietary supplements, or as an ingredient in a blend or formulation of dietary ingredients used in a dietary supplement.[x] To be considered offered for sale, the dietary ingredient must have been physically offered for sale at a retail establishment, through a product catalog or pricing list, or through other means of advertising and promotion.[y]

Documentation would likely be sufficient if it clearly shows that the marketing took place in the United States, the identity and form of the marketed ingredient, and whether the ingredient was marketed as a dietary ingredient.[z] If there is sufficient evidence of marketing prior to Oct. 15, 1994, then a NDI notification is not required prior to marketing of the ingredient in a dietary supplement product. However, if no such evidence is present, the ingredient is considered "new" and a NDI notification is required.

7.2.3.2 New dietary ingredients

FDA must be notified prior to the introduction of any new dietary ingredients into the marketplace. The Office of Nutritional Products, Labeling and Dietary Supplements (ONPLDS) must receive this notification at least 75 days before the NDI is introduced into interstate commerce.[aa] This notification includes a dossier, which is a comprehensive summary of the ingredient that includes a description of the chemistry of the ingredient, its manufacturing process, its intended use and dosage level, and a safety assessment that supports the intended use and dosage level.[bb] FDA will put the NDI notification, with the exception of any trade secret or confidential information, "on public display" 90 days after receipt of the submission.[cc]

FDA customarily will acknowledge and provide the date of receipt of the NDI notification, which is considered the filing date. The NDI may not be introduced into interstate commerce until 75 days after the filing date. However, even though the

[x]Section IV, A.6, Draft Guidance for Industry: Dietary Supplements: New Dietary Ingredient Notifications and Related Issues, July 2011., http://www.fda.gov/Food/GuidanceRegulation/GuidanceDocumentsRegulatoryInformation/DietarySupplements/ucm257563.htm.

[y]Idem. Examples of documentation that would be sufficient to show a marketing date of prior to October 15, 1994 include sales records, manufacturing records, commercial invoices, magazine advertisements, mail order catalogs, or sales brochures. Idem. at Section IV, A.8.

[z]Idem. at Section IV, A.6–A.9.

[aa]21 U.S.C. § 350b(a)(2) (2016).

[bb]*Supra* fn. 24 at Section II. Safety evidence can be in the form of human clinical studies or animal studies and may also include other in vitro studies to support a nontoxicity conclusion of the subject ingredient.

[cc]21 U.S.C. § 350b(a)(2) (2016).

product may be introduced after 75 days, there is no guarantee that FDA agrees there is sufficient information to presume reasonable safety.

In Jul. 2011, FDA issued a draft guidance on NDI notifications and related issues to much industry criticism because the safety procedures outlined seemed too similar to those of food additives.[dd] Despite the backlash, the agency continues to review submitted NDI notifications in light of the 2011 draft guidance, which could be one of the reasons for the high rejection rate. After many conversations with industry, FDA agreed to reconsider the draft guidance and received a substantial number of comments from industry. The revised guidance was slated for publication by 2015, although this failed to materialize. When this is available, it is hoped it will provide industry with additional insight into FDA's position and review process and that the NDI acceptance rate will improve.

An important exemption from the NDI notification requirement is if the NDI has been present in the food supply as an article used for food in a form in which the food has not been chemically altered. That is, the ingredient is exempt from the NDI notification requirements if it meets two conditions: not only must the exact ingredient be present in the food supply, but it also must be in a form that is not chemically altered.[ee] In discussing chemical alteration, FDA notes that minor loss of volatile components, dehydration, lyophilization, milling, and formation of a tincture or a solution in water, a slurry, a powder, or a solid in suspension do not chemically alter an ingredient.[ff] On the other hand, examples of processes that FDA will likely consider to result in chemical alteration of the ingredient include, but are not limited to, processes that make or break chemical bonds, unless the bonds created by the process are reversed when the ingredient is dissolved in water (e.g., creation of a soluble salt) or during ingestion; use of solvents other than water or aqueous ethanol (tincture) to make an extract; and changing the manufacturing method for an ingredient such that the chemical composition is significantly different (e.g., changes that alter the composition of materials used to make the ingredient, use of a different solvent, use of a chromatographic matrix instead of a passive filter).[gg]

7.2.4 SPECIFIC CONSTRAINTS ON FORMULATION

7.2.4.1 Upper limits and safety limits

Sometimes too much of a good thing can be bad. Consumption of certain ingredients at levels higher than advised can become harmful to the body or have dangerous side effects. Therefore, the intake amount of certain ingredients should be limited for safety reasons. While FDA has set forth limits of some ingredients, other limits come from industry and the scientific community. Thus, it must be ensured not only that the ingredient may be legally incorporated into the product, but also that it is present

[dd]See, for example, *supra* fn. 24
[ee]21 U.S.C. § 350b(a)(1) (2016).
[ff]*Supra* fn. 24 at Section III, B.3.
[gg]*Supra* fn. 24 at Section III, B.4.

at a safe amount. Marketing a product whose dosage is unsafe will render a product adulterated.[hh]

Vitamins and minerals, including but not limited to vitamin A, vitamin C, vitamin D, calcium, and iron, have established Upper Tolerable Intake Level (UL) above which daily intake should not exceed. Surpassing this threshold increases the risk of adverse events. Of important note, a UL applies to the total daily intake, not just the amount present in the product. Therefore, daily consumption of vitamins or minerals from other sources should be kept in mind.

The GRAS status of an ingredient may also hinge on specific levels of use. FDA sometimes specifically sets forth in the applicable regulation that an ingredient may only account for up to a certain percentage of the product. In other cases, the GRAS determination, (regardless of whether it has been submitted to FDA), itself specifies at what levels the ingredient may be used. If an ingredient is present in an amount that exceeds its GRAS designation, the ingredient is no longer considered GRAS and the manufacturer must have another basis to establish safety of the ingredient's inclusion in the product.

Finally, some ingredients are widely recognized as being harmful if consumed in excess and therefore tend to be limited in a product to avoid additional risk of adverse events. For example, caffeine consumption of 400 mg/day for healthy adults is generally considered to be without dangerous, negative side effects; exceeding this amount could pose a potential health hazard.[ii] Therefore, manufacturers limit the amount of caffeine to avoid adverse reactions by users, and oftentimes include additional warnings disclosing the amount of caffeine (for more on voluntary warning statements, see Section 7.2.4.5).

Ultimately, it must be kept in mind that it is the manufacturer's responsibility to ensure the product's safety, and adhering to recognized inclusion safety limits is a good first step.

7.2.4.2 Banned ingredients

Some ingredients are subject to a blanket ban in certain products. A well-known banned ingredient is ephedrine, which FDA prohibited from inclusion in dietary supplements in 2004.[jj] Including a banned ingredient automatically renders the product adulterated. Therefore, it is vital to always confirm not only that the ingredient is appropriate for the product, but that it is also eligible for inclusion. Of note, some ingredients may be added in the United States, but are prohibited or restricted in other countries (e.g., yohimbe).

7.2.4.3 Ingredients that are also used in drug products

Ingredients that were sold as drugs (both over-the-counter and prescription) prior to inclusion in a supplement may not be sold as a dietary supplement as the inclusion of

[hh]This applies to the per serving dosage, as well as the total daily consumption of the product.

[ii]FDA to Investigate Added Caffeine, May 3, 2013. http://www.fda.gov/ForConsumers/ConsumerUpdates/ucm350570.htm.

[jj]FDA Issues Regulation Prohibiting Sale of Dietary Supplements Containing Ephedrine Alkaloids and Reiterates Its Advice That Consumers Stop Using These Products, February 6, 2004. http://www.fda.gov/NewsEvents/Newsroom/PressAnnouncements/2004/ucm108242.htm.

such ingredients automatically renders the product a "drug."[kk] Presence of a drug active ingredient in a dietary supplement would cause the product to be adulterated, as the safety is unknown, as well as misbranded because it is not labeled correctly (since it is regulated as a drug). A famous ingredient with this dilemma is red yeast rice. In 1998, FDA initiated action against a red yeast rice product due to it containing monacolin K, a constituent of red yeast rice extract that is more commonly known as lovastatin. Lovastatin is the active pharmaceutical in prescription products for lowering cholesterol. After a court battle, FDA prevailed and was allowed to regulate the product as a drug. Since then, red yeast rice dietary supplements have still been marketed, albeit with a negligible monacolin K content and typically without cholesterol structure/function claims. FDA has continued to bring enforcement actions against those that fail to limit the presence of monacolin K.[ll]

7.2.4.4 Required warning statements

A product may require mandated warnings based on its composition. The allergen declaration may be the most commonly included mandatory warning. Under the Food Allergen Labeling and Consumer Protection Act of 2004 (FALCPA), products that contain one of the eight major allergens (milk, egg, fish, crustacean shellfish, tree nuts, wheat, peanuts, and soybeans) are required to declare the presence of the allergen.[mm] If the name of the allergen is part of the ingredient name (example: soy lecithin), then the declaration requirement is satisfied just by listing the ingredient. If not, the presence of the allergen must be declared directly underneath the ingredient statement. Unlike the FALCPA warning, which is included if any amount of allergen is present, other warnings are based on the product exceeding stated limits. Under FDA regulations, products that exceed certain levels of iron are legally required to have a specific accidental overdose warning statement.[nn] Another example is protein products where more than half of the calories come from protein content; the required warning varies with the types of claims being made for the product.[oo] Failure to include a required warning will render a product misbranded.

State labeling requirements should also be consulted, as some states require standardized warning statements. The most commonly known is California's Prop 65 warning.[pp] Under Prop 65 rules, California has created a list of certain compounds as being connected with cancer and/or reproductive toxicity.[qq] If a product contains any of these compounds, it must bear a warning on its label disclosing that it contains something the

[kk] 21 U.S.C. § 321(ff)(3) (2016).

[ll] See FDA Warns Consumers to Avoid Red Yeast Rice Products Promoted on Internet as Treatments for High Cholesterol Products Found to Contain Unauthorized Drug, August 9, 2007. http://www.fda.gov/NewsEvents/Newsroom/PressAnnouncements/2007/ucm108962.htm.

[mm] Guidance for Industry: Questions and Answers Regarding Food Allergens, Including the Food Allergen Labeling and Consumer Protection Act of 2004 (Edition 4); Final Guidance, October 2006. http://www.fda.gov/Food/GuidanceRegulation/GuidanceDocumentsRegulatoryInformation/ucm059116.htm.

[nn] See 21 C.F.R. § 101.17(e) (2016).

[oo] See 21 C.F.R. § 101.17(d) (2016).

[pp] See Proposition 65, available at: http://oehha.ca.gov/prop65.html.

[qq] Chemicals Known to the State to Cause Cancer or Reproductive Toxicity, last updated December 4, 2015. http://oehha.ca.gov/prop65/prop65_list/files/P65single120415.pdf.

State of California recognizes as causing cancer and/or reproductive toxicity. In food products, the most common compounds that trigger a Prop 65 warning are selenium, cadmium, lead, and mercury. Failure to bear the warning leads not only to regulatory enforcement action, but to consumer class actions as well. While some compounds have a zero tolerance (any amount present requires a warning), other have so-called "safe harbors" where a warning must be included only if the presence exceeds a certain amount. Commercial laboratories offer testing for specific Prop 65 compounds.

7.2.4.5 Industry recommended warning statements

Even if not explicitly required by FDA, a product may still be misbranded if it does not include appropriate warnings. Every product must bear adequate directions for use, which encompasses not only serving directions but also warnings about known side effects and/or interactions. Some dietary ingredients have been linked to severe adverse reactions, while others have been shown to cause health hazards when combined with certain medications and medical conditions. For example, kava has been linked to severe liver disease.[rr] Other ingredients known to interact with drugs include, but are not limited to, St. John's wort, ginseng, and ginkgo biloba.[ss] Potential side effects and interactions with the product's ingredients should be researched and appropriately warned against. Industry organizations, such as the American Herbal Product Association, may provide safety guidance and standardized warnings for some ingredients.[tt]

Importantly, care should be taken for populations such as pregnant or nursing women, children, the elderly, and those with preexisting conditions. Including a recommendation to consult a physician prior to use for these individuals and any other relevant warnings is advisable.

7.2.4.6 Liquid dietary supplements

There is a presumption that beverages are conventional food products and not dietary supplements. According to FDA, simply adding a dietary ingredient to a conventional beverage is insufficient to transform the product into a supplement.[uu] Several criteria are examined when classifying a product as a conventional beverage or liquid dietary supplement, including the product label, composition, claims, and marketing.[vv] FDA details these criteria in its guidance to industry titled "Distinguishing Liquid Dietary Supplements from Beverages."[ww]

[rr]Consumer Advisory: Kava-Containing Dietary Supplements May be Associated with Severe Liver Injury, March 25, 2002. http://www.fda.gov/Food/ResourcesForYou/Consumers/ucm085482.htm.

[ss]Avoiding Drug Interactions, November 28, 2008. http://www.fda.gov/ForConsumers/ConsumerUpdates/ucm096386.htm#supplements.

[tt]See, for example, AHPA Code of Ethics & Business Conduct, http://www.ahpa.org/AboutUs/AHPASPolicies/CodeofEthics.aspx; and AHPA Guidance Policies, http://www.ahpa.org/AboutUs/AHPASPolicies/GuidancePolicies.aspx.

[uu]Section II.B, Guidance for Industry: Distinguishing Liquid Dietary Supplements from Beverages, January 2014. http://www.fda.gov/downloads/Food/GuidanceRegulation/GuidanceDocumentsRegulatoryInformation/DietarySupplements/UCM381220.pdf.

[vv]Idem.

[ww]Idem.

7.3 PRODUCT MANUFACTURING

7.3.1 FACILITY REGISTRATION

Section 415 of the FD&C Act requires that the owner, operator, or agent in charge of a domestic or foreign facility that manufacturers, processes, packs, or holds food (including dietary supplements) for human or animal consumption in the United States register with the FDA, unless an exception applies. "Food" is uniquely defined in regulations specifically applicable to food facility registration to exempt pesticides and FCS (including packaging materials) from the definition.[xx] Therefore, facilities that manufacture, process, pack, or hold only pesticides or FCS are not required to register with the FDA.

Also exempt from the requirement of facility registration are: (1) a foreign facility, if food from the facility undergoes further manufacturing/processing (including packaging) by another facility outside of the United States; (2) farms; (3) retail food establishments; (4) restaurants; (5) nonprofit food establishments; (6) fishing vessels; (7) facilities regulated exclusively and throughout the entire facility by the US Department of Agriculture; and (8) private residences of individuals even though food may be manufactured, processed, packed, or held in them.[yy]

Food facilities must be registered with the FDA before manufacturing, processing, packing, or holding of food commences. Thereafter, registrants must notify the agency within 60 calendar days of any changes to the facility's registration. In addition, registrations must be renewed biennially during the period beginning Oct. 1 and ending on Dec. 31 of each even-numbered year. Failure to register a facility or submit a biennial renewal is a prohibited act under the FD&C Act, and violators may be subject to enforcement actions, which may include injunction or prosecution. If a foreign facility fails to register, food from the facility that is offered for import into the United States is subject to refusal.

Information required to be submitted in a food facility registration includes: (1) the name, address, and phone number of the facility; (2) the name, address, and phone number of the parent company, if applicable; (3) the name, address, and phone number of the owner, operator, and agent in charge; (4) an email address for the contact person at the facility, or the US agent in the case of a foreign facility; (5) a list of all trade names the facility uses; (6) assurance that the FDA will be permitted to inspect the facility at times and in the manner permitted by the FD&C Act; and (7) certification that the information submitted is truthful and accurate.[zz]

The FDA Food Safety Modernization Act of 2011 provided the FDA with the authority to suspend the registration of a food facility in certain limited circumstances.[aaa] Specifically, if the agency determines that food manufactured, processed, packed, received, or held by a registered food facility has a reasonable probability of causing

[xx]21 C.F.R § 1.227(b)(4) (2016).
[yy]21 C.F.R § 1.226 (2015); 21 C.F.R § 1.227(a)(2) (2016).
[zz]21 C.F.R. § 1.232 (2016).
[aaa]21 U.S.C. § 350d(b) (2016).

serious adverse health consequences or death to humans or animals, FDA may, by order, suspend the registration of a facility that: (1) created, caused, or was otherwise responsible for such reasonable probability; or (2) knew of, or had reason to know of, such reasonable probability, and received, packed, or held such food. If the registration of a food facility is suspended, no person can import or export, or offer to import or export, food from the facility into the United States, or otherwise introduce food into interstate or intrastate commerce from the facility.

Registrants subject to a suspension order will have the opportunity for an informal hearing no later than two business days after the issuance of the suspension order. If during such hearing it is determined that suspension of the registration remains necessary, the registrant will be required to submit a corrective action plan to the FDA explaining measures that will be taken to correct the facility's deficiencies. Once adequate grounds no longer exist to continue the suspension, FDA will vacate the suspension order.

7.3.2 GOOD MANUFACTURING PRACTICES (GMPS)

All food (including dietary supplements) manufactured, packaged, and held in the United States is subject to good manufacturing practice (GMP) regulations. In very general terms, GMP regulations require companies to establish procedures to monitor and control manufacturing processes and facilities to assure the safety of the food supply. GMPs address areas including personnel, physical plant and grounds, equipment and utensils, production and process controls, and holding and distributing activities. While the overall objective of GMPs for foods and dietary supplements is the same, namely to regulate the manufacture of such products in a manner that will help ensure the safety of the food supply, the regulations applicable to each specific food category differ substantially.

The food GMPs can be found in 21 CFR Part 110. The FD&C Act does not specifically authorize the FDA to promulgate GMP regulations for foods. Rather, as its authority for doing so, the FDA relied on section 402(a)(3) of the FD&C Act which specifies that food must not be manufactured under such conditions that render it unfit for consumption, section 402(a)(4) which provides that food may be deemed adulterated if it is prepared, packed, or held under insanitary conditions whereby it may become contaminated with filth or rendered injurious to health, and section 701 which gives the agency the authority to prescribe regulations for the efficient enforcement of the Act. Because FDA's authority to promulgate food GMPs is limited to ensuring that the requirements of sections 402(a)(3) and 402(a)(4) are met, the scope of the food GMPs is limited to prescribing requirements that reasonably relate to controlling insanitary conditions that may contaminate food and render it injurious to health.

The Dietary Supplement Health and Education Act of 1994 amended the FD&C Act to expressly authorize the FDA to promulgate GMPs for dietary supplements.[bbb] While the grant of such authority specifically required that the dietary supplement GMPs be "modeled" after current GMPs for food, the regulations ultimately finalized

[bbb]21 U.S.C. § 342(g) (2016).

go far beyond prescribing requirements to control for insanitary conditions that may contaminate dietary supplements. In fact, the dietary supplement GMPs more closely resemble drug GMPs in many respects, including the production and process control provisions that require the implementation of robust quality control operations to ensure the identity, purity, strength, and composition of manufactured supplements. The dietary supplement GMPs can be found in 21 CFR Part 111.

The FD&C Act expressly provides that dietary supplements are to be deemed "food" under the Act.[ccc] While companies that manufacture, package, label, or hold dietary supplements must comply with the GMP requirements set forth in 21 CFR Part 111, nothing exempts such facilities from also complying with the general food requirements set forth in 21 CFR Part 110. While the FDA included in the final dietary supplement GMP regulations those provisions of the food GMPs that it deemed to be common to dietary supplement manufacturing, it noted in the preamble to the final regulation that there may be operations related to the manufacture of a particular supplement for which certain provisions of Part 110 apply [e.g., dehydrated supplements that rely on the control of moisture consistent with § 110.80(b)(14)].[ddd] In such cases, dietary supplement manufacturers are expected to comply with both Part 111 and the applicable provisions of Part 110.

On Sep. 17, 2015 the FDA issued new final regulations in 21 CFR Part 117 entitled *Current Good Manufacturing Practice, Hazard Analysis, and Risk-Based Preventative Controls for Human Food*.[eee] These regulations did two things: (1) revise the food GMP regulations in 21 CFR Part 110, and (2) establish new Hazard Analysis, and Risk-Based Preventative Controls (HARPC). The compliance date for the new regulations varies depending on the size of the company (Sep. 19, 2016–Sep. 17, 2018). While dietary supplement manufacturers will still be subject to the GMP provisions of Part 117 as discussed previously, they are exempt from the HARPC regulations.

The provisions of 21 CFR Part 110 have now been moved to Part 117, and some additions, deletions, and revisions were made. Most significant are added requirements relating to the prevention of food allergen contamination and cross-contact, and the imposition of minimum training requirements for all employees involved in the manufacturing, processing, packing, or holding of food. 21 CFR Part 110 will be removed after the final compliance date for Part 117.

The new HARPC regulations are similar in concept to the Hazard Analysis and Critical Control Points (HACCP) provisions applicable to seafood and juice producers. Generally, HARPC requires that owners or operators of food facilities prepare and implement written food safety plans that include the following: (1) identify and evaluate the hazards that could affect food manufactured, processed, packed, or held by the facility; (2) identify and implement preventative controls; (3) monitor the performance of those controls; (4) develop corrective action plans if preventative controls are not effective; and (5) verify that preventative controls are effective.[fff] Food

[ccc]21 U.S.C. § 321(ff) (2016).
[ddd]72 Fed. Reg. 34752 at 34764, June 25, 2007.
[eee]See 80 Fed. Reg. 55908, September 17, 2015.
[fff]See 21 C.F.R part 117, Subchapter C.

manufacturers must also conduct various verification activities, including validating that preventative controls are adequate to control identified hazards and verifying that monitoring is being conducted and corrective actions are taken.

Dietary ingredient manufacturers and suppliers are required to comply with the GMP provisions in 21 CFR Part 110. In the preamble to the final dietary supplement GMP regulations, the FDA expressly indicated that 21 CFR Part 111 does not apply to companies that manufacture and supply only dietary ingredients.[ggg] Dietary ingredient manufacturers will be subject to the requirements in 21 CFR Part 117, including the new HARPC regulations, as of the applicable compliance date based on the size of the company.

7.4 PRODUCT CLAIMS

Product claims are regulated by both FDA and the Federal Trade Commission (FTC). A 1971 memorandum of understanding governs the division of responsibilities between the two agencies.[hhh] FDA has primary responsibility for claims appearing on product labeling, including packaging, inserts, and other promotional materials distributed at the point of sale. FTC has primary responsibility for claims made in advertising, including print and broadcast advertisements, infomercials, websites, catalogs, and similar direct marketing materials. This division is enforcement-based only—the rules from both FDA and FTC still apply to all labels, labeling, and advertising.

The most basic requirement is that all advertisements must be truthful, not misleading, and appropriately substantiated. Generally, claims regarding the nutritional value and composition of the product are considered "nutrient content claims" which are regulated by FDA and typically are substantiated with product compositional testing and nutritional data. Efficacy-based claims, however, are typically within the purview of FTC and may fall into several different categories, such as performance claims, comparative claims, superiority claims, and establishment claims. The type of claim dictates the level of substantiation needed, and some claims require highly scientific substantiation, while others may only require nutritional testing. FTC's advertising guide for the supplement industry provides detailed examples of claims, how to identify claims, and ways to determine adequate levels of substantiation.[iii]

[ggg]72 Fed. Reg. 34752 at 34792, June 25, 2007.

[hhh]Memorandum of Understanding Between The Federal Trade Commission and The Food and Drug Administration, May 14, 1971. http://www.fda.gov/AboutFDA/PartnershipsCollaborations/MemorandaofUnderstandingMOUs/DomesticMOUs/ucm115791.htm.

[iii]See Dietary Supplements: An Advertising Guide for Industry, April 2001, https://www.ftc.gov/tips-advice/business-center/guidance/dietary-supplements-advertising-guide-industry#b. FDA also has a guidance that substantially mirrors the FTC guidance. See Guidance for Industry: Substantiation for Dietary Supplement Claims Made Under Section 403(r) (6) of the Federal Food, Drug, and Cosmetic Act, December 2008, http://www.fda.gov/food/guidanceregulation/guidancedocumentsregulatoryinformation/dietarysupplements/ucm073200.htm.

An often overlooked distinction is that of express versus implied claims. Advertisers are not only responsible for express claims, but also for all reasonable interpretations of that claim. Thus language, context, and formatting may individually or altogether convey a different message than a standalone statement. If an advertiser does not have substantiation for an implied message, even if the express claim is adequately documented, then they may be subject to legal action. Enforcement comes not only from FDA and FTC, but also from state attorney generals, consumer lawsuits, and independent self-regulatory bodies such as the National Advertising Division.[jjj]

An overview of common advertising claims is provided later. The parameters of these claims come from FDA, FTC, and even the courts. However, the universal concept to keep in mind is that if an advertisement is not truthful, is misleading, or the substantiation for the claim is inappropriate or insufficient, then the product is subject to enforcement action from government bodies and private citizens.

7.4.1 PRODUCT CLAIMS DEFINED BY REGULATION

7.4.1.1 Structure function claims

Of the many types of claims, the most common claims on dietary supplement products tend to be structure/function claims. Structure/function claims are those that "describe the role of a nutrient or dietary ingredient intended to affect the structure or function in humans or that characterize the documented mechanism by which a nutrient or dietary ingredient acts to maintain such structure or function."[kkk] Both FDA and FTC regulate these claims. Under FDA regulations, the claim must be in accordance with the structure/function final rule (i.e., among other things, it cannot relate to the diagnosis, treatment, cure, prevention, or mitigation of any disease).[lll] FDA also requires a structure/function claim to always be connected to one of the following disclaimers:

§This statement has not been evaluated by the Food and Drug Administration. This product is not intended to diagnose, treat, cure, or prevent any disease.

§These statements have not been evaluated by the Food and Drug Administration. This product is not intended to diagnose, treat, cure, or prevent any disease.

This disclaimer must be prominent and conspicuous; they should be bold, in a box, and no smaller than 1/16th of an inch in height as measured by the lower case "o." The disclaimer should also appear on every panel that bears a claim.

FTC mandates the level of support required for these claims. As health-related claims, they will require competent and reliable scientific evidence.[mmm] This is

[jjj]The National Advertising Division decides advertising challenges brought by itself as well as industry members. Although not legally binding, decisions are typically adhered with, as cases will be referred to FTC if a party does not cooperate.
[kkk]21 C.F.R. § 101.93(f) (2016).
[lll]See 65 Fed. Reg. 1000, February 7, 2000.
[mmm]Section II.B, Dietary Supplements: An Advertising Guide for Industry, April 2001. https://www.ftc.gov/tips-advice/business-center/guidance/dietary-supplements-advertising-guide-industry#b.

defined as "tests, analyses, research, studies, or other evidence based on the expertise of professionals in the relevant area, that have been conducted and evaluated in an objective manner by persons qualified to do so, using procedures generally accepted in the profession to yield accurate and reliable results."[nnn] Although not expressly stated by FTC, the so-called "gold standard" substantiation is a double-blind, placebo-controlled, randomized clinical trial of a healthy population that is adequately powered[ooo] and yields statistically significant results.[ppp] That the study is peer-reviewed and published also adds great weight.

Substantiation should be collected, reviewed, and kept on file prior to disseminating a claim. Marketing a claim without substantiation is not only misbranding under FDA regulations, but also illegal under FTC laws as well.[qqq] At the product development stage, it is crucial to identify, evaluate, and assemble appropriate structure/function claim substantiation. The most important detail is determining the studied dosage. Even if a study meets the "gold standard," it is irrelevant to substantiating the product's claims if the product does not contain at least as much as, if not more than, the amount consumed in the study. It should also be noted that studies on proprietary ingredients typically may not be extrapolated to generic ingredients, and that studies on different formulations (i.e., study on a, b, c, but product contains a, d, e) do not support efficacy of the product formulation.

7.4.1.1.1 403(r)(6) Notifications

Under DSHEA, companies are required to notify FDA of any structure/function claim included with a product and the dietary ingredient it is derived from within 30 days of the product's introduction into commerce.[rrr] The notification may simply be a cover letter with a chart that includes: the name of the product, the dietary ingredient, and the claim(s). The letter must also contain a statement that the company possesses substantiation for the claims, and that the claims are truthful and not misleading. While this is a straightforward premarket requirement, many companies fail to comply with the 403(r)(6) notice requirement. FDA does not have a large enforcement history on this aspect, however, it is a legal requirement and failure to submit is, strictly speaking, breaking of the law.

7.4.1.2 Health claims

In limited instances, a precise health claim has been evaluated by FDA and allowed in marketing if the product meets delineated requirements. In these situations, FDA has reviewed the scientific evidence between a food, food component, or dietary ingredient and a health condition and found that certain claims, called "health claims," may be appropriate. There are three different types of health claims: NLEA Authorized Health Claims, Health Claims Based on Authoritative Statements, and Qualified Health Claims.

[nnn]Idem.
[ooo]The sample size should be sufficiently large so that the power is at least 80%.
[ppp]Statistical significance is when the p value is less than 0.05.
[qqq]See 15 U.S.C. § 52 & 55 (2016).
[rrr]21 U.S.C. § 343(r)(6) (2016).

7.4.1.2.1 NLEA authorized health claims

FDA is authorized to create health claims via regulation under the Nutrition Labeling and Education Act of 1990 (NLEA). These NLEA Authorized Health Claims revolve around a relationship between a risk of disease and a food, food component, or dietary ingredient. Submission of a petition will prompt FDA to review the science and determine if there is an established "significant scientific agreement" on a relationship between the food/ingredient and the health condition. If so, FDA will authorize a claim and set forth in a regulation the criteria a product must meet to bear the claim. An appendix of currently allowed claims is available on FDA's website.[sss]

7.4.1.2.2 Health claims based on authoritative statements

Although somewhat similar in form to NLEA Authorized Health Claims, Health Claims Based on Authoritative Statements have some significant differences. Most importantly, dietary supplements are not eligible to use these claims, although functional foods may. The other significant difference is that the claim is derived from a statement from certain Federal scientific bodies.[ttt] These authoritative statements may be submitted via petition to FDA, who will then evaluate whether the "significant scientific agreement" standard is met by the Federal scientific body's statement. If so, FDA will then publish the requirements for the specific claim, but not officially codify the claim or requirements. FDA has also made available on its website a chart of currently approved Health Claims Based on Authoritative Statements.[uuu]

7.4.1.2.3 Qualified health claims

The final type of health claim is when there is evidence of a relationship between a food, food component, or dietary ingredient and a health condition, but the science does not meet the "significant scientific agreement" standard needed for a NLEA Authorized Health Claim. For these types of petitions, FDA will evaluate the totality and quality of the evidence and determine if the relationship is credible. If so, FDA will not create a regulation allowing the claim, but instead issue a "letter of enforcement discretion" that outlines the conditions under which it thinks the claim may be made. As the claim's name suggests, one of the requirements in the letter of enforcement discretion will be that certain qualifying language is used when the claim is made. FDA's webpage on Qualified Health Claims provides a summary of

[sss]Appendix C: Health Claims, in Guidance for Industry: A Food Labeling Guide, January 2013. http://www.fda.gov/Food/GuidanceRegulation/GuidanceDocumentsRegulatoryInformation/LabelingNutrition/ucm064919.htm.

[ttt]FDA has recognized the following bodies as sources of authoritative statements: National Academy of Sciences; National Institutes of Health; Centers for Disease Control and Prevention; Surgeon General within Department of Health and Human Services; and the Food and Nutrition Service, Food Safety and Inspection Service, and the Agricultural Research Service within the Department of Agriculture. See, for example, Guidance for Industry: Notification of a Health Claim or Nutrient Content Claim Based on an Authoritative Statement of a Scientific Body, June 1998, http://www.fda.gov/Food/GuidanceRegulation/GuidanceDocumentsRegulatoryInformation/ucm056975.htm.

[uuu]*Supra*, fn. 71

claims currently subject to enforcement discretion, as well as those being reviewed and those that have been denied.[vvv]

7.4.1.3 Nutrient content claims

Many common claims are actually strictly defined by regulation. Perhaps the most widely known of these claims would be nutrient content claims such as "low fat" and "good source of calcium." FDA has drafted an easy-to-read overview chart of express nutrient content claims.[www] Another useful FDA chart covers implied nutrient content claims; these "trigger" words signal a nutrient content claim is being made (Table 7.1).

Although the FDA charts are helpful, they do not always accurately reflect all requirements for nutrient content claims. There are also many express and implied nutrient content claims that haven't been discussed here. Whether the claim is appropriate and whether it needs certain disclaimers or triggers other additional FDA requirements differ from product to product based on the nature of the claim and product's composition/nutrition. The regulations should always be consulted to confirm all requirements, as including an inappropriate or inaccurate nutrient content claim may render the product misbranded.

Table 7.1 Select Examples of Implied Nutrient Content Claims

Claim	Requirements
"High," "rich in," or "excellent source of"	Contains 20% or more of the Daily Value (DV) per Reference Amounts Customarily Consumed (RACC). May be used on meals or main dishes to indicate that the product contains a food that meets the definition, but may not be used to describe the meal.
"Good source," "contains," or "provides"	10–19% of the DV per RACC. These terms may be used on meals or main dishes to indicate that the product contains a food that meets the definition but may not be used to describe the meal.
"More," "fortified," "enriched," "added," "extra," or "plus"	10% or more of the DV per RACC than an appropriate reference food. May only be used for vitamins, minerals, protein, dietary fiber, and potassium.

Excerpt from: FDA (2013). Appendix A: Definitions of Nutrient Content Claims, Guidance for Industry: A Food Labeling Guide, FDA, January 2013.[xxx]

[vvv]Qualified Health Claims, last updated November 24, 2014. http://www.fda.gov/Food/IngredientsPackagingLabeling/LabelingNutrition/ucm2006877.htm.

[www]Appendix A: Definitions of Nutrient Content Claims, in Guidance for Industry: A Food Labeling Guide, January 2013. http://www.fda.gov/Food/GuidanceRegulation/GuidanceDocumentsRegulatoryInformation/LabelingNutrition/ucm064911.htm.

[xxx]http://www.fda.gov/Food/GuidanceRegulation/GuidanceDocumentsRegulatoryInformation/LabelingNutrition/ucm064916.htm.

7.4.1.4 Other claims

7.4.1.4.1 "Healthy"

FDA also regulates many other claims. An increasingly more common claim is the term "healthy" and its related variations. FDA has set nutritional parameters on what is considered a "healthy" product," and, although perhaps imperfect as applied to some foods, this definition has been strictly enforced.[yyy] An important caveat is that even implied healthy claims, such as that consuming the product is part of a healthy lifestyle, may require the product to comply with FDA's "healthy" definition (Table 7.2).

7.4.1.4.2 Antioxidant claims

"Antioxidant"-based claims are another common claim subject to FDA regulation. When a claim characterizes the amount of antioxidant in a food, the FDA's requirements[zzz] must be satisfied or the product faces being misbranded:

1. An RDI must be established for each of the nutrients that are the subject of the claim;
2. Each nutrient must have existing scientific evidence of antioxidant activity;
3. The level of each nutrient must be sufficient to meet the definition for "high," "good source," or "more"; and

Table 7.2 Conditions for the Use of "Healthy" for Individual Food Products

	Individual Food
Total Fat	3 g or less of fat per RACC (or per 50 g if RACC is less than 30 g)
Saturated Fat	1 g or less per serving and 15% or less of calories from saturated fat
Cholesterol	Less than 2 mg per RACC and per labeled serving
Sodium	Less than 480 mg sodium per RACC (or per 50 g if RACC is less than 30 g)
Beneficial Nutrients	At least 10% or more of the recommended daily intake per RACC of one or more of vitamin A, vitamin C, calcium, iron, protein, or dietary fiber. Note: If the product is fortified to meet this requirement, the fortification must be in accordance with FDA's Fortification Policy (21 C.F.R. 104.20)

Adapted from: FDA (2013). Appendix B: Additional Requirements for Nutrient Content Claims, Guidance for Industry: A Food Labeling Guide, FDA, January 2013.[aaaa]

[yyy]See FDA Warning Letter to Kind, LLC, March 27, 2015, http://www.fda.gov/ICECI/EnforcementActions/WarningLetters/ucm440942.htm. As in the case of Kind, LLC, FDA notice of violating "healthy" labeling claim requirements often leads to consumer class actions.

[zzz]21 C.F.R. § 101.54(g) (2016).

[aaaa]http://www.fda.gov/Food/GuidanceRegulation/GuidanceDocumentsRegulatoryInformation/LabelingNutrition/ucm064916.htm.

Note: Beta-carotene may be the subject of an antioxidant claim when the level of vitamin A present as beta-carotene in the food is sufficient to qualify for the claim.

4. The name(s) of nutrient(s) that is (are) the subject of the claim is (are) included as part of the claim (e.g., high in antioxidant vitamins C and E).

Under the FTC lens, it is also necessary to have appropriate Oxygen Radical Absorbance Capacity (ORAC) testing to ensure the levels of antioxidants in the product provide antioxidant benefits. If not, the claim may be misleading even if it complies with the FDA requirements.

7.4.1.4.3 "Gluten-free"

Another highly marketed FDA-regulated claim is "gluten free."[bbbb] Although FDA does not certify products as "gluten-free,"[cccc] it has provided extensive guidance[dddd] on what it means to be "gluten-free." Generally speaking, any unavoidable presence of gluten must be less than 20 ppm in order to be labeled as "gluten-free," regardless of whether it is inherently gluten-free or manufactured to be that way. Appropriate product testing must be conducted to ensure this requirement is met, since cross-contact with gluten ingredients may lead to a gluten content exceeding this threshold. A product may not bear the claim "gluten-free" if it has:

- An ingredient that is a gluten-containing grain;
- An ingredient that is made from a gluten-containing grain and that has not been processed to remove gluten; or
- An ingredient that is made from a gluten-containing grain and that has been processed to remove gluten, if the use of that ingredient contains 20 parts per million (ppm) or more gluten.

7.4.1.4.4 "Organic"

One final example of regulated terms is "organic." The USDA handles the National Organic Program and is charged with certifying manufacturers and regulating labeling of "organic" products. Briefly, there are three "tiers" of organic labeling:

- To be labeled "Organic" or use the USDA seal, the product must be >95% organic;
- To be labeled as "Made With Organic," the product must be >70% organic;
- At less than <70% organic, "organic" may only be declared with the ingredient name in the ingredient statement.

[bbbb]See 21 C.F.R. § 101.91 (2016). In November 2015, FDA proposed a rule relating to "gluten-free" labeling of fermented or hydrolyzed foods, as the 2013 final rule on "gluten-free" labeling did not address these products.

[cccc]There are many independent organizations that will certify products as "gluten-free" and license the use of a specific mark. However, this certification does not relieve the manufacturer of maintaining substantiation that the product conforms to FDA "gluten-free" labeling requirements.

[dddd]See Guidance for Industry: Gluten-Free Labeling of Foods, June 2014, http://www.fda.gov/downloads/Food/GuidanceRegulation/GuidanceDocumentsRegulatoryInformation/UCM402559.pdf.

The National Organic Program is extensive and the USDA has devoted several webpages for industry on organic certification and organic labeling.

7.4.2 LITIGATED CLAIMS THAT MAY BECOME REGULATED

7.4.2.1 "Natural"

Currently, there is no regulation defining "natural." However, FDA has indicated it will not object to use of the term on a product that "does not contain added color, artificial flavors, or synthetic substances."[eeee] The product should also be minimally processed, as high levels of processing likely remove the ingredient/product from its "natural" state. FDA sought comments from industry and consumers on whether it should expressly regulate the term and, if so, how it should be regulated.[ffff] The comment period closed May 10, 2016, and FDA is currently reviewing the submissions and determining its next steps.

With little FDA enforcement, "natural" is widely challenged in court, typically in the form of consumer class actions. While these actions have been widespread for several years, recently courts have begun to question their appropriateness. In California, a popular state to bring "natural" challenges, class actions have begun to run into certification challenges due to the difficulty of discerning damages.[gggg]

7.4.2.2 "Non-GMO"

Similar to natural, there is no regulatory definition of "non-GMO" claims, although there have been several legislative efforts (both federally and at the state level) around regulating the term. However, FDA has provided more oversight in this arena by publishing guidance for industry on how it thinks the term should be used.[hhhh] Although the guidance is currently not enforceable, following the FDA's guidelines is advisable since it helps avoid accusations of misbranding or misleading advertising. The cornerstone of FDA's guidance is to avoid using "genetically modified organism" or "GMO" and instead use terms like "bioengineered" or "genetically engineered," as it is more appropriate terminology. FDA also recommends avoiding terms such as "free" or "non" due to the difficulty in substantiating them, and instead advises companies to use claims that indicate the product/ingredient was not made using bioengineering.

Some independent organizations will, for a fee, apply their own standards and certify that a product is not made with genetically engineered ingredients. However,

[eeee]What is the Meaning of 'Natural' on the Label of Food?, last updated November 10, 2015. http://www.fda.gov/aboutfda/transparency/basics/ucm214868.htm.

[ffff]FDA Requests Comments on Use of the Term "Natural" on Food Labeling, November 10, 2015. http://www.fda.gov/Food/NewsEvents/ConstituentUpdates/ucm471919.htm.

[gggg]See, for example, Brazil v. Dole Packaged Foods, LLC, 12-CV-01831-LHK (N.D. Cal. 2014).

[hhhh]Guidance for Industry: Voluntary Labeling Indicating Whether Foods Have or Have Not Been Derived from Genetically Engineered Plants, November 2015. http://www.fda.gov/food/guidanceregulation/guidancedocumentsregulatoryinformation/ucm059098.htm.

it is important to note that the manufacturer bears the burden of substantiating claims, and therefore these independent certifications may not always be sufficient.

7.5 CONCLUSIONS

As outlined previously, there are a number of regulatory constraints that impact how an innovative functional food product can be marketed in the United States. With an increasing demand from consumers for functional foods and a shift to a healthier lifestyle, it is prudent for companies to consider the regulatory landscape in developing new and innovative products.

REFERENCES

Statutes, Regulations, and Cases

Brazil v. Dole Packaged Foods, LLC, 12-CV-01831-LHK. (N.D. Cal. 2014).

Current Good Manufacturing Practice, Hazard Analysis, and Risk-Based Preventive Controls for Human Food, 80 Fed. Reg. 55908, September 17, 2015.

Current Good Manufacturing Practice, Hazard Analysis, and Risk-Based Preventive Controls for Human Food; Clarification of Compliance Date for Certain Food Establishments, 80 Fed. Reg. 71934, November 18, 2015.

Federal Food, Drug, and Cosmetic Act, 21 U.S.C. 301 *et seq.*

Federal Trade Commission Act, 15 U.S.C. § 41 *et seq.*

Food and Drug Administration; Subchapter A—General, 21 C.F.R. § 1 *et seq.*

Food and Drug Administration; Subchapter B—Food for Human Consumption—21 C.F.R. § 100 *et seq.*

Regulations on Statements Made for Dietary Supplements Concerning the Effect of the Product on the Structure or Function of the Body, 65 Fed. Reg. 1000, February 7, 2000.

Safe Drinking Water and Toxicity Enforcement Act of 1986 (Proposition 65), §§ 25000 to 27001.

Guidance Documents

Dietary Supplements: An Advertising Guide for Industry, April 2001. https://www.ftc.gov/tips-advice/business-center/guidance/dietary-supplements-advertising-guide-industry#b

Draft Guidance for Industry: Ingredients Declared as Evaporated Cane Juice, October 2009. http://www.fda.gov/Food/GuidanceRegulation/GuidanceDocumentsRegulatoryInformation/ucm181491.htm

Draft Guidance for Industry: Dietary Supplements: New Dietary Ingredient Notifications and Related Issues, July 2011. http://www.fda.gov/Food/GuidanceRegulation/GuidanceDocumentsRegulatoryInformation/DietarySupplements/ucm257563.htm

Draft Guidance for Industry: Frequently Asked Questions about Medical Foods; second ed., August 2013. http://www.fda.gov/Food/GuidanceRegulation/GuidanceDocumentsRegulatoryInformation/ucm054048.htm

Guidance for Industry: Notification of a Health Claim or Nutrient Content Claim Based on an Authoritative Statement of a Scientific Body, June 1998. http://www.fda.gov/Food/GuidanceRegulation/GuidanceDocumentsRegulatoryInformation/ucm056975.htm

Guidance for Industry: Structure/Function Claims, Small Entity Compliance Guide, January 2002. http://www.fda.gov/Food/GuidanceRegulation/GuidanceDocumentsRegulatoryInformation/ucm103340.htm

Guidance for Industry: Iron-Containing Supplements and Drugs: Label Warning Statements, October 2003. http://www.fda.gov/Food/GuidanceRegulation/GuidanceDocumentsRegulatoryInformation/ucm073014.htm

Guidance for Industry: *Trans* Fatty Acids in Nutrition Labeling, Nutrient Content Claims, Health Claims, August 2003. http://www.fda.gov/Food/GuidanceRegulation/GuidanceDocumentsRegulatoryInformation/ucm053479.htm

Guidance for Industry: Frequently Asked Questions about GRAS, December 2004. http://www.fda.gov/Food/GuidanceRegulation/GuidanceDocumentsRegulatoryInformation/ucm061846.htm

Guidance for Industry: A Dietary Supplement Labeling Guide, April 2005. http://www.fda.gov/Food/GuidanceRegulation/GuidanceDocumentsRegulatoryInformation/DietarySupplements/ucm2006823.htm

Guidance for Industry: FDA's Implementation of "Qualified Health Claims": Questions and Answers; Final Guidance, May 2006. http://www.fda.gov/Food/GuidanceRegulation/GuidanceDocumentsRegulatoryInformation/ucm053843.htm

Guidance for Industry: Questions and Answers Regarding Food Allergens, Including the Food Allergen Labeling and Consumer Protection Act of 2004 (Edition 4); Final Guidance, October 2006. http://www.fda.gov/Food/GuidanceRegulation/GuidanceDocumentsRegulatoryInformation/ucm059116.htm

Guidance for Industry: Food Labeling; Nutrient Content Claims; Definition for "High Potency" and Definition for "Antioxidant" for Use in Nutrient Content Claims for Dietary Supplements and Conventional Foods, July 2008. http://www.fda.gov/Food/GuidanceRegulation/GuidanceDocumentsRegulatoryInformation/ucm063064.htm

Guidance for Industry: Substantiation for Dietary Supplement Claims Made Under Section 403(r)(6) of the Federal Food, Drug, and Cosmetic Act, December 2008. http://www.fda.gov/food/guidanceregulation/guidancedocumentsregulatoryinformation/dietarysupplements/ucm073200.htm

Guidance for Industry: Evidence-Based Review System for the Scientific Evaluation of Health Claims—Final, January 2009. http://www.fda.gov/Food/GuidanceRegulation/GuidanceDocumentsRegulatoryInformation/ucm073332.htm

Guidance for Industry: Food Labeling: Health Claims; Calcium and Osteoporosis, and Calcium, Vitamin D, and Osteoporosis, May 2009. http://www.fda.gov/Food/GuidanceRegulation/GuidanceDocumentsRegulatoryInformation/ucm152626.htm

Guidance for Industry: Current Good Manufacturing Practice in Manufacturing, Packaging, Labeling, or Holding Operations for Dietary Supplements, December 2010. http://www.fda.gov/Food/GuidanceRegulation/GuidanceDocumentsRegulatoryInformation/ucm238182.htm

Guidance for Industry: Food Additive Petition Expedited Review, October 2010. http://www.fda.gov/Food/GuidanceRegulation/GuidanceDocumentsRegulatoryInformation/ucm224104.htm

Guidance for Industry: Questions and Answers about the Petition Process, April 2011. http://www.fda.gov/food/guidanceregulation/guidancedocumentsregulatoryinformation/ucm253328.htm

Guidance for Industry: A Food Labeling Guide, January 2013. http://www.fda.gov/Food/GuidanceRegulation/GuidanceDocumentsRegulatoryInformation/LabelingNutrition/ucm2006828.htm

Guidance for Industry: Considerations Regarding Substances Added to Foods, Including Beverages and Dietary Supplements, January 2014. http://www.fda.gov/Food/GuidanceRegulation/GuidanceDocumentsRegulatoryInformation/ucm381315.htm

Guidance for Industry: Distinguishing Liquid Dietary Supplements from Beverages, January 2014. http://www.fda.gov/downloads/Food/GuidanceRegulation/GuidanceDocumentsRegulatoryInformation/DietarySupplements/UCM381220.pdf

Guidance for Industry: Gluten-Free Labeling of Foods, June 2014. http://www.fda.gov/downloads/Food/GuidanceRegulation/GuidanceDocumentsRegulatoryInformation/UCM402559.pdf

Guidance for Industry: Questions and Answers Regarding Food Facility Registration (sixth ed.), November 2014. http://www.fda.gov/Food/GuidanceRegulation/GuidanceDocumentsRegulatoryInformation/ucm331959.htm

Guidance for Industry: Voluntary Labeling Indicating Whether Foods Have or Have Not Been Derived from Genetically Engineered Plants, November 2015. http://www.fda.gov/food/guidanceregulation/guidancedocumentsregulatoryinformation/ucm059098.htm

Webpages

Advertising Self-Regulatory Council. http://www.asrcreviews.org

American Herbal Product Association Code of Ethics & Business Conduct. http://www.ahpa.org/AboutUs/AHPASPolicies/CodeofEthics.aspx

American Herbal Product Association Guidance Policies. http://www.ahpa.org/AboutUs/AHPASPolicies/GuidancePolicies.aspx

Avoiding Drug Interactions, November 28, 2008. http://www.fda.gov/ForConsumers/ConsumerUpdates/ucm096386.htm#supplements

Chemicals Known to the State [of California] to Cause Cancer or Reproductive Toxicity, last updated December 4, 2015. http://oehha.ca.gov/prop65/prop65_list/files/P65single120415.pdf

Consumer Advisory: Kava-Containing Dietary Supplements May be Associated with Severe Liver Injury, March 25, 2002. http://www.fda.gov/Food/ResourcesForYou/Consumers/ucm085482.htm

Current Good Manufacturing Practices (cGMPs) for Dietary Supplements, last updated August 10, 2015 http://www.fda.gov/Food/GuidanceRegulation/CGMP/ucm079496.htm

FDA, 2004. Issues Regulation Prohibiting Sale of Dietary Supplements Containing Ephedrine Alkaloids and Reiterates Its Advice That Consumers Stop Using These Products, February 6, 2004. http://www.fda.gov/NewsEvents/Newsroom/PressAnnouncements/2004/ucm108242.htm

FDA Warns Consumers to Avoid Red Yeast Rice Products Promoted on Internet as Treatments for High Cholesterol Products Found to Contain Unauthorized Drug, August 9, 2007. http://www.fda.gov/NewsEvents/Newsroom/PressAnnouncements/2007/ucm108962.htm

FDA to Investigate Added Caffeine, May 3, 2013. http://www.fda.gov/ForConsumers/ConsumerUpdates/ucm350570.htm

FDA Requests Comments on Use of the Term "Natural" on Food Labeling, November 10, 2015. http://www.fda.gov/Food/NewsEvents/ConstituentUpdates/ucm471919.htm

Food Industry, last updated September 9, 2015. http://www.fda.gov/Food/ResourcesForYou/Industry/default.htm

FSMA Rules & Guidance for Industry, last updated November 11, 2015. http://www.fda.gov/Food/GuidanceRegulation/FSMA/ucm253380.htm

FSMA Final Rule for Preventive Controls for Human Food, last updated January 22, 2016. http://www.fda.gov/Food/GuidanceRegulation/FSMA/ucm334115.htm

Memorandum of Understanding Between The Federal Trade Commission and The Food and Drug Administration, May 14, 1971. http://www.fda.gov/AboutFDA/PartnershipsCollaborations/MemorandaofUnderstandingMOUs/DomesticMOUs/ucm115791.htm

New Dietary Ingredients Notifications (NDI), last updated August 10, 2015. http://www.fda.gov/Food/GuidanceRegulation/FoodFacilityRegistration/NewDietaryIngredientsNotificationsNDI/default.htm

Non-GMO Project. http://www.nongmoproject.org/

Organic Agriculture, last updated January 9, 2015. http://www.usda.gov/wps/portal/usda/usdahome?navid=organic-agriculture

Organic Certification and Accreditation. http://www.ams.usda.gov/services/organic-certification

Proposition 65. http://oehha.ca.gov/prop65.html

Qualified Health Claims, last updated November 24, 2014. http://www.fda.gov/Food/IngredientsPackagingLabeling/LabelingNutrition/ucm2006877.htm

Registration of Food Facilities, last updated January 5, 2016. http://www.fda.gov/Food/GuidanceRegulation/FoodFacilityRegistration/default.htm

What is the Meaning of 'Natural' on the Label of Food?, last updated November 10, 2015. http://www.fda.gov/aboutfda/transparency/basics/ucm214868.htm

FDA Warning Letters

FDA Warning Letter to Accera, Inc., December 26, 2013. http://www.fda.gov/ICECI/EnforcementActions/WarningLetters/2013/ucm381320.htm

FDA Warning Letter to Kind, LLC, March 27, 2015. http://www.fda.gov/ICECI/EnforcementActions/WarningLetters/ucm440942.htm

FDA Warning Letter to Metagenics, August 13, 2013. http://www.fda.gov/ICECI/EnforcementActions/WarningLetters/2013/ucm367142.htm

FDA Warning Letters, Webpage. http://www.fda.gov/ICECI/EnforcementActions/WarningLetters/

CHAPTER

Formulations and challenges: a special emphasis on stability and safety evaluations

8

Surashree Sen Gupta, Mahua Ghosh
Department of Chemical Technology, University of Calcutta, Kolkata, India

8.1 INTRODUCTION

Functional foods and nutraceuticals are designed to provide nutrients to individuals that they generally cannot consume in sufficient proportions. Such terms correspond to a wide range of products including isolated bioactive components from natural sources, vitamins, minerals, dietary fibers, fatty acids, herbal extracts, processed foods, beverages, etc. Such food supplements play a major role in enhancing the defensive mechanism of the body, thus helping to prevent opprobrious lifestyle diseases and at the same time manage any sort of preexisting health conditions (Arjmandi and Smith, 2002).

In keeping with today's boisterous way of life, several lifestyle diseases are on the rise (Gupta et al., 2010). The aftermath of this is the arrival of varied tenacious medications and drugs, which are consumed by one and all. These drugs are associated with numerous adverse side effects. On the contrary, consumption of regulated daily doses of natural health-promoting products can effectively prevent such lifestyle diseases. Hence, there is a great demand for the evolution of such health food products, leading to the concept of nutraceuticals, an assemblage of "nutrition" and "pharmaceuticals." However, from the commercialization aspect, it is observed that apart from chemical activity, the basic characteristics of the compounds are more important (Kawakami, 2012).

Divergent varieties of dietary supplements are available today, which include about 50,000 different types of nutraceuticals (CNN, 2013). A massive population consume these food supplements almost daily. However, often such products are found to be ineffective in treating any specific disease and if consumed without any proper consultation, they can lead to unwanted after effects.

Developing New Functional Food and Nutraceutical Products. http://dx.doi.org/10.1016/B978-0-12-802780-6.00008-0

8.2 NUTRACEUTICALS FORMULATION AND DESIGN

Establishment of a definite entity of the nutraceutical is essential for the well-being of an individual whereupon the physicochemical properties of the product can also be improved based on the properties displayed by the nutraceutical compounds. This will result in the development of such a product that will effect better wellness and improvement of health. Certain bioactive food components are directly consumed after isolation and purification from natural sources. Others are often modified structurally, functionally, or in terms of delivery properties to pertain with the market demand and consumer feasibility. Hence, in the present section several formulation strategies adopted for the evolution of nutraceuticals are discussed.

8.2.1 ISOLATION OF NUTRACEUTICALS FROM NATURAL SOURCES

Natural products from certain plants can be isolated as pure compounds or extracts in edible-grade solvents (Duraipandiyan et al., 2006). Such food supplements extracted from natural sources contain various types of specific bioactive compounds that are functional in nature. By means of different analytical methodologies accurate extraction, isolation, and characterization of the active ingredients in nutraceuticals are performed. After that, their safety aspects are evaluated before releasing them to the market for consumption purposes. The two chief difficulties encountered in the entire process include the selection of a proper solvent and the extraction technique for isolation of the active ingredient (Sasidharan et al., 2011). Most importantly, the purification of these compounds is absolutely essential before they can be declared safe for consumption. Several characterization techniques are common today such as high performance liquid chromatography (HPLC), thin layer chromatography (TLC), gas chromatography (GC), fourier transform infra red (FTIR) spectroscopy, nuclear magnetic resonance spectroscopy (NMR), mass spectroscopy (MS), etc. which properly help to determine the level of purity of the product and also its chemical entity.

8.2.2 LIPOSOMAL CARRIER SYSTEMS

Liposomes, indicated by microscopic units which are spherical in nature and basically corpuscles of lipid matter, most commonly of phospholipids, serve as the most common carrier systems for nutraceuticals. They house several common food supplements such as vitamins, minerals, herbal products, antioxidants, etc. with a lipid-soluble nature. These units also play an important role in gearing the pharmacokinetics of the absorption of such nutraceuticals in the body. Important factors controlling the development of these liposomes include shape and size of the liposomes, their charge distribution, and their chemical makeup (Padamwar and Pokharkar, 2006). An important advantage of this type of nutraceutical delivery system is that it can essentially cross the membrane barriers of the body, especially as it is structurally designed in such a way as to be similar to biological membranes (Padamwar and Pokharkar, 2006).

8.2.3 MICROSPONGES AND NANOSPONGES

A group of novel delivery systems in the market today are the micro and nanosponges. They contain micro or nanospheres, which are porous in nature. An added advantage of these units is their large internal surface area, which enhances their loading capacity for the nutraceuticals. These porous polymeric systems have found application for accommodating genetically modified melanin, a biological nutraceutical responsible for skin pigmentation. This was observed to distribute the pigment uniformly on the skin surface, thus incurring protection of the skin from harmful UV radiation. Other applications like resveratrol and its stability, three-dimensional polymers of cyclodextrin, etc. were also developed using the micro/nanosponge models (Patravale and Mandawgade, 2008; Ansari et al., 2011). These complexes are gaining popularity due to their excellent and timely release properties, permeation nature, stability, and ease of application of the nutraceuticals (Ansari et al., 2011).

8.2.4 CYCLODEXTRIN COMPLEXATION

Cyclodextrins or their derivatives have acted as excellent delivery systems over the last decade. The greatest advantage of these units is the improved bioavailability of the nutraceuticals (Gonnet et al., 2010; Tonnesen et al., 2002; Yuan et al., 2008, 2012). Different types of vitamins can complex with such molecules, thereby improving their functional properties on application (Kim et al., 2010).

8.2.5 BIODEGRADABLE HYDROGELS

Biodegradable hydrogels have found primary applications in a variety of pharmaceutical industries. Of late attempts have been made to use them for nutraceutical formulations. The dominant effect of these hydrogels is their protective nature for oxidation-prone or sensitive nutraceuticals. This in turn induces the retention of the nutraceuticals in their nascent state, before they are consumed. Due to the thermolabile nature of vitamins, for vitamin formulations hydrogel technology is most commonly adopted. Other studies involving the use of hydrogels have been reported for protecting riboflavin, soy isoflavones, milk proteins, etc. (Maltais et al., 2009; Livney, 2010).

8.2.6 NANOTECHNOLOGY-BASED FORMULATIONS FOR NUTRACEUTICAL DELIVERY

Nanotechnology is the term of the day to prepare delivery systems for insoluble nutraceuticals (Mathew et al., 2012). It is used to embellish the solubility of such compounds in water, enhance their absorption in the body, and formulate several dosages. Furthermore, the bioavailability of the nutraceuticals becomes appreciable due to their targeted release in the gastrointestinal tract. However, it is unfortunate that such nanoformulations of nutraceuticals have not yet made an appreciable impact in the industrial arena. It is still under research consideration (Sen Gupta and Ghosh, 2012b, 2014b, 2015). Nanosuspensions and nanoemulsions have also found applications by dissolution of the nutraceuticals. This contrivance of nutraceuticals remains in a liquid form and finds applications in both oral deliveries and combination with edible food items. Both nanosuspensions and nanoemulsions dissolve hydrophobic nutraceuticals and effectively improve their solubility and bioavailability (Gonnet et al., 2010; Li et al., 2012; Mitri et al., 2011).

Nanostructured lipid carriers consist of a lipid grid into which the nutraceuticals are embedded. Due to their size, shape, and structures, these units show quick absorption into the body via the membrane, thus delivering the nutraceuticals at the appropriate site and at the right time (Teeranachaideekul et al., 2007). Nanomicelles or nanovehicles are another group of nutraceutical delivery units. They are used to treat several ailments such as neurological and cardiovascular diseases (Tan et al., 2012). Several phytonutrients have been encapsulated and then dispatched at the appropriate site (Sen Gupta and Ghosh, 2012a, 2014b, 2015). On being released at the targeted site these nutraceuticals have been shown to display several health-aiding effects. Furthermore, due to the controlled release nature of these nanocapsules, the toxicity generated on consuming multiple nutraceuticals in a dose is considerably minimized. Moreover, these nanocapsules provide substantial protection of the encapsulated nutrients against temperature, pH, or pressure-related degradation (Shpigelman et al., 2010).

8.2.7 SOLID DISPERSIONS

Solid dispersions are another means of formulating nutraceuticals to achieve targeted delivery of these to the desired locale. Here the bioactive compounds are dispersed uniformly in a polymer complex, which is basically lipophobic in nature. This technique is highly suitable for mildly water-soluble food supplements that display limited absorption in the body (Nepal et al., 2010; Sen Gupta and Ghosh 2014a). A common bioactive product, resveratrol, has been developed in the form of solid dispersion, which considerably enhanced the targeted release properties (Weigel et al., 2013).

8.3 CHALLENGES IN THE FORMULATION OF FOOD SUPPLEMENTS

The major challenge faced by the nutraceutical industry is the incorporation of the food supplement into daily consumable products. This limits their acceleration to the market level. Furthermore the chemical composition of the nutraceuticals also limits their use in various formulations. In the present section the problems faced in formulating nutraceuticals are briefly discussed.

8.3.1 POOR AQUEOUS SOLUBILITY

Most lipophilic bioactive compounds are insoluble in an aqueous medium. This makes it difficult for them to be absorbed by the body. Hence the beneficial qualities of the nutraceuticals remain mostly unutilized. Some such products include ω-3 fatty acid rich oils, phytosterols, phytostanols, carotenoids, oil-soluble vitamins, curcumin, and so on (Yi et al., 2014). This prevents their usage in foods or beverages, as most of these are not water-soluble. This poses one of the biggest challenges concerning today's high price of nutraceuticals, as considerable methodologies need to be put into practice to make the lipid-based nutraceuticals soluble in aqueous food products. This effectively increases the cost of production of edible-grade nutraceuticals, making them an expensive buy. Hence cost-effective technologies need to be improvised to enhance the solubility of nutraceuticals. Thus considerable research has already been pursued in that direction and the aqueous solubility of nutraceuticals has been improved by means of specialized technologies (Salvia-Trujillo et al., 2013; Sen Gupta and Ghosh, 2012b, 2014b, 2015).

8.3.2 HIGH MELTING POINT OF NUTRACEUTICALS

Compounds such as phytosterols, fatty alcohols, and carotenoids are crystalline at normal room temperature. This makes it an arduous job to incorporate such products into foods especially at room temperature. If forcibly they are introduced in foods, the stability and shelf life of such food items are markedly deteriorated, resulting in disagreeable appearance and obnoxious odor and mouthfeel, thus in effect affecting the market value and consumer demand of such food items. Additionally, such food items also have lower bioavailability, further limiting their utility.

In such cases, generally the nutraceuticals are dissolved in suitable edible-grade solvents to transform them to a liquid phase. They are then incorporated in foods. They may also be introduced as suspended nanocrystals in the respective food item in a manner that does not hamper their shelf life or affect their appearance and quality (McClements, 2012).

8.3.3 CHEMICAL INSTABILITY

The biggest challenge of the food and nutraceutical industry is the instability of the hydrophilic and lipophilic bioactive compounds which results in the quick degradation of these food items under normal storage conditions (Rao et al., 2013). This makes it extremely difficult to market, store, or transport such products for extended periods of time. This problem is especially seen for compounds like ω-3 fatty acid rich oils such as fish oils, flaxseed oil, cod liver oils; carotenoids or lycopene or curcumin; and so on (McClements and Decker, 2000; Boon et al., 2010; Waraho et al., 2011; Sen Gupta and Ghosh, 2013; Heger et al., 2014). The extent of the chemical degradation is completely dependent on the composition of the bioactive product, the environmental conditions such as temperature, pH, pressure, etc., or the presence of metals or other such oxidation-promoting agents. For such compounds, development of nanoscale products is essential to protect them from degradation.

8.3.4 PROBIOTIC COMPOUNDS

In the case of the development of probiotics special selective bacterial strains are necessary. Food products based on these specialized compounds are very popular today. However, a great challenge today is the selection of the proper strain followed by their incorporation in foods. Application of any mismatching bacterial cultures or toxic ones in addition to their negligent handling can cause disastrous consequences.

8.4 RISKS AND SAFETY

The first step of selecting a suitable nutraceutical entails thorough study to assess its safety upon consumption by humans or animals. Common problems that can be encountered are allergic reactions and toxicity. Before marketing a particular nutraceutical, proper research regarding the potential harm that can be caused by it, especially if consumed in unregulated doses, must be conducted. Special care must be taken for products that have been developed by genetic or structural modification techniques. The primary objective for a safety assessment is to acquire suitable knowledge regarding a new nutraceutical product or its modified form, before launching it in the market. This knowledge could be utilized by suitable authorities to impose decisions and regulations on the consumption of the nutraceuticals.

An assessment of the safety criteria or nutritional impact starts sequentially from the identification of the characteristics causing the potential hazard. Thereafter the subsequent consequences and the risk posed by each of these hazard characteristics need to be evaluated. Hence, apart from knowledge of the scientific quality of these products, a proper assay is definitely necessary to assess and manage any associated risk factors. Based on the efficient judgement of all the risks that can be encountered, a decision needs to be taken regarding the application to be selected for proper treatment/management of the risk, to ensure safety.

In the case of some nutraceuticals and dietary supplements, often it is not possible to accurately pinpoint the exact cause of a risk or hazard (Garcia-Alonso, 2010). Hence in such cases it is mandatory to perform several trials and based on the level of uncertainty, the data collected is used to estimate the nature and intensity of the risk. Often long-term preservation or exposure to the environment can also deteriorate the quality of a nutraceutical. Different suitable techniques are adopted to formulate them in such a manner as to enhance their shelf lives (Friel and Lederman, 2006). Hence in terms of nutraceuticals, the term "absolute safety" is not very appropriate. Any such product, no matter how well tested, can be harmful if consumed in unregulated doses. Certain safety measures widely followed during the development of nutraceutical products are discussed herein.

8.4.1 CHARACTERIZATION OF THE MOLECULES

It is essential to have a sound knowledge regarding the molecular composition of a nutraceutical, especially if it is genetically or structurally modified. This is the first step to acquire relevant information regarding the chemical composition of the product and evaluate any risks associated with it. Furthermore, if a DNA analysis is performed, then any unintended side effects associated with the food product would also be detected (Macdonald, 2012).

8.4.2 ADDITION OF CHEMICAL INHIBITORS DURING FORMULATION

Oxidation of many sensitive nutraceuticals can be avoided by addition of certain antioxidants or chelating agents. The chelating agents will complex with transition metals and the antioxidant will scavenge free radicals, thus effectively preventing deterioration of the food supplements (Gülçin et al., 2012; McClements and Decker, 2000; Sen Gupta and Ghosh, 2013; Waraho et al., 2011). However, they should be added at a definite level, beyond which they themselves may act as prooxidants. Some such chelating agents/antioxidants commonly used include EDTA, BHA, TBHQ, etc. (Matalanis et al., 2012; Sen Gupta and Ghosh, 2015; Zhang et al., 2014).

8.4.3 CONTROL OF DEGRADATION DUE TO LIGHT SCATTERING AND ABSORPTION

Certain bioactive molecules undergo auto-oxidation when exposed to UV or visible radiation. Carotenoids, resveratrol, curcumin, and ω-3 fatty acids are some of the common examples. The stability of such compounds can be enhanced by micro or nanoformulating them or by introducing certain compounds during their formulation so that additives can absorb, scatter, or deflect all sorts of radiation without affecting the nutraceutical (Davidov-Pardo and McClements, 2015).

8.4.4 CONTROLLED OR TARGETED RELEASE

The controlled or targeted delivery of nutraceuticals is a major step to prevent any perverse or toxic effects, especially related to the overconsumption of a particular food supplement. Furthermore, by this technique the site specificity for release of the bioactive molecules can be controlled. This fact also establishes that the bioactive product reaches the desired location in an intact manner (Jones and McClements, 2010; McClements, 2015; Nagaich et al., 2014; Shewan and Stokes, 2013; Sen Gupta and Ghosh, 2014b, 2015; Sivakumar et al., 2013). An added advantage of this kind of delivery system is that it boosts the mouthfeel and appearance of the final product, which further adds to the consumer appeal.

8.4.5 REGULATORY MEASURES

Based on the studies performed in the laboratories, different regulatory bodies impose certain rules regarding the safety measures and protocols to be followed by nutraceutical manufacturers. Based on these norms, a nutraceutical can be deemed as suitable for consumption only after it has passed the set criteria. These aspects of protection are based on the assessed risk or the risk hypotheses of the formulated compound (Evans et al., 2006). The primary focus of these policies is that consumption or application of nutraceuticals should not lead to any kind of adverse health effects on humans or animals. The environmental and ecological impacts are also considered. When disposed of, the nutraceuticals or the byproducts of nutraceutical processing should not be the cause of any kind of environmental pollution or hazard.

8.5 CONCLUSIONS

Today many food supplements in the market contain active ingredients necessary to promote sound health. However, one needs to be conscious of the undesired side effects, especially at the time when a new product is being consumed. No products should be taken without proper knowledge of the chemical composition and their effects on the body. It is absolutely essential to ensure that the concerned dietary supplements are fully tested and purified. Any preexisting health conditions or diseases of the individual should also be taken under consideration. Unregulated consumption of any such products can lead to unprecedented health effects that can be fatal. Consumption of such products in excessive amounts also raises some health risk. Special care should be taken in respect of children, the aged, and sick people.

However, the increasing interest in the development and fabrication of nutraceuticals has led to the development of several types of nascent and structured bioactives. Special delivery systems are designed to deliver these food-grade ingredients to the required site. The composition and structure of these delivery systems are carefully tailored to ensure that the complete functional benefits of the nutraceuticals can be obtained. The properties of these products should also be evaluated before releasing them to the market, to make certain that improved physicochemical stability, dispersibility, bioavailability, and restrained release are achieved. The future holds ample prospects for reaching new heights with respect to the development of novel nutraceuticals. At the same time, research should focus on the development of better measures of ascertaining the stability and safety of the new products. Subsequently these products can make headway from laboratories to industries and from there, directly to the markets to fulfill consumer demands.

REFERENCES

Ansari, K.A., Vavia, P.R., Trotta, F., Cavalli, R., 2011. Cyclodextrin-based nanosponges for delivery of resveratrol. In vitro characterisation, stability, cytotoxicity and permeation study. Am. Assoc. Pharm. Sci. 12, 279–286.

Arjmandi, B.H., Smith, B.J., 2002. Soy may prevent bone loss in menopausal women. J. Nutr. Biochem. 13, 130–137.

Boon, C.S., McClements, D.J., Weiss, J., Decker, E.A., 2010. Factors influencing the chemical stability of carotenoids in foods. Crit. Rev. Food Sci. Nutr. 50 (6), 515–532.

CNN, 2013. Half of Americans Use Supplements. Park, Madison.

Davidov-Pardo, G., McClements, D.J., 2015. Nutraceutical delivery systems: resveratrol encapsulation in grape seed oil nanoemulsions formed by spontaneous emulsification. Food Chem. 167, 205–212.

Duraipandiyan, V., Ayyanar, M., Ignacimuthu, S., 2006. Antimicrobial activity of some ethnomedicinal plants used by Paliyar tribe from Tamil Nadu, India. BMC Complementary Altern. Med. 6, 35–41.

Evans, J., Wood, G., Miller, A., 2006. The risk assessment–policy gap: an example from the UK contaminated land regime. Environ. Int. 32, 1066–1071.

Friel, H., Lederman, H., 2006. A nutritional supplement for influenza A (H5N1) infection in humans. Med. Hypotheses 67, 578–587.

Garcia-Alonso, M., 2010. Current challenges in environmental risk assessment: the assessment of unintended effects of GM crops on nontarget organisms. IOBC/WPRS Bull. 52, 57–63.

Gonnet, M., Lethuaut, L., Boury, F., 2010. New trends in encapsulation of liposoluble vitamins. J. Control. Release 146, 276–290.

Gülçin, I., Elmastaş, M., Aboul-Enein, H.Y., 2012. Antioxidant activity of clove oil—a powerful antioxidant source. Arab. J. Chem. 5 (4), 489–499.

Gupta, S.C., Kim, J.H., Prasad, S., Aggarwal, B.B., 2010. Regulation of survival, proliferation, invasion, angiogenesis, and metastasis of tumor cells through modulation of inflammatory pathways by nutraceuticals. Cancer Metastasis Rev. 29, 405–434.

Heger, M., van Golen, R.F., Broekgaarden, M., Michel, M.C., 2014. The molecular basis for the pharmacokinetics and pharmacodynamics of curcumin and its metabolites in relation to cancers. Pharmacol. Rev. 66 (1), 222–307.

Jones, O.G., McClements, D.J., 2010. Functional biopolymer particles: design, fabrication, and applications. Compr. Rev. Food Sci. Food Safety 9 (4), 374–397.

Kawakami, K., 2012. Modification of physicochemical characteristics of active pharmaceutical ingredients and application of supersaturable dosage forms for improving bioavailability of poorly absorbed drugs. Adv. Drug Deliv. Rev. 64, 480–495.

Kim, S.O., Youn, J.Y., Kim, K.M., Kang, K.C., Pyo, H.B., Lee, S.J., 2010. Characterization of an inclusion complex of 7-dehydrocholesterol and cyclodextrin. J. Ind. Eng. Chem. 16, 119–121.

Li, Y., Zheng, J., Xiao, H., McClements, D.J., 2012. Nanoemulsion-based delivery systems for poorly water soluble bioactive compounds: influence of formulation parameters on polymethoxyflavone crystallization. Food Hydrocoll. 27, 517–528.

Livney, Y.D., 2010. Milk proteins as vehicles for bioactives. Curr. Opinion Colloid Interface Sci. 15, 73–83.

Macdonald, P., 2012. Developing workable regulatory frameworks for the environmental release of transgenic plants. Collect. Biosafety Rev. 6, 126–159.

Maltais, A., Remondetto, G.E., Subirade, M., 2009. Soy protein cold-set hydrogels as controlled delivery devices for nutraceutical compounds. Food Hydrocoll. 23, 1647–1653.

Matalanis, A., Decker, E.A., McClements, D.J., 2012. Inhibition of lipid oxidation by encapsulation of emulsion droplets within hydrogel microspheres. Food Chem. 132 (2), 766–772.

Mathew, A., Aravind, A., Brahatheeswaran, D., Fukuda, T., Nagaoka, T., Hasumura, T., Iwai, S., Morimoto, H., Yoshida, Y., Maekawa, T., Venugopal, K., Sakthikumar, D., 2012. Amyloid-binding aptamer conjugated curcumin-PLGA nanoparticle for potential use in Alzheimer's Disease. Bionanoscience 2 (2), 83–93.

McClements, D.J., 2012. Crystals and crystallization in oil-in-water emulsions: implications for emulsion-based delivery systems. Adv. Colloid Interface Sci. 174, 1–30.

McClements, D.J., 2015. Nanoscale nutrient delivery systems for food applications: improving bioactive dispersibility, stability, and bioavailability. J. Food Sci. 80 (7), N1602–N1611.

McClements, D.J., Decker, E.A., 2000. Lipid oxidation in oil-in-water emulsions: impact of molecular environment on chemical reactions in heterogeneous food systems. J. Food Sci. 65 (8), 1270–1282.

Mitri, K., Shegokar, R., Gohla, S., Anselmi, C., Muller, R.H., 2011. Lutein nanocrystals as antioxidant formulation for oral and dermal delivery. Int. J. Pharm. 420, 141–146.

Nagaich, U., Pal, A.K., Bharti, C., Gulati, N., 2014. Formulation and evaluation of nutraceutical tablet using herbal drugs by direct compression method. J. Drug Deliv. Therapeut. 4 (2), 47–51.

Nepal, P.R., Han, H., Choi, H., 2010. Enhancement of solubility and dissolution of Coenzyme Q10 using solid dispersion formulation. Int. J. Pharm. 383, 147–153.

Padamwar, M., Pokharkar, V.B., 2006. Development of vitamin loaded topical liposomal formulation using factorial design approach: drug deposition and stability. Int. J. Pharm. 320, 37–44.

Patravale, V.B., Mandawgade, S.D., 2008. Novel cosmetic delivery systems: an application update. Int. J. Cosmetic Sci. 30, 19–33.

Rao, J.J., Decker, E.A., Xiao, H., McClements, D.J., 2013. Nutraceutical nanoemulsions: influence of carrier oil composition (digestible versus indigestible oil) on β-carotene bioavailability. J. Sci. Food Agric. 93 (13), 3175–3183.

Salvia-Trujillo, L., Qian, C., Martin-Belloso, O., McClements, D.J., 2013. Influence of particle size on lipid digestion and beta-carotene bioaccessibility in emulsions and nanoemulsions. Food Chem. 141 (2), 1472–1480.

Sasidharan, S., Chen, Y., Saravanan, D., Sundram, K.M., Yoga Latha, L., 2011. Extraction, isolation and characterization of bioactive compounds from plants' extracts. Afr. J. Tradit. Complement. Altern. Med. 8 (1), 1–10.

Sen Gupta, S., Ghosh, M., 2012a. Microencapsulation of conjugated linolenic acid-rich pomegranate seed oil by an emulsion method. Food Sci. Technol. Int. 18, 549–558.

Sen Gupta, S., Ghosh, M., 2012b. In vitro study of anti-oxidative effects of β-carotene and α-lipoic acid for nanocapsulated lipids. LWT-Food Sci. Technol. 49, 131–138.

Sen Gupta, S., Ghosh, M., 2013. In vitro antioxidative evaluation of α- and β-carotene, isolated from crude palm oil. J. Anal. Methods Chem. 2013, 1–10.

Sen Gupta, S., Ghosh, M., 2014a. Formulation development and process parameter optimization of lipid nanoemulsions using an alginate-protein stabilizer. J. Food Sci. Technol. 52 (5), 2544–2557.

Sen Gupta, S., Ghosh, M., 2014b. Preparation and characterisation of protein based nanocapsules of bioactive lipids. J. Food Eng. 121, 64–72.

Sen Gupta, S., Ghosh, M., 2015. Synthesis, characterization, stability evaluation and release kinetics of fiber-encapsulated carotene nanocapsules. Grasas Aceites 66 (4), 1–13.

Shewan, H.M., Stokes, J.R., 2013. Review of techniques to manufacture micro-hydrogel particles for the food industry and their applications. J. Food Engr. 119 (4), 781–792.

Shpigelman, A., Israeli, G., Livney, Y.D., 2010. Thermally-induced protein–polyphenol co-assemblies: beta lactoglobulin-based nanocomplexes as protective nanovehicles for EGCG. Food Hydrocoll. 24, 735–743.

Sivakumar, B., Aswathy, R.G., Nagaoka, Y., Iwai, S., Suzuki, M., Venugopal, K., Kato, K., Yoshida, Y., Maekawa, T., Sakthikumar, D., 2013. Aptamer conjugated theragnostic multifunctional magnetic nanoparticles as a nanoplatform for pancreatic cancer therapy. RSC Adv. 3, 20579–20598.

Tan, B., Liu, Y., Chang, K., Lim, B.K.W., Chiu, G.N.C., 2012. Perorally active nanomicellar formulation of quercetin in the treatment of lung cancer. Inter. J. of Nanomedicine 7, 651–661.

Teeranachaideekul, V., Souto, E.B., Junyaprasert, V.B., Muller, R.H., 2007. Cetyl-palmitate based NLC for topical delivery of Coenzyme Q10—development, physicochemical characterization and in vitro release studies. Eur. J. Pharm. Biopharm. 67, 141–148.

Tonnesen, H.H., Masson, M., Loftsson, T., 2002. Studies of curcumin and curcuminoids. XXVII. Cyclodextrin complexation: solubility, chemical and photochemical stability. Int. J. Pharm. 244, 127–135.

Waraho, T., McClements, D.J., Decker, E.A., 2011. Mechanisms of lipid oxidation in food dispersions. Trends Food Sci. Technol. 22 (1), 3–13.

Weigel, L.A., Mauer, L.J., Edgar, K.J., Taylor, L.S., 2013. Crystallization of amorphous solid dispersions of resveratrol during preparation and storage—impact of different polymers. J. Pharm. Sci. 102, 171–184.

Yi, J., Li, Y., Zhong, F., Yokoyama, W., 2014. The physicochemical stability and *in vitro* bioaccessibility of beta-carotene in oil-in-water sodium caseinate emulsions. Food Hydrocoll. 35, 19–27.

Yuan, C., Jin, Z., Xu, X., Zhuang, H., Shen, W., 2008. Preparation and stability of the inclusion complex of astaxanthin with hydroxypropyl-b-cyclodextrin. Food Chem. 109, 264–268.

Yuan, C., Jin, Z., Xu, X., 2012. Inclusion complex of astaxanthin with hydroxypropyl-b-cyclodextrin: UV, FTIR, HNMR and molecular modeling studies. Carbohydr. Polym. 89, 492–496.

Zhang, Z., Decker, E.A., McClements, D.J., 2014. Encapsulation, protection, and release of polyunsaturated lipids using biopolymer-based hydrogel particles. Food Res. Int. 64, 520–526.

CHAPTER

9 Nutraceutical Formulations and Challenges

Raj K. Keservani*, Rajesh K. Kesharwani, Anil K. Sharma†, Surya P. Gautam‡, Santosh K. Verma‡**

**School of Pharmaceutical Sciences, Rajiv Gandhi Proudyogiki Vishwavidyalaya, Bhopal, India; **Department of Biotechnology, NIET, NIMS University, Shobha Nagar, Jaipur, India; †Department of Pharmaceutics, Delhi Institute of Pharmaceutical Sciences and Research, New Delhi, India; ‡CT Institute of Pharmaceutical Sciences, Sahapur Campus, Jalandhar, India*

9.1 INTRODUCTION

The term "nutraceutical" was introduced as a hybrid of nutrition and pharmaceutical in 1989 by Dr. Stephen DeFelice, Chairperson of the Foundation for Innovation in Medicine. A nutraceutical is any material that may be considered a food or component of a food and offers medical or health advantages, encompassing prophylaxis and cure of diseases (Brower, 1998; Kalra, 2003; Biesalski, 2001; Keservani et al., 2010). Functional foods are any food or food ingredient that may provide a health benefit beyond the traditional nutrients it contains (Keservani et al., 2015d). The safety of what is proposed and promising nutritional and therapeutic influences have generated an appreciable curiosity in nutraceuticals and functional foods. This has rendered the nutraceutical and functional food industry in a special place to focus on the benefits of their terminal users, that is, the consumers.

Nutraceuticals offer several advantages, but the current review emphasizes the role of nutraceuticals in the maintenance of health. Nutraceuticals are believed to provide physiological benefit or immunity against ensuing diseases/disorders such as cardiovascular disorders (Rissanen et al., 2003; Choudhary and Tomer, 2013), obesity (Boozer et al., 2001; Kaur et al., 2015), diabetes (Singh et al., 2012), cancer (Temple and Gladwin, 2003; Ahmad et al., 2015), stress, Parkinson's and Alzheimer's diseases (Latif et al., 2007; Keservani et al., 2015a), neurological diseases (Keservani et al., 2015b), pulmonary and respiratory disorder (Keservani et al., 2015c), and may function as immunomodulators (Keservani et al., 2015d).

Developing New Functional Food and Nutraceutical Products. http://dx.doi.org/10.1016/B978-0-12-802780-6.00009-2

Another category of nutraceuticals that have quickly growm in the past two decades are the "probiotics" (i.e., health-promoting bacteria); with a global market value of $32.6 billion in 2014 (Cook et al., 2012; Hafez et al., 2013). The oral delivery of probiotics is hampered by the poor stability of the bacteria in the GIT (gastrointestinal tract) and consequent loss of viability under the influence of high acidity and concentrations of bile salts. Preformed growth factors and cytokines in colostrum upon absorption into the bloodstream have been shown to facilitate the function of bone marrow stem cells. Furthermore, the proteins in colostrum stimulate the innate immune system, leading to the activation of stem cells (Tokuyama et al., 1990; Stelwagen et al., 2009; Rusu et al., 2010; Krissansen, 2007; Davison and Diment, 2010; Davis et al., 2007; Shytle et al., 2006).

Activities such as influential activation, mobilization, and differentiation of progenitor cells that dwell within specialized tissues throughout the body, particularly the bone marrow, fall under continued tissue and organ repair functions and are indicators of sound health. Functional foods may increase regenerative potential via induction of immune cell pathways (Bachstetter et al., 2010; Mikirova et al., 2009). This chapter focuses on the challenges and opportunities associated with the product formulation of nutraceuticals, stability and safety of nutraceuticals, Indian and global markets, current scenario, and future prospects.

9.2 NUTRACEUTICAL FORMULATION

Although it may be easy for a formulator to develop a pharmaceutical form (tablets, powders, capsules, suppositories, etc.) containing bioactive food, it is quite challenging to obtain a satisfactory bioavailability for such nutraceuticals. The bioavailability is often jeopardized by the low solubility, stability, and/or permeability of the bioactive component in the GIT. For example, in the case of curcumin, extremely low serum levels, limited tissue distribution, apparent rapid metabolism, and short circulation half-life are the underlying causes of its low oral bioavailability (Anand et al., 2007; Kesharwani et al., 2015). Another example is the probiotics encapsulation technology (or PET), which has emerged recently as a strategy for delivery of probiotics. Conditions that maintain cell viability like biomaterial selection, solvent type and toxicity, and choice of proper technology are of paramount concern in formulating probiotics. Natural and/or synthetic polymers can be used as biomaterials. Factors to be considered include: (1) physicochemical characteristics (chemical makeup, morphology, mechanical strength, stability in GI fluids); (2) toxicity; and (3) manufacturing and sterilization processes (Gbassi and Vandamme, 2012). The most common biomaterials used for probiotics encapsulation are alginate, chitosan, carrageenan, gelatin, whey proteins, cellulose acetate phthalate, locust bean gum, and starches (Chen et al., 2012).

An alternative approach would be to use solid dispersions and cyclodextrin complexes, which have shown promise in maximizing the effective surface area available for GI fluids so as to enhance dissolution rate and oral absorption (Yallapu et al., 2010a,b; Abu-Dahab et al., 2013). The advent of nanotechnology for pharmaceutical applications has opened a new avenue for stability, solubility, and/or permeability enhancement of problematic nutraceuticals (Yallapu et al., 2010c, 2012, 2013). The goal is targeted toward health and wellness after the mean age of the people is prolonged. Out of a proposed population of about 400 million it is predicted there would be approximately 142 million Americans over the age of 50 by halfway through the 21st century (http://www.fimdefelice.org/archives/arc.researchact.html).

The introduction of generics to the market may lower the price of a few nutraceuticals, yet people's reliance on such products and their enhancing outreach suggest that the growth of the market shall stay stable. Presently various nutritional supplements are available in the market; some of them are listed in Table 9.1.

There has been observed an appreciable ability as a delivery vehicle with controlled release capacity when submicron-level cargo-loaded emulsions have been re-encapsulated into macroscale food matrices, although the technology needs to be optimized. The encapsulation method and emulsion preparation, shelf life, and system quality assessment have been found to be interrelated. There is evidence of the procedure being employed to enhance product quality through improvement in the technique of emulsion preparation (Panagiotou and Fisher, 2012). An incorporation efficiency of up to 97% has been achieved by ionotropic gelation employing furcellaran beads to encapsulate β-carotene, obtained from sea buckthorn juice (Laos et al., 2007). Attention to exploiting cyclodextrins and cyclodextrin complexes for molecular entrapment of lipophilic bioactive cores is in process particularly in applications where other customary approaches do not execute well, or where the ultimate utility may justify the price of such a costly substance. Most of the marketable uses for cyclodextrins have been for flavor entrapment and packaging films (Szente and Szejtli, 2004). The outstanding solubility in water, good gel forming, film forming, and emulsifying features render the proteins the opportunity to be used as encapsulants (Kim and Moore, 1995; Hogan et al., 2001). Lipids are often employed as secondary coating compounds used on primary microcapsules or applied to powdered bioactive cores to enhance their moisture resistance characteristics (Wu et al., 2000).

Liquid emulsions are also an alternative set of vehicles for delivery of foods (Appelqvist et al., 2007; McClements et al., 2007). The lipids and lipid-soluble bioactive agents may be delivered by formulating them as oil-in-water emulsions. The homogenization of a mixture of either an oil or an oil having a lipid-soluble bioactive agent, with an aqueous solution having the encapsulating substance yields kinetically stable oil-in-water emulsions. Microemulsions, which are spontaneously prepared, that are thermodynamically stable may be loaded with nutraceuticals and applied as delivery systems. Water-soluble vitamins may be encapsulated in double emulsions. Fechner et al. (2007) demonstrated that vitamin B12 entrapped in the inner phase of an oil/water/oil emulsion stabilized by caseinate–dextran conjugates, in the place of pure protein, decreased the release of the vitamin in an acidic environment. The stable emulsion preparations or spray-dried emulsions are usually employed as delivery systems for lipid-soluble vitamins (e.g., vitamins A, D, E, and K) and provitamin A (β-carotene). The lipid-soluble bioactive compounds are often delivered as emulsions (McClements et al., 2007). Emulsion-based processes and spray drying are at present the strategies most generally used for microencapsulation and delivery of functional

Table 9.1 List of Nutraceutical Products and Manufacturers

Product	Category	Contents	Manufacturer	References
Calcirol D-3	Calcium supplement	Calcium and vitamins	Cadilla Healthcare Limited, Ahmedabad, India	Chauhan et al. (2013); Nelson (1999)
GRD	Nutritional supplement	Proteins, vitamins, minerals, and carbohydrates	Zydus Cadila Ltd., Ahmedabad, India	Chauhan et al. (2013); Nelson (1999)
Coral calcium	Calcium supplement	Calcium and trace minerals	Nature's Answer, Hauppauge, NY, USA	Chauhan et al. (2013); Nelson (1999)
Proteinex	Protein supplement	Predigested proteins, vitamins, minerals, and carbohydrates	Pfizer Ltd., Mumbai, India	Chauhan et al. (2013); Nelson (1999)
Amiriprash (Gold)	Good immunomodulator	Chyawannprash Avaleha, Swarnabhasma, and RasSindur	UapPharma Pvt. Ltd.	Chauhan et al. (2013)
Omega woman	Immune supplement	Antioxidants, vitamins, and phytochemicals (e.g., lycopene and resveratrol)	Wassen, Surrey, UK	Chauhan et al. (2013); Nelson (1999)
Chyawanprash	Immune booster	Amla, ashwagandha, and pippali	Dabur India Ltd.	Chauhan et al. (2013)
Celestial healthtone	Immune booster	Dry fruit extract	Celestial Biolabs Limited	Chauhan et al. (2013); Nelson (1999)
Orisett	Glucose management	Low molecular weight chitosan oligosaccharide bioactives derived from crab shells	NutraGenesis, LLC, Brattleboro, VT	(http://www.nutraceuticals-world.com)
Benolea (EFLA943)	Cardiovascular health	Benolea olive leaf extract	Frutarom USA	(http://www.nutraceuticals-world.com)
Appleactiv	Joint health, cardiovascular health	Dried apple peel powder	AppleActives and Leahy Orchards	(http://www.nutraceuticals-world.com)
Mushroom optimizer	Immune health	Mushrooms, polysaccharides, and folic acid	Jarrow Formulas, Los Angeles, CA, USA	Madley-Wright (2003)
Appetite intercept	Appetite suppressant	Caffeine, tyrosine, and phenylalanine	Natrol, Chatsworth, CA, USA	Madley-Wright (2003)

fatty acids into food (Sanguansri and Augustin, 2001; McClements et al., 2007). The use of alginate–chitosan microcapsules has also been shown to increase the mechanical strength of the capsules to endure in vitro digestion (Urbanska et al., 2007).

Targeted drug delivery has been achieved by using liposomes (Merisko-Liversidge et al., 2003; Rabinow, 2004, 2005; Panagiotou and Fisher, 2008, 2011; Panagiotou et al., 2009; Liversidge, 2011; Lasic, 1997). The selective oral delivery of dietary and nutritional supplements is now being executed by exploiting the adaptability of liposomes (Panagiotou and Fisher, 2012; Gomez-Hens and Fernandez-Romero, 2006; Mozafari et al., 2008) to refine the bio-efficacy of those agents that have demonstrated poor absorption and bioavailability rates while administered via conventional tablets and capsules. Additionally, entrapment of nutrients confined in liposomes offers a very efficient way of circumventing the degradation by the gastric system, and may also bypass recognition by the immune system, that is, cells of the reticulo-endothelial system (RES) (Mozafari and Mortazavi, 2005; Bender, 2003).

Hypertension may be prevented, its onset delayed, its severity reduced, treated, and controlled by optimal nutrition, nutraceuticals, vitamins, antioxidants, minerals, weight loss, exercise, stopping of smoking, and limiting alcohol and caffeine intake, in addition to other lifestyle alterations. Nutrients and nutraceuticals with calcium channel blocking activity (thereby, antihypertensive action) include α-lipoic acid, magnesium, vitamin B6 (pyridoxine), vitamin C, nacetyl cysteine, hawthorn, celery, ω-3 fatty acids, etc. (Houston, 2005).

Flavonoids are broadly prevalent in onion, endives, cruciferous vegetables, black grapes, red wine, grapefruits, apples, cherries, and berries (Hollman et al., 1996). Flavanoids in plants available as flavones (containing the flavonoid apigenin found in chamomile), flavanones (hesperidin—citrus fruits; silybin—milk thistle), and flavonols (tea—quercetin; kaempferol and rutin—grapefruit; rutin—buckwheat; ginkgo flavonglycosides—ginkgo) (Majoa et al., 2005) play a major role in curing the cardiovascular diseases (Cook and Samman, 1996; Hollman et al., 1999). Flavonoids inhibit the angiotensin-converting enzyme (ACE) that elevates blood pressure, inhibit the "suicide" enzyme cyclooxygenase that fragments prostaglandins, and they avert platelet stickiness and therefore platelet aggregation. Flavonoids also guard the vascular system and strengthen capillaries that convey oxygen and vital nutrients to all cells. Flavonoids suppress the enzymes that make estrogen, hence lowering the threat of estrogen-induced cancers in some (Pietinen et al., 2001; Dai et al., 2002; Boccardo et al., 2004; McCann et al., 2004) but not all studies (Kilkkinen et al., 2004; Zeleniuch-Jacquotte et al., 2004) and a prospective study was equivocal (den Tonkelaar et al., 2001).

9.3 CHALLENGES AND OPPORTUNITIES

Nutraceuticals appear to be discernible in a variety of aspects from functional foods which challenge the regulatory concepts of food and drugs (Kwak and Jukes, 2001). For nutraceutical industries, two threats are evident: regulatory ambiguity, and reliability of labeling claims. Regulatory vagueness occurs both nationally, providing the regulatory swamp in which the nutraceutical industries have survived at present, and in export markets because of the deficiency of international accord on how to label nutraceuticals (Hobbs, 2001). There exist numerous plants with widely varying degrees of the same therapeutic action. For example, in India more than 100 plants are reported to possess antidiabetic and/or hypoglycemic properties (Ajikumarannair and Subramoniam, 2005). When the ethnobotanical approach is utilized for drug discovery, information available from the traditional knowledge holder is shared. It is necessary to honor the intellectual property rights of a given cultural group or tribes or local people of a country where plants are collected based on their ethnomedical knowledge (Baker et al., 1995).

Natural products have major roles in the discovery of novel antibacterial and antifungal agents. For example, the chemical diversity of polyketides and glycopeptides has imparted unique antibacterial activities against multiple organisms. Tygacil (Tigecyclin, GAR-936) is an example of this type of antibiotics. Spices such as turmeric and garlic, vegetables such as bitter gourd, fruits such as grapes and papaya, rice bran, and gooseberry are examples of nutraceuticals. These ingredients of food have biologically active molecules also. For example, curcumin from turmeric has antioxidant, antiinflammatory, and cancer-preventive properties when consumed at reasonable levels. Too much could be harmful. Similarly, sulfur compounds in garlic extract have a hypolipidemic property. Grapes, peanuts, etc. can be considered as nutraceuticals because they contain, among other things, pharmacologically active resveratrol. Lycopene from tomatoes is reported to avert specific kinds of cancer. Herbal moieties which are not edible (not ingredients of a diet) are not nutraceuticals. Many pharma and biotech companies erroneously extended the term nutraceutical even to isolated compounds from wild plants, which are not ingredients of diet (Kamboj, 2000).

Manufacturers in general cannot merely incorporate a nutraceutical into a food while preparing functional foods that have adequate sensory appeal and desired health advantages. The adequate use of microencapsulation for stabilizing nutraceuticals facilitates their efficient delivery via food. Careful planning of the delivery system assists to guard susceptible nutraceuticals from the ambience and processing stresses faced during food manufacture, and checks undesirable interactions of the nutraceutical with elements in the food matrix (Augustin and Sanguansri, 2015).

Microfluidization, or particle collision technology, was effectively employed to prepare yogurt, soya drink, cereal bars, and chocolate fortified with phytosterols (Zawistowski et al., 2003). Stabilization of aqueous suspensions may also be attained by admixing phytosterols with lyophilic colloids (carrageenin, alginate, guar gum, xanthan gum, etc.), stabilizers like microcrystalline cellulose, and emulsifiers like polysorbates, sucrose fatty acid esters, and lecithin (Zawistowski et al., 2003), as well as protein (Moreau, 2004). Lecithin was also exploited to prepare water-dispersible stanols as emulsified micelles with potential uses in fat-free drinks (Ostlund et al., 1999). But, the higher cost of lecithin has largely limited its market uses. Another newer strategy that might address the aforementioned difficulties is to explore nanotechnology. This deals with the application of free phytosterols with very tiny particle sizes (e.g., 300 nm), in forms of nanoparticles and/or nanoemulsions. The application of nanotechnology is increasing in food products. It has been proposed that this technology can boost food processing, packaging, and stability, as well as yield improved flavor and nutrition (Weiss, 2007).

The approach in which a bioactive product is chemically linked to a polymer is well known as polymer conjugation. The conjugates modulate the physicochemical, pharmacokinetic, and therapeutic properties of the therapeutic agent. The water-soluble curcumin polyconjugates showed altered biodistribution and improved anti-cancer efficacy as they combine the dual advantages of enhanced aqueous solubility and polymer-mediated drug internalization (Safavy et al., 2007).

Another challenge to the manufacturer of food substances is the intrinsic "powderiness" of free phytosterols owing to their crystalline state. The particle size of solid components has an outstanding effect on organoleptic traits of foods, like taste and texture. Insoluble particles may add grittiness, and coarseness, which characterize in part the overall "mouth-feel" of foods. For instance, chocolate is reckoned as having a gritty texture when the cocoa powder used in the preparation has a particle size greater than 35 μm. If a minimum of 20% of the cocoa powder particles are greater than 22 μm, the chocolate texture is known as being coarse. One challenge left out is that of spreading awareness among the public to make wise selections with respect to consumed foods. The food souk is ever more varied, with individual selection being the single most significant driving force. Now it may in fact be supposed that individual consumers consume what they opt for. The food industry may play a key role in betterment of the nutritional diversity arising from the food intake patterns of the public (Erickson-Otto, 1998).

9.4 STABILITY AND SAFETY OF NUTRACEUTICALS

Nutraceutical products possess an excellent growth prospect. However, companies must take appropriate steps to optimize, develop, preserve, and protect their intellectual property rights in order to be viable (Lapointe, 2008). The judicious application of nutraceuticals is based on the objective assessment of the clinical evidence and subjective characterization of the risk, merits, economic prices, and probable drug interactions (Mechanick, 2005).

Vitamin E and β-carotene are dietary antioxidants that are present in numerous supplements. Vitamin E is generally regarded as safe, while the safety of β-carotene is heavily debated (Bast and Haenen, 2002). Phenolic materials are broadly distributed in plants. These substances are called vital antioxidants due to their capability to donate a hydrogen atom or an electron in order to form a stable radical intermediate. They check the oxidation of different biological moieties. Indeed, many oilseeds and their byproducts have been studied for phenolic materials for supply of natural antioxidants (Namiki, 1995; Jeong et al., 2004). Iron compounds that are generally recognized as safe (GRAS) have been enlisted elsewhere (Salgueiro et al., 2002). Water-soluble iron substances are more bioavailable but they are prone to accelerate oxidative reactions and possess a disagreeable metallic taste in liquid foods. Therefore, water-soluble iron materials are normally exclusively employed in low-moisture foods.

It is well known that sesame acts in several ways to maintain good health and it has been recognized for countless years that sesame oil is extremely resistant to oxidative degradation (Nagata et al., 1987; Halliwell, 1997). Its noteworthy stability is because of the presence of a huge amount of endogenous antioxidants like sesaminol, sesamol, and α-tocopherol (Niwa and Miyachi, 1986; Fukuda and Namiki, 1988; Abou-Gharbia et al., 2000). The inherent properties of alcohol cosurfactants will affect usage. In particular, alcohol chain length has a significant effect on interfacial disorder, with shorter-chained alcohols increasing interfacial disorder (flexibility) (Tabony et al., 1983; Bansal et al., 1979). Nevertheless, toxicity issues confine the application of efficient cosurfactants like 1-butanol, 2-butanol, and tert-butanol (Attwood, 1994). The commonly employed food-safe cosurfactant is ethanol. However, its application in foods confines product branding because of regulatory requirements, these varying from country to country.

Various clinical trials (phase 1 and phase 2) with reference to curcumin safety have exhibited that curcumin is safe and effective. Doses (500–1200 mg/day) were given in capsule form to the patients and safety was evaluated. Few developed adverse reactions such as diarrhea, headache, skin rashes, and yellow stool. Curcumin has also been found to block the function of drug-metabolizing enzymes such as CYP-450, glutathione S-transferase, and UDP-glucuronosyltransferase (Shehzad et al., 2012). Studies on prolonged supplementation to assess the biological impact of such nutraceutical supplements after regular consumption are in general deficient and investigations on probable adverse effects, accumulation, and toxicity are warranted. Clinical studies with healthy as well as unhealthy volunteers have exhibited huge interindividual variation and absence of uniformity in the findings. This may be attributed to the following variables: (1) differences in the chemical makeup of the nutraceutical evaluated; (2) differences in the dosage form used (pills, capsules, gels, etc.) which may influence the stability and bioavailability of the substances; and (3) physiological condition of the volunteers (Espin et al., 2007).

Herbal and synthetic cosmetic products have been developed to counter the problem of hair loss, yet synthetics have the potential to cause side effects (e.g., local irritation), whilst herbal products are generally safe. Green tea is one food-derived active ingredient with potential as a topical hair grower. Amin et al. (2014) formulated nutraceutical hair tonic(s), which is (are) stable, effective toward hair growth, and safe. Oats serve as a nutraceutical and functional food, having a high amount of well-balanced protein and soluble fiber, energy in the form of carbohydrates and oil, and many vitamins and minerals, hence, it has been known as a healthful and nutritious food. It is a rich supply of antioxidants, phytic acid, and different phenolic materials. The antioxidants are located in the outer stratum of the kernel and also assist to keep the stability of processed oat products and may stabilize oils and fats versus rancidity (Peterson, 2001).

Ramachandran and Nagarajan (2014) studied quality characteristics, nutraceutical profile, and storage stability of an aloe gel–papaya functional beverage blend and found that nutraceutical-rich aloe gel (AG) could be successfully utilized to develop functional fruit beverages with improved quality and shelf life. Attya et al. (2010) reported an unclassical way to determine the level of thermal stability of nutraceuticals in extra virgin olive oil. The current theme of innovation is manifested by the widespread application of hyphenated mass spectrometric techniques based on multistep ion analysis (Attya et al., 2010).

Several human clinical trials (six) have been carried out with AOD9604, covering two intravenous (IV) and two oral dosing pilot studies. Two oral phase IIb investigations with 300 and 500 fat adults were carried out to evaluate efficacy (body mass lowering), safety, and acceptability. In all human studies, AOD9604 was observed to be safe and well accepted (Stier et al., 2013). Amin et al. (2015) studied encapsulation of nanocarotenoids, evaluating stability and safety and the carotenoid nanoparticles were measured by using transmission electron microscopy. The microencapsulation of nanocarotenoids caused oxidation prevention and increased the thermal stability (Amin et al., 2015). Microencapsulation/nanoencapsulation is defined as a technology of packing solids, liquids, or gaseous materials in miniature, sealed capsules that can release their contents at controlled rates under specific conditions (Desai and Park, 2005). These capsules protect the encapsulated product from the light and oxygen and consequently prevent its degradation. Antioxidants can be added to systems where they might degrade, become hazardous by oxidation reactions, and change their original physical properties.

9.5 INDIAN MARKET

The Indian nutraceutical market, worth $1480 million in 2011, might rise to $2731 million in 2016, according to a recent report. As per the report by business research and consulting firm Frost & Sullivan, functional foods will be the fastest mounting class followed by dietary supplements until 2015. The report described that the dietary supplements were the largest group, making up 64% of the nutraceuticals market. This market is governed prima facie by the pharmaceutical field in the form of vitamin and mineral supplements (Brower, 1998; Shashank, 2006; Scarlett, 1998). Nutraceuticals are priced high with wider profit margins than traditional foods, yielding a great motivation for companies to come into this market (Agriculture and Agri-Food Canada, 2001). The ensuing research will result in a new generation of foods that will definitely lie at the interface between food and drugs, which will turn out to be progressively more permeable (Andlauer and Furst, 2002).

9.6 GLOBAL REGULATORY MARKET

As per the study, the global nutraceutical market was anticipated to be $149.5 billion in 2011 with the United States, Europe, and Japan being the largest regional markets, making up approximately 93% of the global nutraceutical needs (Brower, 1998;

Shashank, 2006; Scarlett, 1998). The way to control and marketing of nutraceuticals is remarkably diverse globally. This is largely because of the challenges in categorizing such products, lack of an appropriate regulatory class for these hybrid products, and differing opinions on what is assumed to be satisfactory scientific confirmation to wind up their functionality (Kotilainen et al., 2006). At this stage, there are no regulations and no regulatory procedures which define and overtly address nutraceuticals. There is no sign of if and when specific regulations for nutraceuticals will come into existence.

Some Asian countries such as South Korea, Malaysia, Indonesia, Thailand, the Philippines, and Singapore have approved phytosterol-containing foods. The main marketed food product of such type in the Pacific Rim countries is milk for heart health by Nestlé (Zawistowski, 2008). On the contrary, under EU and US regulations, the daily consumption of phytosterols may be divided across one, two, or three servings of food (De Figueiredo and Lajolo, 2008).

9.7 CURRENT SCENARIO

The Food Safety and Standard Rules, 2011 have come into force in India, effective from May 5, 2011. The Food Safety and Standards Authority has also provided regulations about licensing and registration of food businesses, packing and labeling, food products standards and additives, etc. These acts, rules, and regulations have been executed from Aug. 5, 2011. Hence, there is a single legislation and specified authority to control the manufacture, sale, and distribution of nutraceuticals, functional foods, and dietary supplements in India. Nevertheless, because of the absence of simplicity of certain regulations for registration of nutraceuticals and permitted additives, entrepreneurs purporting to introduce nutraceuticals in India are still experiencing challenges. To surmount these obstacles, it would be essential to amend Schedule K of the Drugs and Cosmetics Rules, 1945 to offer specific immunity to nutraceuticals, dietary supplements, and health supplements from the scope of the Drugs and Cosmetics Act, 1940 and Rules, 1945. It is also imperative to have specific regulations for brand approval, approval of claims, permitted additives, amount of vitamins and minerals, etc. for nutraceuticals as it is essential to take care of this section as an independent and unique unit under the Food Safety and Standards Act, 2006 (Bhowmik et al., 2013).

9.8 FUTURE PROSPECTS

Nutraceuticals have gained significant focus due to their presumed safety and promising nutritional and therapeutic outcomes. Pharmaceutical and nutritional companies are aware of the monetary success availing those providing advantage for the many health-seeking consumers and the altering patterns leading to a propagation of these value-added products purported to address conditions from heart health to cancer (Rajasekaran et al., 2008). It is apparent that the prospects of nutraceuticals have already spurred the scientific community into discovering new compounds that guarantee to lengthen healthy life (Andlauer and Furst, 2002). Numerous scientists consider that enzymes manifest another fascinating front line in nutraceuticals. Enzymes have been underutilized so far. They are going to be a potential area in the future. Fermentation technology employing microbes to generate new food products also appears promising (Pandey et al., 2010). Nutraceuticals have constituted a sharply growing area of attention for research, product development, consumer interest, and regulatory endeavors in recent years. Nutraceuticals offer a unique intersection of the pharmaceutical and food industries with a broad scope (Palthur et al., 2010).

9.9 CONCLUSIONS

It was the progress in deciphering the connection between nutrition and health, usually at the molecular level, that resulted in the concept of "nutraceuticals" as a practical and new approach to attain optimal health and probably lower the danger of disease. A number of industries manufacture and market the nutraceuticals, where the side effects (in particular when consumed in high quantities) of these nutraceuticals are not accounted for or generally not documented. In order to provide scientific knowledge pertaining to nutraceuticals, the general public should be made aware, where recommended daily doses of these nutraceuticals should be known to every consumer. The major concerns about the nutraceutical product are how to prepare it, its stability in different conditions, and safety. Scientists and researchers are preparing nutraceutical products because of the demand in the Indian and international markets. Presently nutraceuticals are drawing attention as being beneficial in treating diseases like obesity, diabetes, cancer, coronary heart disease, osteoporosis, and other chronic and degenerative diseases like Parkinson's and Alzheimer's diseases. Thus it can be concluded here that nutraceutical products are greatly needed by people worldwide for the management and prevention of diseases.

REFERENCES

Abou-Gharbia, H.A., Shehata, A.A.Y., Shaidi, F., 2000. Effects of processing on oxidative stability and lipid classes of sesame oil. Food Res. Int. 33, 331–340.

Abu-Dahab, R., Odeh, F., Ismail, S., Azzam, H., AlBawab, A., 2013. Preparation, characterization and antiproliferative activity of thymoquinone-beta-cyclodextrin self assembling nanoparticles. Pharmazie 68, 939–944.

Agriculture and Agri-Food Canada (AAFC), 2001. Potential Benefits of Functional Foods and Nutraceuticals to the Agri-Food Industry in Canada. AAFC, Canada.

Ahmad, A., Ginnebaugh, K.R., Li, Y., Padhye, S.B., Sarkar, F.H., 2015. Molecular targets of naturopathy in cancer research: bridge to modern medicine. Nutrients 7, 321–334.

Ajikumarannair, S., Subramoniam, A., 2005. Indian medicinal plants with anti-diabetes properties. In: Khan, I.R., Khanum, A., Khan, A.A. (Eds.), Modern and Alternative Medicine for Diabetes. Ukaaz Publications, Hyderabad, India, pp. 43–193.

Amin, J., Simamora, E.L.P., Anwar, E., Djajadisastra, J., 2014. Green tea (*Camellia sinensis*, L.) ethanolic extract as hair tonic in nutraceutical: physical stability, hair growth activity on rats, and safety test. Int. J. Pharm. Pharma. Sci. 6 (5), 94–99.

Amin, A.A., Salama, M.F.S Effat I.S., Mahmoud, K.F., 2015. Encapsulation of nano carotenoids; evaluation of stability and safety. Int. J. Curr. Microbiol. App. Sci 4 (2), 1017–1029.

Anand, P., Kunnumakkara, A.B., Newman, R.A., Aggarwal, B.B., 2007. Bioavailability of curcumin: problems and promises. Mol. Pharm. 4, 807–818.

Andlauer, W., Furst, P., 2002. Nutraceuticals: a piece of history, present status and outlook. Food Res. Int. 35, 171–176.

Appelqvist, I.A.M., Golding, M., Vreeker, R., Zuidam, N.J., 2007. Emulsion as delivery systems in foods. In: Lakkis, J.M. (Ed.), Encapsulation and Controlled Release Technologies in Food Systems. Blackwell, Ames, IA, pp. 1–80.

Attwood, D., 1994. Microemulsions as drug delivery systems. Marcel Dekker, New York.

Attya, M., Benabdelkamel, H., Perri, E., Russo, A., Sindona, G., 2010. Effects of conventional heating on the stability of major olive oil phenolic compounds by tandem mass spectrometry and isotope dilution assay. Molecules 15, 8734–8746.

Augustin, M.A., Sanguansri, L., 2015. Challenges and solutions to incorporation of nutraceuticals in foods. Annu. Rev. Food Sci. Technol. 6, 463–477.

Bachstetter, A.D., Jernberg, J., Schlunk, A., Vila, J.L., Hudson, C., Cole, M.J., et al., 2010. Spirulina promotes stem cell genesis and protects against LPS induced declines in deural stem cell proliferation. PLoS ONE 5 (5), e10496.

Baker, J.T., Borris, R.P., Carte, B., Cordell, G.A., Soejarto, D.D., Cragg, G.M., Gupta, M.P., Iwu, M.M., Madulid, D.R., Tyler, V.E., 1995. Natural products drug discovery and development: new perspectives on international collaboration. J. Nat. Prod. 58, 1325–1357.

Bansal, V., Chinnaswamy, K., Ramachandran, C., Shah, D., 1979. J. Colloid Interface Sci., 524–537.

Bast, A., Haenen, G.R.M.M., 2002. The toxicity of antioxidants and their metabolites. Environ. Toxicol. Pharmacol. 11, 251–258.

Bender, D.A., 2003. Nutritional Biochemistry of the Vitamins. Cambridge University Press, Cambridge, UK.

Bhowmik, D., Gopinath, H., Kumar, B.P., Duraivel, S., Sampath, K.P., 2013. KumarNutraceutical– a bright scope and opportunity of Indian healthcare market. Pharma Innov. 1 (11), 29–41.

Biesalski, H.K., 2001. Nutraceuticals: the link between nutrition and medicine. In: Kramer, K., Hoppe, P.P., Packer, L. (Eds.), Nutraceuticals in Health and Disease Prevention. Marcel Deckker Inc, New York, pp. 1–26.

Boccardo, F., Lunardi, G., Guglielmini, P., Parodi, M., Murialdo, R., Schettini, G., Rubagotti, A., 2004. Serum enterolactone levels and the risk of breast cancer in women with palpable cysts. Eur. J. Cancer 40 (1), 84–89.

Boozer, C.N., Nasser, J.A., Heymsfield, S.B., 2001. An herbal supplement containing ma huang-guarana for weight loss: a randomized, double-blind trial. Int. J. Obes. Relat. Metab. Disord. 25, 316–324.

Brower, V., 1998. Nutraceuticals: poised for a healthy slice of the healthcare market? Nat. Biotechnol. 16, 728–731.

Chauhan, B., Kumar, G., Kalam, N., Ansari, S.H., 2013. Current concepts and prospects of herbal nutraceutical: a review. J. Adv. Pharm. Tech. Res. 4 (1), 4–8.

Chen, S., Zhao, Q., Ferguson, L.R., Shu, Q., Weir, I., Garg, S., 2012. Development of a novel probiotic delivery system based on microencapsulation with protectants. Appl. Microbiol. Biotechnol. 93, 1447–1457.

Choudhary, M., Tomer, V., 2013. Cardioprotective effect of nutraceuticals—the emerging evidences. Proc. Indian Natl. Sci. Acad. 79 (4), 985–996.

Cook, N.C., Samman, S., 1996. Flavonoids—chemistry, metabolism, cardioprotective effects, and dietary sources. J. Nutr. Biochem. 7, 66–76.

Cook, M.T., Tzortzis, G., Charalampopoulos, D., Khutoryanskiy, V.V., 2012. Microencapsulation of probiotics for gastrointestinal delivery. J. Control Rel. 162, 56–67.

Dai, Q., Franke, A.A., Jin, F., Shu, X.O., Hebert, J.R., Custer, L.J., Cheng, J., Gao, Y.T., Zheng, W., 2002. Urinary excretion of phytoestrogens and risk of breast cancer among Chinese women in Shanghai. Cancer Epidemiol. Biomarkers Prev. 11, 815–821.

Davis, P.F., Greenhill, N.S., Rowan, A.M., Schollum, L.M., 2007. The safety of New Zealand bovine colostrum: nutritional and physiological evaluation in rats. Food Chem. Toxicol. 45 (2), 229–236.

Davison, G., Diment, B.C., 2010. Bovine colostrum supplementation attenuates the decrease of salivary lysozyme and enhances the recovery of neutrophil function after prolonged exercise. Brit. J. Nutr. 103 (10), 1425–1432.

De Figueiredo, M.C., Lajolo, F.M., 2008. Supplements and functional foods legislation in Brazil. In: Bagchi, D. (Ed.), Nutraceutical and Functional Food Regulations in the United States and Around the World. Academic Press, New York, pp. 349–364.

den Tonkelaar, I., Keinan-Boker, L., Veer, P.V., Arts, C.J., Adlercreutz, H., Thijssen, J.H., Peeters, P.H., 2001. Urinary phytoestrogens and postmenopausal breast cancer risk. Cancer Epidemiol. Biomarkers Prev. 10 (3), 223–228.

Desai, K.G.H., Park, H.J., 2005. Preparation of cross-linked chitosan microspheres by spray drying: effect of cross-linking agent on the properties of spray dried microspheres. J. Microencapsul. 22 (4), 377–395.

Erickson-Otto, P., 1998. Phytonutrients: putting a dent in disease. Food Prod. Des. 8, 16–23.

Espin, J.C., Garcia-Conesa, M.T., Tomas-Barberan, F.A., 2007. Nutraceuticals: facts and fiction. Phytochemistry 68 (22–24), 2986–3008.

Fechner, C., Knoth, A., Scherze, I., Muschiolik, G., 2007. Stability and release of double-emulsions stabilised by caseinsate-dextran conjugates. Food Hydrocoll. 21, 943–951.

Fukuda, Y., Namiki, M., 1988. Recent studies on sesame seed and oil. Nippon Shokuhin Kogyo Gakkaishi 35, 552–562.

Gbassi, G.K., Vandamme, T., 2012. Probiotic encapsulation technology: from microencapsulation to release into the gut. Pharmaceutics 4, 149–163.

Gomez-Hens, A., Fernandez-Romero, J.M., 2006. Analytical methods for the control of liposomal delivery systems. Trends Anal. Chem. 25, 167–178.

Hafez, M.M., Maghrabi, I.A., Zaki, N.M., 2013. Toward an alternative therapeutic approach for skin infections: antagonistic activity of lactobacilli against antibiotic-resistant staphylococcus aureus and pseudomonas aeruginosa. Probiotics Antimicrob. Proteins 5 (3), 216–226.

Halliwell, B., 1997. Antioxidants and human diseases. Nutr. Rev. 55, S44–S52.

Hobbs, J.E., 2001. Developing Supply Chains for Nutraceuticals and Functional Foods: Opportunities and Challenges. Institute of Nutraceuticals and Functional Foods, Centre for Research in the Economics of Agrifood, Université Laval, Québec.

Hogan, S.A., McNamee, B.F., O'Riordan, E.D., O'Sullivan, M., 2001. Microencapsulating properties of whey protein concentrate 75. J. Food Sci. 66, 675–680.

Hollman, P.C.H., Hertog, M.G.L., Katan, M.B., 1996. Analysis and health effects of flavonoids. Food Chem. 57, 43–46.

Hollman, P.C.H., Feskens, E.J., Katan, M.B., 1999. Tea flavonols in cardiovascular disease and cancer epidemiology. Proc. Soc. Exper. Biol. Med. 220, 198–202.

Houston, M.C., 2005. Nutraceuticals, vitamins, antioxidants, and minerals in the prevention and treatment of hypertension. Prog. Cardiovas. Dis. 47, 396–449.

Jeong, S.M., Kim, S.Y., Kim, D.R., Nam, K.C., Ahn, D.U., Lee, S.C., 2004. Effects of seed roasting conditions on the antioxidant activity of defatted sesame meal extracts. J. Food. Sci. 69, 377–381.

Kalra, E.K., 2003. Nutraceutical—definition and introduction. AAPS Pharm. Sci. 5, 1–2.

Kamboj, V.K., 2000. Herbal medicine. Curr. Sci. 78, 35–39.

Kaur, G., Mukundan, S., Wani, V., Kumar, M.S., 2015. Nutraceuticals in the management and prevention of metabolic syndrome. Austin J. Pharmacol. Ther. 3 (1), 1063.

Keservani, R.K., Kesharwani, R.K., Vyas, N., Jain, S., Raghuvanshi, R., Sharma, A.K., 2010. Nutraceutical and functional food as future food: a review. Der Pharmacia. Lett. 2, 106–116.

Keservani, R.K., Kesharwani, R.K., Sharma, A.K., Ahmad, M.F., 2015a. Stress, Parkinson's, and Alzheimer's disease: role of dietary supplements. In: Bagchi, D., Preuss, H.G., Swaroop, A. (Eds.), Nutraceuticals and Functional Foods in Human Health and Disease Prevention. CRC Press, Taylor and Francis, Abingdon, UK, pp. 241–254.

Keservani, R.K., Singh, S., Singh, V., Kesharwani, R.K., Sharma, A.K., 2015b. Nutraceuticals and functional foods in the prevention of mental disorders. In: Bagchi, D., Preuss, H.G., Swaroop, A. (Eds.), Nutraceuticals and Functional Foods in Human Health and Disease Prevention. CRC Press, Taylor and Francis, Abingdon, UK, pp. 255–270.

Keservani, R.K., Kesharwani, R.K., Sharma, A.K., 2015c. Pulmonary and respiratory health: antioxidants and nutraceuticals. In: Bagchi, D., Preuss, H.G., Swaroop, A. (Eds.), Nutraceuticals and Functional Foods in Human Health and Disease Prevention. CRC Press, Taylor and Francis, Abingdon, UK, pp. 279–298.

Keservani, R.K., Kesharwani, R.K., Sharma, A.K., Jarouliya, U., 2015d. Dietary supplements, nutraceutical and functional foods in immune response (immunomodulators). In: Bagchi, D., Preuss, H.G., Swaroop, A. (Eds.), Nutraceuticals and Functional Foods in Human Health and Disease Prevention. CRC Press, Taylor and Francis, Abingdon, UK, pp. 343–360.

Kesharwani, R.K., Srivastava, V., Singh, P., Rizvi, S.I., Adeppa, K., Misra, K., 2015. A novel approach for overcoming drug resistance in breast cancer chemotherapy by targeting new synthetic curcumin analogues against aldehyde dehydrogenase 1 (ALDH1A1) and glycogen synthase kinase-3 β (GSK-3β). Appl. Biochem. Biotechnol. 176 (7), 1996–2017.

Kilkkinen, A., Virtamo, J., Vartiainen, E., Sankila, R., Virtanen, M.J., Adlercreutz, H., Pietinen, P., 2004. Serum enterolactone concentration is not associated with breast cancer risk in a nested case–control study. Int. J. Cancer 108 (2), 277–280.

Kim, Y.D., Moore, C.V., 1995. Encapsulating properties of several food proteins. IFT Annual Meeting Poster, p. 193.

Kotilainen, L., Rajalahti, R., Ragasa, C., Pehu, E., 2006. Health Enhancing Foods: Opportunities for Strengthening the Sector in Developing Countries. International Bank for Reconstruction and Development, Washington, DC.

Krissansen, G.W., 2007. Emerging health properties of whey proteins and their clinical implications. J. Am. Coll. Nutr. 26 (6), 713S–723S.

Kwak, N.S., Jukes, D.J., 2001. Functional foods. Part 1: the development of a regulatory concept. Food Control 12, 99–107.

Laos, K., Lougas, T., Mandmets, A., Vokk, R., 2007. Encapsulation of β-carotene from sea buckthorn (*Hippophae rhamnoides* L.) juice in furcellaran beads. Innov. Food Sci. Emerg. Technol. 8, 395–398.

Lapointe, S., 2008. Food for thought: IP protection for nutraceuticals and functional foods. Health Law Can. 28 (3–4), 101–111.

Lasic, D.D., 1997. Liposomes in Gene Delivery. CRC Press LLC, Boca Raton, FL.

Latif, S., Anwar, F., Ashraf, M., Gilani, A.H., 2007. Moringa oleifera: a food plant with multiple medicinal uses. Phytother. Res. 21, 17–25.

Liversidge, G.G., 2011. Controlled release and nanotechnologies: recent advances and future opportunities. Drug Dev. Deliv. 11, 1–7.

Madley-Wright, R., 2003. New products. Nutraceutical World 6, 66–71.

Majoa, D.D., Guardiaa, M.L., Tripolia, E., Giammancoa, S., Finotti, E., 2005. Flavonoids as inhibitors of lipid peroxidation in membranes. Food Res. Int. 38, 1161–1166.

McCann, S.E., Muti, P., Vito, D., Edge, S.B., Trevisan, M., Freudenheim, J.L., 2004. Dietary lignan intakes and risk of pre- and postmenopausal breast cancer. Int. J. Cancer 111 (3), 440–443.

McClements, D.J., Decker, E.A., Weiss, J., 2007. Emulsion based delivery systems for lipophilic bioactive components. J. Food Sci. 72, R109–R124.

Mechanick, J.I., 2005. The rational use of dietary supplements and nutraceuticals in clinical medicine. Mt Sinai J. Med. 72 (3), 161–165.

Merisko-Liversidge, E., Liversidge, G., Cooper, E., 2003. Nanosizing: a formulation approach for poorly-water-soluble compounds. Eur. J. Pharm. Sci. 18, 113–123.

Mikirova, N., Jackson, J., Hunninghake, R., Kenyon, J., Chan, K., Swindlehurst, C., et al., 2009. Circulating endothelial progenitor cells: a new approach to anti-aging medicine? J. Transl. Med. 7 (1), 106.

Moreau, R.A., 2004. Plant sterols in functional foods. In: Dutta, P.C. (Ed.), Phytosterols as Functional Food Components and Nutraceuticals. Marcel Dekker, New York, pp. 317–345.

Mozafari, M.R., Mortazavi, S.M., 2005. Nanoliposomes: From Fundamentals to Recent Developments. Trafford Publishing Ltd, Oxford, UK.

Mozafari, M.R., Johnson, C., Hatziantoniou, S., Demetzos, C., 2008. Nanoliposomes and their applications in food nanotechnology. J. Liposome Res. 18 (4), 309–327.

Nagata, M., Osawa, T., Namiki, M., 1987. Stereochemical structures of antioxidative sesaminol. Agric. Biol. Chem. 51, 1285–1289.

Namiki, M., 1995. The chemistry and physiological functions of sesame. Food. Rev. Int. 11, 281–329.

Nelson, N.J., 1999. Purple carrots, margarine laced with wood pulp? Nutraceuticals move into the supermarket. J. Natl. Cancer Inst. 91 (9), 755–757.

Niwa, Y., Miyachi, Y., 1986. Antioxidant action of natural health products and Chinese herbs. Inflammation 10, 79–91.

Ostlund, Jr., R.E., Spilburg, C.A., Stenson, W.F., 1999. Sitostanol administration in lecithin micelles potently reduces cholesterol absorption in humans. Am. J. Clin. Nutr. 79, 826–831.

Palthur, M.P., Palthur, S.S.S., Chitta, S.K., 2010. Nutraceuticals: concept and regulatory scenario. Int. J. Pharm. Pharma. Sci. 2 (2), 14–20.

Panagiotou, T., Fisher, R.J., 2008. Form nano-particles via controlled crystallization: a "bottom-up" approach. Chem. Eng. Prog. 104 (10), 33–39.

Panagiotou, T., Fisher, R.J., 2011. Enhanced transport capabilities via nanotechnologies: impacting bio-efficacy, controlled release strategies and novel chaperones. J. Drug Deliv., 1–14, ID 902403.

Panagiotou, T., Fisher, R.J., 2012. Improving product quality with entrapped stable emulsions: from theory to industrial application. Challenges 3, 1–39.
Panagiotou, T., Mesite, S., Fisher, R.J., 2009. Production of norfloxacin nano-suspensions using microfluidics reaction technology (MRT) through solvent/anti-solvent crystallization. Ind. Eng. Chem. Res. 48 (4), 1761–1771.
Pandey, M., Verma, R.K., Saraf, S.A., 2010. Nutraceuticals: new era of medicine and health. Asian J. Pharmaceut. Clin. Res. 3 (1), 11–15.
Peterson, D.M., 2001. Oat antioxidants. J. Cereal Sci. 33 (2), 115–129.
Pietinen, P., Stumpf, K., Männistö, S., Kataja, V., Uusitupa, M., Adlercreutz, H., 2001. Serum enterolactone and risk of breast cancer: a case–control study in Eastern Finland. Cancer Epidemiol. Biomarkers Prev. 10, 33.
Rabinow, B.E., 2004. Nanosuspensions in drug delivery. Nat. Rev. Drug Discov. 3 (9), 785–796, 2004.
Rabinow, B.E., 2005. Pharmacokinetics of nanosuspensions, proceedings of the nanotechnology for drug delivery conference, Philadelphia, PA.
Rajasekaran, A., Sivagnanam, G., Xavier, R., 2008. Nutraceuticals as therapeutic agents: a review. Res. J. Pharm. Tech. 1 (4), 328–340.
Ramachandran, P., Nagarajan, S., 2014. Quality characteristics, nutraceutical profile, and storage stability of aloe gel-papaya functional beverage blend. Int. J. Food Sci. 847013, 1–7.
Rissanen, T.H., Voutilainen, S., Virtanen, J.K., Venho, B., Vanharanta, M., Mursu, J., Salonen, J.T., 2003. Low intake of fruits, berries and vegetables is associated with excess mortality in men: the Kuopio Ischaemic Heart Disease risk factor (KIHD) study. J. Nutr. 133, 199–204.
Rusu, D., Drouin, R.J., Pouliot, Y., Gauthier, S., Poubelle, P.E., 2010. A bovine whey protein extract stimulates human neutrophils to generate bioactive IL-1Ra through a NF-kappaB- and MAPK-dependent mechanism. J. Nutr. 140 (2), 382–391.
Safavy, A., Raisch, K.P., Mantena, S., Sanford, L.L., Sham, S.W., et al., 2007. Design and development of water-soluble curcumin conjugates as potential anticancer agents. J. Med. Chem. 50, 6284–6288.
Salgueiro, M.J., Zubillaga, M., Lysionek, A., Caro, R., Weill, R., Boccio, J., 2002. Fortification strategies to combat zinc and iron deficiency. Nutr. Rev. 60 (2), 52–58.
Sanguansri, L., Augustin, M.A., 2001. Encapsulation of food ingredients. WO 01/74175A1.
Scarlett, T., 1998. How modernized is FDA now? FDLI Update 3, 1.
Shashank, B., 2006. Capacity building in the Indian food industry: opportunities and challenges. In: Proc. Ind. Symp., CFTRI Mysore (India), June 23–25, 2006.
Shehzad, A., Khan, S., Sup Lee, Y., 2012. Curcumin molecular targets in obesity and obesity-related cancers. Future Oncol. 8, 179–190.
Shytle, R.D.J.E., Tan, J., Vila, J., Cole, M., Sanberg, C.D., Sanberg, P.R., Bickford, P.C., 2006. Oxidative stress of neural, hematopoietic, and stem cells: protection by natural compounds. Rejuvenation Res. 10 (2), 173–178.
Singh, U., Singh, S., Kochhar, A., 2012. Therapeutic potential of antidiabetic nutraceuticals. Phytopharmacology 2 (1), 144–169.
Stelwagen, K., Carpenter, E., Haigh, B., Hodgkinson, A., Wheeler, T.T., 2009. Immune components of bovine colostrum and milk. J. Anim. Sci. 87 (13), 3–9.
Stier, H., Vos, E., Kenley, D., 2013. Safety and tolerability of the hexadecapeptide AOD9604 in humans. J. Endocrinol. Metab. 3 (1–2), 7–15.
Szente, L., Szejtli, J., 2004. Cyclodextrins as food ingredients. Trends Food Sci. Techno. 15, 137–142.

Tabony, J., Llor, A., Drifford, M., 1983. Colloid Polym. Sci., 938–946.
Temple, W.J., Gladwin, K.K., 2003. Fruits, vegetables, and the prevention of cancer: research challenges. Nutrition 19, 467–470.
Tokuyama, H., Tokuyama, Y., Migita, S., 1990. Isolation of two new proteins from bovine colostrums which stimulate epidermal growth factor-dependent colony formation of NRK-49F cells. Growth Factors 3 (2), 105–114.
Urbanska, A.M., Bhathena, J., Prakash, S., 2007. Live encapsulated Lactobacillus acidophilus cells in yoghurt for therapeutic oral delivery: preparation and in-vitro analysis of alginate-chitosan microcapsules. Can. J. Physiol. Pharm. 85, 884–893.
Weiss, J., 2007. Nanotechnological food applications—with a focus on novel delivery system for bioactives. In: Nanotechnology Short Course. IFT, Chicago, IL.
Wu, W.-H., Roe, W.S., Gimino, V.G., Seriburi, V., Martin, D.E., Knapp, S.E., 2000. Lowmelt encapsulation with high laurate canola oil. US 6,153,236.
Yallapu, M.M., Jaggi, M., Chauhan, S.C., 2010a. beta-cyclodextrin-curcumin self-assembly enhances curcumin delivery in prostate cancer cells. Colloids Surf. B Biointerfaces 79, 113–125.
Yallapu, M.M., Jaggi, M., Chauhan, S.C., 2010b. Poly(Î²-cyclodextrin)/curcumin self-assembly: a novel approach to improve curcumin delivery and its therapeutic efficacy in prostate cancer cells. Macromol. Biosci. 10, 1141–1151.
Yallapu, M.M., Jaggi, M., Chauhan, S.C., 2010c. Scope of nanotechnology in ovarian cancer therapeutics. J. Ovarian Res. 3, 19.
Yallapu, M.M., Jaggi, M., Chauhan, S.C., 2012. Curcumin nanoformulations: a future nanomedicine for cancer. Drug Discov. Today 17, 71–80.
Yallapu, M.M., Jaggi, M., Chauhan, S.C., 2013. Curcumin nanomedicine: a road to cancer therapeutics. Curr. Pharm. Des. 19, 1994–2010.
Zawistowski, J., 2008. Regulations of functional foods in selected Asian countries in the Pacific Rim. In: Bagchi, D. (Ed.), Nutraceutical and Functional Food Regulations in the United States and Around the World. Academic Press, New York, pp. 365–401.
Zawistowski, J., Kitts, D.D., Stoynov, N., Du, K., 2003. Thermal oxidation behavior of a phytosterol mixture in different oil systems. 94th American Oil Chemists' Society Annual Meeting, 4–7 May, Kansas City, MI.
Zeleniuch-Jacquotte, A., Adlercreutz, H., Shore, R.E., Koenig, K.L., Kato, I., Arslan, A.A., Toniolo, P., 2004. Circulating enterolactone and risk of breast cancer: a prospective study in New York. Br. J. Cancer. 91 (1), 99–105.

CHAPTER

Sourcing, supply chain, and manufacturing of nutraceutical and functional foods

10

Sourcing Datta
Supply Chain Manager at eBay, San Jose, CA, United States

10.1 INTRODUCTION STATEMENT

The supply chain has had such an important effect on organizations that it has made companies millions and pushed them ahead of other companies and in some cases irregular supply chain processes have forced companies to go out of business. In the food and beverage industry, a constant flow of material is very important for a company's success. As rightly put by Thomas Friedman:

> *Supply chains cannot tolerate even 24 hours of disruption. So if you lose your place in the supply chain because of wild behavior you could lose a lot. It would be like pouring cement down one of your oil wells. (Supply Chain Quotes, 2016)*

10.2 SCOPE AND OPPORTUNITY

The start of the development of a new product is very interesting for a company. Before starting manufacturing of a new product, the internal team researches which product they would want to make, market conditions are determined, and current situations are analyzed. Based on the analysis, the internal team works with a separate SWOT team to determine which nutraceutical product the company would look to make. It might so happen that the one year is favorable for products that would cure arthritis and the next year could be more favorable for curing diabetes.

As per the National Institute of Diabetes and Digestive and Kidney Diseases, in 2015, 31.2% of adults in the United States have a body mass index (BMI) less than 24.9 and hence are considered underweight or normal weight (Health Statistics Information, 2015).

However, it's frightening when we see that the other 68.8% are overweight. Out of the 68.8%, 33.1% have BMIs from 25 to 29.9 and thus belong to the category of

Developing New Functional Food and Nutraceutical Products. http://dx.doi.org/10.1016/B978-0-12-802780-6.00010-9

overweight. 35.7% have BMIs equal to or greater than 30 and are considered to be obese. At present, stress and anxiety are not helping the current situation and only making it worse. Add to this the scenario of diabetes.

There could be other substitutes to medicines, which alone could have a cure to various problems. In order to control diabetes and weight gain, certain ingredients are either used to create different forms of drinks or even sold in powder form. These new forms of food have various advantages: (1) ease of intake, (2) efficacy in weight control, and (3) cost of product. These products are traditionally much lower in cost than other medicines and thus have a greater advantage when it comes to customers deciding which options to take.

The greatest advantage of nutraceutical products is the aftermarket. They could either be sold directly to the customer as end products or also could be sold to other health drink manufacturing companies as semifinished products.

10.3 MANUFACTURING

10.3.1 PRECONDITIONS FOR MANUFACTURING OF NUTRACEUTICALS/FUNCTIONAL FOODS

If the raw materials are produced outside of United States soil, then it's very important to keep in a mind a few important points. The details are described next:

1. One needs to ensure preapproved US Customs clearance for importing goods from outside. Preapproved customs clearance is beneficial since it helps in smooth movement in the product flow once the production is set and the manufacturing process is in a steady state of operation.
2. It is important to know foreign policies, rules, and regulations so the imports of products are not stopped and there are no unforeseen issues in the middle of the cycle.
3. It is very important to know the list of prohibited items in the product that are not allowed in the United States.
4. It is important to do the cost analysis of producing locally versus producing in another country and then bringing in the finished goods.
5. Customs duty could be an added factor to the cost since products are imported.
6. Check for restrictions on imports—not only on the type of food but on the form as well, for example, restrictions vary whether the product is in liquid or powder or solid form.
7. The available capacity of the vendor to produce both quantity and quality along with a defined response time.
8. The total time required from placing an order to receiving the semifinished goods from the overseas supplier.
9. The shipment time from overseas and local vendors and the difference between the two.
10. The total life of the imported versus domestic items. Items that are imported must have a life longer than the domestic product.

11. The following permissions/clearances are also required:
    a. Pollution clearance from the EPA (Environmental Protection Agency).
    b. License is also required from the DSHEA (Dietary Supplement Health and Education Act).
    c. It is important to have permission from the USFDA (United States Food & Drug Administration) for good manufacturing practices.
    d. For better acceptability of the product in the market, permission and clearance are required to use the logo of the NSF (National Supplemental Foundation).
    e. The manufacturing unit along with all employee registration with the IRS for federal tax and state department taxes are also mandatory.
    f. In order to employ and engage any employee, permission from the INS (Immigration and Naturalization Service) Department is mandatory.
    g. Obtain permission from the Labor Department for W2, that is, Authorization to work or a Work Authorization Permit.
    h. Similar permission and clearance levels are required from other countries where the final product would be sold.

10.3.2 ACTIVITIES OF VARIOUS DEPARTMENTS FOR MANUFACTURING OF PRODUCT

The different departments in the organization are responsible for different actions and detailed descriptions of each of the departments are given. The different tasks are:

planning,
procurement/buyer,
quality check for all incoming materials,
raw material warehousing,
manufacturing/execution,
process control and online quality check,
quality check before final packing,
packaging,
distribution to warehouse,
manpower and safety measures, and
effective pollution control measures.

10.3.3 PLANNING

After assessing the proper demand in terms of quality as well as quantity required to sell to various health companies, a raw materials breakup is required. The supply chain activities will start from this point. The raw material planning is one of the most important criteria for both cost control and quality output and needs to be performed in a very streamlined manner. Planning needs to have the following details captured:

1. Quantity of each item/raw material and the total quantity.
2. Shelf life for each item.
3. Storage facility.
4. Lead time of each item if they need to be procured from different location/facility.
5. Material handling facility (it becomes important since quality measures and control can be achieved through proper material handling steps).
6. Availability of storage bin or racks.
7. Thorough analysis of safety of environment and employee, good health care control, and preventive measures are a must or it might lead to factory shutdown (impact on factory shutdown time must be also captured).
8. Insurance facilities for material, environment, and manpower.

10.3.4 PROCUREMENT/BUYER

The buyer will now start the activities for procurement of all materials based on information received from the planning team. The buyer will perform the following functions after discussion with the planning team:

1. Assessment of all materials required as informed by the planning team.
2. Analysis of economic order quantity (EOQ) for all material.
3. Identify multiple vendors and assess the rating of each vendor about supply, quality, delivery, and response.
4. Analysis of lead time for each vendor and commodity.
5. Ensure material is accompanied with quality/safety/health test report from approved laboratories and verify all parameters are within standard limit and if quality issues arise, steps need to be specified for correction if quality standards from vendors are not met as mentioned in the relevant purchase order.
6. Penalty/demurrage clause in case of delay or nondelivery.
7. Arbitration clause in case of dispute.
8. Indication of delivery mode and insurance applicability during in-transit time.
9. Ensuring each material's shelf life period.
10. Placeholder in the purchase order about type and modes of packaging if necessary.

10.3.5 QUALITY/QUANTITY CHECK

The quality department must have a manual for checking the incoming material. An example of a detailed checklist that could be used is as follows:

1. A manual/handbook must be prepared for all incoming material, where the detailed procedure related to each parameter, process, and method of checking must be clearly noted.
2. As soon as the incoming material is reported to the factory/warehouse location, immediately the verification needs to be started by the security personnel, and it should start from checking whether the material is accompanying the purchase

order. The quality, safety, and health report cards along with the requisite quantity as mentioned in the purchase order must be thoroughly checked.
3. The quality department then comes into the picture. The quality control personnel will physically inspect the material and thereafter either randomly should draw a sample or draw as per the quality control manual guidelines and should perform an extensive and thorough check based on various parameters.
4. If the incoming material meets all parameters as laid down in the manual, the quality control team will give the go-ahead to the warehouse team for product storage.
5. However, if the material does not meet the specification the same should be returned to the vendor with a rejection note together with the reason for such a rejection.

10.3.6 WAREHOUSING

Before storing the material, the warehousing department will ensure the following compliances are met:

1. Ensure the material has been received as per the proper purchase order.
2. Ensure arrangements for the proper storage of material.
3. Material should be stored in a way such that material handling will be easier and handling distance must be at a minimum.
4. Ensure proper physical verification from time to time.
5. All materials must be tagged and it must be ensured that they are visible at all times in the warehouse.
6. All coverage insurance at all times must cover the entire risk and value of any materials.
7. As soon as product is received, the insurance should kick in to avoid any items being without insurance.
8. Due precaution must be taken for storing hazardous, toxic chemicals along with highly inflammable and temperature-controlled items.
9. Personnel from the storage department must take due care for all requirements needed for safety for both products as well as other employees.
10. One must ensure items are stored in a proper place, which avoids absorption of moisture.
11. The material can only be stored after receiving due clearance from the quality control team.
12. The responsibility of this department is also to ensure the inventory levels of various incoming materials are managed. The level may be fixed based on various inputs, for example, total lead time, EOQ, transit time for each material, inventory holding cost.
13. Identify the slow and nonmoving items.
14. Document the procedure for handling slow and nonmoving/obsolescence items and final disposal of the same.
15. Inform the procurement/buyer department if any material reaches MOL (minimum ordering level).

10.4 MANUFACTURING/EXECUTION

After receiving the details of quantity and product from the planning department, the manufacturing department will start the activities of manufacturing the final product from the raw materials. The department will ensure the manufacturing process as follows:

1. Create manufacturing process handbooks for all products along with process flow diagrams with detailed instructions.
2. Inform the quality control personnel engaged in line quality control checks at various stages of manufacturing.
3. Ensure safety measures in terms of both health and avoidance of any accidents for any and all people engaged for manufacturing activities.
4. Ensure medical checkups for all factory personnel at fixed intervals of weeks/months.
5. It is critical to keep communication open with both upstream and downstream departments. Information needs to flow downstream to the final quality control team regularly before the material is forwarded to the packaging department for final packing.
6. A logbook or plant utilization report is a must that will prove/capture the usage of machineries.
7. Similarly the manpower utilization report will also prove the utilization of manpower in the respective working zone.
8. Control and keeping a proper count of semifinished goods at various stages are the responsibilities of the manufacturing department.

10.4.1 QUALITY CONTROL MEASURES

The quality control department will complete the following to ensure quality:

1. Create quality control manuals for all products and make it available at all times. Examination/verification should be done according to the guideline of the manual to check the quality of all manufactured finished products before final packaging.
2. Update the product knowledge handbook. This is key and also very important for quality control department personnel, since both input and output may vary from time to time due to new developments or changes in the current scenario.
3. Communication must be sent to the packaging department for packing after the quality control department approves quality.

10.4.2 PACKAGING, WAREHOUSING, AND DISTRIBUTION

After getting confirmation of the final product's availability for packaging, the packaging department will start packaging the items as per the guidelines of the planning department. The packaging department will also ensure the following:

1. Perform detailed packaging responsibilities as per the planning department's requirement according to the sizes and type of packaging.
2. Ensure the availability of stock required for packaging.
3. Make sure all relevant information is contained in the package in order to avoid noncompliance issues as per statutory requirements.
4. Quality is very important and every effort needs to be taken so that the packages don't get returned from the customers' end on grounds of poor packaging quality.
5. Inventory holding is also an important factor and frequent changes of design/input in the package will make the existing packaging boxes and items out of design or place for products that are packed.
6. One must ensure that the packaging cover will protect the items at least up to the end of the life of the product.
7. Coordinate closely with the marketing department to understand the life, size, and design of packaging items.
8. Provide a high level of quality and standard on the type and quality of packaging since these are also important to avoid in-transit damages.
9. Ensure the minimum necessary inventory to avoid obsolescence and nonmoving stock.
10. Fix various inventory holding parameters in order to avoid blocking of working capital and avoid chances of obsolescence.
11. Ensure the correct steps are laid down so that the finished products are transferred to the finished goods warehousing department, waiting for final dispatch to the end customer.
12. The responsibility of the finished goods warehousing department is to now arrange all the steps to dispatch goods as per the "dispatch plan" to the various customers.
13. The important issue is to deliver the goods to the customers' end in good condition and at the right time. For delivering the products in good condition, it is also important to ensure the safety and security of the product and it is important to have the necessary transit insurance policy.
14. This team needs to ensure that proper checks and controls have been performed after filling the various packs and before dispatching the products to the warehousing department for storing.

10.4.3 MANPOWER AND SAFETY MEASURES

The deployment of manpower is also a very important factor in the manufacturing area. The following checklist will ensure proper compliance along with the best utilization of manpower in relation to the various cost centers:

1. The manpower must be properly deployed to ensure the manufacturing process will function properly. At the same time it is also very important to identify and reduce bottlenecks in the system.

2. It is not always the best approach to run machines at maximum output. The best option is to reduce bottlenecks so that the best system output is achieved.
3. Safety procedures are of paramount importance to avoid any contamination to the food and also decrease any chance of accidents. Extra care must be put in place around the accident zone, where the risks of damages/accidents are high.
4. Since the workforce could be working in a chemical plant, as per statutory requirements, it is necessary to provide onsite medical support and health checks at regular intervals for the entire manufacturing force that is related directly or indirectly to production.
5. Another safety precaution is a must: protection against fumes that might be generated.

10.4.4 POLLUTION CONTROL MEASURES

Before the start of production, it is mandatory for the manufacturing unit to obtain necessary clearances from the appropriate authorities. The EPA looks into all pollution-causing elements like smoke, discharge of chemicals, polluted water, etc. The following checklist is important for any manufacturing unit to meet these compliance issues:

1. Need to ensure all equipment and measures are already installed and obtained in order to comply as per the requirement of the pollution control guidelines and it is very important to be within the tolerance limit of the EPA.
2. It is very important to have a detailed and documented procedure to comply with the pollution control guidelines.
3. Have a procedure for frequent checks by an in-house safety department from time to time to ensure no violations have occurred.
4. The renewal process for EPA clearances needs to be in place since EPA clearance is for a time-bound period.
5. Often times pollution clearance is also state specific. Hence, the proper procedure documenting the necessary steps must be documented in order to receive clearance.

10.4.5 RISK FACTORS FROM MANUFACTURING

There are a number of risks involved in every step in a manufacturing setup and there are even higher risks for that of a chemical or food-processing factory. Let's look in detail at the risks involved:

1. The State may not give any permission to set up any chemical/food factory in a place as per the choice of the requester. The selection of the place may be made in consultation with regulatory bodies as per existing guidelines. The clearance is often country specific and sometimes even state specific. This could have a tremendous impact on the organization. It could increase costs and also might make it difficult to capture clients.

2. If major raw materials are imported from overseas countries, the import process is entirely dependent on the country's import policy.
3. Since major raw materials are imported, bilateral foreign policies are equally important between both the countries: the country who is exporting and the country who is importing.
4. For any factory, pollution control clearance is mandatory; however, in the case of a food/beverage factory additional clearance is required from the regulatory authority of the Food and Beverage Control authority. Hence, clearance from both the EPA and DSHEA in the United States is mandatory. These are very process-oriented and time-bound processes and could take a lot of time, effort, and money. The time increases more when clearance needs to be obtained for other countries as well.
5. Safety is of paramount importance and apart from just having onsite medical facilities for employees, often times it is necessary to have doctors 24×7 around the facility and mini safety and medical units are needed onsite.
6. Regular and employee-specific medical checks could be mandatory for all personnel engaged directly or indirectly with the manufacturing activities. This often increases cost.

10.5 OUTSOURCING

10.5.1 CONDITIONS FOR OUTSOURCING

"Make versus buy" has always been a very hard decision to finalize for any company. There are various factors that determine the decision and in that time, money and ability to enter the market are some critical factors to note.

The nutraceutical companies who buy raw materials or semifinished goods from another company tend to look into the following to determine the viability of make versus buy.

10.5.2 PRICING

Pricing is one of the four most important criteria for a decision. Pricing needs to be competitive and often, other factors remaining the same, the vendor with the lowest price wins the deal.

10.5.3 QUALITY

Quality is of paramount importance when it comes to food and liquid. The quality of the product can make or break a product's future in less than a few months.

10.5.4 QUANTITY

A constant source of quantity is very important for a product or products to be truly successful in the market. It is very important to determine the right amount in the right proportion for the right market.

10.5.5 CONSISTENT SUPPLY

The supply of raw materials needs to be constant and not fluctuate.

10.5.6 OTHER FACTORS FOR OUTSOURCING

1. Manufacturing source:
   In most cases, companies look to get product directly from the manufacturing source. There are a number of reasons for this. Generally quality is better, the price is cheaper, and it is easier to determine the capability of the vendor when the product or products are sourced at the manufacturing site.
2. Multiple sources of raw material/semifinished product:
   Often times, a nutraceutical company would like to have multiple sources of the same product. Apart from quality and price control, it also lets them manage their vendors effectively and an impact on one site and vendor is minimized when there is an option of sourcing the same products from another.
3. Match the correct requirements:
   Matching the correct technical requirements is the key to quality outsourcing. For example, in the past certain nutraceutical companies have heavily outsourced from China, but the semifinished products didn't match the requirements of the company. Also counterfeit products have been used at certain times, the quality has not been met, and prohibited substances have been used.
4. Variable resources and local management:
   Nutraceutical companies often mitigate risks by taking two strategies—have various sources for the same raw material and also send a group of their own employees to manage the raw material at the vendor site. Quality control could greatly improve due to this.

   At certain times, when it is critical, sometimes the companies create a smaller divisional office close to the vendor site to manage both quality and cost. This also gives two other major advantages:
   a. Product knowledge: Close proximity to the vendors helps in gaining more in-depth knowledge about the raw materials and semifinished goods.
   b. Create one's own formula: Close proximity to vendors also increases product knowledge, which helps the nutraceutical company to determine whether the product will be passed to the United States as a raw material or semifinished good. In some cases, depending on the requirement, the local division becomes adept at creating either of the forms depending on the demand from the headquarters.
5. Control the content of raw material:
   There are very strict guidelines governing the content of nutraceutical foods. Most natural and herbal products have strict parts per million (PPM) requirements and changes to either the content or the ratio of contents could be detrimental.

For example: Green tea imported from outside the country containing arsenic and cadmium beyond the allowed parts per million (ppm) has been found before and also substances like butanol had been used to increase the extraction. This causes a harmful reaction. The contamination level approved by regulatory bodies should always be followed and regulatory bodies specially look for critical heavy metals such as arsenic, mercury, lead, and cadmium. Also only approved solvents should be used for extraction purposes.

6. Shipment cost:
   Shipment cost and freight cost are very important and they could break the backbone for importing products from other countries. For example, Mexico could be a key location in the future since shipping costs and labor costs are becoming increasingly high in China.
7. Raw material hunting:
   Getting the right raw material, and getting good-quality material with the right specification are extremely important. Add constant supply to all the aforementioned factors. Quality raw material is extremely important.
8. Safety stock:
   Keeping safety stock and calculating the right value for safety stock are critical. Because of the nature of the industry and the high demand of nutraceutical food, often times companies keep an extra 20–30% in stock.
9. Temperature/climate conditions:
   Temperature and climate conditions play a very important role. Three important criteria during raw material processing are:
   a. Temperature and variation of temperature. In addition frequent climate changes and changes in humidity affect the products greatly.
   b. Problems with pesticides—issues with rodents.
   c. Contamination during storage—contamination shouldn't affect the materials.

   In most cases, a tropical climate is ideal for growth of numerous raw materials. India and China hold the lead in that with tea and coffee for example. In the United States, Florida has a few advantages due to the tropical weather. Numerous products are manufactured in Florida and raw materials are aplenty, for example, coconut.
10. Customs:
    Customs could be a very important factor for outsourcing of products. Not all products are allowed in the United States from all countries, so this is a challenge that can be both product specific and country specific.

10.6 SUPPLY CHAIN MANAGEMENT

10.6.1 VENDOR MANAGEMENT

Vendor management is very important in the supply chain of the functional food industry. The vendor management process has gone through a number of changes over the years but the basic principle has remained fairly consistent for a number of years. A basic outline of the process is captured next:

1. Determine which product the company is looking to develop.
2. The vendor management team works with the internal technical team to determine the requirements from the vendor.
3. Internal analysis to finalize the requirement and type of vendor required is performed.
4. Check to determine if an existing vendor who has a good rating could be used for new product development.
5. Also other vendors should be analyzed and a group of vendors could be shortlisted for providing the raw materials.
6. A request for quote (RFQ) process is initiated and a request for proposal (RFP) is initiated and completed.
7. Two or three vendors are chosen and either an existing vendor or a new vendor who matches the quality, quantity, and pricing set forth by the company is chosen as the primary vendor and a few others who closely match the requirements are chosen as secondary vendors.
8. Create agreements with the vendors.
9. Provide current and forecasted requirements to the primary and secondary vendors.
10. Perform value metrics to assess the performance of secondary vendors and slowly bring the secondary vendors up to speed.
11. Perform quarterly business reviews with vendors to determine their performance.
12. Make sure the vendor is able to provide a constant supply while maintaining the price and quality (Institute of Supply Chain Management, 2016).

10.7 INVENTORY MANAGEMENT

Inventory management is described under warehouse management and the process doesn't change much. But nowadays functional manufacturing companies sometimes let the vendor manage the entire inventory. Often times vendors have raw materials in their local manufacturing hubs, which are in close proximity to the company. This helps the vendor to manage seasonality and uncharacteristic demand. Letting the vendor manage the inventory also helps in following government rules and regulations since the manufacturer can do an audit and perform toxicity level checks and solvent checks on the vendor inventory and their own inventory that is managed by vendors. However, this is not universal and in a number of cases, a manufacturer likes to manage its own inventory. The two important factors are storage and maintaining effective safety stocks. Storage becomes very important since storage containers can't have excess salinity, maintain a controlled temperature, and perfect humidity for storing products (Institute of Supply Chain Management, 2016).

10.7.1 CAN NUTRACEUTICAL COMPANIES CONTROL THE SUPPLY CHAIN AND EMERGE AS LEADERS?

Nutraceutical companies can definitely control the supply chain and emerge as strong winners and they can do so mainly because of the reasons described below:

1. Patents
   Patents are one of the key factors to capture market growth and revenue for years to come. Patents are given a very high value in the market and patents have the power of generating higher market share and better product pricing.
2. Clinical study
   Clinical study goes hand in hand with patents. More clinical study proves the efficacy of the product and that helps in receiving patents, which in turn are used to control the market.
3. Power of money and controlling the money flow
   Let's take the example of a particular raw material. The cost of making this supplement in India and China is comparatively cheap while the final selling price in the USA is significantly higher. So often times, the raw material is acquired in the Asian subcontinent and converted to semifinished goods. The semifinished goods are then brought to the United States and then more research and development is done and the product is improved. It is then sold at a much higher price. At the same time, keeping a high price is also important to manage quality and import duties, manage quality standardization, import, research and development, marketing, and positioning of the product.
4. Effective management of raw materials, production of semifinished goods and vendors
   The motto of the effective supply chain is: producing the right quantity in the right time in the right proportion. This is key for the nutraceutical companies as well. Having a local office in the source location of a raw material helps in effective raw material management, production of semifinished goods, and effective shipping of the goods to the United States. A local office or presence helps in managing the vendor better and the United States office can then determine more downstream functionalities and can concentrate more on selling the final product effectively. Effective management of raw materials would mean:
   a. identify products that are going to be a part of semifinished goods,
   b. source raw material where it is cultivated and manufactured,
   c. source material right at the site and be able to match the requirements, and
   d. produce semifinished goods and not finished goods.

If required, having a small office at the source would help in determining what level of semifinished goods should be sent to the United States. (Note: A complete and finished product could have more tax implications and the difference in price from finished product to semifinished product could vary by as much as 75%.)

10.7.2 FACTORS FOR FUTURE SUCCESS

In order to be successful in the future, it is imperative that the raw material supplier, the manufacturer of functional food, and the final seller all form a close bond with each other and work towards the success of the industry. All three players play very critical roles in the growth of the nutraceutical industry. Trust is one of the key factors and will continue to be key in the future.

Trade beyond borders has made trust more important, as captured in the Food Processing article (Trade beyond Borders, 2015).

Another very important factor in the entire scheme of things is the importance of urbanization. Increases in disposable incomes and urbanization are moving people towards nutraceutical food. Working families, changes in lifestyle, and smaller families are some of the factors forcing people to look at avenues of food which they didn't think about before—easy and quick to make but healthy—and that's where the importance of nutraceutical and functional food comes into the picture. Demand is growing and supply needs to grow in order to meet the demand—a happy problem to have for all the nutraceutical and functional food companies of the world.

10.8 CONCLUSIONS

This chapter focused on nutraceutical and functional foods. Most of the raw materials are obtained from the Asian subcontinent currently and specifically from China and India. The original concept of nutraceutical foods was actually derived from Japan and Japan was known to be able to cure a number of diseases by mixing traditional food with nontraditional food. However, after the growth in nutraceutical products, a number of products are now manufactured in Malaysia, Australia, Indonesia, and New Zealand. These countries could hold a key position in the future in terms of providing raw material or semifinished products. The concept of nutraceutical product is not new. It was developed and nurtured by people who stayed in different countries for centuries. Ayurveda was available in India for a long time and was a part of normal life for medicinal value. A number of natural products like turmeric have been brought to the United States and have been converted to drugs, research and development have been performed to ensure the quality of the final product; this was then tested and made into a functional product. One more fairly recent development has been the production of raw materials in different countries—for example turmeric that was originally developed only in India now comes from different locations and it is cultivated in other south Asian countries. Most of the products are distributed across the globe and final product development along with marketing and product launch are taking place in the United States and to some extent in Europe (approximately 80% in the United States and 20% in Europe, respectively). Some other scattered products are also developed in New Zealand and Australia and parts of Africa like Congo. After the initial development of the products, they are sent to the United States for performing efficacy and safety checks, and more research and development is performed on the raw materials in order to make the product market ready for selling. Due to the high cost of raw material and purification of the materials in the United States, Mexico is used as another place for purifying the raw materials and sending them to the United States. Another great advantage of Mexico is its proximity to the United States and that helps in lowering shipping costs.

In order to be successful, there needs to be enormous confidence in the viability of the product. Outsourcing the product manufacturing to another low-cost country is beneficial for cost but also sometimes risky. Often times, raw material is procured from multiple sources and that helps to take away the dependency on one source. After the raw material is gathered and semifinished products are manufactured, products are brought to the United States and quality checks and controls are performed. One of the key successes of the products will be the actual determination of the authenticity of the source and validation of authenticity. The United States performs a detailed investigation before bringing any product into the market.

A good example of research and development, product development, and marketing is captured in the following paragraph.

A very important raw material grows in a remote village in India and China and the manufacturing cost of converting the abundant raw material into a semifinished product is extremely low. The semifinished goods are then brought to the United States and purified and the cost for the entire operation is extremely cost effective. The product is then sold at a much higher price in the United States.

REFERENCES

Health Statistics Information. http://www.niddk.nih.gov/health-information/health-statistics/Pages/overweight-obesity-statistics.aspx

Institute of Supply Chain Management. https://www.instituteforsupplymanagement.org

Supply Chain Quotes. http://www.brainyquote.com/quotes/keywords/supply_chain.html

Trade beyond Borders. Examining the global supply chain. Food Processing. http://www.foodprocessing.com/articles/2015/examining-the-global-supply-chain/

CHAPTER

Raw materials analysis and quality control

11

Varuzhan Sarkisyan*, Vladimir Bessonov, Alla Kochetkova***

**Laboratory of Food Biotechnology and Foods for Special Uses, Federal Research Centre of Nutrition and Biotechnology, Moscow, Russia; **Laboratory of Food Chemistry, Federal Research Centre of Nutrition and Biotechnology, Moscow, Russia*

11.1 INTRODUCTION

Raw materials are the sources of nutrients on which the development of the functional foods and nutraceuticals is based. The main criterion for the applicability of a particular type of raw material is the availability of its complete and accurate specifications from the standpoint of quality, safety, and authenticity. This specification includes information as to a wide variety of indicators, including the chemical composition and physical-chemical properties, structure, sensory characteristics, and, in some cases, origins.

Given that the majority of the functional foods and nutraceuticals are used not only in the country where they are produced but also in the world market, it is appropriate to use internationally recognized analytical methods. Hence, only these techniques and methods of analysis of raw materials are the subject matter of this chapter. However, it should be noted that newer, more accurate and sensitive analytical methods are constantly being developed, the description of which is beyond the scope of this chapter. Detailed description of such analytical tools can be found in the books dealing entirely with the issues of analytical chemistry for foods (Kennedy and Lloyd, 1995; Nollet and Toldra, 2015; Pomeranz, 2013; Sun and MyiLibrary, 2008).

The characteristics of the raw materials obtained by the generally accepted methods for analysis allow the food producers to carry out a selection of the most secure raw materials of adequate quality that meet their technological needs. These characteristics are also critical for consumers to select a particular product on the market.

The term "raw material" is very broad and is regularly updated with the development of new and more promising sources of nutrients. The systematic classification of raw materials is given here to simplify their description.

11.2 CLASSIFICATION OF RAW MATERIALS

All the existing classifications of raw materials are conditional and are mainly used to make the selection of a specific source of nutrients simple. In this chapter, raw materials are classified according to the source of their origin (Fig. 11.1)

Developing New Functional Food and Nutraceutical Products. http://dx.doi.org/10.1016/B978-0-12-802780-6.00011-0

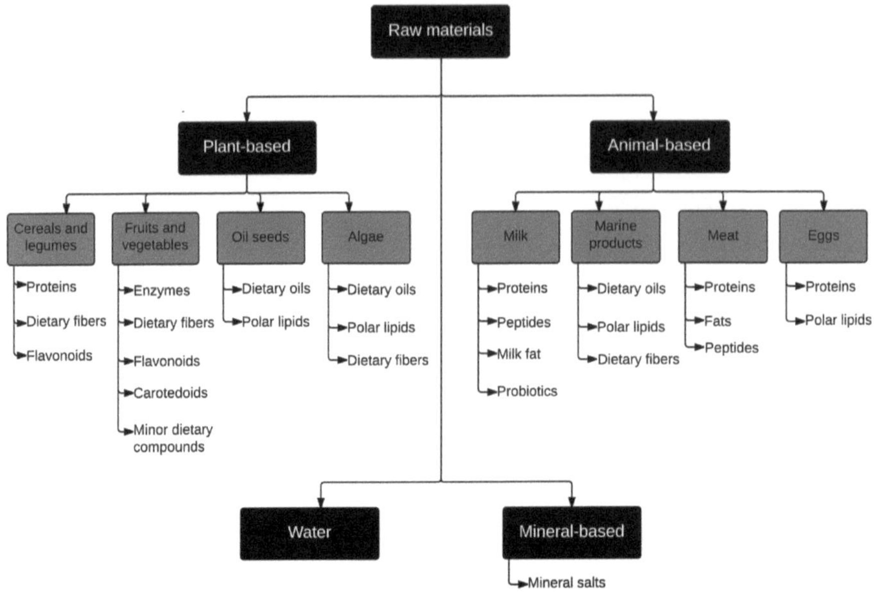

FIGURE 11.1 Classification of Raw Materials and their Main Nutritive Compounds

As is shown in Fig. 11.1, raw materials can be subdivided into four groups based on the source of their origin: water, mineral raw materials, animal-based raw materials, and plant-based raw materials. The origin of a raw material determines the whole complex of characteristics that a certain group possesses, and also determines the ways in which the samples are prepared and the appropriate methods of analysis.

The main sources of animal-based raw materials are meat, milk, seafood, and eggs. The main sources of plant-based raw materials are cereals, leguminous plants, vegetables, and fruits as well as oil-bearing plants. These groups of raw materials are used either as they naturally occur or as a source of biologically active substances. Animal-based raw materials are often used as a source of macronutrients, such as proteins, fats, carbohydrates, specific amino acids, and fatty acids.

Drinking water, as a raw material for the production of functional foods and nutraceuticals, has a special role when used for technological purposes. Water is one of the most widely used raw materials, although it is not always taken as such.

Animal-based raw materials also constitute a separate group of raw materials—a source of indispensable food components that play a key role in the vital functions of human organisms (potassium, calcium, magnesium, sodium, iron, and many others).

11.3 PREREQUISITES FOR RAW MATERIALS ANALYSIS

Prior to the analysis, it is important to ascertain that the main prerequisites are met, such as adequate laboratory conditions for the study, and that the rules for appropriate sample selection are met. The matrices of raw material for food are very complex systems, and it is often a very difficult task to isolate the required component. With this consideration in mind, the sampling of the raw materials must be done in accordance with the international methods. For instance, the sampling of cereals, pulses, and milled products should be conducted according to ISO, 2009b, *ISO 24333:2009—Cereals and cereal products—Sampling*, 2009). This sampling protocol covers manual or mechanical sampling up to a depth of 3 meters and mechanical sampling up to 12 meters. As mentioned in the standard, samples should be taken jointly by representatives of the buyer and seller or by a jointly appointed sampling superintendent, and samples shall be as representative as possible of the lots from which they are taken. Given that the composition of a lot is seldom uniform, a sufficient number of increments shall be taken and carefully mixed, thereby giving a bulk sample from which the laboratory samples are obtained by successive divisions or otherwise. Special care should be taken to ensure that all the sampling apparatus is clean, dry, and free from contaminants. Sampling will also be carried out in a way that protects the samples and their containers, as well as the sampling instruments, from contamination with rain, dust, etc.

These recommendations, in general, are applicable for the analysis of all raw materials, but every standard on sampling has its own special requirements. For example, the standard for the sampling of animal and vegetable fats and oils (ISO, 2001, *ISO 5555:2001—Animal and vegetable fats and oils—Sampling*, 2001) in addition to general requirements, describes procedures for examination of samples from vertical cylindrical land tanks, ships' tanks, weigh tanks, small packing units (including consumer packs), and other storage vessels. The sampling techniques among them vary in methods for providing a homogeneous sample for analysis (to exclude fractionation of oil, water content, etc.). The guidelines for the sampling of milk and milk products (ISO, 2008, *ISO 707:2008 (IDF 50:2008)—Milk and milk products—Guidance on sampling*, 2008) additionally determine the storage temperatures used before and during the transportation of the samples and pay special attention to sampling equipment and the sequence of the analysis (microbiological, chemical, physical, and sensory examinations).

It is very important to follow the appropriate sampling procedure to ensure that there are no significant variations during the analysis of the raw materials.

In addition, all the analyses must be done in laboratories that meet the requirements of good laboratory practice (GLP), which is a quality system concerned with the organizational process and the conditions under which nonclinical health and environmental safety studies are planned, performed, monitored, recorded, archived, and reported to ensure the uniformity, consistency, reliability, reproducibility, quality, and integrity of test results (*ENV/MC/CHEM(98)17*, 1997). The GLP standard describes principles of organization:

1. Test facility organization and personnel (test facility management responsibilities, study director's responsibilities, principal investigator's responsibilities, study personnel's responsibilities).
2. Quality assurance program (responsibilities of the quality assurance personnel).
3. Facilities (test system facilities, facilities for handling test and reference items, archive facilities, waste disposal).

4. Apparatus, materials, and reagents.
5. Test systems (physical/chemical, biological).
6. Test and reference items (receipt, handling, sampling, and storage, characterization).
7. Standard operating procedures.
8. Performance of the study (study plan, conduct of the study).
9. Reporting of study results.
10. Storage and retention of records and materials.

Only when the aforementioned conditions for sampling and laboratory practice are satisfied is it possible to obtain accurate and reproducible results of the analysis of a raw material for food products.

11.4 BASICS OF ANALYTICAL METHODS

There are a number of methods of analysis for food products. This chapter will deal with a brief description of the main principles of the research methods. Detailed information of the methods of measurements can be found elsewhere (Cordella and Moussa, 2002; Martinez et al., 2003; Reid et al., 2006).

The methods of measurements can be classified into the following main groups based on the physical principle underlying the research: gravimetry, refractometry, spectroscopy, mass spectrometry, chromatography, stable isotope analysis, enzymatic, immunological, and DNA methods.

Gravimetry is one of the basic methods of studying the composition of raw materials, which deals with the quantitative determination of an analyte based on the mass of the solid components in a sample. These methods have a significantly limited usage due to their nonspecificity. However, they are considered decisive in defining several qualitative indices of the raw material, such as the amount of moisture and certain solid components (depending on the sample preparation).

Refractometry is based on the measurement of the refractive index of the substance, that is, the ratio of the speed of light in a vacuum to its speed in a substance. This speed is related to the temperature, composition, concentration, and purity of the substance. The index of refraction depends on the density of a liquid and thus on the concentration of the solute in the solution. Measurements are usually made at 20°C using the sodium D-line as the light source.

There are four widely used spectroscopic techniques: UV-Vis spectrophotometry, infrared spectroscopy (IRS), nuclear magnetic resonance (NMR) spectroscopy, and atomic adsorption spectroscopy. The basic principle of all spectroscopic methods is in the measuring of the difference between the properties of light (in IR, Vis, UV range, etc.) before and after its contact with the testing sample. UV-Vis spectroscopy is one of the most inexpensive and simple methods that are accepted in analytical chemistry and might be used in the majority of analytical laboratories.

Mass spectrometry is a method based on ion formation from compounds followed by the steps of their separation and detection. The result of this procedure is the information of the mass-to-charge ratio (*m/z*) and an abundance of individual ions formed during the analysis. This allows for qualitative and quantitative characterization of the composition of the sample.

The aforementioned methods might be used directly or as part of more complicated chromatographic techniques.

Chromatography is a process based on the multiple repetitions of absorption and desorption of a sample while moving in a stream of a mobile phase (eluent) along a stationary phase (sorbent). Separation of the complex mixtures by chromatographic methods is based on the different affinity of the components to the eluent and sorbent. During the chromatography, the eluent containing the sample is moving through the stationary phase. Normally the stationary phase is a material with a developed surface, and the mobile phase is a flow of gas or liquid that is filtered through a sorbent bed.

Chromatographic techniques are usually classified by the nature of the stationary and the mobile phases. The main types are thin layer chromatography (TLC), systems consisting of a planar solid phase and a liquid mobile phase; gas chromatography (GC), systems with a gaseous mobile phase; and liquid chromatography (LC), systems with a solid stationary phase and a column of a solid stationary phase (Cserháti and Forgács, 2005).

After the chromatographic separation of the components, they might be analyzed (detected) by various methods including spectroscopy (in various ranges), refractometry, mass spectrometry, etc. The informative value of these measurements can be enhanced by in-line coupling of different detectors. This procedure allows describing several physical properties of the separated analyte simultaneously in a single chromatography run.

The fermentative methods of food analysis may be classified into three groups according to their purpose: analysis determining the enzyme activity level in raw food; the preparation of a raw sample for a further study; and enzyme-linked immunosorbent assays (ELISA).

The analysis determining the enzyme activity level in raw food is devised to evaluate the quality of the raw material; the usage of the technological processing of the raw material; and other indices connected with the presence of specific fermentative activity.

As a tool for sample preparation, the fermentative methods are mainly used for a specific separation of certain components by means of hydrolysis, such as monosaccharides from polysaccharides and fatty acids from triglycerides. The hydrolysates are then analyzed spectrophotometrically or by the chromatographic methods.

ELISA assays are based on the colorimetric measurement of binding of antigens (analyte from testing material) to antibodies (specific to the antigen) resulting in the formation of a colored complex. The binding reaction is conducted with the antigen or analyte immobilized on the solid surface of a well plate. This assay allows one to quantitatively characterize the presence of the analyte. There are several modifications of the ELISA method that differ from each other by the sequence of antigen and antibody immobilization. The main ones are direct, indirect, and sandwich procedures.

11.5 ANALYSIS OF THE RAW MATERIALS

The principles of analysis of raw materials from various sources are similar in many ways but they significantly differ in the sample preparation, the reason being the necessity of maximum recovery of the test material from the matrix. The main generally accepted methods of analysis of the chemical composition of raw materials are given in Table 11.1. The classification of the methods of analysis is given in accordance with the above-mentioned classification of the raw materials.

11.5.1 WATER ANALYSIS

In the analysis of water, the main markers of quality are the content of minerals (calcium, magnesium, chlorides, sulfates, general salinity), the content of the salts of heavy metals, and microbiological markers. In this specific case, the most informative method of analysis is AAS which allows identifying and providing a highly accurate qualitative characteristic of the mineral composition of water used for technological purposes.

Thus, the relevant standard (ISO, 2003a, *ISO 15586:2003—Water quality—Determination of trace elements using atomic absorption spectrometry with graphite furnace*, 2003) includes principles and procedures for the determination of trace levels of: Ag, Al, As, Cd, Co, Cr, Cu, Fe, Mn, Mo, Ni, Pb, Sb, Se, Tl, V, and Zn in surface water, ground water, drinking water, waste water, and sediments, using atomic absorption spectrometry with electrothermal atomization in a graphite furnace. The method is applicable to the determination of low concentrations of elements. Water samples are preserved by acid treatment, filtered and preserved by the addition of acid, or digested. Sediment samples are digested. A small subsample of a sample solution is injected into a graphite furnace of an atomic absorption spectrometer. The furnace is electrically heated. By increasing the temperature stepwise, the sample is dried, pyrolyzed, and atomized. Atomic absorption spectrometry is based on the ability of free atoms to absorb light. A light source emits light specific to a certain element (or elements). When the light beam passes through the atom cloud in the heated graphite furnace, the light is selectively absorbed by atoms of the chosen element(s). The decrease in light intensity is measured with a detector at a specific wavelength. The concentration of an element in a sample is determined by comparing the absorbance of the sample with the absorbance of calibration solutions. If necessary, interferences may be overcome by adding a matrix modifier to the samples before analysis, or by performing the calibration with the standard addition technique.

Very frequently (ISO, 1986, *ISO 7980:1986—Water quality—Determination of calcium and magnesium—Atomic absorption spectrometric method*, 1986) is used which specifies a method for the determination of dissolved calcium and magnesium by flame atomic absorption spectrometry. It is intended for the analysis of raw and drinking waters and can be used for waters having calcium content of up to 50 mg/L and magnesium content of up to 5 mg/L. For samples containing higher concentrations of calcium or magnesium, a smaller volume of the sample must be taken for the analysis.

Table 11.1 General Methods of Quality Analysis

Raw material	Analyte	Method	Principle	Notes
Water	Calcium	ISO 7980	AAS	
	Chloride	ISO 9297	Titrimetry	
	Iron, dissolved	ISO 6332	Spectrophotometry	
	Sulfates	ISO 9280	Gravimetry	
	Arsenic	ISO 17294-2; 15586	ICP-MS; GF-AAS	
Animal-based materials				
Meat	Fat	ISO 1443	Gravimetry	
	Lead	AOAC 934.07	Colorimetry	
	Tin	AOAC 985.16	AAS	
Milk	Iron	AOAC 999.11	AAS	
	Melamine	ISO/TS 15495	LC-MS/MS	
	Free fatty acids	ISO 1740; IDF 6	Titrimetry	The fat acidity is expressed in millimoles per 100 g of fat
	Peroxide value	AOAC 965.33	Titrimetry	Should be conducted shielded from direct light source
	Milk fat purity	ISO 17678; IDF 202	Gas chromatography	
Fish and fish oils	Histamine	AOAC 977.13	Fluorometry	
	Mercury	AOAC 977.15	FAA	
	Fatty acid composition	ISO 12966	Gas chromatography	
	Peroxide value	ISO 3960	Titrimetry using iso-octane	
	Acid value	ISO 3960	Titrimetry	
	P-anisidine value	AOCS Cd 18-90	Spectrometry	
Plant-based materials				
Cereals and legumes	Protein	ISO 1871:2009	Titrimetry, Kjeldahl digestion	Does not take into account the nitrogen from nitrates and nitrites
	Ash	ISO 2171	Gravimetry	
	Moisture	AOAC 925.09	Gravimetry	Vacuum oven
	Fat, crude	AOAC 945.38F; 920.39C	Gravimetry	Ether extraction
Oil seeds	Cocoa shell	AOAC 968.10, 970.23	Spiral vessel count, stone cell count	
	Fat	AOAC 963.15	Gravimetry	Soxhlet extraction
	Free fatty acids	ISO 660	Titrimetry	
	Unsaponifiable matter	ISO 3596	Titrimetry	

When using the air/acetylene flame and the dilution factor 1 in 10, the optimum range is 3–50 mg/L for calcium and 0.9–5 mg/L for magnesium. The principle of the method is measurement by flame atomic absorption spectrometry after adding lanthanum chloride (if an air/acetylene flame is used) or cesium chloride (if a nitrous oxide/acetylene flame is used) to reduce interferences. For calcium, the absorbance is measured at 422.7 nm and for magnesium at 285.2 nm.

However, in addition to AAS the spectrometric methods of analysis for metals are used, namely, (ISO, 1988, *ISO 6332:1988—Water quality—Determination of iron—Spectrometric method using 1,10-phenanthroline*, 1988), that specifies a 1.10-phenanthroline spectrometric method for the determination of iron in the water. Procedures are described for the determination of dissolved and undissolved iron. The methods are applicable to the determination of iron concentrations between 0.01 and 5 mg/L. Iron concentrations above 5 mg/L may be determined after suitable dilution of the sample. The methods involve addition of 1,10-phenanthroline solution to a test portion and photometric measurement of the orange-red complex at a wavelength of about 510 nm.

Another highly effective method of water analysis is (ISO, 2003b, *ISO 17294-2:2003—Water quality—Application of inductively coupled plasma mass spectrometry (ICP-MS)—Part 2: Determination of 62 elements*, 2003), that describes the determination of 62 elements in water. The 62 elements determined include almost all elements in the environmental analysis. The halogens, noble gases, hydrogen, nitrogen, and oxygen are exceptions. The filtered acidified samples (nitric acid) are analyzed to obtain dissolved elements and the whole sample is digested with nitric acid (ISO 15587-2) or aqua regia (ISO 15587-1) to produce total elements. The standard warns that digestion for the total elements procedure may only partially dissolve some elements like silicon, aluminum, tin, tungsten, zirconium, and antimony. The technique is based on the measurement of specific isotopes of elements.

11.5.2 ANIMAL- AND PLANT-BASED MATERIALS

The analysis of the main macro components in animal- and plant-based raw materials generally converges on determining the content of proteins, fats, carbohydrates, water, and ash. A widely accepted principle of determining the total protein is described in (ISO, 2009a, *ISO 1871:2009—Food and feed products—General guidelines for the determination of nitrogen by the Kjeldahl method*, 2009), that provides general guidelines for the determination of nitrogen by the Kjeldahl method. It applies to food and feed products containing nitrogenous compounds that can be directly determined by the Kjeldahl method. The principle of this method is digestion of a test portion with concentrated sulfuric acid in the presence of catalysts to convert the organic nitrogen into ammonium sulfate. Excess sodium hydroxide is added to the cooled digest to release the ammonia. The released ammonia is distilled into an excess of boric acid solution and then titrated with a standard solution of sulfuric or hydrochloric acid. The nitrogen content is calculated from the quantity of ammonia produced.

Then, the content of the protein in the sample is calculated based on the content of the total nitrogen using the specific coefficients of recalculation for each type of product.

This method is not specific to nitrogen (organic/inorganic and protein/nonprotein), therefore additional methods of analysis are used for certain raw materials to detect the possible existence of nonprotein nitrogen in the raw material. For example, (ISO/TS, 2010, *ISO/TS 15495:2010 (IDF RM 230:2010)—Milk, milk products and infant formulae—Guidelines for the quantitative determination of melamine and cyanuric acid by LC-MS/MS*, 2010) gives guidelines for the quantitative determination of melamine and cyanuric acid content in milk, powdered milk products, and infant formulae by electrospray ionization liquid chromatography tandem mass spectrometry (LC-MS/MS).

Moreover, in the case of seafood, an important factor in the evaluation of its quality is the determination of the presence of specific types of biogenic amines, namely histamine, the content of which is determined in accordance with (ISO, 2004, *ISO 1740:2004 (IDF 6:2004)—Milkfat products and butter—Determination of fat acidity (Reference method)*, 2004), which specifies a method for the determination of histamine in seafood. The principle of this method is the extraction of the probe with 75% (v/v) methanol. The extract is passed through an ion exchange column. The *o*-phthaldialdehyde solution is added to the eluate to form fluorescent histamine derivatives. The fluorescent intensity of the derivatives is measured by a fluorometer, and the histamine is quantified using external standards.

The content of fat in a raw material is usually determined gravimetrically after its extraction from the raw material, for example, by (ISO, 1973, *ISO 1443:1973—Meat and meat products—Determination of total fat content*, 1973) that specifies a reference method for the determination of the total fat content of meat and meat products. The principle of this method is in boiling of the sample with dilute hydrochloric acid to free the occluded and bound lipid fractions, filtration of the resulting mass, drying, and extraction with n-hexane or light petroleum of the fat retained on the filter.

Besides the total content of fat, one of the fundamental criteria of fat quality is the composition of fatty acids which is determined in accordance with (ISO, 2015, *ISO 12966-4:2015—Animal and vegetable fats and oils—Gas chromatography of fatty acid methyl esters—Part 4: Determination by capillary gas chromatography*, 2015) that gives an overview of the gas chromatographic determination of fatty acids, free and bound, in animal and vegetable fats and oils following their conversion to fatty acid methyl esters (FAMEs). The GLC of FAMEs is applicable to all natural and synthetic mixtures of tri-, di-, and monoacylglycerols, to fatty acid esters, free fatty acids, soaps, and other fatty compounds. With this suite of standards, FAMEs from C4 to C26 can be determined, including saturated fatty acid methyl esters, cis- and trans-monounsaturated fatty acid methyl esters, and cis- and trans-polyunsaturated fatty acid methyl esters.

In some cases, certain fat fractions have to be examined to evaluate the quality of a raw material. Thus, for instance, (ISO, 2000b, *ISO 3596:2000—Animal and vegetable fats and oils—Determination of unsaponifiable matter—Method using diethyl ether extraction*, 2000) specifies a method using diethyl ether extraction for determination of the unsaponifiable matter content of animal and vegetable fats and oils. This method is not applicable to waxes and, moreover, gives approximate results with certain fats of high unsaponifiable matter content, for example with fats derived from marine animals. A method given in (ISO, 2000a, *ISO 18609:2000—Animal and vegetable fats and oils—Determination of unsaponifiable matter—Method using hexane extraction*, 2000) may be used when the climatic conditions, or regulations, do not permit the use of diethyl ether. The principle of this method is that the fat or oil sample is saponified by boiling under reflux with an ethanolic potassium hydroxide solution. Then the unsaponifiable matter is extracted from the soap solution by diethyl ether. The solvent is evaporated and the residue is weighed after drying.

When evaluating the quality of fat in the raw food, it is essential to determine the content of the primary and secondary products of lipid oxidation, as well as the progress of the hydrolysis processes.

The main marker of the progress of the lipid hydrolysis is the content of the free fatty acids which is determined titrimetrically, such as in accordance with (ISO, 2004, *ISO 1740:2004 (IDF 6:2004)—Milkfat products and butter—Determination of fat acidity (Reference method)*, 2004) that specifies a method for the determination of the acidity of the fat contained in milk fat products and in butter. In the particular case of butter, the fat is first separated from the melted butter by centrifuging. In an oven, the melted milk fat product or fat from butter is filtered through a filter paper. The filtrate is dissolved in a mixture of propan-2-ol and light petroleum and then titrated with a tetra-n-butylammonium hydroxide standard solution using thymol blue as indicator.

A marker of the existence of the primary products of oxidation is the peroxide value which is determined, for instance, in accordance with AOAC 965.33, a standard that applies to vegetable oils and animal fats and established the method of determination of the peroxide number in vegetable oils and animal fats of various levels of purity in the range from 0.1 to 45 mmol of active oxygen per 1 kg of oil or fat. The method is based on the interaction of the products of oxidation of the vegetable oils and animal fats (peroxides and hydroperoxides) with potassium iodide in the solution of acetic acid and iso-octane or chloroform, with a subsequent quantitative determination of the released iodine using sodium thiosulphate solution by the titrimetric method.

The secondary products of oxidation are aldehydes and ketones, the content of which may be determined in accordance with AOCS Official Method Cd 18-90 that determines the amount of aldehydes (principally 2-alkenals and 2,4-dienals) in animal and vegetable fats and oils, by reaction in an acetic acid solution of the aldehydic compounds in an oil and *p*-anisidine, and then measuring the absorbance at 350 nm.

The total carbohydrates are calculated by the difference between the net weight of the raw material and the overall weight of its protein, fat, moisture, and ash. However, the analysis of the chemical composition of a raw material is not limited to the analysis of the macro components only. The determination of the content of specific biologically active compounds has a great significance in the evaluation of the quality of a raw material (Table 11.2).

Table 11.2 Special Components Analysis Methods

Analyte	Method	Principal	Notes
Total dietary fibers	AOAC 2009.01	Enzymatic gravimetry	
Insoluble dietary fibers	AOAC 991.42 AACC Intl 32-20.01	Enzymatic gravimetry	Quantitation lost for inulin, resistant starch, polydextrose, and resistant maltodextrins. Refer to specific methods
Soluble dietary fibers	AOAC 993.19	Enzymatic gravimetry	Quantitation lost for inulin, resistant starch, polydextrose, and resistant maltodextrins. Refer to specific methods
(1→3)(1→4) Beta-D-glucans	AOAC 995.16	Enzymatic gravimetry	
Fructans	AOAC 997.08	Enzymatic gravimetry HPAEC-PAD	Oligofructoses, inulin, hydrolyzed inulin, polyfructoses, fructooligosaccharides (applicable to added fructans)
Protein efficiency ratio (PER)	AOAC 960.48	Rat bioassay	
Gluten	ELISA R5 Mendez	Immunoassay	Overestimates barley hordein content in barley-contaminated foods
Vitamin A	AOAC 974.29	Colorimetry	
Riboflavin	AOAC 970.65	Fluorometry	
Vitamin B_{12}	AOAC 952.20	Microbioassay	
Vitamin C	AOAC 967.22	Microfluorometry	
Vitamin D	AOAC 936.14	Rat bioassay	Result in international units
Vitamin E	AOAC 971.30	Colorimetry	

One of the groups of analytes belonging to this category is the substances the composition of which requires additional research after the analysis by the standard methods. A typical case is the differentiation of the fiber fractions and their separate classes from the total content of carbohydrates in a sample.

The analysis of the content and composition of dietary fibers (carbohydrates that are resistant to digestion and absorption in the human small intestine with complete or partial fermentation in the large intestine) is carried out by enzymatic gravimetry methods. Despite lengthy discussions, there is not a unanimous opinion about the methods of determining dietary fibers.

Following Codex Alimentarius documents, the AOAC 2009.01 method should be used for measuring total dietary fiber recovered by ethanol precipitation of the digested sample, and nondigestible oligosaccharides and polysaccharides that escape precipitation by ethanol, to fully account for the undigested polysaccharides and oligosaccharides. According to this standard, the prepared sample containing dietary fibers should be digested in a mixture of pancreatic amylase and amyloglucosidase in a sodium maleate buffer. Then, after inactivation of the enzymes, it should be digested with protease followed by ethanol precipitation. The mass of the filtered precipitate, containing high molecular weight dietary fibers (HMWDF), then is measured gravimetrically after drying. The supernatant, containing low molecular weight dietary fibers (LMWDF), is analyzed by means of HPLC. Total dietary fibers are measured as the sum of HMWDF and LMWDF.

The determination of soluble dietary fibers (SDF) and insoluble dietary fibers (IDF) for the purpose of food labeling as per Codex Alimentarius documents should be measured by the AOAC 993.19 and AOAC 991.42 methods accordingly. Both methods are based on the same enzymatic gravimetry principle but unlike the 2009.01 method this procedure includes digestion of samples with heat-stable α-amylase, followed by digestion with amyloglucosidase. After digestion, the IDF and SDF fractions are separated by filtration and then analyzed gravimetrically.

These methods have several limitations such as quantitation loss for inulin, resistant starch, polydextrose, and resistant maltodextrins. But these components might be analyzed by specific methods mentioned in the Codex standards.

For example, (1–3)(1–4) beta-D-glucans should be analyzed according to the AOAC 995.16 method that involves two enzymes. Lichenase hydrolyzes the molecule to oligosaccharides and β-glucosidase hydrolyzes the oligosaccharides to glucose. Glucose is then analyzed spectrophotometrically as a colored substance which is produced by treatment with a glucose oxidase/peroxidase reagent. Another example of a specific method for dietary fibers is the AOAC 997.08 method for determining the concentration of fructans. This method is based on a two-stage enzymatic extraction of fructans from the matrix using amyloglucosidase and inulinase. The concentration of the fructans, calculated as the sum of glucose and fructose after first and second hydrolysis (corrected with the free glucose and fructose level in the initial extract), is measured using high-performance anion-exchange chromatography with pulsed amperometric detection (HPAEC-PAD).

A separate group of methods includes the analysis of the content of essential nutrients, the share of which in the human diet is strongly regulated. An example of this is the analysis of the contents of vitamins in raw material: information which is of significant importance for the production of most functional foods and nutraceuticals.

Vitamin A according to Codex standard 234 should be measured using the AOAC 974.29 calorimetric method. The principle of measurement is based on the formation of a blue complex between antimony trichloride or trifluoroacetic acid and retinol in chloroform, which is measured at 620 nm. The concentration of vitamin A is calculated then using a standard calibration curve.

AOAC International Method 936.14 describes the procedure for measuring vitamin D by means of rat bioassay. According to this method, weanling rats are depleted of vitamin D over a 16–25-day depletion course and, after depletion, standard vitamin D doses and test samples are fed for 7–12 days. Recalcification of the tibia or distal end of a radius or ulna is determined by staining newly calcified areas with silver nitrate. The length and width of the calcification line determine the concentration of vitamin D in international units.

Vitamin E should be measured using AOAC Official Method 971.30 that includes saponification of a sample followed by quantitation of α-tocopherol by TLC.

Most of the water-soluble vitamins are measured using microbiological assays, for example, the AOAC 952.20 method measuring the concentration of vitamin B12 might be mentioned. This method is based on *L. delbrueckii* growth response that is sensitive to vitamin B12 at concentrations approaching 1.0 pg m/L of assay growth media. This sensitivity is sufficient to quantitate vitamin B12 in foods containing less than 0.5 μg 100 g^{-1}. According to this method, vitamin B12 active compounds should be extracted in a phosphate buffer containing a reducing agent such as metabisulfite or ascorbic acid to protect the cobalamins throughout the extraction. Standard microbiological assay techniques are used to quantitate the vitamin B12 (AOAC Method 960.46).

Nevertheless, there are some exceptions like the AOAC 967.22 method for measuring vitamin C (total) by the microfluorometric method. According to this method the steps for vitamin C analysis are as follows: extraction with the metaphosphoric acid–glacial acetic acid extractant used with the 2,6-dichloroindophenol assay (Method 967.21) followed by quinoxaline fluorescent derivative formation. Fluorescence then is determined at Ex $\lambda = 364$ nm, Em $\lambda = 440$ nm.

The third group of bioactive substances include the components the existence of which, even in trace amounts, in raw food materials is inadmissible. Examples of such substances are the allergens: lactose, gluten, milk protein, peanuts, and many others.

One of the methods approved by Codex Alimentarius is the ELISA R5 method, that is a sandwich enzyme immunoassay for the quantitative analysis of contaminations by prolamins from wheat (gliadin), rye (secalin), and barley (hordein) in raw products like flours (buckwheat, rice, corn, oats, teff) and spices as well as in processed food like noodles, ready-to-serve meals, bakery products, sausages, beverages, and ice cream. This method is based on binding of R5 monoclonal antibodies to the potentially celiac toxic epitope GLU-GLU-PRO-PHE-PRO (QQPFP) which is presented in all fractions of wheat gliadin, barley hordein, and rye secalin. The limit of quantification for the ELISA R5 method is 2.5 ppm gliadin, corresponding to 5 ppm gluten while materials which contain less than 20 ppm of gluten can be labeled as "gluten free."

11.6 AUTHENTICATION METHODS

It is not an easy task to determine the adulteration of food products, which explains the absence of internationally accepted procedures for revealing the adulteration. In such cases, the analysis converges on the selection of the theoretical information about the standard composition of the raw material and the comparison of the latter with the experimental data for every specific item under study.

There are three general types of adulteration in food materials: addition or replacement of undeclared species, false declaration of raw material origin, and false declaration of production system (Table 11.3).

Table 11.3 Adulterations and Authentication Methods

	Type of Adulteration		
	Addition of or Replacement by Undeclared Species	**False Declaration of Origin**	**False Declaration of Production System**
Adulteration purposes	- Lowering price - Masking unfavorable properties - Enhancing weak sensory properties - Increasing volume/ weight - Interfering with analysis results	- Lowering price - False declaration of geographical origin - False declaration of source species	- Lowering price - Expansion of user market
Authentication methods	HPLC, IRS, NMR, enzymatic, GC, spectrophotometry, etc.	PCR, ELISA, IRMS, IRS, MS, etc.	Regular intrusive inspection, IRMS, HPLC, AAS, etc.

IRMS, Isotope-ratio mass spectrometry; IRS, Infrared spectroscopy.

The most widely spread type of adulteration is addition or replacement of undeclared species. This type of adulteration is commonly done with the purpose of lowering the price of the final food product, masking the unfavorable properties, enhancing the weak sensory properties, increasing the volume/weight of the product, and interfering with the results of the analysis. In this case, the adulterant is any undeclared additive that interferes with the quality of the results of the analysis.

Common examples are the addition of sugar to honey, the addition of melamine to milk, and the replacement of milk fat by vegetable oils in dairy products.

The authenticity testing of honey is a good example illustrating the possible complications that might occur under this type of adulteration. The official AOAC method (998.12) for authentication of honey is based on measurement of the ratio of stable isotopes (IRMS) of carbon value (^{12}C and ^{13}C) for protein isolated from honey, that is considered to be a standard to which the stable carbon isotope ratio value of the whole honey is compared. In unadulterated honey samples the difference between ratio values should not be significant, whereas there will be a significant difference for honey adulterated by sugars from C-4 plants (from corn or cane). This official method has several limitations, the main one being that it is impossible to recognize the presence of sugars from C-3 plants (beet, rice, wheat, etc.).

Another yet nonofficial IRMS method might be used to overcome these limitations (Elflein and Raezke, 2008). As it is described by the authors of the method, it is based on the separation of sugars from honey by HPLC with their further carbon isotopes analysis. This method allows comparing the $^{13/12}C$ isotopes ratios of different sugars (eg, fructose and glucose) pairwise that are assumed as markers of the presence of sugars from C-3 plants in honey. The combination of the official and LC-IRMS methods will help to identify honey with added sugar from both C-3 and C-4 plants.

False declaration of raw materials' origin as well as in the previous case is used for lowering the price of raw materials production and involves a false declaration of geographical origin and biological origin of the product.

Common examples are false declaration of geographical origin of milk, false declaration of biological origin of protein, and false declaration of absence of genetically modified organisms (GMO).

In the case of the investigation of false declaration of geographical and biological origin, IRMS and ELISA methods should be used accordingly. The general principles of these methods are similar to the corresponding authentication methods mentioned above. Details are available elsewhere (Ballin, 2010; Fronza et al., 2002).

The presence of GMO in raw materials is an issue strictly controlled by national authorities around the world. There are two methods most used for detecting genetic modifications: ELISA and polymerase chain reaction (PCR). In this case, the ELISA method is designed to specifically identify the individual transgenic proteins. Although the ELISA method is rather simple to use and not time consuming it has a significant drawback associated with possible denaturation of the target protein during technological processing of the raw materials leading to a false negative result.

More precise and reliable results might be obtained by the PCR method. The main idea of PCR is in selective amplification of special fragments of DNA (typically gene cassette promoters) extracted from raw material followed by gel electrophoresis separation of the amplified DNA fragments (Gachet et al., 1998; Waiblinger and Grohmann, 2014). Analysis of the gene cassette promoter region allows detecting the presence or absence of any genetic modification in raw materials.

The third type of adulteration associated with false declaration of the production system cannot always be identified by analytical methods. For example, it is impossible to determine if the raw material has been produced under Kosher or Halal conditions unless local authorities conduct an intrusive inspection. However, there is a vast variety of methods for authentication of organic foods production, including sensory analysis, metabolites profiling, major and trace element analysis, volatile fingerprinting, stable isotopes analysis, etc. (Capuano et al., 2013).

11.7 CONCLUSIONS

A survey of the principles underlying the analytical methods of research of raw materials leads to the conclusion that, despite a well-devised system of methods for the supervision of the quality of the experiments and the rules of sampling, the list of internationally accepted methods of analysis is rather limited.

Most of the commonly accepted methods fall far behind the advanced scientific developments in the field of analytical chemistry, therefore, they are time-consuming, and frequently the results are less accurate. The same is true about the principles of the determination of raw material adulteration.

These circumstances determine the current tendency in the development of methods of analysis for raw food materials, namely the extension of the list of the commonly accepted methods as well as the extension of the public sources of raw food materials which will provide an effective way of determining adulteration in the production quality and safe foods and nutraceuticals.

REFERENCES

Ballin, N., 2010. Authentication of meat and meat products. Meat Sci. 86, 577–587.

Capuano, E., Boerrigter-Eenling, R., van der Veer, G., van Ruth, S.M., 2013. Analytical authentication of organic products: an overview of markers. J. Sci. Food Agric. 93, 12–28.

Cordella, C., Moussa, I., 2002. Recent developments in food characterization and adulteration detection: technique-oriented perspectives. J. Agric. Food Chem. 50, 1751–1764.

Cserháti, T., Forgács, E., 2005. Chromatography in authenticity and traceability tests of vegetable oils and dairy products: a review. Biomed. Chromatogr. 19, 183–190.

Elflein, L., Raezke, K.-P., 2008. Improved detection of honey adulteration by measuring differences between 13C/12C stable carbon isotope ratios of protein and sugar compounds with a combination of elemental analyzer—isotope ratio mass spectrometry and liquid chromatography—isotope ratio mass spectrometry (δ13C-EA/LC-IRMS). Apidologie 39, 574–587.

ENV/MC/CHEM(98)17, 1997. OECD series on principles of good laboratory practice and compliance monitoring.

Fronza, G., Fuganti, C., Schmidt, H.-L., Werner, R., 2002. The δ18O-value of the *p*-OH group of L-tyrosine permits the assignment of its origin to plant or animal sources. Eur. Food Res. Technol. 215, 55–58.

Gachet, E., Martin, G.G., Vigneau, F., Meyer, G., 1998. Detection of genetically modified organisms (GMOs) by PCR: a brief review of methodologies available. Trends Food Sci. Technol. 9, 380–388.

ISO 1443:1973—Meat and meat products—Determination of total fat content, 1973.

ISO 7980:1986;1;—Water quality—Determination of calcium and magnesium—Atomic absorption spectrometric method, 1986.

ISO 6332:1988—Water quality—Determination of iron—Spectrometric method using 1,10-phenanthroline, 1988.

ISO 18609:2000—Animal and vegetable fats and oils—Determination of unsaponifiable matter—Method using hexane extraction, 2000.

ISO 3596:2000—Animal and vegetable fats and oils—Determination of unsaponifiable matter—Method using diethyl ether extraction, 2000.

ISO 5555:2001—Animal and vegetable fats and oils—Sampling, 2001.

ISO 15586:2003—Water quality—Determination of trace elements using atomic absorption spectrometry with graphite furnace, 2003.

ISO 17294-2:2003—Water quality—Application of inductively coupled plasma mass spectrometry (ICP-MS)—Part 2: Determination of 62 elements, 2003.

ISO 1740:2004 (IDF 6:2004)—Milkfat products and butter—Determination of fat acidity (Reference method), 2004.

ISO 707:2008 (IDF 50:2008)—Milk and milk products—Guidance on sampling, 2008.
ISO 1871:2009—Food and feed products—General guidelines for the determination of nitrogen by the Kjeldahl method, 2009.
ISO 24333:2009—Cereals and cereal products—Sampling, 2009.
ISO 12966-4:2015—Animal and vegetable fats and oils—Gas chromatography of fatty acid methyl esters—Part 4: Determination by capillary gas chromatography, 2015.
ISO/TS 15495:2010 (IDF RM 230:2010)—Milk, milk products and infant formulae—Guidelines for the quantitative determination of melamine and cyanuric acid by LC-MS/MS, 2010.
Kennedy, J., Lloyd, L., 1995. Analytical chemistry of foods. Blackie Academic and Professional, Glasgow, UK.
Martinez, I., Aursand, M., Erikson, U., 2003. Destructive and non-destructive analytical techniques for authentication and composition analyses of foodstuffs. Trends Food Sci. Technol. 14, 489–498.
Nollet, L.M.L., Toldra, F. (Eds.), 2015. Handbook of Food Analysis. third ed. CRC Press, Boca Raton, FL.
Pomeranz, Y., Meloan, C.E., 2000. Food Analysis—Theory and Practice, third ed. Aspen Publishers Inc., Gaithersburg, Maryland.
Reid, L., O'Donnell, C., Downey, G., 2006. Recent technological advances for the determination of food authenticity. Trends Food Sci. Technol. 17, 344–353.
Sun, D.W., MyiLibrary, 2008. Modern Techniques for Food Authentication. Annals of Physics Academic Press, Oxford, UK.
Waiblinger, H.-U., Grohmann, L., 2014. Guidelines for validation of DNA extraction methods applied in subsequent PCR analysis of food and feed products for the presence of genetically modified material. J. für Verbraucherschutz und Leb. 9, 183–190.

CHAPTER

12 Strategize the research investigations: pre-clinical and clinical evaluations

Victor Tutelyan*, Alla Kochetkova, Vasily Isakov†, Elena Smirnova**, Varuzhan Sarkisyan****

**Laboratory of Enzymology of Nutrition, Federal Research Centre of Nutrition and Biotechnology, Moscow, Russia; **Laboratory of Food Biotechnology and Foods for Special Uses, Federal Research Centre of Nutrition and Biotechnology, Moscow, Russia; †Department of Gastroenterology and Hepatology, Federal Research Centre of Nutrition and Biotechnology, Moscow, Russia*

12.1 INTRODUCTION

The development of functional foods and nutraceuticals has the potential to eliminate the deficiency of essential nutrients in the human diet and to lead to their enrichment by way of components that exhibit proven biological activity in organs and organ systems. The practice of recent years has shown that these areas of the food industry are very promising but include numerous challenges.

First and foremost is the problem of obtaining reliable evidence of the efficacy of functional foods and nutraceuticals together with their safety for human health. A series of preclinical and clinical investigations need to be carried out to obtain this evidence.

Preclinical and clinical research investigations, the most valuable stages in new product development, serve as the only reliable evidence of safety and efficacy of functional foods and nutraceuticals. These investigations provide the most comprehensive and unbiased data that can be used in different countries, despite significant variation in their regulations.

The general sequence of safety and efficacy studies is shown in Fig. 12.1.

According to Fig. 12.1 there are three main cases in which to conduct preclinical and clinical research investigations:

First case: neither safety nor efficacy data is available. In this case, both toxicological and efficacy evaluation studies need to be conducted.

Second case: safety data is available, but efficacy data is unavailable. In this case preclinical and clinical efficacy studies need to be conducted.

Third case: both safety and efficacy data are available, but there is a lack of information on the appropriate food matrix for active ingredients. In this case, only clinical studies of the functional food or nutraceutical need to be conducted.

Developing New Functional Food and Nutraceutical Products. http://dx.doi.org/10.1016/B978-0-12-802780-6.00012-2

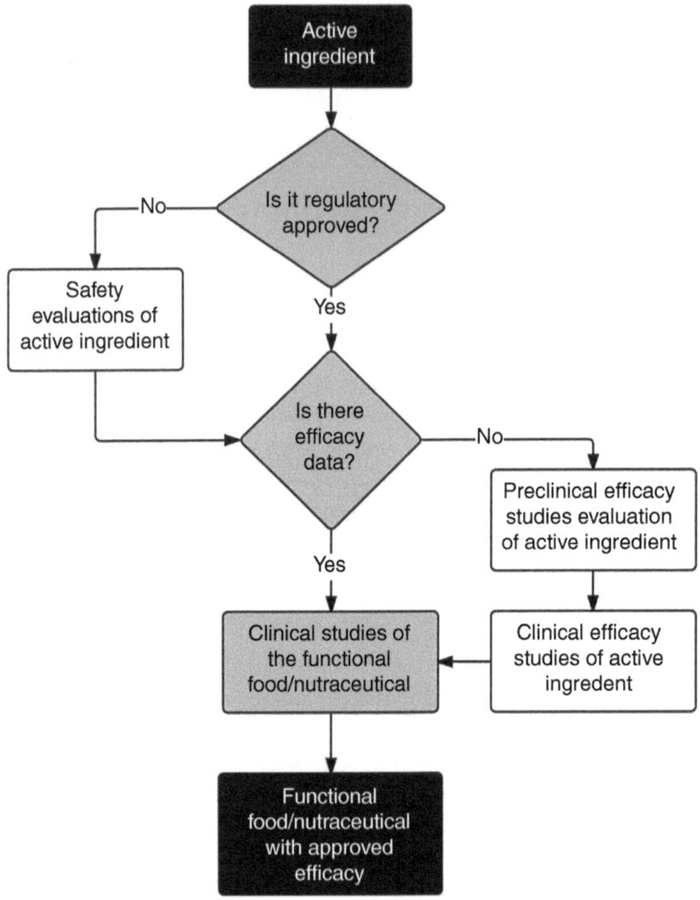

FIGURE 12.1 General Strategy of Preclinical and Clinical General Sequence of Safety and Efficacy Studies

The scope of the experiments varies depending on the available and most up-to-date scientific data on the safety and efficacy of the food compounds under study. If there is a lack of scientific data on the safety of certain food compounds, additional research should be carefully carried out to examine these overlooked areas.

Preclinical and clinical research investigations generally can be divided into two types of research: toxicological investigations and efficacy evaluation studies. There are numerous methods that can be used in both of these types of studies (Table 12.1).

Preclinical studies include in silico, in vitro, and in vivo levels of research. The in silico level is an application of combinatorial chemistry methods for virtual screening of potentially active components in relation to the targets. The most used methods, in this case, are molecular docking, quantitative structure–activity relationships (QSAR), and pharmacopoeia mapping (Raunio, 2011). The outcome from such

Table 12.1 General Review of Preclinical and Clinical Investigation Methods (AbuMweis et al., 2010; Chatzigeorgiou et al., 2009; Ekins et al., 2007; Savoie, 1994; Yao et al., 2013)

Type of Research		Aim of Usage	Experiment Duration	Throughput	Cost	Main Drawbacks
Preclinical	In silico	Hypothesis on MOA, SAR, QSAR	Low	High–ultra high	Low	• Metabolic pathways cannot be considered • Only targets with a known structure can be examined
	In vitro	Toxicology, MOA	Low	Medium	Medium	• Metabolic pathways cannot be considered
				High–ultra high	High	• Metabolic pathways cannot be considered • Expensive equipment is required
	In vivo	Toxicology, hypothesis on MOA, dose–response, biomarkers	Medium	Medium	Medium–high	• Dose exposure to humans may not be clearly recognized • Differences in metabolism between species should be considered
Clinical	Cohort studies	Examining associations between nutrient consumption and disease, generating hypothesis on MoA, possible biomarkers	High	Low	High	• A large number of subjects need to be studied for a long time • Only widespread diseases can be studied • Only diseases with long latency can be studied • Biased due to differential loss to follow-up
	Case-control studies		Medium	Low	Medium	• Only the relative risk might be measured • Study is limited to one endpoint measurement • Exposure information is fallible
	Randomized controlled clinical trials	Examining causation between nutrient consumption and disease, testing hypothesis on MoA, possible biomarkers	High	Low	Very high	• Time consuming and resource intensive • Randomized group may not be representative of the whole population • Dropouts may be substantial

MOA, Mechanism of action; SAR, structure–activity relationship; QSAR, quantitative structure–activity relationship; MoA mode of action.

studies can be used to develop hypotheses on the chemical properties of nutraceuticals, their mechanism of action (MOA), and how they can be used to determine potentially effective compounds or the most appropriate biological targets.

Though in silico methods cannot appropriately describe absorption, distribution, metabolism, and excretion (ADME) of nutraceuticals, they are widely used because they are relatively inexpensive, less time consuming, and highly productive.

The in vitro studies include a wide variety of approaches, such as testing in isolated tissues or organs, cell-based assays, or receptor or enzyme binding tests. In vitro studies are used as a first stage, in both toxicological and efficacy evaluation studies. Results from these studies provide insights on the MOA, ADME, toxicology, possible biomarkers, and endpoints, as well as reveal the dose–response relationship (Dayan, 1998). In vitro methods might be conducted in a high-throughput manner, by using automated systems that significantly increase productivity but also noticeably increase the research cost (Lang et al., 2006).

The in vivo study is the final stage of preclinical research that provides the most important information on toxicity and possible efficacy of the functional food or nutraceutical. This stage includes tests in whole animals, usually rats and mice. Results of these studies serve as a basis for making decisions on the safety of a food compound for clinical trials and also provide the information upon which to develop a hypothesis on the mode of action (MoA) to validate proposed biomarkers and dose–response relationships. Although in vitro and animal approaches provide useful information, only human clinical trials provide definitive evidence of linkages between food consumption and health and disease.

The overall effectiveness of investigations is based on appropriate research strategies that combine the aforementioned methods in a consistent sequence. General strategies for the safety investigations and efficacy evaluation studies are shown in Figs. 12.2 and 12.3 accordingly.

12.2 TOXICOLOGICAL INVESTIGATIONS

The most substantial and accepted way to conduct toxicological research is the “tiered approach,” since it uses less complex tests to obtain sufficient data for risk assessment (EFSA, 2011; FDA, 2007; R1.2.3156-13, 2014). This approach is designed to address four general toxicological aspects such as:

1. Toxicokinetics;
2. Toxicity (subchronic, chronic toxicity, and carcinogenicity);
3. Genotoxicity;
4. Reproductive and developmental toxicity.

There are three levels for toxicological investigations within the scope of the tiered approach. Tier 1 is mostly used to determine if the compound under examination adsorbs or demonstrates toxicity or genotoxicity. If it does then Tier 2 testing is required to specify toxicological effects during long-term experiments. Tier 3 testing

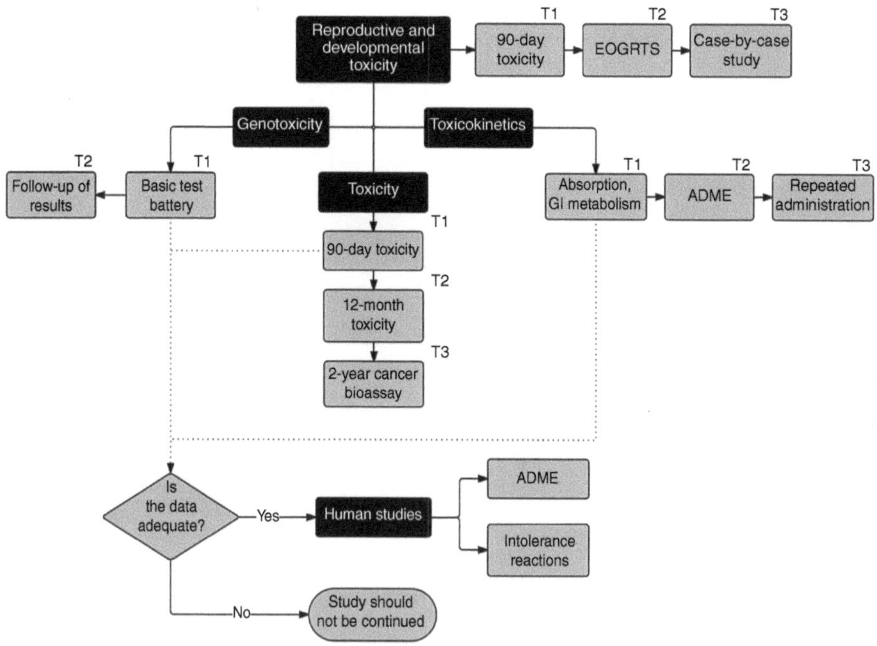

FIGURE 12.2 Tiered Approach for Toxicological Evaluation

T1, T2, and *T3* Tiers 1–3 accordingly; *ADME*, absorption, distribution, metabolism, excretion; *EOGRTS*, extended one-generation reproductive toxicity study.

is performed on an individual basis considering all the available data to clarify specific endpoints that are needed for further investigation of Tier 1 and 2 findings.

In this chapter the tiered approach for food components is described structurally to help understand the main points of the different stages for each type of study.

12.2.1 TOXICOKINETICS STUDIES

Toxicokinetics, that is, ADME, is an important part of research in health risk assessment that provides information on species and doses for toxicity testing through the comparison of internal doses in experimental animals and humans (Creton et al., 2009). However, it should be noted that the toxicokinetic processes may become saturated at doses higher than those expected to be relevant to human exposure, which can result in toxicity that would not be relevant to the intended use of a compound (Slikker et al., 2004).

12.2.1.1 Tier 1: absorption studies

The aim of the absorption studies is to establish the potential of the food compound, its metabolites, or breakdown products, to be absorbed from the gastrointestinal tract.

Generally used methods: in vitro, in vivo, and ex vivo absorption and bioavailability methods, including methods of determining physicochemical properties such as

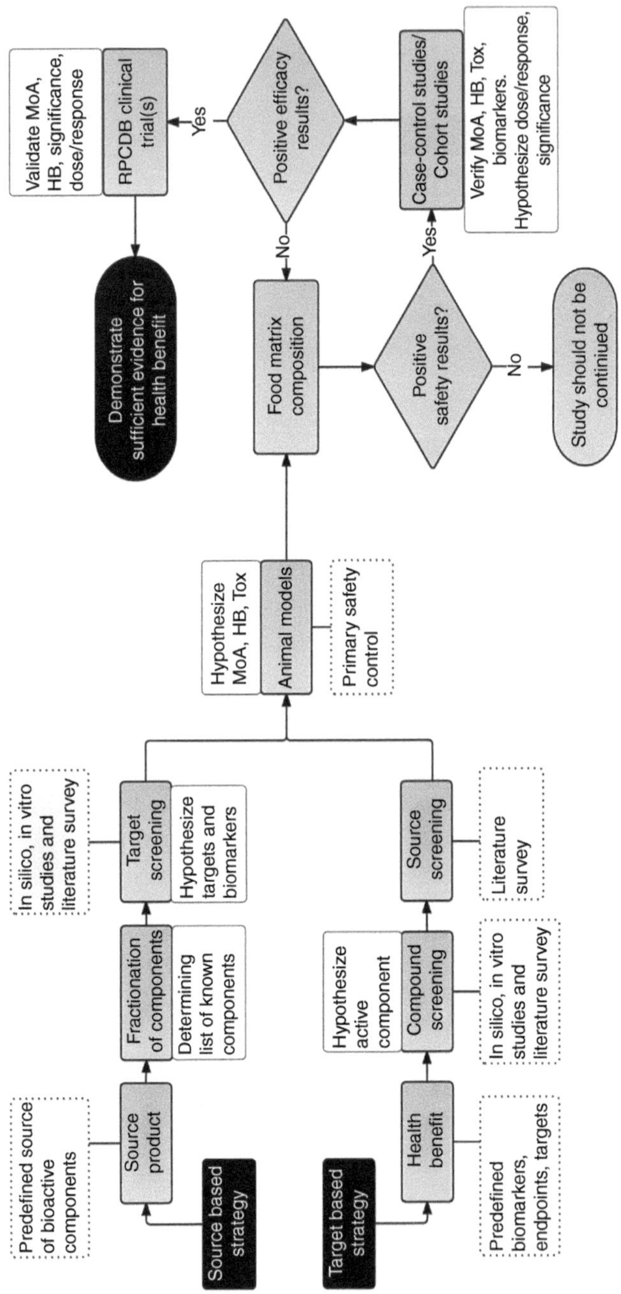

FIGURE 12.3 Strategies for Efficacy Evaluation

Tox, toxicology; *HB*,health benefit; *MoA*, mode of action; *RPCDB*, randomized, placebo-controlled double-blind.

hydro- and lipo-philicity, molecular weight, and the ionization constant; the in vitro Ussing chamber (Ussing and Zerahn, 1951) and the inverted sac (Kato et al., 2004) are gastrointestinal metabolism models.

Interpretation of the results: if a negligible absorption is demonstrated on Tier 1, further studies on higher tiers may not be conducted. Otherwise higher-tier investigations should be carried out.

12.2.1.2 Tier 2: single dose distribution, metabolism, and excretion studies

The aim of the studies is to assess ADME of food compounds through the use of in vivo experiments.

Generally used methods are described in OECD TG 417. These methods are focused on the assessment of a compound's ADME along with the definition of single dose toxicokinetic parameters, such as area under curve (AUC), maximal concentration (C_{max}), and half-life time ($t_{1/2}$). In vitro data and dose-dependency assays to determine the effects on various enzymes, liver subcellular fractions, or cell lines may also be useful in this tier.

Interpretation of the results: Tier 2 studies are expected to provide a proposed metabolic pathway, potential species and sex differences, possible dose-dependent changes in toxicokinetic parameters, rates of clearance, and bioaccumulation potential.

12.2.1.3 Tier 3: repeated administration toxicokinetic studies

The aim of the studies is to investigate the possible mechanism of slow or limited excretion leading to bioaccumulation, by means of repeated administration.

Generally used methods are in vivo repeated dose toxicokinetic studies to help predict the ADME of a food compound.

Interpretation of the results: Tier 3 studies are expected to provide possible age, physiological state, and disease state dependent toxicokinetic parameters differences.

12.2.2 SUBCHRONIC, CHRONIC, AND CARCINOGENICITY TOXICITY STUDIES

The toxicity studies of peroral administration of food compounds are used to provide data on biochemical changes in body fluids (urine, blood) and histopathological changes in organs. Such investigations include subchronic and chronic toxicity studies, as well as carcinogenicity studies to establish the highest dose level of the food compound and target organs and tissues. The toxicity studies may reveal the need for additional immunological or neurotoxic studies.

12.2.2.1 Tier 1: subchronic toxicity studies

The aim of the studies is to indicate target organs and tissues and to identify chemicals with potentially neurotoxic, immunological, reproductive organ effects, or endocrine-mediated effects.

Generally used methods: 90-day toxicity test according to OECD TG 408 in rodents; repeated dose 28-day test as described in OECD TG 407.

Dose levels of the food compound should be selected considering any existing toxicity and toxicokinetic or other relevant data. The highest dose level should be chosen with the aim of inducing toxic effects but not death or severe suffering.

Interpretation of the results: Tier 1 toxicity studies are expected to identify the no observed adverse effect level (NOAEL) or a lower confidence bound of the benchmark dose (BMDL) of food compounds. Extended investigations are needed if any indicators have been observed in the OECD TG 407 test.

12.2.2.2 Tier 2: chronic toxicity studies

The aim of the studies is to confirm effects observed in subchronic studies or reveal effects not evident in subchronic studies.

Generally used methods: 12-month toxicity and carcinogenicity tests according to OECD TG 452 and OECD TG 451, respectively.

Interpretation of the results: Tier 2 toxicity studies are expected to investigate the MoA or mechanism of toxicity or carcinogenicity.

12.2.2.3 Tier 3: carcinogenicity studies

The aim of the studies is to evaluate carcinogenic effects during the majority of the normal life span of the testing animals. Neurotoxicity, immunotoxicity, or endocrine-mediated effects may also be considered.

Generally used methods: 24-month tests for rats and 18-month tests for mice according to OECD TG 451.

Interpretation of the results: Tier 2 toxicity studies are expected to determine the relevance of effects observed in the test species for humans.

12.2.3 GENOTOXICITY

The goal of genotoxicity studies is to identify genetic alterations in somatic or germ cells caused by the food compounds. These alterations may affect cardiovascular, immune, nervous, and reproductive systems.

Both in vitro and in vivo tests are required to provide sufficient information on mechanisms and modes of genotoxic action of food compounds, considering the effect of metabolites and breakdown products of the compound.

12.2.3.1 Tier 1: in vitro genotoxicity studies

The aim of the studies is in vitro screening to determine potential genotoxicity of the test components.

Generally used methods: bacterial reverse mutation assay according to OECD TG 471, in vitro mammalian cell micronucleus test according to OECD TG 487.

Interpretation of the results: if all in vitro endpoints are negative, it can be concluded that the substance is not a genotoxic hazard. In the case of inconclusive results from Tier 1 studies, it may be appropriate to conduct further testing in vitro to resolve the situation.

In the case of positive results from the Tier 1 studies, further in vivo studies should be conducted on Tier 2.

12.2.3.2 Tier 2: in vivo genotoxicity studies

The aim of the studies is to identify substances which could potentially be genotoxic in humans and their MOA and MoA.

Generally used methods: in vivo Comet assay, in vivo micronucleus, and a transgenic rodent test (OECD TG 474 and OECD TG 488 accordingly).

Interpretation of the results: if the results of Tier 2 studies are negative, then it can be concluded that the substance is not an in vivo genotoxic, otherwise it can be concluded that the substance is an in vivo genotoxic.

12.2.4 REPRODUCTIVE AND DEVELOPMENTAL TOXICITY

Reproductive toxicity and developmental toxicity studies are long-term experiments, including a wide range of tests on the influence of food components on the reproductive ability of an organism and the development of its progeny. In contrast to other toxicological studies, these studies require assessment of the second generation.

12.2.4.1 Tier 1: toxicokinetics and 90-day subchronic toxicity studies

The aim of the studies is to characterize bioavailability and toxicity to reproductive systems caused by the food compound.

Generally used methods: absorption study and repeated dose 90-day oral toxicity study (OECD TG 408) focusing on toxicity to reproductive organs.

Interpretation of the results: if Tier 1 studies show sufficient absorption or toxicity to the reproductive system in subchronic experiment Tier 2, reproductive and developmental toxicity studies should be considered. If negligible absorption is shown at Tier 2, testing need not be performed.

12.2.4.2 Tier 2: reproductive and developmental toxicity studies

The aim of the studies is to identify possible lethal, teratogenic, or other toxic effects on the embryo and fetus caused by food components and their influence on male and female libido and fertility, as well as on the female's ability to carry pregnancy.

Generally used methods: prenatal developmental toxicity studies according to OECD TG 414 and extended one-generation reproduction toxicity study (EOGRTS) as described in OECD TG 443.

Interpretation of the results: Tier 2 reproductive and developmental toxicity studies are expected to determine reproductive and developmental endpoints. Extended investigations on Tier 3 are needed if any indicators have been observed.

12.2.4.3 Tier 3: extended reproductive and developmental toxicity studies

Tier 3 studies are used to investigate modes of toxic action and additional developmental and reproductive toxicity of food compounds, such as developmental neurotoxicity. These studies are conducted on a case-by-case basis based on the study results from the Tier 2 stage.

12.2.5 HUMAN STUDIES

As is shown in Fig. 12.1, a favorable outcome of Tier 1 toxicokinetic, genotoxicity, and 90-day toxicity studies is the prerequisite for human safety studies. These studies include physical examination, blood chemistry, hematology, urine analysis, and organ function tests. Clinically controlled assessment for single or short-term repeated administration of the food component to volunteers is used to conduct absorption, metabolism, distribution, and elimination studies, as well as tolerance studies.

Toxicokinetic studies of food compounds in volunteers would help to extrapolate the in vitro and in vivo investigations' results to humans. Food intolerance tests are used to investigate such symptoms as headaches and gastrointestinal discomfort, that cannot be studied by way of animal experiments.

12.3 EFFICACY EVALUATION

In contrast to toxicological studies, efficacy evaluation is a simple task, both by the amount of possible experimental setups and by the interpretation of further results.

As shown in Fig. 12.2, the preclinical stage of efficacy evaluation may be conducted by the following two strategies: target based and source based. The target-based strategy is an approach that is widely used in drug development procedure. In this case, the researcher starts from the predefined health benefits (HB), biomarkers, and biological targets. The aim of this type of research is to characterize active compounds that might be used as functional components for nutrition. The source-based approach can be used when the whole range of compounds is known, but the targets or biomarkers are unknown (e.g., in studies of the components of traditional medicine) (Narayana, 2010). Using a source-based strategy, the researcher starts from known biological sources (herbal or animal origin) with a known composition and proposed health benefit. The aim of this type of research is to find appropriate biological targets and to propose a mechanism of action and possible biomarkers.

Both of these strategies involve identical in vivo experiments that are focused on the primary validation of proposed activities of compounds.

Appropriate food matrix selection for the functional compound is an important and usually quite complicated task in functional food design. The food matrix acts as a delivery system and environment for bioactive compounds that significantly influences their bioavailability and stability.

Modification of the stability and bioavailability of epigallocatechin-3-gallate (EGCG) is an excellent example to illustrate the importance of the food matrix. EGCG is the main polyphenol in green tea that has shown various health effects as an antioxidant, exhibiting antiinflammatory and antiatherogenic properties. In addition, there are a number of research papers showing that EGCG has strong inhibitory activity against alpha-amylase and alpha-glucosidase in vitro. At the same time, the overwhelming majority of in vivo experiments and clinical trials show a small or an unfavorable inhibitory activity against carbohydrate hydrolyzing enzymes (Liu et al., 2014; Zheng et al., 2013). This difference might be caused by low bioavailability and/or lack of stability of EGCG. As described in Mereles and Hunstein (2011), the oxidative stability of EGCG is highly correlated with the pH of the food matrix, the presence of metal ions such as Ca^{2+} or Mg^{2+}, the concentration of oxygen, and storage temperature. The presence of protein, vitamin C, ω-3 fatty acids, and presence of food are some of the factors affecting the bioavailability of EGCG.

Despite this, in most of the afore-mentioned preclinical and clinical investigations only aqueous solutions of EGCG or green tea water extracts were used without considering the influence of environment on its stability and bioavailability. A subsequent research, using different food matrices for EGCG, showed an increase of glucose-lowering effect (Ortsäter et al., 2012; Wolfram et al., 2005), which was not seen with the aqueous solution. In light of the aforementioned factors, a study on the influence of the food matrix on the bioavailability of an active ingredient is a significant step in testing its efficacy as a component of functional foods.

12.3.1 PRECLINICAL RESEARCH

As in the case of toxicological studies, clinical investigations require prerequisites. In the case of efficacy evaluations these prerequisites are in silico, in vitro, and in vivo studies that are not regulative, but may significantly simplify further clinical research.

In silico screening studies, despite their limitations, are widely used in the design of new functional foods. Analysis of several programs for in silico studies is presented in the work of Cross with coauthors (Cross et al., 2009). The most widespread application of these methods for functional food design is the screening of new antidiabetic components.

A recent example of the screening of antidiabetic components is a study by Laura Guasch and coauthors who identify partial agonism of PPARγ for a variety of natural extracts that exhibit antidiabetic activity. For this purpose the authors used a virtual screening method based on two structure-based pharmacophores, protein–ligand docking, and an electrostatic/shape similarity analysis. Screening of the initial set of more than 29,000 components led to predicting 22 new molecules as potential PPARγ partial agonists. Furthermore, their results allowed making a hypothesis on the MOA of the tested natural extracts for their antidiabetic properties, that is, the insulin-stimulated glucose uptake is augmented through the action on PPARγ (Guasch et al., 2013). Such results obtained from in silico studies, providing information on a large list of compounds, should subsequently be confirmed in in vitro and in vivo tests.

In vitro methods that are used for functional products assessment are similar to methods which are used in investigations of new medications, which can be illustrated by taking the example of the prebiotic inulin. Wang and Gibson measured the in vitro fermentability of inulin by measuring the bacterial end-product formation in batch cultures. Their experiments showed that inulin was preferentially utilized by the *Bifidobacterium* at normal pH, while potentially pathogenic species such as *E. coli* and *Clostridium* were maintained at low levels (Wang and Gibson, 1993).

The same bifidogenic effect has been proved in terms of in vivo investigations in a study using rats, which were fed diets containing inulin. The bifidogenic activity of inulin was confirmed by an increase of the bifidobacteria count, short-chain fatty acids production, P-galactosidase and fructohydrolase activities, and also by a decrease in intestinal pH. This experiment revealed the dose-dependent manner of the inulin action and its MoA (Andrieux et al., 1991).

12.3.2 CLINICAL TRIALS

Evaluation of the clinical efficacy of functional food products is not a simple task because the clinical effect of food significantly differs from that of medication. Sometimes the clinical efficacy of functional products takes longer to develop an effect than in the case of pharmaceuticals.

Research investigations into the efficacy of functional food and nutraceuticals should be based on substantial evidence to ascertain a causal relationship between food consumption and a specific health benefit. The most common approach determining causation and its separation from other potential factors is the evaluation of the scientific results using criteria proposed by Bradford Hill (Golan and Linn, 2015). According to Dr. Hill, evidence from research investigations must satisfy nine criteria to be able to prove a causative relationship between the proposed active compound and a particular disease. They are:

1. Strength—evidence should be strong, that is, statistically significant.
2. Consistency—evidence should be consistent and confirmed in other studies.
3. Specificity—results should demonstrate a predictable relationship between the bioactive component and health effects.
4. Temporality—health effects should always be preceded by the treatment.
5. Dose-response relationship—increase in the dose of the bioactive component should have an impact on health effect.
6. Plausibility—the effect of the bioactive component should be characterized by a plausible mechanism of action.
7. Coherence—the association between bioactive components and outcomes should be compatible with the existing scientific knowledge.
8. Experiment—it should be possible to subject it to experimental evidence.
9. Analogy—the effect of similar factors should be considered.

These criteria are the evidentiary basis for evaluating the effectiveness of functional foods and nutraceuticals.

The goal of the study of functional foods is to learn their medical effects, including the rate of remission, recovery time, the incidence of complications of any underlying disease, as well as normalization or improvement of the nutritional status of the patient.

Randomized controlled studies with parallel-group or with cross-over design are usually used to evaluate the clinical efficacy of functional foods and nutraceuticals. It is a substantially complicated task to conduct studies with a double-blind design because of their volume and sensory properties. Producing a "placebo product" having sensory properties of the test product, but no medicinal properties (i.e., without components that are associated with the therapeutic effect), is extremely difficult or even impossible in some cases.

An impartial assessment of the therapeutic effect of a functional food or nutraceutical is a challenging task because they are often not the only or even the main treatment. Treatment of gluten enteropathy and similar diseases are the few exceptions. Usually, they are used as an addition to the medical therapy or as a component of a complex treatment protocol, using surgical, medicinal, or other therapies of the disease. In this regard, groups of patients (basic and control) in the study of the effectiveness of the functional food must be homogeneous not only by age, sex, severity of the underlying disease, and other standard indicators, but also by the method for treatment of the disease (set of medications, their dosage, forms of surgical interventions, and others), and if the trial is performed in patients, by whether they have received any medical treatment.

Specific attention should be paid to the nutrition of patients under any such study. The basic diet of both groups of patients should not differ in calories, macronutrient composition, and multiplicity of meals and, in general, must comply with the physiological needs for nutrients and energy for a patient according to his or her sex, age, and level of physical activity. One of the standard diets that are regulated on the national level can be used in both groups of patients as another alternative. Basic calorie diet and nutrient composition should be adjusted (reduced) in the main study group in which patients receive functional food products based on the caloric and macronutrient composition of the particular product under study.

Normalization or improvement (change of parameters toward normal levels) of the patient's nutritional status is one of the most important criteria for the functional food or nutraceutical beneficial health effect. Nutritional status is as follows:

1. The actual nutrition and body composition (evaluated using bio-impedancometry or X-ray densitometry);
2. Resting energy expenditure (assessed by indirect calorimetry with the calculation of the oxidation rate of proteins, fats, and carbohydrates, using intermediate indicators of nonprotein energy expenditure and nonprotein respiratory quotient);
3. Biochemical blood parameters characterizing the metabolism of
   a. proteins (total protein, urea, creatinine, uric acid, albumin);
   b. lipids (lipidogram);
   c. carbohydrates (glucose, glycated hemoglobin, insulin, homeostatic model assessment);
   d. vitamin status indicators (vitamins A, C, E, B6);
   e. mineral and trace elements (sodium, potassium, magnesium, calcium, zinc, selenium, chromium, iodine).

It should be noted that a beneficial health effect of the same functional food product can be observed in a variety of diseases. This is not surprising since functional foods tend to realize their effect on the level of "shunting" disturbed metabolic pathways or correction of a nutrient insufficiency that is typical for many diseases. However, this does not mean that the resulting beneficial health effect of a dietary product in one disease may be mechanistically extrapolated into another, even if the nature of the nutrients' metabolic abnormalities in these diseases is similar. Therefore, clinical efficacy studies of functional foods or nutraceuticals for each disease or even in a disease's different clinical forms are required.

The study of functional foods in irritable bowel syndrome (IBS) could be considered as an example. This disease occurs in two main clinical forms: diarrhea-predominant and constipation-predominant. The main clinical parameters that characterize IBS are reduced quality of life, intestinal disorders, and abdominal pain. Normalization or disappearance of these parameters will be the disease remission or recovery markers, in this case. Patients diagnosed with IBS with a clinical form recognized according to standard criteria (Longstreth et al., 2006) are randomized into two groups. They both receive a basic diet. The experimental group additionally receives the functional product with the basic diet adjusted for essential nutrients taking into account the chemical composition of the functional product. The Bristol stool chart should measure stool characteristics after 0, 3, 7, and 14 days of the studies. The intensity of abdominal pain should be measured based on the visual analogue scale or validated questionnaires and indices (such as the index of the severity of IBS BEST) (Spiegel et al., 2006). The quality of life is also evaluated by general or specialized questionnaires, for example, IBS-QOL (Hahn et al., 1997). A sufficient amount of efficacy studies on inulin-enriched foods in IBS has been done according to these criteria (Collado Yurrita et al., 2014; Isakov et al., 2013). Foods with preventive action are a specific category of functional foods. The systematic (regular) use of these foods by all age groups of healthy people reduces the risk of human diseases.

The subject of these research studies is the preventive effect of a product. The preventive effect is usually confirmed by lower rates of the disease in the treated group compared to those of the control group that did not receive the functional food product. Comparative randomized controlled prospective studies with parallel and cross-over designs are widely used in this case. The feature of studying the effectiveness of foods with preventive action is that the preventive effect (unlike the therapeutic effect) is achieved over a longer period of time (sometimes over decades).

Therefore, there is broad acceptance for studies on the preventive effect of functional foods in which the preventive effect is evaluated by the surrogate marker but not by decreasing the incidence of disease. By way of example, epidemiological studies reveal that the incidence of hypertensive disease is closely related to excessive consumption of sodium. Moreover, it has been shown that this relationship is dose-responsive. An increase in the ratio of salt to 5 g/day (2000 mg sodium) is associated with an increase in risk of cardiovascular diseases by 17% and stroke by 23% (Strazzullo et al., 2009).

Thus, it would be appropriate to evaluate the preventive effect of functional foods with a reduced amount of sodium as compared to a conventional product in a cohort study. Two fairly large groups of several thousand people, who would be randomly assigned to the control group, receiving the traditional diet, and the trial group, should be formed when using the classic criteria of efficacy. In the trial group, the participants would receive a preventive food with reduced sodium content instead of the usual for several years. It would be possible to talk about the preventive effect if the difference in the incidence of cardiovascular disease between the groups became significant at some point.

This type of study, however, can be designed in another way. It is known that determining sodium intake with food using the questionnaire method or a food diary is difficult. Nevertheless, daily urinary sodium excretion accurately reflects the level of its consumption; that is why this surrogate marker is used in epidemiological studies (Takahashi et al., 2006).

Hence it appears that if the functional product's consumption leads to a significant reduction in urinary sodium excretion (even in a small group of patients) then the functional product exhibits a health-beneficial effect.

The main disadvantage of this latter kind of research is that one cannot use it to assess the magnitude of the preventive effect, that is, it is impossible to answer the question: "How long should a person consume food with preventive properties to show the preventive effect?" However, this disadvantage is common to all the studies, including studies with the use of medications in which surrogate markers are used. For example, it is known that the preventive effect of statins on cardiovascular disease risk is associated with lower blood cholesterol. Thus, by assigning statins and determining the proper dose, the physician is guided by this surrogate marker. Notwithstanding that, it has been found in long-term prospective studies that continuous use of statins for several years is required to decrease the risk of cardiovascular diseases significantly (Ebrahim et al., 2014). Therefore, if there are studies demonstrating a strong association between a surrogate marker (or its numeric value) and incidence of the disease, this marker can be considered as indirect evidence of a preventive effect.

12.4 CONCLUSIONS

Preclinical and clinical research investigations are a complicated process that include studies from the molecular to entire organism levels with the general aim to provide significant data on the safety and efficacy of new types of functional foods and nutraceuticals. Every stage of the investigations has its difficulties and limitations that should be carefully considered to avoid undesirable outcomes.

Strategies for preclinical and clinical investigations alluded to in this chapter are an overall guideline to help optimize the choice of appropriate analytical methods and clinical approaches. They will substantially decrease the cost and duration of experiments and will increase the overall impact of the research.

REFERENCES

AbuMweis, S.S., Jew, S., Jones, P.J.H., 2010. Optimizing clinical trial design for assessing the efficacy of functional foods. Nutr. Rev. 68, 485–499.

Andrieux, C., Lory, S., Dufour-Lescoat, C., de Baynast, R., Szylit, O., 1991. Physiological effects of inulin in germ-free rats and in heteroxenic rats inoculated with a human flora. Food Hydrocoll. 5, 49–56.

Chatzigeorgiou, A., Chatzigeorgiou, A., Halapas, A., Kalafatakis, K., Kamper, E., 2009. The use of animal models in the study of diabetes mellitus. In Vivo 23, 245–258.

Collado Yurrita, L., San Mauro Martín, I., Ciudad-Cabañas, M.J., Calle-Purón, M.E., Hernández Cabria, M., 2014. Effectiveness of inulin intake on indicators of chronic constipation; a meta-analysis of controlled randomized clinical trials. Nutr. Hosp. 30, 244–252.

Creton, S., Billington, R., Davies, W., Dent, M.P., Hawksworth, G.M., Parry, S., Travis, K.Z., 2009. Application of toxicokinetics to improve chemical risk assessment: implications for the use of animals. Regul. Toxicol. Pharmacol. 55, 291–299.

Cross, J.B., Thompson, D.C., Rai, B.K., Baber, J.C., Fan, K.Y., Hu, Y., Humblet, C., 2009. Comparison of several molecular docking programs: pose prediction and virtual screening accuracy. J. Chem. Inf. Model. 49, 1455–1474.

Dayan, A.D., 1998. Drug Discovery and Evaluation edited by HG Vogel and WH Vogel. Hum. Exp. Toxicol. 17, 591.

Ebrahim, S., Taylor, F.C., Brindle, P., 2014. Statins for the primary prevention of cardiovascular disease. BMJ 348, g280.

EFSA, 2011. Draft guidance for submission for food additive evaluations 1. EFSA J. 10, 1–57.

Ekins, S., Mestres, J., Testa, B., 2007. In silico pharmacology for drug discovery: applications to targets and beyond. Br. J. Pharmacol. 152, 21–37.

FDA, 2007. Redbook 2000: Guidance for industry and other stakeholders: Toxicological principles for the safety assessment of food ingredients: Redbook 2000: Page 1 of 4, Redbook. Hum. Stud. 3, 1–286.

Golan, D., Linn, S., 2015. From statistical associations to scientific causality. Harefuah 154, 389–393, 403.

Guasch, L., Sala, E., Mulero, M., Valls, C., Salvadó, M.J., Pujadas, G., Garcia-Vallvé, S., 2013. Identification of PPARgamma partial agonists of natural origin (II): in silico prediction in natural extracts with known antidiabetic activity. PLoS One 8, e55889.

Hahn, B.A., Kirchdoerfer, L.J., Fullerton, S., Mayer, E., 1997. Evaluation of a new quality of life questionnaire for patients with irritable bowel syndrome. Aliment. Pharmacol. Ther. 11, 547–552.

Isakov, V., Pilipenko, V., Shakhovskaya, A., Tutelyan, V., 2013. Efficacy of inulin enriched yogurt on bowel habits in patients with irritable bowel syndrome with constipation: a pilot study. FASEB J. 27, lb426.

Kato, T., Hayashi, Y., Inoue, K., Yuasa, H., 2004. Functional characterization of the carrier-mediated transport system for glycerol in everted sacs of the rat small intestine. Biol. Pharm. Bull. 27, 1826–1830.

Lang, P., Yeow, K., Nichols, A., Scheer, A., 2006. Cellular imaging in drug discovery. Nat. Rev. Drug Discov. 9, 343–356.

Liu, C.-Y., Huang, C.-J., Huang, L.-H., Chen, I.-J., Chiu, J.-P., Hsu, C.-H., 2014. Effects of green tea extract on insulin resistance and glucagon-like peptide 1 in patients with type 2 diabetes and lipid abnormalities: a randomized, double-blinded, and placebo-controlled trial. PLoS One 9, e91163.

Longstreth, G.F., Thompson, W.G., Chey, W.D., Houghton, L.A., Mearin, F., Spiller, R.C., 2006. Functional bowel disorders. Gastroenterology 130, 1480–1491.

Mereles, D., Hunstein, W., 2011. Epigallocatechin-3-gallate (EGCG) for clinical trials: more pitfalls than promises? Int. J. Mol. Sci. 12, 5592–5603.

Narayana, A.D.B., 2010. Reverse pharmacology for developing functional foods/herbal supplements: approaches, framework and case studies. In: Smith, J., Charter, E. (Eds.), Functional Food Product Development. Wiley-Blackwell, Oxford, pp. 244–255.

Ortsäter, H., Grankvist, N., Wolfram, S., Kuehn, N., Sjöholm, A., 2012. Diet supplementation with green tea extract epigallocatechin gallate prevents progression to glucose intolerance in db/db mice. Nutr. Metab. 9, 11.

R1.2.3156-13, 2014. Evaluation of the toxicity and hazards of chemical substances and their mixtures to human health. Russia.

Raunio, H., 2011. In silico toxicology – non-testing methods. Front. Pharmacol. 2, 33.

Savoie, L., 1994. Digestion and absorption of food: usefulness and limitations of in vitro models. Can. J. Physiol. Pharmacol. 72, 407–414.

Slikker, W., Andersen, M.E., Bogdanffy, M.S., Bus, J.S., Cohen, S.D., Conolly, R.B., David, R.M., Doerrer, N.G., Dorman, D.C., Gaylor, D.W., Hattis, D., Rogers, J.M., Setzer, R.W., Swenberg, J.A., Wallace, K., 2004. Dose-dependent transitions in mechanisms of toxicity. Toxicol. Appl. Pharmacol. 201, 203–225.

Spiegel, B.M.R., Mayer, E., Bolus, R., Chang, L., 2006. Development and initial validation of a concise point-of-care IBS severity index: the 4-item "BEST" questionnaire. Gastroenterology 130, S1040.

Strazzullo, P., D'Elia, L., Kandala, N.-B., Cappuccio, F.P., 2009. Salt intake, stroke, and cardiovascular disease: meta-analysis of prospective studies. BMJ 339, b4567.

Takahashi, Y., Sasaki, S., Okubo, S., Hayashi, M., Tsugane, S., 2006. Blood pressure change in a free-living population-based dietary modification study in Japan. J. Hypertens. 24, 451–458.

Ussing, H.H., Zerahn, K., 1951. Active transport of sodium as the source of electric current in the short-circuited isolated frog skin. Acta Physiol. Scand. 23, 110–127.

Wang, X., Gibson, G.R., 1993. Effects of the in vitro fermentation of oligofructose and inulin by bacteria growing in the human large intestine. J. Appl. Bacteriol. 75, 373–380.

Wolfram, S., Raederstorff, D., Wang, Y., Teixeira, S.R., Elste, V., Weber, P., 2005. TEAVIGO (epigallocatechin gallate) supplementation prevents obesity in rodents by reducing adipose tissue mass. Ann. Nutr. Metab. 49, 54–63.

Yao, C.K., Gibson, P.R., Shepherd, S.J., 2013. Design of clinical trials evaluating dietary interventions in patients with functional gastrointestinal disorders. Am. J. Gastroenterol. 108, 748–758.

Zheng, X.-X., Xu, Y.-L., Li, S.-H., Hui, R., Wu, Y.-J., Huang, X.-H., 2013. Effects of green tea catechins with or without caffeine on glycemic control in adults: a meta-analysis of randomized controlled trials. Am. J. Clin. Nutr. 97, 750–762.

CHAPTER

Concepts of flavor creation in novel nutraceuticals and functional food formulations

13

Shane T. McDonald

Principal Flavor Chemist, Kalsec Inc., Kalamazoo, MI, United States

13.1 INTRODUCTION

While most of this book is dedicated to developing foods and nutraceuticals that deliver health benefits to the consumer, this chapter discusses how to develop those foods so they taste good. The flavorings themselves probably do not deliver health benefits (unless the flavorings have health implications in their own right, such as certain spices).

We will start by discussing the importance of flavor in these products. This leads into a basic review of flavor chemistry. Unique flavor challenges to flavoring functional foods and nutraceuticals follow, as well as ways to overcome those challenges. Most developers of functional foods will choose to partner with a flavor company in flavoring their products, so we will conclude the chapter with how to work with a flavor house.

13.2 WHY IS FLAVOR IMPORTANT IN NUTRACEUTICALS AND FUNCTIONAL FOODS?

Functional foods are first and foremost foods, and they compete with conventional foods (and competitors' functional foods) for a share of the public's stomach. Despite the added value of heath functionality, consumers are unlikely to buy functional foods over the long term if the foods don't taste good (Reineccius, 2000). A 2011 consumer trending survey on functional foods conducted by the International Food Information Council found that the most important factors affecting Americans' decisions to consume functional foods were price and taste (Food Insight, 2011).

Functional foods should not be studied as one homogeneous group, but as separate products within the various food categories. Increasing a functional food's functionality, such as increasing bioactive compounds, may cause off-flavors that cause

Developing New Functional Food and Nutraceutical Products. http://dx.doi.org/10.1016/B978-0-12-802780-6.00013-4

aversive customer reaction due to taste (Siro et al., 2008). A study on off-flavor in juice with specific health claims (Tuorila and Cardello, 2002) found that even with knowledge of the health claims, consumers' degree of liking and likelihood of consumption decreased with increasing severity of the flavor defect. No support was found for the theory that a slight off-flavor was perceived as a positive marker of health benefits to the consumer.

Even in medical foods, where the consumer may be taking the product under doctor's prescription, flavor still is very important. Flavor has several benefits besides making food taste good including stimulation of digestion as well as the physiological well-being of the person (Reineccius, 2000).

Flavors are added to functional foods:

1. To impart flavor to an otherwise bland product.
2. To modify or complement an existing flavor base.
3. To disguise or cover undesirable flavor attributes or off-notes, either by blocking, masking, or otherwise tricking the taste buds into not recognizing the off-flavor.
4. A combination of 1, 2, or 3 (Keskar and Igou, 2011).

13.3 WHAT IS FLAVOR?

All of the senses play into food choice. The senses can be considered "gatekeepers," guiding the selection of food as acceptable for consumption as opposed to that which should be rejected (Delahunty, 2011). All of the senses take part in food selection, but this chapter will concentrate on taste, smell, and the tactile sensations, especially "chemesthesis" (discussed later). According to Hall (1968), "flavor is the sum of those characteristics of any material taken in the mouth, perceived principally by the senses of taste and smell, and also the general pain and tactile receptors in the mouth, as received and interpreted by the brain." The word can either refer to the biological perception of food being taken into the mouth, or to an attribute of the food being perceived (Fisher and Scott, 1997).

There are five widely recognized tastes: sweet, salty, bitter, sour, and umami. There are probably evolutionary bases for taste discrimination. Salty taste detects salt, which is a necessary nutrient. Sweet indicates sugars, which are sources of energy. Umami may be an indicator of protein. On the other hand, sour may indicate spoilage, and bitter may be a warning of toxic compounds.

Food contains certain volatile chemicals. When these escape the food and are inhaled through the nose (orthonasal), some of these compounds stimulate receptors associated with the olfactory nerve and are perceived as aroma. When food is ingested, the taste buds detect the basic tastes. At the same time, air containing volatile compounds is swept into the back of the mouth and into the nasal cavity (retronasal), where some compounds are detected by the olfactory nerve. The largest contribution to the diversity of flavor sensation is the sense of smell. As the taste and smell occur at the same time, people confuse taste with smell. The difference can be seen by eating foods with the nose pinched closed, therefore blocking the nasal passage. Familiar foods will not be recognizable and it is even possible to confuse apples with onions (Delahunty, 2011)!

Chemethesis is the tactile response of some compounds which are capable of triggering the pain receptors and mechanoreceptors (light touch receptors). Examples include the "heat" or pungency of chili peppers (from the compounds capsaicinoids), the cooling of mint from menthol, the tingling of sanshools in Szechuan pepper or from carbon dioxide in carbonated beverages, and the numbing of eugenol in clove oil. Besides chemesthesis, there are some other feeling factors such as metallic (from metal ions), astringency (mouth drying from tannins, for example), and soapiness caused by the salts of long-chain fatty acids.

13.4 LEGAL DEFINITIONS

In the United States, the principal regulatory agency is the Food and Drug Administration (FDA), part of the Department of Health and Human Services. Probably the only time the FDA has defined the term "natural" is in the codex of Federal Regulations (CFR):

> *The term natural flavor or natural flavorings means the essential oil, oleoresin, essence or extractive, protein hydrolysate, distillate, or any product of roasting, heating or enzymolysis, which contains the flavoring constituents derived from a spice, fruit or fruit juice, vegetable or vegetable juice, edible yeast, herb, bark, bud, root, leaf, or similar plant material, meat, seafood, poultry, eggs, dairy products, or fermentation products thereof, whose significant function in food is flavoring rather than nutrition.*

13.4.1 21 CFR101.22(A)(3)

There are a lot more details that follow, but basically, if you start with a natural substance and modify it using an approved process, and the resulting product is for flavor not food, then that product can be considered a natural flavor. "Artificial flavor" essentially means a flavor that doesn't meet those standards. The term "natural flavor" on a label can encompass all sorts of materials, from spice extracts to natural flavor chemicals, and reactions. The exact composition of the flavor is not disclosed for proprietary reasons.

Several other definitions are of importance. When the flavor house ships a flavoring to a customer (such as a functional foods manufacturer, but not the consumer), the label must declare nonflavoring materials such as carriers (vegetable oil, propylene glycol, ethyl alcohol, etc.). The title of the flavoring must reflect the nature of the flavor. A "Natural Strawberry Flavor" contains flavoring materials derived solely from strawberries. A "Strawberry Flavor WONF" (with other natural flavors) means that the natural flavor contains flavor derived from strawberries as well as other natural flavorings that "simulates, resembles, or reinforces" the characterizing flavor (in this case, strawberry). While not defined by the FDA, a "Strawberry Type Flavor" tastes like strawberries but does not contain any flavor derived from strawberries. Note that the natural flavor designation is not affected by the carrier or any nonflavor items in the flavoring. Onion oil, which is a natural flavoring, diluted in propylene glycol (PG), which is synthetic, is still a natural flavor. The PG is listed separately as it is not covered under flavor regulations (FDA, 2000, 21 CFR 101.22).

The flavor manufacturer needs to supply documentation such as the label and the component range breakdown so the customer can label their products correctly. The food application in which the flavor is used will also affect how the finished product is labeled.

Besides the FDA, two other US agencies are involved in flavor regulation. The United States Department of Agriculture (USDA) is involved in flavors containing a certain amount of meat, or when the flavors are used in products containing meat. They also operate the National Organic Program (NOP). The Tax and Trade Bureau (TTB) of the Treasury Department regulates products going into alcoholic beverages.

Another important concept is "Generally Recognized as Safe," or GRAS. The FDA introduced the concept of GRAS as a type of "positive list." The idea was that foods long in use should not have to be tested again. In 1960, the United States flavor industry trade group Flavors and Extract Manufacturers Association (FEMA) instituted an independent (but FEMA-sponsored) group of experts (FEXPAN, or the "Expert Panel") to evaluate additional flavoring materials for safety. FEMA members that wanted a flavoring substance to be FEMA GRAS would submit information to the panel, who would determine if the substance (chemical or botanical) could be used safely in flavors and under what conditions. The substance received a "FEMA number," which means the flavoring substance can be considered GRAS if used as specified. For instance, the fruity flavor chemical ethyl-2-methyl butyrate (apple aroma) has a FEMA number of 2443. FEMA GRAS allows for the use of this chemical in certain applications at certain maximum levels. For example, it can be used safely in alcoholic beverages up to 8.18 ppm, baked goods up to 46.70, soft candy up to 30.56, and chewing gum up to 500 ppm. Some flavorings do not have FEMA GRAS numbers, but are allowable in food as natural flavor by FDA regulation. For instance, amyris oil (woody) has no FEMA number, but is regulated by 21 CFR 172.510 (Burdock, 2010).

The FDA has refused to define "natural" in any other case, but has given this guidance:

> *The agency will maintain its current policy ... not to restrict the use of the term "natural" except for added color, synthetic substances, and flavors as provided in [21 C.F.R.] § 101.22. Additionally, the agency will maintain its policy ... regarding the use of "natural," as meaning that nothing artificial or synthetic (including all color additives regardless of source) has been included in, or has been added to, a food that would not normally be expected to be in the food. (Long et al., 2011; Federal Registrar, 1993)*

Be aware that other countries have their own regulations. It is necessary to let the flavor house know if your products are being sold outside of the country.

13.5 FLAVOR CHEMISTRY

Chemicals can be either volatile or nonvolatile (or somewhere in between). Chemicals that cause taste sensations tend to be nonvolatile, and also water-soluble. For example, sucrose is dissolved in the saliva or beverage and is detected by certain taste buds, to be perceived as sweet. A certain stereochemical arrangement of functional groups on the particular molecule will fit (bind) into the receptor in the taste bud to cause the sweet perception. While sucrose is the standard for sweetness, other molecules are detected as sweet, and many give a much stronger sweetness response, such as aspartame and saccharine (called High Intensity Sweeteners or HIS). Similarly, bitter receptors require molecules with a certain arrangement to elicit bitter response. Unfortunately, similarities between the sweetness receptors and the bitterness receptors result in molecules, especially many HIS, being both sweet and bitter (Lindsay, 2008). Umami is elicited most notably by the sodium salt of L-glutamic acid (MSG) and 5′ nucleotides, such as 5′-inosine monophosphate and 5′-guanosine monophosphate.

Bitterness, sweetness, and umami are all due to these molecules interacting with G-protein receptors (Rawson and Li, 2004). Saltiness and sour perception are associated with ion channels. Sour is associated with the hydrogen protons of acids. Salt taste is elicited by the movement of sodium ions, although some other monovalent cations can also generate salty response, such as lithium and potassium.

Chemesthesis can be triggered by nonvolatile and volatile compounds. Examples of nonvolatile pungent (hot) chemesthetic agents include capsaicinoids from chili peppers; gingerols, shogaols, and paradols from ginger; and piperine from black and white pepper. Sanshools are tingling compounds from Szechuan peppers and spilanthol from jambu. Incidentally, Szechuan peppers, chili peppers, and black and white peppers are not related.

Volatile chemesthetic compounds include the pungent compounds cinnamaldehyde (from cinnamon and cassia), isothiocyanate (mustard and horseradish), and eugenol (numbing compound from cloves).

Most of what is perceived as flavor arises from volatile compounds that impact the sense of smell. This encompasses a huge variety of various types of compounds: phenolics, carbonyls, acids, esters, heterocyclic, and "aromatic" (the organic chemistry meaning, as typical of a benzene ring) compounds, sulfur- and nitrogen-containing chemicals, and more. The only characteristics they have in common are that they are volatile to some extent and are able to elicit a response from olfactory receptors. Note that there is a lot of variation in the strength of flavor compounds. Vanillin, the characterizing compound in vanilla extract, has a recognition threshold of 4 ppm, while para-menthene-8-thiol, called the grapefruit mercaptan, has a threshold of 10 ppt (Burdock, 2010)!

13.6 SOURCES OF FLAVORING MATERIALS

Early in the history of flavor creation, only natural extracts were available. The main sources were the extracts of spices and herbs, along with numerous other botanicals. Many parts of the plants could be used, including flowers, buds, seeds, bark, exudates, and rhizomes. A limited number of animal-sourced flavorings were also available (rarely if at all used in flavors today), such as musk from musk deer and castoreum from beavers.

Plant extracts are separated from the nonflavorful portion of the plant material by either solvents or distillation. Ethyl alcohol is a great solvent and the resulting extract can be used as is (without the removal of the ethanol) to form a tincture. Other organic solvents can also be used, such as hexane or acetone. However, these solvents must be removed via vacuum to below a federally permitted level, and the resulting flavor extracts are called oleoresins. If supercritical carbon dioxide is used to extract the materials, the CO_2 is released and there is no residual solvent. The solvent extracts have the complete flavor of the source material, including volatile and nonvolatile components. Flavorings can be separated from the source material using distillation, creating essential oils that have the aroma of the source material but lack the nonvolatile portion of the flavors. For example, black pepper oleoresin (solvent extracted) has the aroma of black pepper, as well as the pungent "bite" from piperine, and is pigmented and viscous. Black pepper essential oil, however, smells like black pepper but is a clear, thin oil which does not contain piperine. Other methods of distillation extraction involve spinning cone technology, which creates aqueous extracts.

The development of organic chemistry led to the creation of many synthetic chemicals for use in the flavor industry. This included synthetic versions of naturally occurring flavor chemicals, such as vanillin derived from petroleum or wood pulp. Also, chemicals not found in food or even in nature were found for flavor use. One such compound is ethyl methyl phenyl glycidate, or "aldehyde C16," used for "red pop," a strawberry, or raspberry-like flavor.

In the 1980s, artificial flavors became less popular and natural flavors became more popular. Organic chemists were able to make many of the same chemicals found in nature using methods that followed the FDA definition of natural flavor (discussed earlier). The availability of such flavor chemicals has increased dramatically since then, so it is now possible to create most of the flavors as natural that were once only available artificially.

Another source of natural flavors is reaction flavors. This is the formation of flavorful materials from generally low or nonflavorful precursors. For example, the amino acids phenylalanine and leucine, heated under specific conditions in the presence of a reducing sugar (fructose, dextrose, xylose, etc.), can form a reasonable chocolate flavor. This technique is very popular for creating meat-type flavors without animal-derived ingredients, perfect for vegetarian foods. Another process is pyrolysis, good for creating smoke and grill flavors.

There continue to be advances in fermentation, enzyme, and extraction technologies to offer natural flavorings. All of the flavor extracts mentioned earlier are available for a flavor creator (called a flavorist or flavor chemist) to make many natural flavors for use in many applications.

13.7 FORMS OF FLAVORS

Flavors can be considered to have two parts: the flavoring moiety, and the carrier (or delivery system). The flavor portion consists of the materials listed previously, and is most commonly lipophilic (oil-like). Its function is to simulate the named flavor, to maintain its flavor through processing, and to enhance the product's flavor impression and its acceptability (Fischetti, 2014).

The carrier has several roles: it keeps the flavor portion in solution, adds color if any, regulates flavor strength, gives the flavor physical fixation (to survive processing), inhibits chemical reactions, may act as a preservative, is a vehicle for presenting the flavor, and determines the form of the flavor (Fischetti, 2014).

Liquid flavors may be either oil-soluble, water-soluble, or emulsions. Oil-soluble carriers include vegetable oils and synthetic oils, such as mid-chain triglycerides or triacetin. They work well for oil-soluble applications and many sauces.

Water-soluble carriers include natural products like ethyl alcohol, water, and glycerine. Water may have poor solvent ability for many of the nonpolar flavor compounds, and supports microbial growth. Ethanol is a great solvent for flavors, but there are concerns for Halal use (Islamic food law), flammability, and for use in products targeted for children. Ethanol can also react to some flavor constituents, and is unsuitable to high-temperature applications where the ethanol and the flavor may vapor off. Glycerine is a better use for all of those reasons, but is viscous and not a particularly good solvent. PG is the most common synthetic hydrophilic carrier in use. However, it is synthetic and also will react with some flavoring substances.

Emulsions are more or less stable dispersions of flavorings in a carrier, typically a hydrophobic flavor load dispersed in a hydrophilic phase, usually water. To accomplish this, some type of emulsifier must be used. Common emulsifiers include gum arabic, gum acacia, encapsulating starches (usually modified food starches), or quillaja extract; along with the application of some type of shear. Coarse emulsions commonly used for baking have large particle sizes of the flavor-containing oil. Beverages require small particle sizes, typically requiring high pressures in a piston homogenizer (like the kind used to make homogenized milk). Because of the water content in the emulsion, there is the possibility of microbial growth. This can be suppressed by adding an acidulant to reduce the pH, commonly citric acid. The water activity can be lowered by the addition of salt, sugar, glycerine, or PG. Finally, an antimicrobial preservative may be added, including potassium sorbate or sodium benzoate. Another option is sterilization followed by aseptic packaging.

Spray drying is a way to make an emulsion that will not support microbial growth because the water is removed. The oil flavor load is dispersed in a mixture of water, emulsifiers, and carriers (maltodextrin is popular). The emulsion gets heated and pumped into a heated chamber through some type of atomizer. The water is vapored off of the emulsion, leaving the oil droplets surrounded by dried emulsifier and carrier, which "protects" the oil load. This method is very commonly used to make a powder that resists the loss of the flavor compounds by volatilization. Powders can also be made via plating. There, the oil load is blended onto a carrier such as maltodextrin, salt, or dextrose. There is no protection of the oil load; however, if packaged properly and used promptly, it represents an inexpensive way to create powdered flavors.

13.8 CHALLENGES TO FLAVORING FUNCTIONAL FOODS AND NUTRACEUTICALS

Functional foods are foods, so flavoring them should be similar to flavoring other processed foods. However, the active ingredients may cause off-flavors that need to be masked or diminished. For example:

Foods with soy protein may have a beany flavor that many people find objectionable (Reineccius, 2000; Hazen, 2003).

Vitamins have a meaty brothy flavor. Many functional ingredients are bitter. Other functional ingredients are prone to flavor deterioration, such as carotenoids (oxidation), thiamin (degradation to meaty or vitamin notes), and fish oils (oxidation) (Reineccius, 2000).

Base interactions are also a factor. Added fiber in foods effectively binds flavors, as do proteins and starches. Flavors that work well in full fat foods may not taste good in a low or no fat system.

Many functional food items contain various herbs and spices that contain bitter, astringent, grassy, and other flavor defects.

Mineral formulations can be metallic (Hazen, 2003).

Flavor chemicals are by nature often reactive. Flavorists know that and can compensate in regular food. However, they may be unfamiliar with the reactive characteristics of the herbal blends in the food (Reineccius, 2000).

Nutraceuticals, while not food, will most likely have bitter, metallic, and other off-flavors. Also, they may have characteristics unlike food, like high or low pH or long shelf lives, and may need to be registered with the relevant health authorities (Wright, 2011).

13.9 BASE STRATEGIES

It is better to eliminate/reduce/mask the off-flavors in the base to get the base as neutral in flavor as possible. Just overpowering the defects with flavor is expensive and the high levels of flavors are likely to cause other defects (Hazen, 2003). The best place to start is by using high-quality ingredients in the base that have minimal flavor defects. Examples include soy protein with low beany flavor, or blond pysllium. Alternatively, the base can be further processed to remove off-flavors. It does no good to start your new product with an avoidable off-flavor (Reineccius, 2000). A totally flavorless base presents other problems. It is difficult to provide all of the flavor by means of flavor chemicals without the product tasting imbalanced and artificial. Sweetness, saltiness, and other taste effects must be balanced (Hatchwell, 2006).

One word of caution: product developers who have been working on a base for a long time may overlook, get accustomed to, or even enjoy, the flavors and off-flavors in their bases. It may be a good idea to have someone not closely associated with the product, such as nontechnical coworkers, or even friends and family, taste the product and get their opinion. You can also taste the base away from the laboratory, such as at home or in the break room.

13.10 STRATEGIES FOR THE FINISHED PRODUCT

If the best efforts to make an unobjectionable base still leave undesirable flavors, the next step is to work with the flavor company to mask or otherwise obviate the off-flavors.

Woo (2015) described five different strategies for reducing bitterness perception or other off-flavors and tastes: avoid, remove, blend, mask, and block. As mentioned earlier in base strategies, the best practice to avoid flavor defects in the base is by avoiding the problem by selecting high-quality raw materials with little or no off-flavors. Selection of processing conditions to minimize flavor defects (high thermal treatment may cause flavor defects) is also important. Microbial contamination may also cause off-flavors and should be controlled.

If the bitterness compound cannot be avoided, such as if they are inherent in the base or the active ingredients, the compounds can be removed by physical, chemical, or microbial/enzymatic means. Encapsulation so the active ingredient can be delivered but not tasted is a physical strategy. Filtration may be employed to remove bitter compounds. Some negative tastants can be destroyed or modified via chemical means (such as oxidation). Alternately, the compounds can be destroyed or modified into a nonbitter form by enzymes or fermentation. For instance, a bitter compound may be in the form of a conjugate with a carbohydrate moiety. The molecule may be hydrolyzed and the aglycone rendered flavorless.

Blending may be an option. Ingredients with high levels of bitterness can be blended with larger quantities of ingredients with low levels of bitterness so that the final product has an acceptable taste. Masking and blocking strategies are described next.

13.11 BITTER BLOCKING

Ley (2008) reviews various methods for blocking bitterness. One approach is to prevent the bitter molecules from contacting the taste buds. This includes fats and starches which can coat the tongue, and cyclodextrines which can sequester the molecules until the complex dissolves in the acidic environment of the stomach.

A more sophisticated approach is to utilize molecules that prevent binding onto the bitter receptors. There are 25 identified types of bitter receptors, detecting various classes of bitter compounds (McGregor, 2011). The large number reflects the importance of bitter perception by our ancestors. Being unable to detect a bitter and potentially toxic substance was a distinct evolutionary disadvantage. This causes a problem for employing bitter blockers in food. The blocker has to specifically block the receptor for the bitter compound found in the food. If the blocker affects the wrong receptor, it won't help your product. Also, many of these blockers are not food grade (they are mainly used for pharmaceuticals) or not naturally derived, so may not be appropriate for your all-natural functional food.

Adenosine 5′-monophophate (AMP) is capable of reducing taste cell activation by bitter molecules. Many other compounds are being tested by measuring their effect on bitter receptors or by sensory studies. They are wide in variety, such as neodiosmine taurine, L-thanine, and ferulic acid. Some spices and herbs can reduce bitterness, for example, cassia, cumin, ginger, and red pepper (Hatchwell, 2006). As there are such a wide variety of bitter compounds, no single bitter blocker usually suffices. A bitter blocking compound system is usually developed on a case-by-case basis, and requires the skills and experience of a flavor house to be developed.

13.12 MASKING

Masking is the addition of flavors to minimize the detectability of off-flavors. The methods include: taste modifying by providing other sensations, taste suppression by competing with bitter receptors (blocking), or accentuating other flavors (taste enhancing) (Gascon, 2011). Blocking was discussed earlier.

Sweetness often is enough to mask bitter flavors, since bitter and sweet receptors are similar in structure. However, just adding more sucrose may not be the answer, especially for dietary foods. Low-calorie HIS including glycyrrhizin and neohesperidin dihydrochalcone can be used. Nonsweet masking agents include miraculin. A protein isolated from the berries of *Synsepalum dulcificum,* it has the curious effect of making sour foods taste sweet. Personal experience has demonstrated that the protein can make biting into a lemon slice seem like sweet lemonade. Other compounds used for masking include the common flavoring maltol and furanones such as "strawberry furanone" (4-hydroxy-2,5-dimethyl-3(2H)-furanone) and methylcyclopentenolone (MCP), which are all related to the thermal decomposition of sugars, and some amino acids such as alanine and glycine.

Maltol, vanillin, and MCP all smell "sweet," that is, they possess aromas that are associated with sweet foods (think ice cream, sweet bakery items, and maple syrup). Added to a food with a bitter component, the sweet aroma will trick the mind into ignoring the bitter component as it is not congruent with the sweet taste and aroma.

Vanillin has a desirable flavor and is very good at rounding out the flavor of a base, improving the mouthfeel of the food (Lindsay, 2008). An example of vanillin used in a commercial product is in a natural retail sugar substitute. The substitute is made using a blend of the natural HIS stevia, along with greater quantities of the low-sweetness sugar alcohol erythritol. The blend and its constituents were compared to sugar in a sensory panel at Kalsec, Inc. (unpublished data). The results depicted

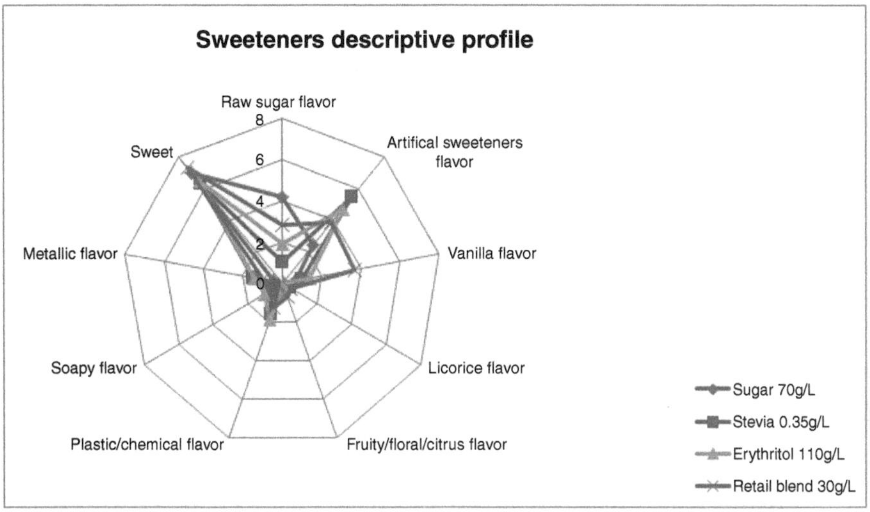

FIGURE 13.1 Retail Blend—Natural Sugar Substitute Flavor Blend

in Fig. 13.1 show that the sensory profile of the blend is closer to that of sugar than stevia and erythritol by themselves. Note, however, that the blend has a considerable peak identified as "vanilla flavor" which is not typical of sugar and is not found in the stevia or erythritol. Probably vanillin was added to the blend to mask an off-flavor and improve mouthfeel.

Salt and potentiation using MSG or other high-umami products may work as well for savory products.

Once the perception of off-flavors in the base has been reduced, flavorings can be added to the base. Since eliminating all off-flavors in a base is impractical, a good approach is to layer the flavor sensations, altering the temporal profile and confusing the senses. Adding cayenne pepper, mint, and salt adds tactual sensations which are more pervasive and linger longer than bitterness, making the flavor more acceptable. Vanillin works in sweet applications (Hatchwell, 2006).

The flavors added may be congruent or contrasting. Congruent flavors have some of the same taste characteristics of the perception you want to mask. For example, grapefruit and dark chocolate are both bitter, so some bitterness from the functional food will not detract from the flavor of the product. Citrus flavors can be used for sour tastes. Mint combinations work with alkaline defects, and metallic defects are compatible with grape and lemon-lime flavors (Reineccius, 2000).

Contrasting flavors, on the other hand, have little of the off-flavor note. The addition of "brown, sweet" flavor compounds, such as maltol and vanilla, or flavors such as maple or honey, have aromas associated with sweet-tasting products. This reinforces the sweetness of the functional food and deemphasizes any bitterness. Other options are to add spices, acids, or pungent spices such as capsicum to add impact to the flavor of the product and decrease the perception of the bitterness.

13.13 CHOICE OF FLAVORINGS

Medicinal properties of many herbs and spices and their extracts have been reported. For example, dill is carminative (prevents intestinal gas), stomachic (improves stomach function), and antipyretic (reduces fever). Garlic has anticholesterol effects, poppy seeds are an expectorant, and thyme is an antispasmodic (Kurian, 2012). Plus, many have antioxidative and antimicrobial properties (Kurian, 2012). Culinary spices and herbs have favorable taste qualities, but many active compounds for nutraceutical effects are bitter, astringent, acrid, and acidic (Peter, 2012). So, spices and herbs and their extracts can be used for their flavoring, if the flavor is favorable to the base and the extracts don't interfere with the functionality of the products. Oleoresins are solvent extracted, so there may be resistance to the solvents employed even if under acceptable residual levels. Alternative extraction methods include distillation (essential oils), carbon dioxide extraction (supercritical or subcritical), expeller pressed, or alcohol/water extraction.

Reaction flavors are considered natural flavor in the United States (if they conform to the FDA definition of natural flavor). Be aware of any carriers such as maltodextrin that may or may not be acceptable. The reactants are usually not disclosed to the customer.

Many flavorings are made in part with flavor chemicals, which are not disclosed to the customer for proprietary reasons. Flavorings are often in a carrier, such as oils, PG, ethyl alcohol, or maltodextrin, which are generally disclosed to the customer but may or may not be disclosed to the consumer because they are considered a "processing aid" or an "incidental additive." The same goes for emulsifiers employed to solubilize the flavor for aqueous or hydrophilic applications. It is necessary to let the flavor company know what ingredients are acceptable or unacceptable when requesting the flavor.

The desired form of the flavor is dependent on the application. Water-soluble or oil-soluble liquids are the most popular, depending on compatibility to the base. Ethanol-based liquid flavors are a poor choice for applications that undergo heating, such as hard candies. Plating flavors onto carriers (maltodextrin, salt, starch, dextrose, etc.) is an inexpensive way to deliver flavors for dry products, but does not protect the flavor from environmental effects such as volatilization or oxidation. Spray drying is much better for applications with those concerns, but can be expensive or too dilute. Emulsions are popular when a product needs a water-dispersible flavor. Strength, microbial stability, and potentially objectionable ingredients can be a barrier to using them.

13.14 PROCESSING AND OTHER CONSIDERATIONS

The flavoring system, including masking and blocking agents, must be compatible with the base and the processing. Thermal conditions are deleterious to many flavors, and may also cause flavor changes in the base. For instance, Maillard browning may add sweet or brothy flavors. Heat may also degrade unsaturated fatty acids. Adding flavor in open systems allows for flavor loss as most flavor compounds are volatiles. pH, either inherent in the base (citrus juices) or from the addition of acids (for flavor, microbial stability, etc.), will affect the flavor. Citrus flavors work best in acid environments, chocolate is better in neutral conditions. Air contact has the potential to increase oxidation for the flavor and the functional ingredients.

13.15 LABELLING AND OTHER MARKETING CONSIDERATIONS

Although natural flavors are essentially the same on a molecular basis as artificial ones, marketing considerations often dictate natural flavors. Note that the regulations for natural flavors can be quite different in the United States as compared to Europe, Japan, and other countries. Even in the United States, there are three agencies with flavor-regulating responsibilities (the FDA is the major one, but the USDA and TTB all have different regulations concerning natural flavors). Also, be very aware that natural flavors are not the same as "all natural." There is no definition for "all natural" in the United States. Natural claims are interpreted by each customer. Some influential companies have natural regulations that attain a quasiregulatory status, such as Whole Foods. Also, "certified organic," under the USDA's National Organic Program, does allow certain synthetic ingredients.

Religious programs bring additional burdens. Kosher is based on Jewish dietary law. Halal is based on Islamic law. The two have similarities but are not the same. Other considerations include allergens, genetically modified organisms, vegan/vegetarian, among others.

13.16 WORKING WITH A FLAVOR HOUSE

Keskar and Igou (2011) listed a number of common problems faced with flavoring nutraceuticals and functional foods. They include:

1. Not having a plain base to use to flavor.
2. Failing to cover off-flavor notes first.
3. Overflavoring the base.
4. Not using dual-function ingredients (nutraceutical ingredients that have positive influences on the product, such as agave nectar, with is a source of inulin and sweetness).
5. Flavor function restrictions (a flavor that works well in one product may not be suitable in another).
6. Vitamin and mineral issues.
7. High-protein products.

Flavoring functional foods and nutraceuticals is tricky, and must be done on a case-by-case basis. The best approach by far is to partner with a flavor house during all phases of product development. The tendency is to want to "perfect" the base first, and then bring in the flavor house at the last stages of development. This greatly reduces the probability of a successful product launch. The flavor house has valuable skills in reducing off-flavors in your base, which needs to be accomplished before the base is flavored.

When it comes to flavoring your base, either you can request samples from a flavor house and do your own selecting and application, or you can leverage the resources of the flavor house to do this for you.

13.17 IF YOU WANT TO DO YOUR OWN FLAVORING OF BASES

If you decide to do your own flavoring, request samples from companies with proven experience and success in the area of flavoring functional foods. Be aware that no company, even the large multinational ones, is equally good at all applications and types of flavors; they all have their strengths and weaknesses. Also, many companies are "sweet" flavor companies, that is, they primarily make flavors that are complemented by sugar, such as fruit flavors and brown sweet flavors. Others are savory companies, which make flavors that are complemented by salt, such as meat and meat substitutes and vegetable flavors. Companies are often stronger in specific types of flavors, such as dairy flavors or flavors for beverages. Select flavor houses whose strengths are in the areas of your products.

Make sure the company knows what your applications are, pertinent labeling restrictions (natural, Kosher, non-GMO, etc.), volume, cost, and target flavor profiles. With enough information, they can make educated guesses on flavors that can meet your needs. Pay attention to their recommended dosage levels; adding too much flavor can be as bad as or worse than too little.

A big mistake many customers make is inappropriately screening flavors. Many just sniff a bottle of liquids, or taste a pinch of powdered flavors to determine which ones they want to select for additional work. This is undoubtedly a horrible way to screen flavors. As mentioned earlier, the functional ingredients can greatly affect the perception of a flavor. The base can also. Fat, starch, protein, gums, acidity, and other constituents and properties can all affect the flavor. Processing can also make a huge difference, especially high heat. A flavor that smells OK out of a bottle may or may not be the same in the base after processing. The best way to screen flavors is in the same base and under the same processing conditions as the product will be made. If that is impractical, try a model system. Examples include citric acid and water for a beverage base (carbonated water for carbonated drinks) or salt and MSG for savory applications. In this example, thermally process the flavored water. Don't rely on a "golden tongue" (a single expert) but objective panels such as sensory panels.

It may be tempting to skip this intensive screening of flavors for the sniff or pinch test due to time or cost restraints. However, consider all of the time and resources required to develop a product, scale it up in a plant, and for packaging, marketing, and delivery.

You want the very best flavors you can obtain (for the money) for the best chance for your product to be successful. You don't want your product to fail because the best flavor is sitting unused on your lab bench due to invalid screening procedures. Remember, no matter how healthy your product is, its success will probably be due to its taste.

13.18 PARTNERING WITH A FLAVOR COMPANY

A better approach is to partner with a flavor house. As mentioned, every functional food will require a customized flavor solution. As a proprietor of a new product, you may be very cautious on the information you give out to vendors. Have your partner sign a "nondisclosure agreement" (NDA) so that their technical staff can talk openly with your company's technical staff.

Again, make sure the company you are going to partner with has proven expertise in functional foods of the form you are producing. The very large multinational companies have awesome resources to rely upon. The cost of their flavors reflects those resources. Also, the very large flavor companies only want very large customers. A smaller or start-up company may find a better working relationship with a small or medium-sized flavor house.

Start working with a partner company early in the development process. It is disheartening as a flavorist to receive a base with a strong flavor defect and be expected to somehow overcome that defect with flavor. The experts at the flavor company can work with you for a low-flavor base (see the earlier sections on ingredient selection, masking, and bitter blocking). It is always better to start out with a base with very low flavor defects before adding the final flavors.

Be open with your ingredients, especially the functional ingredients. Functional ingredients may have unusual chemical properties and may react with common flavorings. Processing information is vital. Be up-front with labelling, regulatory, and marketing requirements. Cost of course is another big consideration. Be aware that the price of the flavor is less important than the "cost in use." You are better off with 0.02% of a $50/kg flavor than 1% of a $5/kg flavor. Many procurement departments have trouble with this concept. The flavor company doesn't want to spend a lot of resources to make the perfect flavoring system to find out that it costs too much or can't contain glycerin. The flavor company also needs to know timing. When do they need to deliver the solution? What is the expected volume? The flavor company needs to know how much resource they can afford to expend on your project.

Give the flavor house your unflavored base and detailed ideas on processing and the desired flavor profile. That way flavors can be evaluated in the proper base.

Leverage their regulatory, sensory, and analytical capabilities (within reason). Invite their technical staff to your company for a firsthand look at your product and process. Many companies love to invite their customer's technical staff to work with the flavorists and applications personnel at the flavor house. That way, flavor systems can be designed by fast iterations in the lab rather than spending days or weeks shipping products back and forth and waiting for feedback.

This should all be the best opportunity to get the desired flavor profiles for your product. Remember, though, that all of this service costs the flavor company money. They usually get paid back only when you buy their flavors. The resources expended are reflected in the prices you pay for the flavoring systems. It is tempting to take the flavor produced by a resource-heavy company and shop it out to a low-cost flavor company for them to match. Not only is this not particularly ethical, it will be very unlikely that that company will want to work with your company in the future. And any problems caused by the third-party flavor are not their problem.

Knowing your products, the unique features of your functional foods, and working closely and openly with a reputable flavor company with proven expertise in your type of products will give you the best chance for successful functional foods and nutraceuticals.

REFERENCES

Burdock, G., 2010. Fenaroli's Handbook of Flavor Ingredients, sixth ed. CRC Press, Boca Raton, FL.

Delahunty, C., 2011. The sensory systems and food palatability. In: Lanham-New, S., Macdonald, I., Roche, H. (Eds.), Nutrition and Metabolism. second ed. Wiley Blackwell, Oxford, (Chapter 9).

FDA, 2000. 21 CFR 101.22(a)(3). Foods: labeling of spices, flavoring, colorings and chemical preservatives.

Federal Registrar, 1993. Food labelling. 2407 in 58 Federal Register 2302 January 6, 1993. (Notice).

Fischetti, F. Jr., 2014. The science of artificial flavor creation. Presentation at the Society of Flavor Chemists Symposium 2014: The state of the industry after sixty years. October 8–9, 2014, Jersey City, NJ.

Fisher, C., Scott, T., 1997. Introduction—problems in flavor research. In: Food Flavors: Biology and Chemistry. RSC Paperbacks, Cambridge, UK (Chapter 1).

Food Insight, 2011. 2011 Functional foods/foods for health consumer trending survey. http://www.foodinsight.org/2011_Functional_Foods_Foods_For_Health_Consumer_Trending_Survey

Gascon, M., 2011. Masking agents for use in foods. In: Taylor, A., Hort, J. (Eds.), Modifying Flavor in Food. CRC Press, Boca Raton, FL, (Chapter 13).

Hall, R., 1968. Food flavors: benefits and problems. Food Technol. 22, 1388.

Hatchwell, L., 2006. Flavor applications. In: Mosicano, G. (Ed.), Successful Flavors. From Formulation to QC to Applications and Beyond. Allured Books, Carol Stream, IL, pp. 85–108.

Hazen, C., 2003. Secrets of masking flavors. Food Product Design. http://www.foodproductdesign.com/articles/2003/11/secrets-of-masking-flavors.aspx

Keskar, A., Igou, W., 2011. Flavoring of nutraceuticals. In: Pathak, Y. (Ed.), Handbook of Nutraceuticals. Volume 2. Scale up, Automation, and Processing. CRC Press, Boca Raton, FL (Chapter 19).

Kurian, A., 2012. Health benefits of herbs and spices. Peter, K. (Ed.), Handbook of Herbs and Spices, vol. 2, second ed. Woodhead Publishing, Cambridge, UK, (Chapter 4).

Ley, J., 2008. Masking bitter taste by molecules. Chem. Percept. 1, 58–77.
Lindsay, R., 2008. Flavors. In: Damodaran, S., Parkin, K., Fennema, O. (Eds.), Fennema's Food Chemistry. fourth ed. CRC Press, Boca Raton, FL, (Chapter 10).
Long, J., Park-Suk, H., Tyler, L., 2011. Does the FDA have a definition for "natural" or not? The National Law Review. Posted on December 17, 2011.
McGregor, R., 2011. Bitter blockers in foods and pharmaceuticals. In: Taylor, A., Hort, J. (Eds.), Modifying Flavor in Food. CRC Press, Boca Raton, FL, (Chapter 12).
Peter, K., 2012. Introduction to herbs and spices: medicinal uses and sustainable production. Peter, K. (Ed.), Handbook of Herbs and Spices, vol. 2, second ed. Woodhead Publishing, Cambridge, UK, (Chapter 1).
Rawson, N., Li, X., 2004. The cellular basis of flavor perception: taste and aroma. In: Taylor, A., Roberts, D. (Eds.), Flavor Perception. Blackwell, Oxford, (Chapter 2).
Reineccius, G., 2000. Flavoring systems for functional foods. In: Schmidl, M., Labuza, T. (Eds.), Essentials of Functional Foods. Aspen Publications, Silver Spring, MD, (Chapter 4).
Siro, I., Kapolna, E., Kapolna, B., Lugasi, A., 2008. Functional food. Product development, marketing, and consumer acceptance—a review. Appetite 51, 456–467.
Tuorila, H., Cardello, A., 2002. Consumer responses to an off-flavor in juice in the presence of specific health claims. Food Qual. Prefer. 13, 561–569.
Woo, A., 2015. A taste modulation workshop. Presented 11/4/15 to Kalsec© Inc of Kalamazoo, MI. W2O Food Innovation. Chicago, IL.
Wright, J., 2011. Applications, second ed. Flavor CreationAllured Books, Carol Stream, IL, (Chapter 11).

CHAPTER

Product design for bread through mind genomics and cognitive economics

14

Howard Moskowitz*, Stephen Rappaport, David Moskowitz*, Sebastiano Porretta†, Brigitte Velema*, Livia Rossi*, Margie Atwater***

**Mind Genomics Associates, White Plains, NY, United States; **Stephen D. Rappaport Consulting LLC, Norwalk, CT, United States; †Experimental Station for the Food Preserving Industry, Parma, Italy*

14.1 A SHORT BACKGROUND ABOUT MIND GENOMICS AND COGNITIVE ECONOMICS

The traditional scientific method teaches us to understand variables by isolation, by determining how a single variable performs when all other variables are held at a single level for each. This "one-at-a-time" approach works well for many phenomena, but fails to show the dynamics of a compound system when variables influence each other. In our daily lives it is mixtures which rule, mixtures comprising variables at different levels. It may be possible to understand part of our daily reality by isolating one variable and studying it, but we're likely to get a different and perhaps somewhat more ecologically valid picture by studying known mixtures of variables in a systematic way.

The foregoing paragraph, sterile as it might seem, opens us up to the wonderful world of subjective experience, specifically for this chapter the experience of thinking about food products. We can mix ingredients to make a product; cooks and food manufacturers do it all the time. We can mix messages to make an advertisement; companies do that all the time. And the strategy works; companies sell complex food products, mixtures of ingredients, and companies advertise these foods through commercial messages, that is, mixtures of words, images, sounds and emotions.

Mind Genomics, at its very simplest level, is the study of these mixtures, the study of everyday stimuli, to understand the rules of choice, to discover what is important to a person and what is not (Moskowitz et al., 2005a, 2005b). Rather than forcing the person to evaluate stimuli that vary only along one dimension, Mind Genomics studies mixtures of the variables, typically mixtures of messages, in an effort to develop rules of experience.

Developing New Functional Food and Nutraceutical Products. http://dx.doi.org/10.1016/B978-0-12-802780-6.00014-6

This chapter looks at some of the rules emerging when we investigate responses to messages about artisanal, European-style bread. The approach described in this chapter can be expanded to virtually any food product, and the nature of the variables, the messages about the bread, can be changed to fit as closely as possible to the nature of the food that is studied.

14.2 A SHORT INTRODUCTION TO MIND GENOMICS

Mind Genomics is an experimental science which tries to understand the world of everyday life and experience. Mind Genomics studies life-relevant topics which are important to people, such as artisanal bread, the example in this chapter. The topic need not be extraordinarily important, such as, the reasons for good and evil. Rather, the topic has to be germane to consumer researchers.

Mind Genomics begins at the granular level, understanding the mental dynamics of small bits of reality, called topic areas, approached through a microscope, in the style of a pointillist painter, who creates the picture by aggregating little dots of paint. This "from the ground up" pointillist approach contrasts with the typical top-down approach of the aerial photo, which classifies the entire world in a set of macro-clusters. With those macro-clusters it is difficult to know what to do with a specific topic because the typical approaches work at generalities, not specifics (Claritas, 1999; Wells, 1975).

Mind Genomics determines which aspects of the topic drive interest, or not; how people differ in the way they respond to these aspects of the topic; and how these differences cluster to form segments, called mind-sets. Lastly, Mind Genomics is concerned with assigning people to a relevant mind-set segment, which is accomplished through one of two techniques, either by personal interaction (personal viewpoint identifier) or by tagging customer databases (digital viewpoint identifier).

Table 14.1 briefly outlines the aspects of Mind Genomics.

In many cases, the history of "pricing research" suggests that interest in a product or an aspect of a product and price covary; people say that they will pay more for what they like. Mind Genomics along with Cognitive Economics allow us to understand this covariation at a deeper level. When we examine the patterns of interest and dollar values for each element, we develop a different picture, uncovering the trade-offs that people make between features and value. Table 14.2 shows a 2x2 matrix of interest and price, four combinations, which allow the economics narrative to be better understood.

We label the four cells using colloquial language to describe the appeal of the combinations of interest and price:

1. Profit Opportunity—elements which are considered desirable for which people are willing to pay a higher price. These are the messages that consumers "absolutely desire."
2. Must Haves—elements which drive interest but are not worth a premium. These are messages that are expected, the "cost of entry."

Table 14.1 A Synopsis of Mind Genomics

The objectives of mind genomics are:
To understand reality from the "ground up," topic by topic
From the point of view of the person's experience
In a way which leads to quantification about what is important, for each person
So that general rules about each topic of the everyday may emerge from the empirical study
With the goal of creating an archival science, empirical, and functionally useful for business and social good
Thus leading to the science of that topic, a science founded in human experience and perception
The tools of Mind Genomics comprise:
The raw materials of the topic (silos, i.e., categories of interest, and elements, e.g., specifics within the silos, which may include, but are not limited to, attributes, benefits, features, functions, or emotions)
Experimental design to create test stimuli, groups of 3–4 elements called "vignettes"
Study respondents are the test people, subjects, aspects of whose "minds" are to be understood through the experiments
Rating questions, which allow respondents to communicate their reactions to the test stimuli
Technology, that is, computers, to present the test stimuli, acquire the ratings, calculate the statistics, create the report for the user, and develop the segment assignment tools
The work products of Mind Genomics are:
A database for each topic area showing which specific "elements" or ideas drive interest, are valued highly, or affect one's emotions
Mind-set segments (viewpoints) for the topic area, based upon what elements are important
Viewpoint identifier tool, to determine the segment membership of an individual
The ability to sequence a person's mind on a large array of topic areas, using an array of viewpoint identifiers
The practical outcomes of Mind Genomics are:
A micro-science of a topic area, that is, what works, what doesn't work
How the topic area divides, based upon the minds of people (deeper mind-truths)
Tools (interactive viewpoint identifiers, look-alike modeling) to drive desired behaviors (purchase, compliance, interest, perceived value)

3. Not Worth Paying For—elements which are neutral or slightly negative which people expect to pay more for when present. People don't want to spend extra for a message which does not interest them; it's a "bad deal."
4. Irrelevant—elements that markedly detract from interest and for which people are unwilling to pay or feel that it costs too much. These are messages people discount right away that can quickly stop them from considering a product, service, or an experience.

Table 14.2 The 2×2 Matrix of Interest in a Product Element (e.g., Communication Element about Artisanal Bread) and Willingness to Pay for That Element (Price)

	Low interest—Low or Negative Impact	**High interest—High Positive Impact**
High price—dollar value above the average of all elements	Not worth paying for	Profit opportunity
Low price—dollar value below the average of all elements	Irrelevant	Must have

The matrix in Table 14.2 makes an important point: Just because an aspect of a product or service is pleasing to a person does not mean that particular aspect will command a higher price. There may be some covariation between liking and price, but the covariation is not perfect by any means. The nature of such covariation may vary by product, and by the nature of the person doing the evaluation. That covariation, and indeed the hedonics (liking/disliking) and Cognitive Economics (price willing to pay), are all topics for empirical investigation presented here.

14.3 PRAGMATICS—DOING THE MIND GENOMICS/ COGNITIVE ECONOMICS STUDY

The Mind Genomics approach involves the following steps, taken from Table 14.1 (Moskowitz et al., 2001):

1. Develop elements, that is, raw materials that later will be incorporated into vignettes. These elements are simple, stand-alone messages which pertain to the topic of interest, here artisanal bread. Elements are organized into categories, called silos. The silos span a range, from descriptions of the sensory response to the product, to health features, to emotions, respectively. Each silo, or general group, comprises six elements.
2. Experimental design to create test stimuli, so-called vignettes, which are combinations of 3–4 elements, one each from three or four silos, but never two or more elements from a single silo. The rationale behind the experimental design has been presented by Gofman and Moskowitz (2012) and by Moskowitz and Gofman (2004). In marketing research the approach using experimental design is known as conjoint analysis (Green and Srinivasan, 1990). Conjoint measurement has also been used to identify the features of new to the world products (Mahanna et al., 2009). Experimental design has been extensively explicated in authoritative texts, for example, Box et al.'s (1978) book, *Statistics for Experimenters*.
3. Rating questions, which allow respondents, the tested people, to communicate their reactions to the test stimuli on the basis of interest (question 1), and price willing to pay (question 2).

4. Recruiting the test people whose "minds" are to be understood through the science.
5. Running the experiments using technology, that is, a computer to present the test stimuli (vignettes) and acquire the ratings.
6. Data analysis utilizing regression, clustering, and discriminant function analysis to create the mind-set segments and a system for creating the viewpoint identifiers (Green and Krieger, 1991).

14.4 THE RAW MATERIALS

These comprise silos, that is, groups which themselves comprise related elements. For this particular study we created six silos, each silo comprising six elements. The Mind Genomics project covers a wide range of ideas and messages across the 36 elements. Table 14.3 shows the silos and elements.

1. The six silos, denoted with capital letters A–F, concern six different aspects of artisanal, European-style bread. The elements are denoted by the numbers 1–6. This alphanumeric system provides a standard bookkeeping method, enabling the analysis to refer to each individual element.
2. Silo A refers to elements having to do with Europe, bread heritage, and purity. Silo B deals with country or region of Origin. Silo C: Ingredients. Silo D: Satisfaction. Silo E: History and Recipes. Silo F: Sensory Emotion.

Table 14.3 Silos and Elements

	Silo A: Europe, Heritage, Purity
A1	It is stunning how delicious the bread is in Europe
A2	European bread has that crispy crust and is nice and chewy and is less dense than American bread
A3	The Europeans only use basic ingredients in their bread—they produce it fresh daily, that's why it is so delicious
A4	Get more conscious of what is in our food—bread should be a source of purest nutrition, but here in the United States it is not
A5	Honest food doesn't need any taste enhancers—no preservatives, GMO ingredients, sugar, or additives
A6	Bread is one of the most important foods we consume every day—it should come from the best ingredients
	Silo B: Origin
B1	From France
B2	From Italy
B3	From the Netherlands
B4	From Germany
B5	From Spain
B6	From the best bread countries in Europe
	Silo C: Ingredients

(Continued)

Table 14.3 Silos and Elements (*cont.*)

C1	Dough made with only local European ingredients—GMO free
C2	Bread variations from artisanal bakeries made by professional bakers
C3	Artisanal bread recipes without any sugar, taste enhancers, colorants, GMO ingredients, or preservatives
C4	Bringing the purest bread variants fresh to your table every day
C5	Your everyday bread made from ancient, pure local recipes
C6	Freshly baked every day in the United States
	Silo D: Satisfaction
D1	Eating pure and healthy every day
D2	Honest, delicious, and fresh food for me and my family
D3	My daily bread providing the best nutrition
D4	The original recipes guarantee the best taste and health
D5	Benefits long-term health as it is free of GMO ingredients, artificial additives, and sugar
D6	Easy way to make healthy and nutritious breakfast or lunch for me and my family
	Silo E: History, Recipe
E1	Recipes just using simple ingredients such as water, dough, yeast, and craftsmanship
E2	We selected just the best recipes from a time when making bread was not industrialized
E3	We produce the dough in the country of origin, which brings the real taste to the United States
E4	Made from European pure ingredients
E5	Bread that has been a nutritious staple in Europe for centuries
E6	Delivered to you fresh every day
	Silo F: Sensory Emotion
F1	It tastes like coming home
F2	It smells and tastes like homemade
F3	Your every morning joy
F4	It tastes like being on holiday every day
F5	It tastes healthy
F6	Crispy crust tells you that it was freshly baked, nice and chewy bite

3. The elements are written in simple English, in a way that would be readable and engaging for a consumer. Respondents are accustomed to reading information about products in ads, on packages, and via other forms of communication. To the degree that we can present the information in a normal, "nonclinical" fashion, we will gather responses to stimuli which represent the type of messages and mixtures of messages one receives daily from commercial sources seeking to describe a product, to sell a product, or both.

14.5 THE EXPERIMENTAL DESIGN

Conventional science works by isolating one variable at a time. As mentioned previously, Mind Genomics combines variables, that is, elements, in a systematic fashion so that each element appears in combination with other elements, simulating the typical compound stimulus one encounters in everyday life. Statisticians provide the experimental design, ensuring that the elements appear independently of each other, in a statistical sense, and that each element appears in 5 of the 48 vignettes (combinations). Thus each element is absent from 43 of the 48 vignettes.

Each respondent tests a unique set of 48 vignettes, constituting a full experimental design. The Mind Genomics strategy begins with a single experimental design, dictating the composition of all 48 vignettes, using that single design as the base. The base design is *permuted*, by changing the order of elements in each silo. That change, a permutation, creates a new set of 48 combinations, a set which maintains the deep structure of the experimental design. Each new design is isomorphic (same structure) as the basic design, but with different combinations (Moskowitz and Gofman, 2004; Gofman and Moskowitz, 2010). With 152 respondents who ultimately participated we end up with 152 unique sets of combinations, all permutations of the basic design.

Table 14.4 shows an example of part of the full experimental design for one respondent, explicating the first six vignettes or combinations of elements.

Table 14.4 Experimental Design—Composition of the First Six Vignettes

	Vig1	Vig2	Vig3	Vig4	Vig5	Vig6
Silo A	Absent	Absent	Absent	A1	A4	A4
Silo B	B2	B1	B1	Absent	B2	Absent
Silo C	Absent	Absent	C5	C3	C1	C3
Silo D	Absent	D5	D6	D3	D2	Absent
Silo E	E2	E6	E3	Absent	Absent	E5
Silo F	F6	F5	Absent	Absent	Absent	F2
Example of binary coding for Silo F (sensory emotion)						
1 = present, 0 = absent						
F1	0	0	0	0	0	0
F2	0	0	0	0	0	1
F3	0	0	0	0	0	0
F4	0	0	0	0	0	0
F5	0	1	0	0	0	0
F6	1	0	0	0	0	0
Rating						
Q1—interest on 9-point scale	9	7	2	6	6	8
Q2—selected price "number" on scale	2	4	1	2	2	2
Top 3 (INT), rating 1–6 = 0, 7–9 = 100	100	100	0	0	0	100
Actual price	3.99	5.79	1.99	3.99	3.99	3.99

14.6 THE RATING QUESTIONS

The respondent communicates feelings through the rating questions. This study required the respondent to rate his feelings in two ways for each of the 48 vignettes. The first rating allows the respondent to report how he feels about the artisanal bread, based upon the vignette, that is, an overall evaluation of interest in the bread. The second rating allows the respondent to tell us how much he would pay for the bread.

The two questions appear next. The dollar values are presented to the respondent in irregular order, preventing the respondent from repeating the same keystroke for the two questions. That is, the irregular order forces the respondent to think a bit more for the second question:

Question 1: Based on the description you see, how interested are you in this bread idea?
1 = Not interested at all ...... 9 = would be extremely interested
Question 2: Based on this screen only... How much would you pay?
1 = \$1.99; 2 = \$3.99; 3 = \$0; 4 = \$5.79; 5 = \$2.89; 6 = \$9.49; 7 = \$4.89;
8 = \$7.99; 9 = \$6.69

14.7 EXECUTING THE INTERVIEW

In its current incarnation Mind Genomics works with Internet-based studies, allowing respondents to participate either at home/in the office or an Internet-enabled café. In some cases the studies can in fact be run in special locations, as long as an Internet connection is available. This includes schools, doctor's offices, hospitals, and so on.

The Mind Genomics study begins with an orientation page, shown in Fig. 14.1. The orientation page describes the background of the study, without however divulging too much information. The introduction sets the stage for what the study aims at, in general. The specifics will be filled in through the elements.

You will be seeing several screens describing different aspects of artisanal European bread

You will be asked 2 questions for each description and asked to rate them on a 1-9 scale.
Any of the 9 numbers can be used to rate it.

Here are the 2 questions you will be asked on each creen:
1) Based on the description you see, how interested are you in this bread idea?
1=Not interested at all ...... 9=would be extremely interested

2) Based on this screen only... How much would you pay?
1=\$1.99; 2=\$3.99; 3=\$0; 4=\$5.79; 5=\$2.89; 6=\$9.49; 7=\$4.89; 8=\$7.99; 9=\$6.69

Some screens may seem repetitive but they are all in fact unique.
Please do not use the back button after you click a response.

Thank you for your time. Your feedback is important to us.

FIGURE 14.1 The Orientation Page

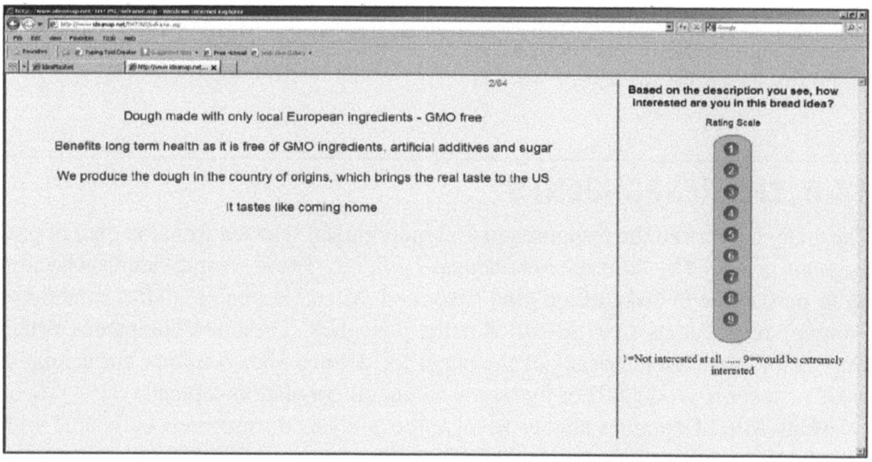

FIGURE 14.2 A Vignette With the First Rating Scale (Interest in Bread) on the Right Side

The remainder of the orientation presents details about the study, such as the rating questions, and the fact that the vignettes (screens) are all unique. The experimental design dictates vignettes comprising 3–4 elements, with the same element appearing 5 times across 48 vignettes. It is perfectly normal for a respondent to feel that he has already seen the vignette, because he recognizes the element. Up-front reassurance about the uniqueness of the vignettes suffices to reduce anxiety and possible irritation.

The 48 vignettes, already unique, are presented in a randomized order. Figs. 14.2 and 14.3 show these vignettes. One vignette remains on the screen until the respondent

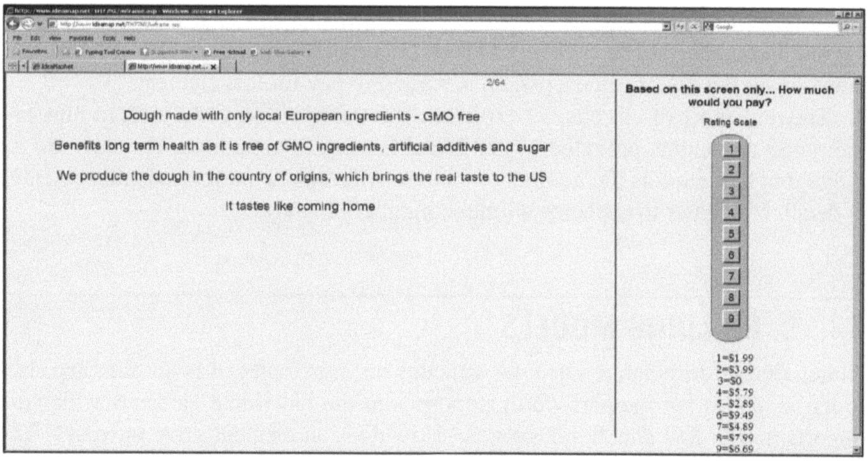

FIGURE 14.3 The Same Vignette With the Second Rating Scale (Price) on the Right Side

answers both questions. Once answered, the screen will automatically give way to the next vignette in the randomized list, again remaining on the screen until the two rating questions are answered.

14.8 THE RESPONDENTS

The study comprised the responses of 152 individuals, selected from the pool of participants provided by Amazon Mechanical Turk, Inc. These are individuals who sign up to participate in tasks of the kind described. Mind Genomics studies sometimes "source" respondents from so-called panel providers. These are companies which can "tailor" the characteristics of the panel for defined specifications, for example, users of specific products. For this study a general population sufficed.

Many Mind Genomics studies involve the analysis of responses by people with specific characteristics, such as age, gender, and even general attitude toward a topic. It is important to collect this information directly from the respondents at the time of the Mind Genomics interview. To better understand the respondents, Mind Genomics devotes the second half of the interview to a short self-profiling questionnaire. The questionnaire comprises a set of classification questions which allow the researcher to learn more about who the respondent "is" in terms of demographics (e.g., age, gender, etc.), psychographics (e.g., attitudes, beliefs), and behaviors (e.g., what the respondent does in terms of purchase and consumption).

14.9 LEVELS OF LEARNING

The Mind Genomics enterprise allows for a number of different analyses, beginning with simple relations between the presence/absence of elements and responses (interest, dollar value), moving to the identification of different groups in the population based upon newly emergent "mind-sets" (mind-set segmentation), and then moving beyond those analyses to uncover the relation between a person's interest in an element versus the dollar value a person is willing to pay for that element.

The remainder of this chapter explicates a number of these analyses, to illustrate the range of insights provided by the Mind Genomics enterprise. The data are far richer, but space limits the number of analyses that can be performed and presented in detail. It is better to explicate a limited number of analyses.

14.10 BUILDING MODELS

Although each respondent rated 48 vignettes on 2 attributes, it is not the responses to the vignettes but the part-worth responses to the individual elements which are important. The first simple question is "How does an element drive interest?" The second simple question is "how does an element drive value, that is, dollar worth?"

We analyze the data at the level of each respondent. This is known as a "within-subjects analysis." The individual pattern of responses to the 36 elements is thus the basic unit of analysis and the basic information. The conclusions are based on the aggregation of individual data from the appropriate respondents, that is, those who fit a specific criterion for inclusion in the average.

The workhorse analytic tool is OLS (ordinary least-squares) regression, often referred to as "curve fitting." OLS regression deconstructs an individual's responses to the 48 vignettes into the separate, part-worth responses of the components, the elements. The mathematics behind OLS regression is well accepted in statistical circles, and available in most off-the-shelf statistics packages for the personal computer.

Mind Genomics creates three individual-level models or equations for these data: Persuasion, Interest, and Dollar, respectively. Since each respondent evaluated 48 vignettes, with the 48 vignettes constituting an experimental design, it is straightforward to run 152 regression equations for each of the 3 models, hence altogether 3 regression equations for each respondent:

1. Persuasion (PER): We use the rating on the 9-point scale as the dependent variable. OLS analysis deconstructs the rating as the sum of individual contributions of elements on a 9-point scale. The Persuasion model predicts the rating on the 9-point scale when each element is incorporated into the vignette. The equation for the Persuasion model is expressed as: 9-Point rating = $k_0 + k_1$(Element A1) + k_2(Element A2) … k_{36}(Element F6). The additive constant, k_0, shows the number of rating points to be expected in the absence of elements. The additive constant is a theoretical number, that is, a purely estimated value, because all vignettes comprised 3–4 elements. We use k_0 as a baseline. In turn, each coefficient, k_1–k_{36}, shows the number of rating points on the 9-point scale to be obtained by inserting the element into the vignette.
2. Interest (INT): We follow the conventions of political polling, sociology, and market research which place people into groups for analysis. We transform the rating scale, converting it into a binary scale. Ratings of 1–6 are transformed to 0 to denote no interest in the vignette. Ratings of 7–9 are transformed to 100 to denote interest in the vignette. The cut-point, between 6 and 7, is arbitrary, having evolved from 30 years of research.
3. Dollar: We substitute the dollar value corresponding to each rating point. Thus the rating of "1" on the dollar scale (question 2) corresponds to $1.99, whereas the rating of 2 on the dollar scale corresponds to $3.99, and so forth. We then apply the OLS regression to the dollar ratings of the 48 vignettes for each respondent, to generate a dollar model: Dollar = k_1(Element A1) … k_{36}(Element F6). There is no additive constant in the dollar model, based upon the working hypothesis that in the absence of any element in the vignette there is no way to understand the value of the product.
4. Some technical notes:
   a. To each value of the dependent variables, whether the 9-point scale (Persuasion model), the 0/100 binary value (interest model), or the dollar

value, we add a small random number ($<10^{-5}$). That small number does not affect the coefficients or impact value, but it ensures that the OLS regression will not "crash" when the respondent assigns the same number to every vignette.

b. Whether or not we incorporate an additive constant into the regression makes no difference to the relative values of the coefficients. The coefficients estimated in the presence of the additive constant correlate perfectly with the coefficients estimated in the absence of the additive constant.

14.11 THE INTEREST AND DOLLAR MODELS FOR THE TOTAL PANEL

The data from Mind Genomics studies of this type are very rich. We need not go into detail to explicate each element. Rather, we will look at some of the highlights:

1. Table 14.5 shows the average values for the coefficients or impacts, both for the Interest (INT) Model and for the Dollar Model. The two sets of coefficients correlate highly; what a person likes is typically what a person feels to be valuable. However, there will be exceptions to that general rule.
2. The basic interest (additive constant) is 3, close to 0. The basic interest in artisanal bread is low. *It will be the elements which must do the work to raise interest, and the dollar values will tell us how they are valued.*
3. Profit Opportunity: Recall that elements in this group are characterized by high interest and high (above-average) dollar values. These elements have impact values of 12 or greater, and a dollar value greater than the average of $1.00. Here are the top three elements with great profit opportunity (Table 14.6).
4. What makes a "winner?": Honest, artisanal products, without anything added, generate the highest level of interest and price. The coefficients are 19 and 18 for the top two elements in Table 14.6, meaning that when one of these elements is presented in a vignette we get an additional 18–19% of the respondents voting 7–9 on a vignette which incorporates the element. The vignette goes from a near zero value of 3 to 22, with the addition of just one of the two elements. The total panel values these elements roughly 20–50% greater than the average.
5. Must Haves: Must Have messages combine high interest with average or below average dollar values. These are the elements that people expect. These elements add richness, offering assurance that there is substance underlying the respondents' keen interest as reflected in the Profit Opportunity elements. Table 14.7 shows the two Must Haves.
6. Not Worth Paying For: These elements combine low interest ratings with high dollar values. When they are present in a communication these elements may prompt the buyer to say: "That's too much money to pay for something I don't

Table 14.5 Interest and Price (Dollar) Models for the Total Set of 152 Respondents, Based on the Average From the 152 Individual Models

	Total Panel	INT	Price
	Average value	8.4	*1.00*
	Additive constant	3	($)
A5	Honest food doesn't need any taste enhancers—no preservatives, GMO ingredients, sugar, or additives	19	*1.53*
C3	Artisanal bread recipes without any sugar, taste enhancers, colorants, GMO ingredients, or preservatives	18	*1.46*
C4	Bringing the purest bread variants fresh to your table every day	12	*1.26*
A3	The Europeans only use basic ingredients in their bread—they produce it fresh daily, that's why it is so delicious	13	*1.25*
C1	Dough made with only local European ingredients—GMO free	16	*1.21*
D5	Benefits long-term health as it is free of GMO ingredients, artificial additives, and sugar	13	*1.18*
C5	Your everyday bread made from ancient, pure local recipes	9	*1.18*
B1	From France	3	*1.14*
A2	European bread has that crispy crust and is nice and chewy and is less dense than American bread	14	*1.10*
F6	Crispy crust tells you that it was freshly baked, nice and chewy bite	13	*1.10*
E5	Bread that has been a nutritious staple in Europe for centuries	9	*1.10*
B2	From Italy	5	1.09
C2	Bread variations from artisanal bakeries made by professional bakers	14	1.08
D6	Easy way to make healthy and nutritious breakfast or lunch for me and my family	7	1.07
A1	It is stunning how delicious the bread is in Europe	5	1.04
A6	Bread is one of the most important foods we consume everyday—it should come from the best ingredients	10	1.02
B4	From Germany	6	1.02
E2	We selected just the best recipes from a time when making bread was not industrialized	8	1.01

(Continued)

Table 14.5 Interest and Price (Dollar) Models for the Total Set of 152 Respondents, Based on the Average From the 152 Individual Models (*cont.*)

	Total Panel	INT	Price
E1	Recipes just using simple ingredients such as water, dough, yeast, and craftsmanship	10	0.98
E6	Delivered to you fresh every day	9	0.98
E3	We produce the dough in the country of origin, which brings the real taste to the United States	10	0.97
C6	Freshly baked every day in the United States	8	0.93
F4	It tastes like being on holiday every day	6	0.93
D4	The original recipes guarantee the best taste and health	4	0.93
F2	It smells and taste like homemade	9	0.89
B6	From the best bread countries in Europe	7	0.88
F3	Your every morning joy	4	0.88
E4	Made from European pure ingredients	7	0.86
F1	It tastes like coming home	4	0.79
D1	Eating pure and healthy every day	8	0.77
A4	Get more conscious of what is in our food—bread should be a source of purest nutrition, but here in the United States it is not	7	0.77
F5	It tastes healthy	5	0.77
D3	My daily bread providing the best nutrition	5	0.77
B3	From the Netherlands	2	0.71
B5	From Spain	0	0.71
D2	Honest, delicious, and fresh food for me and my family	5	0.70

Table is sorted by the coefficients from the price model.

Table 14.6 Elements With Great Profit Opportunity (High Interest, and Commands a High Price)

	Element	INT	Price
A5	Honest food doesn't need any taste enhancers—no preservatives, GMO ingredients, sugar, or additives	19	$1.53
C3	Artisanal bread recipes without any sugar, taste enhancers, colorants, GMO ingredients, or preservatives	18	$1.46
C1	Dough made with only local European ingredients—GMO free	16	$1.21

Table 14.7 The Must Have Elements (High Interest, Average or Below Average Dollar Values)

	Element	INT	Price
E1	Recipes just using simple ingredients such as water, dough, yeast, and craftsmanship	10	$0.98
E3	We produce the dough in the country of origin, which brings the real taste to the United States	10	$0.97

Table 14.8 The Not Worth Paying For Elements (Low Interest, High Dollar Values)

	Element	INT	Price
B4	From Germany	6	$1.02
B2	From Italy	5	$1.09
B1	From France	3	$1.14

Table 14.9 The Irrelevant Elements (Low Interest, Low Dollar Values)

	Element	INT	Price
D4	The original recipes guarantee the best taste and health	4	$0.93
F3	Your every morning joy	4	$0.88
F1	It tastes like coming home	4	$0.79

care much about." Respondents were uninterested in breads actually imported from Germany, Italy, and France, and apparently wary of the price premium that they would expect to pay (Table 14.8).

7. Irrelevant: Fifteen of the 36 elements are considered Irrelevant by the respondents, indicating that they were not interested in them and valued them below the average. Here are three of the lower-scoring irrelevant elements. They are somewhat flowery and do not resonate with the respondents (Table 14.9).

14.12 THE EMERGENT COGNITIVE ECONOMICS NARRATIVE

Analyzing the patterns of the coefficients and organizing them into the four categories—Profit Opportunity, Must Haves, Not Worth Paying For, and Irrelevant—creates and shapes a narrative for communicating with people about artisanal bread.

The patterns enable advertisers or marketers to select a group of messages that raise interest while simultaneously conforming to what people feel is worth paying for. The narrative also shows us which messages are likely to send signals to consumers that a certain artisanal bread may not be "worth it" and even signals those messages that will turn a person "off" and dissuade them from considering purchase.

14.13 FROM TOTAL PANEL TO KEY SUBGROUPS

Mind Genomics allows us to compare the response patterns of different groups, and even to compare individual corresponding parameters. It does so by creating models at the individual respondent level, models whose corresponding parameters can be averaged arithmetically to estimate the central tendency. We move beyond the general conclusion that Group A differs from Group B, to a precise profile of the nature of the differences. All analyses comparing groups look for narratives by which to understand these groups, and in a more profound sense "understand how nature works." Mind Genomics refines that search for narrative by studying the patterns of impacts across elements, and actual impact values of single elements, that is, patterns and points.

As an example, consider the strongest-performing elements (by Interest, question 1). Table 14.10 shows all elements with INT (interest) values above 10, for the 72 respondents who are age 29 or younger, and the 80 respondents who are age 30 or older. Table 14.10 also shows the price coefficients for the same element.

1. The additive constants are the same for the INT model, 2–3, meaning that by itself bread is not interesting. It's what comes along, the elements, which generates interest in the bread.
2. Concerning interest, for the overall evaluation of feeling asked in question 1, younger respondents show higher responsivity to the elements, whereas older respondents show lower responsivity. We see this by the size of the impact values, the coefficients attached to the elements. The younger respondents show higher impact values compared with the older respondents. Sometimes the differences are dramatic.
3. Generally, the differences between young and old are differences of degree. There are no inversions, wherein the older respondents are negative to an idea which appeals to the younger respondents. What the younger respondents like the older respondents like as well, but not to the same degree.
4. When it comes to setting a price for the bread we see flip-flops. Sometimes the younger respondents will pay more for an element (e.g., *Crispy crust tells you that it was freshly baked, nice and chewy bite*), whereas sometimes the older respondents will pay more (*European bread has that crispy crust and is nice and chewy and is less dense than American bread*). There is no clear pattern here.
5. Except for the differences in the magnitudes of the impact values for the INT model, we see no clear differences between the age groups. The differences which do emerge do not "tell a story." There is no clear narrative here, no simple story by which nature is illumed.

Table 14.10 Interest and Price (Dollar) Models for the Strongest-Performing Elements for Two Age Groups: 29 or Younger, 30 or Older

		INT	INT	Price	Price
		29 or Younger	30 or Older	29 or Younger	30 or Older
	Base Size	72	80	71	80
	Constant	2	3		
A5	Honest food doesn't need any taste enhancers—no preservatives, GMO ingredients, sugar, or additives	23	15	1.55	1.52
F6	Crispy crust tells you that it was freshly baked, nice and chewy bite	21	5	1.23	0.98
C2	Bread variations from artisanal bakeries made by professional bakers	18	11	1.08	1.07
A2	European bread has that crispy crust and is nice and chewy and is less dense than American bread	13	14	0.93	1.24
C3	Artisanal bread recipes without any sugar, taste enhancers, colorants, GMO ingredients, or preservatives	13	23	1.50	1.45
F2	It smells and taste like homemade	12	7	0.96	0.84
C1	Dough made with only local European ingredients—GMO free	12	20	1.16	1.25
E5	Bread that has been a nutritious staple in Europe for centuries	11	6	1.21	0.98
E4	Made from European pure ingredients	11	3	0.95	0.78
F4	It taste like being on holiday every day	11	2	0.83	1.02

Table is sorted by the INT model for the younger group, 29 or younger.

14.14 UNDERLYING MIND-SETS IN THE POPULATION UNCOVERED BY MIND-SET SEGMENTATION

Each Mind Genomics study, such as the bread study explicated in this chapter, presents to the researcher an opportunity to understand a small piece of "everyday" reality. Beyond uncovering the relation between elements and responses, Mind Genomics uncovers fundamental differences among people in the way that people process the specific information presented in those elements. These differences are not so much differences among people as they are constellations of ideas which "move together." A person may adopt one constellation of ideas, and be indifferent to, or even reject a competing constellation of ideas for the same topic area. We call these constellations of ideas Mind-sets.

To identify different mind-sets in our topic area of bread we use the statistical procedure of cluster analysis. The "material" on which the clustering takes place is the set of impact values from the questions. For our bread study we will use the 36 Persuasion values for each of our 152 respondents.

1. *The Persuasion or PER model uses the 9-point rating from question 1 as the dependent variable, and the presence/absence of the 36 elements as the independent variables.*
2. *OLS regression enables us to estimate the 36 coefficients and the additive constant.*
3. *We use the 36 coefficients as the inputs to the clustering program.*

The clustering program divides the 152 respondents into a small number of groups or clusters, with the properties that the average profiles of 36 elements across the clusters are very different from each other, whereas the individual profiles within a cluster are similar to each other. The actual computation follows specified steps to achieve the foregoing, explicated in statistics texts (Everitt et al., 2010). The clustering program allocates the 152 respondents into 2, 3, 4, 5, and 6 segments, respectively.

We select that "cluster solution" to satisfy two often mutually clashing objectives:

1. Parsimony: The fewer the number of clusters or segments the better.
2. Interpretability: Each segment must "tell a coherent, simple story." It becomes easier to tell a story when there are many clusters or segments, but having many clusters defeats the objective of parsimony.

Table 14.11 presents the three segments which emerge from the clustering. A two-segment solution is more parsimonious, but not easily understood. The three-segment solution is easier to understand, but still not perfectly interpretable. However, the three-segment solution is sufficiently clear that we can stop here. The left three sets of numbers show the impact or coefficient values for the INT model (overall evaluation, question 1). The right three sets of numbers show the dollar values. Only the strong-performing elements in at least one segment appear in Table 14.11, in order to minimize the size of the table, making it easier to read.

The mind-set segments are mathematical constructs, emerging from the attempt of the clustering algorithm to satisfy specific criteria, without any attempt to "name" the segments. The name comes afterward, emerging from an inspection of the elements which score very high on the INT model (here, impact values for the most part 20 or higher). The segment names come from the elements which are most desired by each segment. The names are not "fixed in stone," but rather represent simplified ways of thinking about these clusters.

Table 14.11 Interest (INT S) and Dollar (DOL S) Segment Models for the Strongest-Performing Elements for the Three Mind-Set Segments

		INT S1	INT S2	INT S3	DOL S1	DOL S2	DOL S3
	Base Size	**30**	**100**	**22**	**30**	**100**	**22**
	Constant	5	−5	35			
A2	European bread has that crispy crust and is nice and chewy and is less dense than American bread	9	**20**	−6	1.09	1.15	0.85
A3	The Europeans only use basic ingredients in their bread—they produce it fresh daily, that's why it is so delicious	−2	**20**	3	0.93	**1.43**	0.83
A5	Honest food doesn't need any taste enhancers—no preservatives, GMO ingredients, sugar, or additives	8	**29**	−15	0.92	**1.77**	1.28
A6	Bread is one of the most important foods we consume every day—it should come from the best ingredients	7	**15**	−7	0.92	1.22	0.23
B1	From France	**27**	−4	0	**1.66**	0.87	**1.65**
B2	From Italy	**16**	4	−5	**1.49**	0.9	**1.38**
B4	From Germany	**23**	0	**13**	**1.46**	0.89	1.01
C1	Dough made with only local European ingredients—GMO free	**16**	**20**	2	**1.52**	1.22	0.71
C2	Bread variations from artisanal bakeries made by professional bakers	**32**	**12**	2	1.28	1.04	0.97
C3	Artisanal bread recipes without any sugar, taste enhancers, colorants, GMO ingredients, or preservatives	**17**	**26**	−16	**1.32**	**1.69**	0.64
C4	Bringing the purest bread variants fresh to your table every day	**21**	**15**	−10	**1.54**	1.16	**1.31**
C5	Your everyday bread made from ancient, pure, local recipes	9	8	**16**	1.26	1.13	**1.32**
D5	Benefits long-term health as it is free of GMO ingredients, artificial additives, and sugar	−12	**22**	10	0.71	**1.41**	0.77

(Continued)

Table 14.11 Interest (INT S) and Dollar (DOL S) Segment Models for the Strongest-Performing Elements for the Three Mind-Set Segments (*cont.*)

		INT S1	INT S2	INT S3	DOL S1	DOL S2	DOL S3
	Base Size	**30**	**100**	**22**	**30**	**100**	**22**
D6	Easy way to make healthy and nutritious breakfast or lunch for me and my family	−7	9	**20**	0.87	1.11	1.16
E3	We produce the dough in the country of origin, which brings the real taste to the United States	**22**	8	6	1.19	0.83	**1.33**
E4	Made from European pure ingredients	**19**	3	8	0.92	0.76	1.23
E5	Bread that has been a nutritious staple in Europe for centuries	3	8	**21**	1.1	0.95	**1.76**
F2	It smells and tastes like homemade	**16**	7	11	1.23	0.76	1.05
F6	Crispy crust tells you that it was freshly baked, nice and chewy bite	**26**	9	**11**	**1.81**	1.01	0.58

Table is sorted by the order of silos and element number.
The value of the bold numbers is to call attention to the elements with high interest levels and dollar values from those with low values.

14.14.1 STRONG-PERFORMING ELEMENTS FROM SEGMENT S1: EUROPEAN, ARTISAN SEEKERS

Bread variations from artisanal bakeries made by professional bakers
From France
Crispy crust tells you that it was freshly baked, nice and chewy bite
From Germany
We produce the dough in the country of origin, which brings the real taste to the United States
Bringing the purest bread variants fresh to your table every day

14.14.2 STRONG-PERFORMING ELEMENTS FROM SEGMENT S2: HEALTHFUL, SIMPLE, FRESH

Honest food doesn't need any taste enhancers—no preservatives, GMO ingredients, sugar, or additives
Artisanal bread recipes without any sugar, taste enhancers, colorants, GMO ingredients, or preservatives

Benefits long-term health as it is free of GMO ingredients, artificial additives, and sugar
Dough made with only local European ingredients—GMO free
European bread has that crispy crust and is nice and chewy and is less dense than American bread
The Europeans only use basic ingredients in their bread—they produce it fresh daily, that's why it is so delicious

14.14.3 STRONG-PERFORMING ELEMENTS FROM SEGMENT S3: BREAD WITH A HISTORY

Bread that has been a nutritious staple in Europe for centuries
Easy way to make healthy and nutritious breakfast or lunch for me and my family
Your everyday bread made from ancient, pure local recipes

14.15 THE RELATION BETWEEN INTEREST (INT) AND DOLLAR VALUE (PRICE) DIFFERS ACROSS SEGMENTS

The scatter plots in Fig. 14.4 show a close relation between interest and dollar value (price) for the total panel, for mind-set Segment 1 (European, artisan seekers) and for mind-set Segment 2 (healthful, high-quality bread). They affirm the expected general pattern between interest and value; we expect that a person will pay more for what the person likes. Whereas a linear relation is intuitive and expected, there is no guarantee, however, that it holds in every situation. As Fig. 14.4 reveals, there is no relation whatsoever between interest and dollar value (price) for mind-set Segment 3 (bread with a history), suggesting that this third segment values the aspects of bread using radically different rules. Segment 3 appears to ascribe value to the more "exotic" aspect of artisanal bread, whether or not they are interested in the bread itself. Segment 3 seems to want to be "impressed" in order to ascribe value to the bread.

14.16 DIFFERENCES AMONG SEGMENTS ARISE FROM THE PATTERNS OF INTEREST AND VALUE

We showed previously that the three segments coalesce around different ideas which give them their identity. Mind Genomics and Cognitive Economics allow us to understand these segments more deeply by grouping the elements within each mind-set according to interest level and value, and thus fitting the elements into the four boxes of our matrix: "*Profit Opportunity,*" "*Must Haves,*" "*Not Worth Paying For,*" and "*Irrelevant,*" respectively. To tell an understandable story, not one which is overwhelming, we will present the elements for one of the four boxes, "Profit

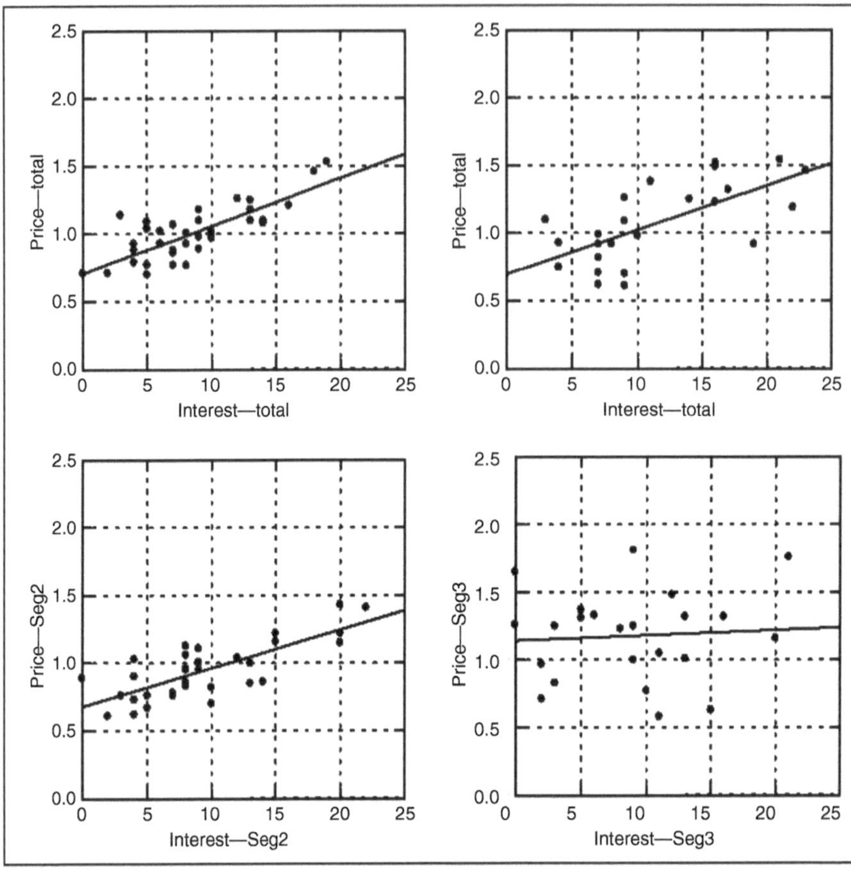

FIGURE 14.4 The Relation Between the Impact Values of 36 Elements for Bread (INT Model, Abscissa) and the Dollar value (Price) of the Element From the Dollar Model (Ordinate)

Each filled circle corresponds to one of the elements.

Opportunity." Recall that these elements in "Profit Opportunity" exhibit the highest interest scores as well as above-average dollar value.

The mind-set segments show specific viewpoints. Each viewpoint tells a different narrative. Table 14.12 shows that the Profit Opportunity elements differ across the three mind-set segments. Even when there is overlap between Segment 1 and Segment 2, for example, elements C1, C2, C3, and C4, their impact values and dollar values differ dramatically. For example, element C2 in Segment 1 has an impact value of 32 and a dollar value of \$1.28. The same element C2 has an impact value of 12 and dollar value of \$1.04 in Segment 2. Each mind-set segment has a different view of what is interesting, and in turn what is worth paying for.

Table 14.12 Profit Opportunities for the Three Mind-Set Segments

		INT	Dollar Value
	Mind-Set Segment 1: European, Artisan, Sensory Seekers		
C2	Bread variations from artisanal bakeries made by professional bakers	32	$1.28
B1	From France	27	$1.66
F6	Crispy crust tells you that it was freshly baked, nice and chewy bite	26	$1.81
B4	From Germany	23	$1.46
E3	We produce the dough in the country of origin, which brings the real taste to the United States	22	$1.19
C4	Bringing the purest bread variants fresh to your table every day	21	$1.54
C3	Artisanal bread recipes without any sugar, taste enhancers, colorants, GMO ingredients, or preservatives	17	$1.32
B2	From Italy	16	$1.49
C1	Dough made with only local European ingredients—GMO free	16	$1.52
F2	It smells and tastes like homemade	16	$1.23
F1	It tastes like coming home	14	$1.25
F4	It tastes like being on holiday every day	11	$1.38
	Mind-Set Segment 2: Healthful, Simple, Fresh		
A5	Honest food doesn't need any taste enhancers—no preservatives, GMO ingredients, sugar, or additives	29	$1.77
C3	Artisanal bread recipes without any sugar, taste enhancers, colorants, GMO ingredients, or preservatives	26	$1.69
D5	Benefits long-term health as it is free of GMO ingredients, artificial additives, and sugar	22	$1.41
A2	European bread has that crispy crust and is nice and chewy and is less dense than American bread	20	$1.15
A3	The Europeans only use basic ingredients in their bread—they produce it fresh daily, that's why it is so delicious	20	$1.43
C1	Dough made with only local European ingredients—GMO free	20	$1.22
A6	Bread is one of the most important foods we consume every day—it should come from the best ingredients	15	$1.22
C4	Bringing the purest bread variants fresh to your table every day	15	$1.16
	Mind-Set Segment 3: Bread With a History		
E5	Bread that has been a nutritious staple in Europe for centuries	21	$1.76
E6	Delivered to you fresh every day	12	$1.48
C5	Your everyday bread made from ancient, pure, local recipes	16	$1.32
D2	Honest, delicious, and fresh food for me and my family	13	$1.32
D6	Easy way to make healthy and nutritious breakfast or lunch for me and my family	20	$1.16

14.17 THE IMPORTANCE OF SENSORY/EMOTIONAL ATTRIBUTES

We often think that the smell of freshly baked bread and other baked goods is central to its interest and appeal. The question to which we now turn is how sensory emotional attributes "work" across the mind-set segments. For example, many of us associate the aroma of bread baking in the oven at home with comfort and emotional connection. So strong is that association that people selling their homes are routinely encouraged to bake bread just before the prospective buyers come into the home to look around.

Examining individual elements shows that their "messaging power" functions differently across the mind-set segments. Consider element F1 (*It tastes like coming home*). F1 is a Profit Opportunity in Mind-set 1. Element F1 is Not Worth Paying For in Mind-set 2. Finally element F1 is Irrelevant in Mind-set 3. This is an important finding that has been replicated across hundreds of Mind Genomics studies; a single element can be highly desirable, expected, a bad trade-off, or a turnoff depending upon the mind-set segment. Inspecting Table 14.12 further shows that some sensory emotional elements may be rated similarly across mind-sets, but the pattern of responses across elements within a mind-set is unique.

When analyzing sensory emotional elements, three different stories emerge. Sensory emotional elements play a major role in mind-set Segment 1, and a minor supporting role in mind-set Segment 3. Sensory emotional elements do not drive interest or price in mind-set Segment 2, or for total panel (Table 14.13).

In summary:

1. Total panel: Sensory emotional elements do not drive interest or price. Almost all are Irrelevant (low interest, low price), and one is seen as Not Worth Paying For (low interest, high price).
2. Mind-set Segment 1 (European, Artisan, Sensory Seekers) is the only segment to rate four of the six sensory emotional elements in the Profit Opportunity group (high interest, high value). Sensory elements matter to this group, and only this group.

Table 14.13 Sensory Elements by Mind-Set Segment and by Group

Element	Total	Mind-Set 1	Mind-Set 2	Mind-Set 3
F1. It tastes like coming home	Irrelevant	Profit	Not worth	Irrelevant
F2. It smells and tastes like homemade	Irrelevant	Profit	Irrelevant	Must have
F3. Your every morning joy	Irrelevant	Irrelevant	Irrelevant	Not worth
F4. It tastes like being on holiday every day	Irrelevant	Profit	Irrelevant	Irrelevant
F5. It tastes healthy	Irrelevant	Irrelevant	Irrelevant	Irrelevant
F6. Crispy crust tells you that it was freshly baked, nice and chewy bite	Not worth	Profit	Irrelevant	Irrelevant

3. Mind-set Segment 2 (Healthful, Simple, and Fresh) is uninterested and unwilling to pay for sensory emotional elements. Nearly all are Irrelevant, with one being rated as Not Worth Paying For.
4. Mind-set Segment 3 (Bread with a History) is mostly uninterested and unwilling to pay for sensory emotional elements. This group responded to element F2's evocation of "smelling and tasting like homemade," but this is an expected aspect for them, fitting into the Must Have category (interested, but Not Worth Paying For).

14.18 MIXING MESSAGES TO RAISE INTEREST AND VALUE: MIND GENOMICS, COGNITIVE ECONOMICS, AND COMMUNICATING ARTISANAL BREAD

Bread makers are in the business of baking bread and selling bread to consumers. One purpose of communication is to create demand, which is often accomplished through such means as commercial communications, for example, advertising. One question nearly every marketer and advertiser wants answered is: "what story does my brand tell and to whom, so that our product will be preferred and purchased, and our brand can maximize sales revenue?"

Findings from Mind Genomics and Cognitive Economics studies provide quantitative guidance for answering these most central questions. The method is to mix elements from the Purchase Opportunity and Must Have groups for each mind-set segment. We present four examples of mixing messages. We will use three elements in each mixture: the two top-scoring elements—those with the highest impact values—from the Profit Opportunity Group, and the top-scoring element from the Must Have group. This mixture includes both the "hot buttons" and also reassurance that an expected aspect of the bread is present. The mixture delivers "sizzle" and "steak."

Recall that the experimental stimuli were constructed from 3–4 elements each from 3–4 different silos A, B, C, D, E, and F, respectively. To mix the messages is a straightforward matter; choose 3–4 top-performing elements, at most one from each silo for each mind-set as well as for the total sample. More than one element from within a single silo is not allowed; the experimental design does not present multiple elements from the same silo in a single vignette. When one element from a silo is present, all others from that silo are absent by design.

By mixing messages as described, coefficients (impact values) of the selected elements can be added together along with the additive constant. This summation is done for the Total panel and for each mind-set segment, respectively. The dollar values of the elements are added independently in a second summation, to compute a "price." Of course this is not necessarily the actual price that would be charged, but here it indicates the relative price that a segment member would pay when specific elements are combined and presented in a vignette.

Adding the impact values to the constant means that when the specified elements are present, we can expect that an additional percentage of respondents—an amount equal to the sum of the values—will rate a vignette as 7, 8, or 9.

Table 14.14 presents four examples of mixing messages.

1. These examples show that, with the exception of mind-set Segment 3, interest in bread is negligible: the additive constants range from –5 to +5.
2. Elements "work" to raise interest across all groups, but they "work harder" in the segments.
3. The elements raised interest in the total panel to 53. The segments do better in driving increased interest. The total panel comprises groups of people with different points of view, so that elements pull in different directions.
4. The segments are more homogeneous. Putting a well-performing element into the vignette, and looking at the performance of that vignette across the segments shows one mind-set segment really liking this combination because the elements are the ones preferred by this segment. Creating the appropriate "optimal" for each mind-set segment brings into play positive elements for that segment only. We see confirmation of this in Table 14.14. The total panel's optimal of 53 thus was exceeded in each of the segments, reaching 63 in mind-set Segment 1 (19% increase) and 83 (32% increase) in mind-set Segments 2 and 3.
5. The same elevation occurs, this time with price. Choosing the best performers for each mind-set segment generates an expected price in that segment which is always higher than what happens when we choose the best for total panel.

As a final exercise in the "Cognitive Economics" of artisanal bread, we can estimate the productivity of messaging, combining the impact value of each element with its dollar value. We operationally define "productivity" as the sum of the interest value times the dollar value, for a group of combined elements. Recall that the interest value is the percentage of respondents who would rate a vignette 7–9, that is, I'd buy the bread. So our "productivity measure" is really a form of *expected value*. Productivity is a measure of sales opportunity. Productivity allows us to compare the computation for the total panel with the computations for each mind-set segment.

Table 14.14 shows the results, and the dramatic increase in "productivity" when we move from a strategy based on total panel to a strategy based on mind-set segments. As Table 14.14 shows, mixing the elements as prescribed, the top two "Profit Opportunity" elements and the top "must have" element, leads to productivity within each segment substantially greater than the productivity of the total panel. The improvements range from 29% to 63%. Marketers and advertisers now have an empirical, scientific basis for selecting messages and segments they wish to target for communications.

Note that this basic sales model could be refined by adding in a conversion factor, known or estimated, for each mind-set segment and the total panel. In addition, marketers or advertisers might allocate their efforts against one, two, or three segments,

Table 14.14 Interest (INT), Price, and Estimated "Sale" (Productivity) for Optimal Vignettes from Total Panel and Mind-Set Segments

	INT	Price
Illustrating With Total Sample		
Additive constant	3	
A5: Honest food doesn't need any taste enhancers—no preservatives, GMO ingredients, sugar, or additives	19	$1.53
C3: Artisanal bread recipes without any sugar, taste enhancers, colorants, GMO ingredients, or preservatives	18	$1.46
E1: Recipes just using simple ingredients such as water, dough, yeast, and craftsmanship	10	$0.98
Sum:	53	$3.97
Total panel sales per 1000 people = 1000 (0.53 × $3.97) = $2104.10		
Illustrating With Mind-Set Segment 1: European, Artisan, Sensory Seekers		
Additive constant	5	
C2: Bread variations from artisanal bakeries made by professional bakers	32	$1.28
B1: From France	27	$1.66
E6: Delivered to you fresh every day	19	$0.98
Sum:	83	$3.92
Segment 1 sales per 1000 people = 1000 (0.83 × $3.92) = $3253.60 (+55%)		
Illustrating With Mind-Set Segment 2: Healthful, High-Quality Bread		
Additive constant	−5	
A5: Honest food doesn't need any taste enhancers—no preservatives, GMO ingredients, sugar, or additives	29	$1.77
C3: Artisanal bread recipes without any sugar, taste enhancers, colorants, GMO ingredients, or preservatives	26	$1.69
D1: Eating pure and healthy every day	13	$0.85
Sum:	63	$4.31
Segment 2 sales per 1000 people = 1000 (0.63 × $4.31) = $2715.30 (+29%)		
Illustrating With Mind-Set Segment 3: Bread With A History		
Additive constant	35	
E5: Bread that has been a nutritious staple in Europe for centuries	21	$1.76
C5: Your everyday bread made from ancient, pure, local recipes	16	$1.32
F2: It smells and tastes like homemade	11	$1.05
Sum:	83	$4.13
Segment 3 sales per 1000 people = 1000 (0.83 × $4.13) = $3427.90 (+63%)		

depending on their goals and objectives. Although enhancements or different targeting scenarios are possible, the core principle is selecting mixtures of messages based on their impact value and dollar value.

14.19 CONCLUSIONS

This study explored artisanal, European-style bread from the standpoint of Mind Genomics and Cognitive Economics, using an experimental design coupled with consumer evaluation and modeling. The objective is to uncover the impact on interest and value of messages, especially sensory emotional elements. The research examined sensory emotional elements in the context of a variety of other elements relevant to artisanal, European-style bread, such as country of origin, ingredients, recipes, heritage, and purity, not in an isolated fashion. Sensory emotional elements selected for the study were not presented by technical jargon, but in the language consumers routinely encounter; evocative mixtures of ideas in communications about bread which appear in advertising, product promotion, or on packages. To ensure that, the sensory emotional elements were selected with the help of marketers and communications professionals.

Sensory emotional elements did not universally drive interest in artisanal European-style bread nor did they drive the perception of value. It was the opposite; each mind-set segment interpreted and responded to different, unique stories about bread. Only mind-set Segment 1, the person interested in the artisanal aspects of bread, showed high interest in and highly valued the warm and comforting sensory emotional attributes. The nonsensory emotional elements drove greater interest and value in the other segments, Segment 2 and Segment 3, respectively.

Mind Genomics and Cognitive Economics provide marketers and advertisers with empirical, scientific guidance when selecting target groups and messages to achieve their business goals: effective marketing, resonant and persuasive communications, and sales success. The approach outlined in this chapter, as noted earlier, is applicable to food products, nonfood products, services, and experiences. The approach, mixing ideas together by design, and deconstructing the contribution of these ideas, finds use in innovation, new product development, and modifying existing products to better meet the needs of consumers.

Mind Genomics and Cognitive Economics are beginning to find creative new applications in the food industry, which reflect new directions of thinking. One new approach is to use the procedures outlined here to *remove ingredients, features, or attributes* that consumers no longer desire, as reported by a major food brand (Cox and Emrick, 2013). They designed a study that included the ingredients in a specific product. After assigning elements to one of the four categories: Profit Opportunity, Must Haves, Not Worth Paying For, and Irrelevant, the brand deleted the Irrelevant ingredients, that is, ingredients with low interest and low dollar values. The brand lowered costs while tuning the product more closely to consumer interests and needs.

Response-based segmentation, dividing people by patterns of response to messages about the specific topic, improves the bottom line by increasing sales and eventually profitability. The segmentation works in two ways:

1. Segmentation identifies groups of people who can be more readily convinced. So we have here an advantage from the realm of marketing.
2. Segmentation reveals groups of people who want only part of the full possible array of features and benefits offered by a product. This identification has monetary implications. To the degree that one can identify segments of individuals who care only about a limited number of product features, one can tailor a more cost-effective offering to that segment. This tailoring is finding use by major food brands, and has been in the forefront of business attention for more than a decade (Gladwell, 2004).

Finally, this journey through Mind Genomics and Cognitive Economics occasionally used the words "narrative" and "story." Narrative refers to the thematic description of the mind-set segments, and story refers to the elements within the segment. Since late 2007, the term "brand storytelling" has gained currency among marketers and advertisers. Most "brand stories" are written from the brand's perspective, not from the perspective of their consumers. It is not what a brand says about itself that matters, but the different narratives and stories people tell about brands. Mind Genomics and Cognitive Economics, focusing as they do on resonant communications, offer a way to discover the consumer stories that brands can use strategically to create products, services, or experiences which match consumer interest. The rich specificity of elements allows realistic, evocative, meaningful stories about products, stories which resound across and within the different mind-sets consumers have toward them. This tight coupling enables brands to serve their consumers as they would like to be served while at the same time improving their chances for business success.

ACKNOWLEDGMENTS

The authors would like to thank Ms. Margie Atwater, Editorial Group at Mind Genomics Advisors, for her ongoing help in preparing this and other manuscripts.

REFERENCES

Box, G.E.P., Hunter, J., Hunter, S., 1978. Statistics for Experimenters. John Wiley, New York.

Claritas, 1999. PRIZM Cluster Snapshots: Getting to Know the 62 Clusters. Claritas Corporation, Ithaca, NY.

Cox, S.A., Emrick, M., 2013. A three-in-one tool: advancing consumer research by applying mind genomics to address product development needs in a challenging environment. Product development needs in a challenging environment. Paper Presented to the Institute of Food Technologists, National Meeting.

Everitt, B.S., Landau, S., Leese, M., Stahl, D., 2010. Cluster Analysis, fifth ed. John Wiley & Sons, Hoboken, NJ.

Gladwell, M., 2004. Choice, happiness, and spaghetti sauce. TED talks. www.ted.com/talks/malcolm_gladwell_on_spaghetti_sauce?language=en

Gofman, A., Moskowitz, H.R., 2010. Application of isomorphic permuted experimental designs in conjoint analysis. J. Sens. Stud. 25, 127–145.

Gofman, A., Moskowitz, H.R., 2012. Rule Developing Experimentation: A Systematic Approach to Understand & Engineer the Consumer Mind. Bentham Science Publishing, Sharjah.

Green, P.E., Krieger, A.M., 1991. Segmenting markets with conjoint measurement. J. Marketing 55, 20–31.

Green, P.E., Srinivasan, V., 1990. Conjoint analysis in marketing: new developments with implications for research and practice. J. Marketing 54, 3–19.

Mahanna, K., Moskowitz, H.R., Lee, S.-Y., 2009. Assessing consumer expectations for food bars by conjoint measurement. J. Sens. Stud. 24, 851–870.

Moskowitz, H.R., Gofman, A., 2004. System and method for performing conjoint measurement. Provisional patent application, 60/538,787, filed January 23, 2004.

Moskowitz, H.R., Gofman, A., Itty, B., Katz, R., Manchaiah, M., Ma, Z., 2001. Rapid, inexpensive, actionable concept generation and optimization: the use and promise of self-authoring conjoint analysis for the food service industry. Food Service Technol. 1, 149–168.

Moskowitz, H.R., German, B., Saguy, I.S., 2005a. Unveiling health attitudes and creating good-for-you foods: the genomics metaphor and consumer innovative web-based technologies. CRC Crit. Rev. Nutr. Food Sci. 45 (3), 191–265.

Moskowitz, H.R., Porretta, S., Silcher, M., 2005b. Concept Research in Food Product Design and Development. Blackwell Professional, Ames, IA.

Wells, W.D., 1975. Psychographics, a critical review. J. Mark. Res. 12, 196–213.

CHAPTER

15 Global market entry regulations for nutraceuticals, functional foods, dietary/food/health supplements

Andrew Shao
Herbalife Nutrition, Los Angeles, CA, United States

15.1 INTRODUCTION

The policy that underlies the regulation for any consumer product should be one that appropriately balances benefit, and consumer access and choice, with safety (risk). This principle should apply regardless of the regulatory category, whether a drug, device, food, supplement, or nutraceutical. Requiring that a new promising treatment for cancer be free of any side effects whatsoever may minimize risk to the public, but such a treatment would likely fail to ever reach the market, thus doing consumers little if any good. In contrast, allowing an untested medical device on the market that provides little benefit but poses high safety concerns, allows for open consumer access, but does not appropriately balance risk.

Starting with market entry, global regulations should aim to strike an appropriate balance between consumer access and risk. For nutraceuticals, functional foods, and supplements, the specific market entry requirements depend on the overall regulatory framework employed within a given country. For most countries, these frameworks are based on a food or drug, including the over the counter (OTC)-based approach. Products regulated under a food-like category tend to be subjected to notification or registration-based systems, whereas products regulated under a drug-like category tend to be subjected to a premarket approval system. The chapter that follows compares and contrasts the different approaches used globally and whether these help to achieve the appropriate access–risk balance.

15.2 MARKET ENTRY REQUIREMENTS

For most countries worldwide, foods (including conventional foods and functional foods) and supplements (including nutraceuticals) are regulated as a category of food. In some regions or countries, there are a specific set of regulations governing

Developing New Functional Food and Nutraceutical Products. http://dx.doi.org/10.1016/B978-0-12-802780-6.00015-8

supplements (e.g., US, EU, ASEAN) and nutraceuticals (India), which stem from a food-based regulatory paradigm. For most of these countries, some form of a notification or registration-based system is required to bring new products to market (Table 15.1). This approach is in contrast to the premarket approval approach required for drugs in most countries, and appropriately fits the category of foods, as these pose inherently low safety risks relative to drugs.

Two key markets in which market entry regulations are currently evolving are India and China. For the latter, market entry regulations were recently revised to allow for a notification-based approach for basic vitamin- and mineral-containing products (China Food Safety Law, inpress). This decision from the China Food and Drug Administration (CFDA) is indicative of substantial progress achieved in a major market to rebalance consumer choice and access with consumer protection. With the safety and health benefits of essential vitamins and minerals well recognized, this new approach calls for a more simplified and cost-effective product notification, rather than the lengthy and onerous registration process which has been in place for years (and remains for other functional food products bearing functional claims).

Recent developments in India have arisen from stagnation in the marketplace resulting in Supreme Court involvement. Due to lack of clarity in the regulations, the Food Safety & Standards Authority of India (FSSAI) had been holding back on the approval of applications for many supplement and functional food products since mid-2013. In late Aug. of 2015 the Bombay Supreme Court ordered FSSAI to cease its backlog of applications and to issue expedited regulations on market access (No.P-15025/SCJ/2015-PA/FSS, 2015).

FSSAI has since issued revised draft regulations that have received public comment and are currently under review.

Table 15.1 Global Notification or Registration for Foods/Supplements

Country/Region	Notification/Registration
Argentina	Registration
ASEAN	Notification or registration
Australia	Notification/listing
Brazil	Registration
Canada	Registration
Chile	Notification
China	Registration or notification
Colombia	Registration
EU	Notification or none
India	Registration
Japan	None
Mexico	Registration
Russia	Registration
USA	Notification or none

15.2.1 POSITIVE AND NEGATIVE INGREDIENT LISTS

The safety and acceptability of new products into the marketplace start first with the ingredients. In most regions of the world, the safety and acceptability of excipients used in a given formula are governed via established food additive requirements. In the United States (US Food and Drug Administration, 2012a), including generally recognized as safe (GRAS) (US Food and Drug Administration, 2012b), and the European Union (European Commission, 2013a), there are well-established food additive regulations, and for those countries which lack their own established approaches, many rely on standards set by the *Codex Alimentarius* (JECFA, 2013). Therefore, when evaluating the entry of new products to the marketplace, most regulators focus their attention on the safety and acceptability of the active or driver ingredients.

In many countries, regulators rely on established formal "positive" and/or "negative" ingredient lists, or derivations thereof to help guide their evaluations. These lists, as their name implies, indicate which ingredients are considered acceptable for use in foods and/or supplements and which are not allowed, including acceptable chemical forms. Some countries will rely on only a positive list, others only a negative list; some countries rely on both and some don't rely on a list but employ a broader approach. The basis for how ingredients are added to these lists varies widely around the world, and the formal role the lists play varies as well. An ingredient may be added to a negative (not allowed in foods) list based on well-established safety issues, for example, a botanical with well-known toxic effects; or an ingredient may be added to the list based on the perception (often ill conceived) or speculation of adverse effects. Ingredients known to be inherently safe, such as vitamins and minerals, will appear on most positive lists, or due to their inherent safety may not appear on a list at all. Most regulatory agencies keep separate lists for botanicals allowable for use in foods and supplements. New botanicals or other novel ingredients may also be allowed and added to positive lists on a case-by-case basis following their formal evaluation. Whether an ingredient is considered new or novel depends on the specific policies and regulations of the given country or region.

In Australia, the therapeutic goods administration (TGA) regulates supplements under a Complementary Medicines category and maintains a positive list of allowable ingredients, including excipients and actives (Therapeutic Goods Administration, 2007). This list includes vitamins, minerals, and botanicals. China (China Food and Drug Administration, 2009) is an example of a country that utilizes both a positive and a negative list. China's positive list includes ingredients that can be added to foods, drugs, and functional foods, as well as there being a list of banned ingredients. Canada employs a unique monograph system—ingredients which comply with already established monographs have an easier path to market as natural health products (NHPs), while those which have not yet been the subject of a monograph must go through a more formal approval process (Natural Health Products Directorate, 2009).

In the European Union, the European Commission has published a list of vitamin and mineral ingredients allowable in foods and food supplements (European

Commission, 2009). Ingredients which have not been used for human consumption to a significant degree before May 1997 are considered novel foods, requiring the marketer to apply for authorization (European Commission, 2012). There is an online catalogue, which uses a traffic light system to highlight whether the ingredient is novel, not novel, can only, be used in food supplements, or it is not certain yet (European Commission, 2013b).

For other physiologic substances, including botanicals, the regulations in the European Union for acceptable ingredients have not yet been harmonized. As a result, different positive lists have been developed by some individual member states. For example, Italy (Ministero della Salute, 2013), Denmark (Ministerialtidende, 2012), and the Czech Republic (SBÍRKA ZÁKONŮ, 2012) have established their own versions of positive lists of substances that can be added to foods and/or supplements. Russia is an example of a nonmember state with a positive list (Customs Union Commission, 2010).

In the United States, the US FDA maintains a list of ingredients that are GRAS for use in foods (US Food and Drug Administration, 2012b). In many cases, these ingredients are acceptable for use in dietary supplements as well, although the agency does not maintain a specific positive list of ingredients for supplements. Regarding ingredients not allowed for use in supplements, the FDA has published several rulings (US Food and Drug Administration, 2004; Center for Food Safety and Applied Nutrition, 2011) and/or warning letters (US Food and Drug Administration, 2001a, 2001b, 2003) indicating certain ingredients are not acceptable for inclusion in dietary supplements or are unsafe for the public (US Food and Drug Administration, 2002). In the United States, new dietary ingredients (NDI) must be notified to the FDA while "old" dietary ingredients, or grandfathered ingredients (ingredients marketed prior to Oct. 1994), need not be notified. However, no formal list of grandfathered ingredients exists in the US.

In South America, the situation varies from country to country, with some not relying on a list (e.g., Argentina, Chile, Colombia), others a positive list (e.g., Brazil), and others both (e.g., Venezuela).

In ASEAN, the region is aiming to establish a restricted list of substances allowed in traditional medicines and health supplements (ASEAN, 2010). In India, the FSSAI has recently issued draft regulations for food or health supplements, nutraceuticals, foods for special dietary uses, foods for special medical purposes, functional foods, and novel foods (No. 1-4/ Nutraceutical/ FSSAI-2013, 2015). The draft regulation includes definitions of the aforementioned product categories, acceptable ingredients, and claims. Draft positive lists are presented for vitamins and minerals, amino acids, botanicals, animal-derived ingredients, probiotics, prebiotics, nutraceuticals, and food additives.

A brief summary of the different types of lists utilized in different countries or regions is given in Table 15.2.

15.2.2 BOTANICALS

Botanicals can present unique regulatory, quality, and safety challenges and thus tend to be more restricted in many markets. Lack of understanding on the part of

Table 15.2 Countries/Regions Employing Positive and/or Negative Ingredient Lists

Market/Region	Positive List	Negative List	Other
Argentina			
Australia	✓		
Brazil	✓		
Canada			Monograph system
Chile			
China	✓	✓	
Colombia			
India[a]	✓		Novel food
Mexico	✓		
Paraguay			
Russia	✓		
Uruguay			
United States			GRAS, New/Old Dietary Ingredients
Venezuela	✓	✓	
EU	✓		Novel food
Italy	✓		
Denmark	✓		
Czech Republic	✓		
ASEAN			Restricted list

[a]*Draft regulation.*

regulators, lack of transparency in the supply chain, and a less robust evidence base relative to vitamins and minerals serve to undermine the confidence of regulators in this category of ingredients. Thus, the appearance of botanical ingredients on positive lists is quite limited in many countries. In Brazil, the presiding regulatory agency over foods and drugs, National Health Surveillance Agency (ANVISA), only allows water or ethanol-extracted botanicals in foods, and each product is evaluated on a case-by-case basis (National Health Surveillance Agency, 2009). In Mexico, the Federal Commission for the Protection against Sanitary Risk (COFEPRIS) maintains a positive and negative list for botanicals allowed in foods (Federal Commission for the Protection against Sanitary Risk, 2011). The agency has recently proposed to add certain botanical ingredients with a long history of safe use to a negative list, including inulin and aloe (Asociacion Nacional de la Industria de los Suplementos Alimenticios, 2013), based on the belief that these have documented therapeutic uses (and thus are medicines, not acceptable for use in foods). In the European Union, the European Food Safety Authority (EFSA) has published a botanicals compendium (European Food Safety Authority, 2012) which includes a list of botanicals reported to contain toxic, addictive, psychotropic, or other substances of concern. While this document does not have official regulatory standing, it could be interpreted as a form of a negative list for the European Union. However, other EU member states have published and continue to update their own positive and negative botanical lists. Italy maintains a list of permitted botanicals with related health benefits and one of nonpermitted botanicals (Ministero della Salute, 2012). Belgium has one list including permitted, nonpermitted botanicals, and mushrooms with related maximum allowable levels for some botanicals (Belgian Health, 2012). Germany maintains a list, which has been used by the regulators as a guidance document for assessing botanicals in Germany (The Federal Office of Consumer Protection and Food Safety, 2013).

In these countries, whether a botanical appears on a positive or negative list for foods depends largely on if it is perceived as medicinal or not. A plant may be judged to be medicinal if there are scientific data showing an influence on physiological functions. Where applicable an assessment of the minimum dosage for such activity is determined. A number of acceptable scientific sources to support this assessment may be consulted, including compendial monographs. Another check is whether the ingredient is novel, in which case the European Union novel food catalogue is consulted as a reference. Finally, the botanical may be checked against conformity with other approvals, for example, additives and if the product would be considered as safe.

The FSSAI in India has recently proposed new regulations that include draft positive lists for several ingredient categories, including plant or botanical sources for "ASU" (Ayurveda, Siddha, and Unani) ingredients (No. 1-4/ Nutraceutical/ FSSAI-2013, 2015).

Clear challenges for regulators in establishing positive or negative lists include lack of quality standards for botanical-derived ingredients, lack of transparency in the supply chain, and the heterogeneous evidence base for the variety of conditions of use and preparations of botanical ingredients. These issues became the focus of the New York Attorney General in the United States, whose controversial stance on the policy of botanical-containing dietary supplements has had global implications. In early 2015 the New York Attorney General launched an investigation and sued a group of retailers and dietary supplement manufacturers, alleging their botanical products were not authentic (New York Attorney General Bashes Herbal Supplement Purity, 2015). While the US industry trade associations vigorously opposed this action which appeared to be aimed more at policy change than identifying true quality issues (CRN, inpress), the ordeal is reflective of issues with botanical ingredient acceptance globally.

The simplest path for botanical ingredient acceptance is to implement a tiered approach based on history of safe use, quality standards, contemporary evidence of efficacy, and a clear and transparent supply chain. Such an approach would allow for expedited review and market access for those ingredients with the most information for a given preparation, with greater scrutiny afforded to those with less information. Regulators and policy makers from Europe, South America, China, and India are currently contemplating such approaches.

15.2.3 NOTIFICATION VERSUS REGISTRATION

Although there are some differences between countries, a notification-based system usually represents a means of informing the regulatory authorities of a product's introduction to the marketplace or an intent to introduce a new product. Notifications tend to be comprised of information on the manufacturer, product form, formulation, and label and these tend to be subjected to minimal review by regulators. Product ingredients must be derived from a positive list (if one exists), and the entry of the product to the marketplace does not signify or constitute an "approval" or "authorization" from regulators, but they do have the authority to question or object to a notification. The notification results in the addition of the new product to a list which regulators maintain for enforcement purposes. The notification approach tends to be the most efficient for getting new products to market, for both industry and regulators, as minimal premarket resources are necessary. This approach also allows regulators to focus their resources on enforcement of critical postmarket activities, such as good manufacturing practices, and adverse event reporting. Notification is thus a proportionate approach to the low level of risk posed by these categories of food products.

Registration-based approaches differ from notification-based approaches in that they require more detailed information in the form of a registration dossier, intense review by regulators, and a lengthier review process. A typical registration dossier requires much more detailed information than is found in a notification, including details on the specifications of the finished product, evidence supporting the safety and efficacy of the ingredients, a Certificate of Analysis, stability testing (to support shelf-life dating), the product label, a certificate of free sale, and a GMP certification or compliance statement. While this approach allows regulators to closely scrutinize new products, in most countries that require registration, it is a slow and extremely resource-intensive process, ranging from months up to years. A lengthy registration time tends to be due to a lack of resources and the obligation regulators have since a positive registration equates to regulator approval or authorization. Drawbacks of this approach are the length of time to market for products, regardless of the risk they may pose, and even fewer resources for enforcement in other areas. A comparison of notification versus registration appears in Table 15.3.

The European Union (EU) is an example of a region which utilizes a notification approach (European Commission, 2002). While the specific requirements differ between countries, of the 27 member states, 23 have opted to incorporate a notification-based approach for new product entry. Austria, Holland, Sweden, and the UK are among those for which notification for new products is optional.

In the ASEAN, the 10-member country region is in the process of finalizing harmonized regulations for traditional medicines and health supplements (Association of South East Asian Nations, 2012). Currently, both registration and notification-based procedures are utilized, depending on the country (Table 15.4). As with other regions, the specific requirements in each country vary (Association of South East Asian Nations, 2006).

Table 15.3 Notification Versus Registration Submissions for Foods, Supplements, and Nutraceuticals

	Notification	Registration
Ingredients in accordance with applicable positive and negative lists	Required	Required
Applicable good manufacturing practices (GMP)	Manufacturer expected to adhere to	Official certification or statement required with dossier
Label according to regulation	Required	Required
Detailed product specifications	N/A	Required
Certificate of Analysis	N/A	Required
Claim substantiation[a]	Manufacturer expected to have on file	Required
Shelf-life substantiation	N/A or manufacturer expected to have on file	Required
Certificate of free sale	N/A	Required
Implications	Surveillance and control regime	Reduces postsurveillance and control regime
	Helps the authority to process a high volume of products	Requires high government resources
	Immediate access to market	Slow access to market

[a]*In those countries where claims are allowed.*

Regulations were harmonized in 2013 and the intent was to begin implementation by 2015, however this has been met with delays in the region due to some misalignment between ASEAN members. For market entry for health supplements, the region will require a comprehensive product dossier as part of a registration-based process, with the emphasis on quality, safety, and efficacy.

The United States represents a somewhat unique situation where no product notification is required unless the product contains a NDI (US Food and Drug Administration, 2011). If a new product does not contain a NDI, then no notification is required. If a new product contains a NDI, then a notification to the US FDA is required 75 days before the product enters the market. The basis of the notification is substantiation that the manufacturer or marketer believes the NDI is safe in the product under its intended conditions of use (US Food and Drug Administration, 2012c).

In Latin America, the market entry requirements for foods and supplements vary, with notification-based approaches employed in Mexico and Chile, while registration-based approaches are followed in Colombia, Brazil, and Argentina. In countries such as Brazil [ANVISA (National Health Surveillance Agency, 2000)],

Table 15.4 Current Health Supplement Market Entry Requirements for Various ASEAN Member Countries

	Brunei	Cambodia	Indonesia	Malaysia	Philippines	Singapore	Thailand	Vietnam
Premarketing registration	No	Yes	Yes	Yes	Yes	No	Yes	Yes
Time line for evaluation								
< 2 months								✓
2–3 months			✓		✓			
6 months							✓	
6–12 months		✓		✓				

China [CFDA (JECFA, 2013)], and Taiwan [Taiwan Food and Drug Administration—TFDA (Taiwan Food and Drug Administration, 2012)] the regulators require animal and/or human clinical studies as a part of the product registration. This requirement tends to be limited to those products for which efficacy claims are intended.

The CFDA has recently proposed to modify its product registration requirements, to restrict extensive testing and premarket registration for nonvitamin and mineral functional foods, while allowing vitamin and mineral products to access the market via a simplified notification system.

In Australia and Canada, supplements and nutraceuticals are regulated more closely as a drug than food category. The TGA in Australia regulates supplements under a Complementary Medicines category (Therapeutic Goods Administration, 2012). The premarket approval requirements for this category are comparable to registration requirements for foods and supplements in other countries (Therapeutic Goods Administration, 2011), although other requirements, such as Good Manufacturing Practices, tend to be more drug-like than food-like. In Canada, Health Canada (HC) regulates supplements under a NHP category (Health Canada, 2012). Due to the allowance of medicinal or therapeutic claims for NHPs, HC takes a more rigorous approach to new product entry. The licensing process in Canada for NHPs is risk-based, with products posing the highest risk requiring the greatest premarket review by the agency.

While a variety of individual systems exist in various countries and regions worldwide, with a few exceptions the approaches for bringing new food, supplement, or nutraceutical products to market tend to follow three basic approaches. The notification-based approach most optimally balances premarket resources, consumer access, and consumer safety. However, a key aspect to assuring product safety and quality in the marketplace, regardless of the premarket requirements, is robust postmarket surveillance.

REFERENCES

ASEAN, 2010. Guiding principles for inclusion of active substances into the restricted list for traditional medicines and health supplements (TMHS). http://www.asean.org/storage/images/archive/SnC/Guiding_Principles_for_Restricted_List_of_Active_Substances.pdf

Asociacion Nacional de la Industria de los Suplementos Alimenticios, 2013. ANAISA noticias. http://www.anaisa.mx/

Association of South East Asian Nations (ASEAN), 2006. Profile of definition, terminology, and technical requirement of traditional medicines and health supplements among ASEAN member countries. http://www.asean.org/storage/images/archive/20258.pdf

Association of South East Asian Nations (ASEAN), 2012. Harmonization of standards and technical requirements in ASEAN. http://www.asean.org/news/item/harmonization-of-standards-and-technical-requirements-in-asean

Belgian Health, F.C. S. a. E., 2012. Food supplements—enriched foodstuffs. http://www.health.belgium.be/eportal/foodsafety/foodstuffs/foodsupplements/index.htm?fodnlang=en#Plants

Center for Food Safety and Applied Nutrition, C.F. S. A. N., Office of Food Additive Safety, 2011. FDA response to OVOS Natural Health homotaurine petition. Department of Health and Human Services, Food and Drug Administration, Washington, DC.

China Food & Drug Administration, 2009. Regulations on supervision of functional foods (draft for approval).

China Food Safety Law. Article 76 regulation on the registration and notification for health foods. http://www.sda.gov.cn/WS01/CL0780/125503.html

CRN responds to NY Attorney General actions against herbal supplements. http://www.crnusa.org/NYAG/PR.html

Customs Union Commission, 2010. Uniform sanitary epidemiological and hygienic requirements for products subject to sanitary and epidemiological supervision (control). http://ec.europa.eu/food/safety/docs/ia_eu-ru_sps-req_req_san-epi_chap-2_23_en.pdf

European Commission. (2002). Directive 2002/46/EC of the European Parliament and of the Council.

European Commission. (2009). Commission Regulation (EC) No 1170/2009 of 30 November 2009 amending Directive 2002/46/EC of the European Parliament and of Council and Regulation (EC) No 1925/2006 of the European Parliament and of the Council as regards the lists of vitamin and minerals and their forms that can be added to foods, including food supplements. Official Journal of the European Union, L314/36.

European Commission, 2012. Novel foods and novel food ingredients. Food and Feed Safety. http://ec.europa.eu/food/food/biotechnology/novelfood/index_en.htm

European Commission. Food additives and flavourings. http://ec.europa.eu/food/fs/sfp/flav_index_en.html

European Commission. Novel food catalogue—search. Food and Feed Safety. http://ec.europa.eu/food/food/biotechnology/novelfood/nfnetweb/mod_search/index.cfm

European Food Safety Authority, 2012. Compendium of botanicals reported to contain naturally occuring substances of possible concern for human health when used in food and food supplements. EFSA J. 2012, 10.

Federal Commission for the Protection against Sanitary Risk, COFEPRIS, 2011. What is the COFEPRIS? http://www.cofepris.gob.mx/Paginas/Idiomas/Ingles.aspx

Health Canada, 2012. Natural health products. http://www.hc-sc.gc.ca/dhp-mps/prodnatur/index-eng.php

Joint Expert Committee on Food Additives (JECFA), 2013. About Codex. Codex Alimentarius. http://www.codexalimentarius.org/scientific-basis-for-codex/jecfa/en/
Ministerialtidende, 2012. Bekendtgørelse om tilsætning af visse andre stoffer end vitaminer og mineraler til fødevarer. https://www.retsinformation.dk/Forms/R0710.aspx?id=137299
Ministero della Salute, 2012. Disciplina dell'impiego negli integratori alimentari di sostanze e preparati vegetali (G.U. 21-7-2012 serie generale n. 169).
Ministero della Salute. Altri nutrienti e altre sostanze ad effetto nutritivo o fisiologico. http://www.salute.gov.it/imgs/C_17_pagineAree_1268_listaFile_itemName_4_file.pdf>
National Health Surveillance Agency, ANVISA, 2000. Registration of products: manual covering procedures for registration and exemption from registration of imported products. http://www.anvisa.gov.br/eng/food/registration.htm
National Health Surveillance Agency, ANVISA, 2009. Novos Ingredientes Aprovados. http://portal.anvisa.gov.br/wps/content/Anvisa+Portal/Anvisa/Inicio/Alimentos/Assuntos+de+Interesse/Novos+Alimentos+e+Novos+Ingredientes/29bd7700401adec6b403b654e035b7cb
Natural Health Products Directorate, 2009. Compendium of monographs. Drugs and Health Products. http://www.hc-sc.gc.ca/dhp-mps/prodnatur/applications/licen-prod/monograph/index-eng.php
New York Attorney General Bashes Herbal Supplement Purity. Chemical and Engineering News, 93 (7), p. 9, 16 February 2015.
No. 1-4/ Nutraceutical/ FSSAI-2013. Draft Food Safety and Standards (food or health supplements, nutraceuticals, foods for special dietary uses, foods for special medical purposes, functional foods, and novel foods). July 2015.
F. No.P-15025/SCJ/2015-PA/FSSAI. Product Approval Division. 26 August 2015.
SBÍRKA ZÁKONŮ, 2008. Kterou se stanoví požadavky na doplňky stravy a na obohacování potravin.
Taiwan Food & Drug Administration, 2012. Food and Drug Administration, Department of Health. http://www.taiwan.gov.tw/ct.asp?xItem=25613&ctNode=1957&mp=999
The Federal Office of Consumer Protection and Food Safety (BVL), 2010. Entwurf einer Liste für die Kategorie "Pflanzen und Pflanzenteile." http://www.bvl.bund.de/SharedDocs/Downloads/01_Lebensmittel/stoffliste/stoffliste_pflanzen_pflanzenteile.html?nn=1406620
Therapeutic Goods Administration, 2007. Substances that may be used in listed medicines in Australia. http://www.tga.gov.au/pdf/cm-listed-substances.pdf
Therapeutic Goods Administration, 2011. Australian Guidelines for Complementary Medicines (ARGCM) Part II: listed complementary medicines.
Therapeutic Goods Administration, 2012. The regulation of complementary medicines in Australia—an overview: pre-market assessment. http://www.tga.gov.au/industry/cm-basics-regulation-overview.htm#pre
US Food and Drug Administration, 2001a. FDA advises dietary supplement manufacturers to remove Comfrey products from the market.
US Food and Drug Administration, 2001b. Letter to health professionals regarding safety concerns related to the use of botanical products containing aristolochic acid.
US Food and Drug Administration, 2002. Consumer advisory: kava-containing dietary supplements may be associated with severe liver injury.
US Food and Drug Administration, 2003. February 28, 2003 warning letter to Powerhouse Supplements.
US Food and Drug Administration, 2004. Final rule declaring dietary supplements containing ephedrine alkaloids adulterated because they present an unreasonable risk, Fed. Regist. 69(28), pp. 6788–6854.

US Food and Drug Administration, 2011. Draft guidance for industry: dietary supplements: new dietary ingredient notifications and related issues. http://www.fda.gov/regulatoryinformation/guidances/ucm257563.htm

US Food and Drug Administration, 2012a. Food additives. http://www.fda.gov/food/ingredientspackaginglabeling/foodadditivesingredients/ucm2006845.htm

US Food and Drug Administration, 2012b. Generally Recognized as Safe (GRAS). http://www.fda.gov/food/ingredientspackaginglabeling/gras/default.htm

US Food and Drug Administration, 2012c. New dietary ingredients in dietary supplements—background for industry. http://www.fda.gov/Food/DietarySupplements/ucm109764.htm

CHAPTER

US regulation of food label claims

16

James E. Hoadley
EAS Consulting Group, Alexandria, VA, United States

16.1 OVERVIEW

Products marketed in the USA and intended for oral consumption, other than tobacco products and alcoholic beverages, are regulated as either foods or drugs. Dietary supplements are a subcategory of food under the US Federal Food, Drug, and Cosmetic Act (FFD&C Act). The US Food and Drug Administration (FDA) regulates the labelling of all foods other than meat and poultry products that are under US Department of Agriculture (USDA) jurisdiction. The advertising of food falls under the jurisdiction of the Federal Trade Commission (FTC). Because the FDA and FTC enforce different laws, there are some differences in the regulations and policies for food advertising and labeling. This chapter will focus on FDA regulation of food label claims and will not address advertising.

The statutory definition of a *food* in the FFD&C Act is:

an article used for food or drink for man or for other animals, chewing gum, and articles used for components of any such article.

The definition of a *drug* in the FFD&C Act is, in part:

articles intended for use in the diagnosis, cure, mitigation, treatment, or prevention of disease in man or other animals, and articles, other than food, intended to affect the structure or any function of the body of man or other animals.

The labeling[a] of food may not include any information about uses of a food in the cure, mitigation, treatment, or prevention of a disease. Food labels can have information about how a food may affect a structure or function of the body (i.e., a structure/function claim). The FFD&C Act also provides for FDA to authorize food label claims that describe how a food may reduce disease risk (i.e., a health claim). Placing an FDA-authorized health claim on a food label does not cause the intended

[a]A distinction is made between the "label" (on the immediate product container) and "labeling" (material accompanying the product). In this chapter, we use the terms "label" and "labeling" interchangeably to describe the written, printed, or graphic matter on any food, its containers or wrappers, or other material accompanying the food.

Developing New Functional Food and Nutraceutical Products. http://dx.doi.org/10.1016/B978-0-12-802780-6.00016-X

use of the product to be a drug use. However, using a nonauthorized health claim on a label results in the product to be subject to regulation as a drug.

The statutory distinction between what is a food and what is a drug is based to a large extent on how the product is represented for intended use. It is possible for a substance to have both food and drug uses. An intended drug use does not necessarily preclude a product from also having a food use, or vice versa. Baking soda, for example, is useful both as an ingredient in baking (food use) and as a treatment of acid indigestion (drug use). A key phrase in the FFD&C Act drug definition is "intended for use". Product labeling is one obvious way to show intended use of a product. A baking soda label stating that baking soda may be used for relief of acid indigestion shows that the product is intended to be used as a drug.[b] Any intended drug use must be in accordance with the premarket review and approval requirements for drugs, including efficacy and safety standards applicable to drugs. A product label claiming that the product is effective in the mitigation of a disease symptom (e.g., relief from osteoarthritis pain) shows the intended use of the product to be as a drug and therefore it would be regulated as a drug, not as a food. A product labelled as a food may not bear a claim pertaining to use of the product as a drug (i.e., use in the cure, mitigation, treatment, or prevention of any disease), irrespective of how good the evidence is to substantiate the effectiveness of the product for that particular use, unless the product has also been approved for use as a drug.

The 1994 Dietary Supplement Health and Education Act (DSHEA) established dietary supplements as a regulatory subcategory of food, with separate labeling and ingredient safety requirements. FDA uses the term *conventional food* to differentiate between dietary supplements and all other food. The criteria for a product to be a dietary supplement include:

- Not represented as a conventional food or as the sole item of a meal or diet;
- Labelled as a dietary supplement;
- Not be an article already approved as a drug unless the article had been marketed as a dietary supplement prior to its approval as a drug; and
- The dietary ingredients (i.e., the active ingredients) must be vitamins, minerals, herb or other botanicals, amino acids, other substances in the human diet; or a concentrate, metabolite, constituent, extract, or combination of the previous categories.

The FDA regulation of label claims is generally the same for conventional food and for dietary supplements, except there are some restrictions placed on dietary supplements for a few specific types of claims (e.g., fat or cholesterol content claims cannot be made for dietary supplements that have less than 40 calories per serving).

[b]Bicarbonate ion-containing active ingredients, such as sodium bicarbonate (baking soda), are an approved antacid over-the-counter drug for relief of heartburn and acid indigestion (21 CFR 331). There are brands of baking soda marketed in the USA with the labeling required for intended use both as a food and as a drug.

16.2 FDA'S AUTHORITY FOR REGULATING FOOD LABEL CLAIMS

The basis for FDA to regulate food labels originated with the 1938 FFD&C Act. The FFD&C Act was amended by the US Congress several times in the 1990s to make significant changes in how food label claims are regulated. These amendments include: the 1990 Nutrition Labeling and Education Act (NLEA) (Nutrition Labeling and Education, 1990), the 1994 Dietary Supplement Health and Education Act (DSHEA) (Dietary Supplement Health and Education, 1994), and the 1997 Food and Drug Administration Modernization Act (FDAMA) (Food and Drug Administration Modernization, 1997). FDA implements these laws through food labeling regulations found in title 21, Chapter 1, subchapter B, of the Code of Federal Regulations.

The 1990 NLEA amended the FFD&C Act in a number of important ways. Two chief aspects of the NLEA amendments are that they establish mandatory nutrition labeling for nearly all packaged foods sold in the USA and that they establish FDA's authority to regulate food label statements that (1) characterize levels of nutrients in a food, or that (2) characterize relationships of food substances to a disease or health-related condition. In general, information provided by manufacturers on food labels is restricted only to the extent that the information may not be false or misleading. For nutrient content and health-related information, however, NLEA stipulates that statements that characterize the level of a nutrient in a food (nutrient content claims) and statements that characterize a relationship of a nutrient to a disease or health-related condition (health claims) may only be used in food labeling in accordance with relevant FDA regulations. Although label statements such as "High Fiber" (a nutrient content claim) and "Fiber May Reduce the Risk of Heart Disease" (a health claim) may be perceived by many consumers as equivalent health messages, NLEA mandates that FDA regulate nutrient content claims and health claims as separate regulatory categories of food label statements. The 1994 DSHEA added new provisions to the FFD&C Act defining and prescribing the use of structure/function claims in dietary supplement labeling. Structure/function claims describe a role of a nutrient, or other dietary supplement ingredient, in affecting the structure or function of the body (e.g., *calcium builds strong bones*).

The process for FDA authorization of food label health claims has evolved from that initially established under NLEA. The NLEA amendments permitted FDA to issue a health claim-authorizing regulation for a specific substance/disease relationship only after having concluded that the relationship is substantiated by an appropriate level of scientific evidence. The validity requirement for the substantiation of health claims mandated by NLEA requires that FDA assess whether there is "significant scientific agreement" (SSA), among qualified experts, that the totality of scientific evidence supports the dietary substance/disease relationship. The 1997 FDAMA amended the FFD&C Act to provide an alternative to FDA's process for SSA assessment and regulation of health claims. FDAMA provides for a process whereby the substantiation of a health claim can be based on authoritative statements from any US federal scientific agency with official responsibility for public health protection or human nutrition research, or by the National Academy of Sciences, rather than being based only on an FDA assessment of the supporting scientific evidence. FDAMA also provided for the authorization of new authoritative statement-based nutrient content claims. The FDAMA provision for authoritative statement-based health claims and nutrient content claims was intended to expedite the process by which new claims are authorized.

The process for authorizing food label health claims further evolved following a series of court decisions which led to the 2003 *FDA Consumer Health Information for Better Nutrition Initiative* (Food and Drug Administration, 2003a), under which certain "qualified health claims" supported by credible, but not well-established or conclusive, scientific evidence can be used in food labeling based on FDA's exercise of regulatory discretion to allow the use of certain qualified health claims without FDA issuing health claim-authorizing regulations. FDA uses the term "qualified health claim" to refer to health claims for which the scientific evidence does not meet the SSA standard; that include a disclaimer that explains (or qualifies) the level of scientific evidence supporting the claim; and that are the subject of a current FDA regulatory discretion enforcement decision.

16.3 HEALTH CLAIMS

16.3.1 VALIDITY STANDARD OF HEALTH CLAIMS

NLEA mandates that health claims may not be used in food labeling without FDA evaluation of the supporting evidence and issuance of a regulation to authorize the specific health claim. NLEA allows FDA to authorize a health claim only after the agency has determined there to be SSA among qualified scientific experts that the claim is supported by the totality of publicly available scientific evidence. The SSA standard for health claims has two components:

1. there is publicly available supporting scientific evidence from well-designed studies conducted in a manner consistent with generally recognized scientific procedures and principles, and
2. there is SSA among qualified experts that the substance/disease relationship is supported by the totality of the publicly available scientific evidence.

An explanation of FDA's use of the SSA standard to evaluate the supporting scientific evidence for a health claim is available on FDA's Internet website (Food and Drug Administration, 2009). The SSA standard is intended to be a strong but flexible standard that provides a high level of confidence in the validity of a substance/disease relationship (Food and Drug Administration, 2009). NLEA did not require FDA to apply the SSA standard to dietary supplement health claims, rather the agency was given discretion to establish a separate validity standard for dietary supplements. FDA considered a "level playing field" approach to be preferable to multiple health claim standards and, therefore, FDA concluded that the same standard to which conventional food health claims were held (i.e., SSA) was also an appropriate standard for dietary supplement health claims.

16.3.2 GENERAL HEALTH CLAIM REGULATION

The FDA *General Requirements for Health Claims* regulation (21 CFR 101.14) sets out general requirements applicable to all food label health claims. FDA also issues specific authorizing regulations with additional criteria specific to each individual health claim that has met the SSA validity criteria. The general health claim regulation defines a health claim as a statement made on a food label

> *... that expressly, or by implication, characterizes the relationship of any substance to a disease or health-related condition. Implied health claims include any material, such as third party references, written statements (e.g., a brand name including the word "heart"), symbols (e.g., a heart symbol) presented in food labeling that suggests, within the context in which it is presented, that a relationship exists between a substance in the food and a disease or health-related condition. [21 CFR 101.14(a)(1) (Nutrition Labeling and Education, 1990)]*

There are two components to a health claim: the substance and the disease or health-related condition. A food label statement that does not include both components would not be a health claim. The FFD&C Act identifies the substance component of a health claim as any "nutrient which is of the type" required to be listed in the food label nutrition information. However, in that the NLEA also included dietary supplements of herbs or other similar nutritional substances in the health claim provisions of the Act, it is clear that the substance of a health claim was not intended to be limited strictly to "nutrients." The legislative history of NLEA shows that Congress intended that individual foods also could be the subject of health claims. The legislative history of NLEA also makes it clear that a health claim about a food must, at least by implication, be a claim about a substance in the food. When a consumer could reasonably interpret a claim about the relationship of a food to a disease or health-related condition to be an implied claim about a specific substance in the food, that claim would satisfy the "substance" component of a health claim. FDA therefore has defined "substance" as meaning any specific food or a component of a food (e.g., nutrients). Defining "substance" broadly allows authorization of health claims for a wider variety of relationships (e.g., health claims for phytosterols) than would a narrower definition of "substance" encompassing nutrients only. Health claims about consuming a substance at decreased levels (e.g., saturated fat, cholesterol, and heart disease), however, are restricted to the nutrients that are required to be listed in the nutrition information on the food label.

FDA's general health claim regulation includes a requirement that the substance in a health claim contribute taste, aroma, nutritive value, or any other recognized technical effect of a food additive. This requirement impacts the potential for use of health claims for food ingredients promoted for functional attributes other than the traditional food ingredient attributes of taste, aroma, and nutrition. Because FDA recognizes that health benefits may be derived from food ingredients through a variety of processes, FDA has provided for flexibility in determining whether a food substance has nutritional value. "Nutritive value" has been defined in FDA's general health claim regulation as value in sustaining human existence by processes such as promoting growth, replacing loss of essential nutrients, and providing energy and includes assisting in the efficient function of classical nutritional processes and of other metabolic processes necessary for normal maintenance of human existence.

The second component of a health claim, the "disease or health-related condition," means damage to an organ, part, structure, or system of the body such that it does not function properly (e.g., cardiovascular disease), or a state of health leading to such dysfunction (e.g., hypertension). The disease component of a health claim must be one for which the general US population, or an identified population subgroup (e.g., the elderly), is at risk. Diseases resulting from essential nutrient deficiencies (e.g., scurvy or pellagra) are excluded from the diseases that may be the subject of a health claim. A food label statement about a nutrient deficiency disease would not be a health claim. However, so as not to be misleading a food label statement about a nutrient deficiency disease should identify the prevalence of the nutrient deficiency disease in the USA.

The nature of the substance/disease relationships addressed in health claims is that of disease risk reduction. When the intended use of a product is to cure, mitigate, treat, or prevent a disease, the product is considered to be a drug, rather than a food. Health claims cannot describe an effect of the substance to cure, mitigate, treat, or prevent a disease because health claims are only for food labels. For example, a label statement about use of a food or dietary supplement for relief from arthritis pain (e.g., mitigation of a disease symptom) is not a health claim because it infers intent that the product be used as a drug. This principle was upheld in the 2004 Whitaker v. Thompson court decision [353 F.3d 947 (D.C. Cir), cert. denied, 125 S Ct. 310 (2004)], which held that it was constitutionally permissible for FDA to deny a health claim about use of saw palmetto extract dietary supplements for relief of urinary symptoms of benign prostatic hyperplasia on the basis that the claim constituted a drug use for the product.

NLEA prohibited FDA from authorizing the use of health claims for foods that contain any nutrient in an amount that increases the risk of a diet-related disease. As such, FDA has set threshold levels for several nutrients above which a food is disqualified from using health claims [21 CFR 101.14(a) (Food and Drug Administration, 2003a)]. The disqualifying nutrient levels for individual foods are 13 g of total fat, 4 g of saturated fat, 60 mg of cholesterol, and 480 mg of sodium per Reference Amount Customarily Consumed (RACC).[c] Where the actual serving size of a food is greater than the RACC, the disqualifying nutrient levels are determined on the labeled serving size; and the disqualifying nutrient levels for foods with a small RACC (defined as 30 g or less or 2 tablespoons or less) are determined on a per 50 g of food basis. Any food that exceeds one or more nutrient disqualifying levels is prohibited from using any health claim, even though that food may otherwise meet other requirements of a specific health claim. FDA has authority to waive disqualifying nutrient levels in instances where the agency finds that allowing a health claim will

[c]Reference Amounts Customarily Consumed per Eating Occasion (RACC) are used as a reference point for determining the serving sizes used for nutrition labeling and as a uniform reference amount for basing nutrient content claims. RACCs for 139 food product categories are listed in 21 CFR 101.12.

assist consumers in maintaining healthy dietary practices in spite of a high level of a disqualifying nutrient, provided that the label contain a disclosure statement about the nutrient (e.g., "See nutrition information for fat content").

The general health claim regulation also restricts the use of health claims on labels of foods with little nutritional value. Except for dietary supplements, a food may not use a health claim unless the food contains at least 10% of the daily value for one or more of six core nutrients (dietary fiber, protein, vitamin A, vitamin C, calcium, or iron) per RACC. The food must be able to meet the 10% daily value minimum nutrient content requirement prior to any nutrient addition. This provision has come to be known as the "jelly bean rule" because it was intended to restrict the use of health claims by foods with little inherent nutritional value, such as jelly beans. FDA has made exceptions to the "jelly bean rule" in some health claims when a public health benefit warranted.

16.3.3 AUTHORIZED HEALTH CLAIMS

NLEA identified 10 specific nutrient/disease relationships that Congress had considered to be probable health claim topics (Table 16.1). FDA was directed by Congress to evaluate the scientific evidence for these initial 10 health claims. FDA concluded there was a SSA level of substantiation on 5 of the initial 10 nutrient/disease relationships and health claim-authorizing regulations were issued for these 5 health claims. FDA concluded that the available evidence did not warrant authorizing claims for two other nutrient/disease relationships.[d] FDA concluded that the evidence for the remaining three nutrient/disease relationships[e] was inconclusive but that there was SSA for a relationship between diets characterized by the types of foods that are rich sources of these nutrients (i.e., fruits, vegetables, and grain products) and reduced disease risk. FDA therefore authorized three alternative health claims for fruits, vegetables, and grain products that are rich in antioxidant vitamins or in soluble fiber (Table 16.1).

NLEA also directed FDA to establish a procedure through which interested parties could petition FDA to authorize additional health claims. NLEA permitted FDA to authorize a petitioned health claim only when the agency has determined that the claim meets FDA's general requirements for health claims and that the SSA validity standard has been satisfied. NLEA specified that health claim-authorizing regulations issued by the FDA shall:

- describe the substance/disease relationship of the health claim and the significance of the substance in affecting the disease,
- require that the claim be stated in a manner that is an accurate representation of the substance/disease relationship, and
- require that the claim be stated in a manner that enables the public to understand the information provided in the claim and to understand the relative significance of such information in the context of a total daily diet.

[d]The relationships that fell short of SSA were "zinc and immune function in the elderly," and "omega-3 fatty acids and heart disease risk."

[e](1) dietary fiber and cardiovascular disease risk, (2) dietary fiber and cancer risk, and (3) antioxidant vitamins and cancer risk.

Table 16.1 Health Claims from the 1990 Nutrition Labeling and Education Act

Substance–Disease Relationship	NLEA Health Claims
Calcium, vitamin D, and osteoporosis	Foods and supplements that are high in calcium, or high in both calcium and vitamin D, and reduced risk of osteoporosis.[a] *21 CFR 101.72*[b]
Dietary fat and cancer	Foods that are low in total fat,[c] or extra lean fish and game meat and reduced risk of cancer. *21 CFR 101.73*
Sodium and hypertension	Foods and supplements that are low in sodium and reduced risk of high blood pressure. *21 CFR 101.74*
Saturated fat/cholesterol and cardiovascular disease	Foods and supplements that are low in both saturated fat and cholesterol,[c] or extra lean fish and game meat and reduced risk of coronary heart disease. *21 CFR 101.75*
Dietary fiber and cancer	Fruit, vegetable, or grain products that are low fat and are a good source of dietary fiber and reduced risk of cancer. *21 CFR 101.76*
Dietary fiber and cardiovascular disease	Fruit, vegetable, or grain products that are low in total fat, saturated fat, and cholesterol, and contain soluble dietary fiber, and reduced risk of coronary heart disease. *21 CFR 101.77*
Antioxidant vitamins and cancer	Fruits or vegetables that are low fat and a good source of vitamin A, vitamin C, or dietary fiber and reduced risk of cancer. *21 CFR 101.78*
Folic acid and neural tube birth defects	Foods/supplements that are a good source of folate and reduced risk of neural tube birth defects.[d] *21 CFR 101.79*
Zinc and immune function in the elderly	No authorized claim
Omega-3 fatty acids and heart disease	No authorized claim[e]

[a] *Initially authorized as a calcium and osteoporosis claim; the authorizing regulation was amended in 2010 to include an optional "calcium and vitamin D" claim.*
[b] *Food labeling regulations are codified in Title 21 of the Code of Federal Regulations (CFR), Part 101. The regulation authorizing the health claim about the relationship of dietary calcium and osteoporosis risk is in section 72 of this part, that is, 21 CFR 101.72.*
[c] *Although dietary supplements are not explicitly excluded from use of these claims, these claims are not available for many dietary supplement products because the "low" nutrient content claims for total fat, saturated fat, and cholesterol are not defined for "low calorie"/"calorie-free" dietary supplements that are.*
[d] *Authorization of the folate health claim was delayed until 1996 pending resolution of safety questions.*
[e] *Although there was insufficient scientific evidence to support an authorizing regulation, there is a qualified health claim for Omega-3 fatty acids and heart disease.*

Table 16.2 Health Claims Authorized through the Health Claim Petitions Process

Substance–Disease Relationship	Authorized Health Claim
Noncariogenic carbohydrate sweeteners and dental caries	Sugar-free foods sweetened with certain sugar alcohols and other sugar substitutes—xylitol, sorbitol, mannitol, maltitol, isomalt, lactitol, hydrogenated starch hydrolysates, hydrogenated glucose syrups, erythritol, D-tagatose, and sucralose. 21 CFR 101.80
Soluble fiber from certain foods and coronary heart disease	Foods that contain a barley (whole grain or beta-fiber) or oat (oat bran, oatmeal, whole oat flour, or oatrim) ingredient, contain at least 0.75 g beta-glucan fiber per reference amount, and are also low in total fat, saturated fat, and cholesterol.
	Foods that contain psyllium seed husk providing at least 1.7 g soluble fiber per reference amount, and are also low in total fat, saturated fat, and cholesterol. 21 CFR 101.81
Soy protein and coronary heart disease	Foods low in total fat, saturated fat, and cholesterol and that contain at least 6.25 g soy protein per reference amount. 21 CFR 101.82
Plant sterol/stanol esters and coronary heart disease	Foods/supplements that are low in saturated fat and cholesterol and that contain at least 400 mg per reference amount of plant sterols or stanols.[a] 21 CFR 101.83

[a]The authorizing health claim regulation had been amended in 2011; however, enforcement of claim criteria in the amended regulation that are different from the original regulation is being delayed. Check the FDA website for the current status of this health claim.

FDA has authorized new health claims in response to a number of health claim petitions over the past two decades (Table 16.2). The health claims that have been authorized as the result of health claim petitions include claims about sugar substitutes and dental caries; whole oat and barley beta-glucan fiber and heart disease; psyllium husk fiber and heart disease; soy protein and heart disease; and plant sterol/stanol esters and heart disease (Table 16.2).

The health claim authorization procedure established by NLEA involves FDA evaluation of supporting scientific evidence. Upon a determination of SSA that the available scientific evidence substantiates the substance/disease relationship, FDA follows a lengthy *notice and comment* rulemaking to issue an authorizing regulation. The FDAMA provision for authoritative statement-based health claims serves as an alternative to the health claim petition and rulemaking process. Substantiation of an authoritative statement-based health claim derives from a review of scientific evidence conducted by any US federal scientific agency, other than FDA, with official responsibility for public health protection or human nutrition research or by the National Academy of Sciences, where the result of such review is published as an official conclusion of that agency about a dietary substance/disease risk reduction relationship. The same SSA validity standard applies in both processes, but they differ in where the deliberative review of the scientific evidence occurs. The process leading to authorization of an authoritative statement-based health claim begins with submission to FDA of a notice that includes the exact wording to be used in the health claim, the authoritative statement that is the source of the claim, and a balanced representation of the scientific literature relating to the substance/disease relationship to which the claim refers. The new health claim that is the subject of an authoritative statement health claim notification becomes authorized by law 120 days after receipt of the notification by FDA provided FDA has taken no regulatory action during that period to prohibit or modify the claim. FDA uses the 120-day period to determine whether both the notification and the claim meet the requirements established by FDAMA. FDA may issue regulations to prohibit, or to modify, authoritative statement-based health or nutrient content claims. FDA guidelines on the procedure for notifying FDA of intent to use health claims based on an authoritative statement from a scientific body are available on the Internet (Food and Drug Administration, 1998). Table 16.3 lists the health claims which have become authorized through the authoritative statement health claim notification process.

16.3.4 QUALIFIED HEALTH CLAIMS

In 1999, the US Court of Appeals for the DC Circuit issued its decision in *Pearson v. Shalala* (164 F.3d 650, D.C. Cir. 1999) (*Pearson*). The *Pearson* plaintiffs had challenged FDA's application of the SSA standard to dietary supplement health claims and FDA's decision not to authorize health claims for four specific substance/disease relationships: dietary fiber and colorectal cancer; antioxidant vitamins and cancer; omega-3 fatty acids and coronary heart disease; and the claim that 0.8 mg of folic acid in dietary supplement form is more effective in reducing the risk of neural tube birth defects than a lower amount in conventional food form. FDA had denied these four claims after concluding that these claims did not meet the SSA substantiation standard and the claims therefore would be misleading. The court held in *Pearson* that the Constitutional protection of free speech does not permit FDA to reject health claims that the agency determines to be potentially misleading unless the agency can also show that adding a disclaimer would not eliminate the potential deception. The court stated in *Pearson* that the potential for claims that do not meet the SSA validity standard to mislead the consumer could be remedied by including "qualifying" language in the claim which would disclose the level of scientific support for the claim (i.e., disclosure of more information is preferential to suppressing information).

FDA interpreted the Pearson court opinion to suggest that where there was more scientific evidence against a specific claim than there was evidence in support of the claim, the claim would be inherently misleading (rather than merely potentially misleading) and could be prohibited. FDA initially applied a "weight of the evidence" standard, as an alternative to the SSA standard, in evaluating one of the first health claim topics to be reevaluated under *Pearson* (a claim for antioxidant vitamin supplements and cancer risk). FDA concluded that there was more scientific evidence

Table 16.3 Health Claims Authorized through the FDAMA Authoritative Statement Notification Process

Substance–Disease Relationship	**Authorized Health Claim**
Whole grains and heart disease and cancer	Foods with 51% or more whole grain ingredients; dietary fiber content is the measurable marker of whole grain content. There are two versions of the claim: *Diets rich in whole grain foods and other plant foods and low in total fat, saturated fat, and cholesterol may reduce the risk of heart disease and some cancers.* *Diets rich in whole grain foods and other plant foods and low in saturated fat and cholesterol may reduce the risk of heart disease.* FDA Docket no. 1999P-2209, 2003Q-0547
Potassium and high blood pressure and stroke	Foods which are good sources of potassium in addition to being low in total fat, saturated fat, cholesterol, and sodium. *Diets containing foods that are a good source of potassium and that are low in sodium may reduce the risk of high blood pressure and stroke.* FDA Docket no. 2000Q-1582
Fluoridated water and dental caries	Fluoridated bottled water, except those specifically marketed for use by infants. *Drinking fluoridated water may reduce the risk of dental caries [or tooth decay].* FDA Docket no. 2006Q-0418
Dietary fats and heart disease risk	Foods which are low in saturated fat and cholesterol, moderate in fat content, and contain less than 0.5 g/RACC *trans* fat. *Diets low in saturated fat and cholesterol and as low as possible in* trans *fat may reduce the risk of heart disease.* FDA Docket no. 2006Q-0458
Substitution of saturated fat in the diet with unsaturated fatty acids and heart disease risk	Vegetable oil and shortening-containing foods that have a total unsaturated fat content of 80% or more of total fat content. *Replacing saturated fat with similar amounts of unsaturated fats may reduce the risk of heart disease. To achieve this benefit, total daily calories should not increase.* FDA Docket No. 2007Q-0192

against the validity of this claim than there was to support the claim, thus FDA would not permit a qualified health claim for antioxidant vitamins and cancer risk because the weight of available scientific evidence did not support this substance/disease relationship (Emord and Associates, 2001). FDA's interpretation of *Pearson* in the decision to deny a qualified health claim for antioxidant vitamin supplements and cancer was challenged in another lawsuit (*Whitaker v. Thompson*, 248 F. Supp 2d 1 (D.D.C. 2002)) (*Whitaker*). The court's opinion in *Whitaker* was that FDA's "weight of the evidence" standard misinterpreted *Pearson* and that qualified health claims must be permitted if there is any credible scientific evidence in support of a claim. Consistent with *Whitaker*, FDA now uses regulatory discretion to permit qualified health claims where there is any credible scientific evidence to support the claim.

To develop a framework to help consumers obtain accurate, up-to-date, and science-based information about conventional foods and dietary supplements, the FDA convened the *Consumer Health Information for Better Nutrition Initiative* Task Force to establish guidelines for health claims that do not meet the SSA standard (Food and Drug Administration, 2003a). The task force compiled an *Evidence-based Review System for the Evaluation of Health Claims* that establishes the procedures that FDA will use for evaluation of available scientific evidence, in a manner consistent with *Pearson* and *Whitaker*, in its deliberations on when to permit qualified health claims in conventional foods and dietary supplements (Food and Drug Administration, 2003b).

To provide for the use of qualified health claims as directed by *Pearson* and *Whitaker*, FDA uses regulatory discretion in enforcing the requirement that food label health claims must meet the SSA validity standard and be authorized by a health claim regulation. FDA's qualified health claim enforcement discretion decisions are posted on the FDA Internet website rather than codifying the decisions in a regulation. The agency letters announcing an enforcement discretion decision include an explanation of the rationale for the decision and specify the conditions under which the agency will consider the exercise of its enforcement discretion (e.g., the exact wording of the qualifying language). The enforcement discretion decision letters caution that since the qualified health claim topics are areas where the evidence for the claim is incomplete and subject to change, FDA intends to evaluate new information as it becomes available to determine whether the new evidence necessitates a change in the decision. Table 16.4 presents a listing of qualified health claims that have been evaluated by FDA.

16.4 STRUCTURE/FUNCTION CLAIMS

Food and supplement labels may bear claims about how a product may affect structures or functions of the body. The FFD&C Act explicitly recognizes that foods, as well as drugs, can be used to affect the structure or any function of the body. The 1994 DSHEA amended the FFD&C Act to clarify the difference between health claims and structure/function claims made for dietary supplement products. FDA uses the phrase "structure/function claim" to refer to food label statements that describe the role of a nutrient or other dietary supplement ingredient in maintaining normal structure or function in humans (e.g., calcium builds strong bones) or to promote general well-being. Structure/function claims can also characterize the means by which a nutrient or other dietary ingredient acts to maintain a body structure or function (e.g.,

Table 16.4 Qualified Health Claims Permitted by FDA Enforcement Discretion

Atopic dermatitis
100% whey-protein partially hydrolyzed infant formula and reduced risk of atopic dermatitis
Cancer
Selenium and a reduced risk of site-specific cancers
Antioxidant vitamins C and E and reduction in the risk of site-specific cancers
Tomatoes and prostate, ovarian, gastric, and pancreatic cancers
Tomatoes and prostate cancer
Calcium and colon/rectal cancer and calcium and colon/rectal polyps
Green tea and risk of breast cancer and prostate cancer
Selenium and certain cancers
Antioxidant vitamins and risk of certain cancers
Cardiovascular disease
Folic acid, vitamin B6, and vitamin B12 and vascular disease
Nuts and coronary heart disease
Walnuts and coronary heart disease
Omega-3 fatty acids and coronary heart disease
Corn oil and corn oil-containing products and heart disease
Unsaturated fatty acids from canola oil and coronary heart disease
Monounsaturated fatty acids from olive oil and coronary heart disease
Cognitive function
Phosphatidylserine and cognitive dysfunction and dementia
Diabetes
Psyllium husk and type 2 diabetes
Whole grains and type 2 diabetes
Chromium picolinate and a reduced risk of insulin resistance, type 2 diabetes
Hypertension
Calcium and hypertension, pregnancy-induced hypertension, and preeclampsia
Neural tube defects
Folic acid and neural tube defects

The enforcement discretion letters describe the specific conditions under which the claim can be used, including the qualifying language which characterizes the strength of the scientific evidence behind the claim. A compilation of qualified health claims currently permitted is posted on www.fda.gov at a page titled "Qualified Health Claims: Letters of Enforcement Discretion."

antioxidants maintain cell integrity or fiber maintains bowel regularity). Health claims and structure/function claims differ in that there are two elements to a health claim: a substance and a disease or health-related condition, whereas structure/function claims have the first element (a substance) but cannot imply any relationship of the substance to a disease or health-related condition. The context of a statement within the entire label is considered in determining whether a label statement implies a relationship to a disease or a health-related condition. For example, a statement such as "supports the cardiovascular system" does not of itself imply a disease connection. However, this same statement on the label of a product called "CardioCure" would constitute a disease claim within the context of the entire label. A food label statement that makes, or implies, a relationship with a disease or health-related condition is either a health claim or a drug claim, not a structure/function claim. An exception to the prohibition of structure/function claims from referencing a disease is that structure/function claims may describe a benefit related to a classical nutrient deficiency disease (e.g., vitamin C and scurvy) when the statement also tells how widespread such a disease is in the USA. The distinction between a disease claim and a dietary supplement structure/function claim is described in 21 CFR 101.93(g).

Structure/function claims, like all food labeling statements, must be truthful and not misleading. Unlike health claims and nutrient content claims, there is no FFD&C Act provision for FDA's prior review or approval of structure/function claims before their use. Structure/function claims are not regulated by FDA to the extent that health claims and nutrient content claims are, but DSHEA set three conditions for the use of structure/function claims in the labeling of dietary supplements. The first condition is that the manufacturer of a dietary supplement using a structure/function claim has substantiation that the claim is truthful and not misleading. The second condition is that the following statement must be displayed prominently and in boldface type on the label with the structure/function claim: '**This statement has not been evaluated by the Food and Drug Administration. This product is not intended to diagnose, treat, cure, or prevent any disease.**' The third condition is that the dietary supplement manufacturer notify FDA, within 30 days of first marketing a dietary supplement label bearing a structure/function claim, that the statement is being made. FDA examines the dietary supplement structure/function claim notifications only to determine that the claims being made are not claims for an effect related to a disease. FDA does not evaluate the substantiation for, or approve, the structure/function claims in these notifications.

DSHEA's structure/function claim conditions apply specifically to dietary supplement labeling. Structure/function claims, whether used on conventional food labels or dietary supplement labels, need to be adequately substantiated to satisfy the "not false or misleading" criterion that applies to all label information. However, there are no provisions in the FFD&C Act requiring manufacturers using structure/function claims on conventional food labels to include the disclaimer statement or to notify FDA of their structure/function claims. Unlike dietary supplements, structure/function claims made for conventional foods must be about effects that derive from the taste, aroma, or nutritive value of the food or food ingredient that is the subject of the claim. This limitation on use of structure/function claims for conventional foods is based on a federal court ruling that structure/function claims on a conventional food label cause the product to be a drug if the claim promotes the product for an effect that is unrelated to the product's "food attributes" of taste, aroma, and nutritive value [*Nutrilab v. Schweiker*, 713 F. 2d 335 (7th Cir. 1983)]. Table 16.5 compares regulatory differences between health claims and structure/function claims used on conventional foods and dietary supplements.

Table 16.5 Requirements for Health Claims and Structure/Function Claims

	Health Claims—Conventional Food and Dietary Supplements	**Structure/Function Claims—Dietary Supplements**	**Structure/Function Claims—Conventional Food**
Petition FDA for new claims	Yes	No	No
Requires FDA review and approval	Yes	No	No
Must not be false or misleading	Yes	Yes	Yes
Notify FDA within 30 days of marketing product labeled with the claim	No	Yes	No
Include a disclaimer stating the claim has not been evaluated by FDA	No	Yes	No
Each manufacturer using the claim must possess scientific evidence to substantiate the claim is true for their product	No	Yes	Yes
Product subject to disqualifying nutrient levels and minimum nutrient levels	Yes	No	No
Substance must have food attributes such as nutritive value	Yes	No	Yes

Because structure/function claims for dietary supplements must be truthful and not misleading, FDA recommends that manufacturers possess adequate substantiation for all reasonable interpretations of the claims made. FDA recommends applying a standard that is consistent with the Federal Trade Commission standard of "competent and reliable scientific evidence" to substantiate a claim (Bureau of Consumer Protection/Federal Trade Commission, 2001). FDA considers the following factors important to establish whether information would constitute "competent and reliable scientific evidence":

- Does each study or piece of evidence bear a relationship to the specific claim?
- What are the individual study's or evidence's strengths and weaknesses? Consider the type of study, the design of the study, analysis of the results, and peer review.
- If multiple studies exist, do the studies that have the most reliable methodologies suggest a particular outcome?
- If multiple studies exist, what do most studies suggest or find? Does the totality of the evidence agree with the claims?

FDA has provided guidelines on the amount, type, and quality of evidence that FDA recommends a manufacturer have to substantiate a structure/function claim (Food and Drug Administration, 2004).

16.5 NUTRIENT CONTENT CLAIMS

A nutrient content claim is a food label statement that directly, or by implication, characterizes the level of a nutrient in the food. The use of nutrient content claims in food labeling is restricted to statements made in accordance with FDA's authorizing regulations for nutrient content claims. FDA's nutrient content claim regulations define how descriptive terms can be used to characterize the level of a nutrient in a food. The level of a nutrient in a food may be characterized using descriptive terms such as *free*, *high*, and *low*, or the level of a nutrient in one food may be compared to that in another food using descriptive terms such as *more*, *reduced*, and *lite*. The requirements governing the use of nutrient content, claims help ensure that descriptive terms are used consistently for all types of food products and are thus meaningful to consumers. Nutrient content claims include both explicit claims (e.g., *low fat*) and implied claims (e.g., *healthy*). With a few exceptions, only nutrients with an established daily value for nutrition labeling may be the subject of a nutrient content claim. The general principles for the use of nutrient content claims are listed in the Code of Federal Regulations, title 21 part 101, section 13 (21 CFR 101.13).

Nutrient content claims have been defined only for situations where they will be nutritionally relevant. The terms *good source* (defined as 10–19% of daily value per RACC), *high* (defined as at least 20% of daily value per RACC), and *more* (defined as at least 10% of daily value more than that in a reference food) have been defined for nutrients that have a daily value. The requirements for the use of *good source*, *high*, and *more* content claims are listed in 21 CFR 101.54. The terms *free*, *low*, and *reduced* have been defined for use only with calories, total fat, saturated fat, cholesterol, sodium and, in some cases, sugars and salt. These descriptive terms are defined individually for each of the listed nutrients. In addition, there are specific criteria that apply to each individual nutrient as well. For example, the criteria for *sugar free* include a sugar content limit (0.5 g per RACC), a requirement that the label show the calorie profile (e.g., *low calorie*), and trigger levels for other nutrients (such as total fat) that require a disclosure statement. Consequently, the regulations governing these claims should be read carefully before using them on a food label. Authorized nutrient content claims for sugar include *free*, *reduced*, and *no added sugar*; but a *low sugar* nutrient content claim is not permitted. *Light* (or *lite*) is a nutrient content claim that has been defined for certain foods reduced in fat and/or calories, or certain foods reduced in sodium content. The term *light* may also be used outside of a nutrient content claim context in describing organoleptic properties such as color, for example, "light brown sugar." Requirements for the use of *light/lite* nutrient content claims are found in 21 CFR 101.56. Requirements for the use of content claims about calories and sugar are found in 21 CFR 101.60. Requirements for the use of content claims about sodium and salt are found in 21 CFR 101.61. Requirements for the use of content claims about fat, saturated fat, and cholesterol are found in 21 CFR 101.62.

One exception to the rule that only nutrients with established daily values can be the subject of nutrient content claims is a provision that permits the use of quantitative statements (e.g., *100 mg of lycopene*). A quantitative statement may be used in food labeling to describe the amount of a nutrient present in a food, when a daily value has been established for that nutrient, provided that the statement does not use language that characterizes the amount. A statement such as *packed with 100 mg of lycopene* characterizes the amount of lycopene in the food by suggesting this to be a significant amount of lycopene. Because there is not an established daily value for lycopene, the *packed with 100 mg of lycopene* statement would not be permitted. The terms *contains* and *provides* are defined as nutrient content claim descriptors synonymous with *good source*. However, when *contains* or *provides* are used with a quantitative statement about a nutrient for which there is not an established daily value, the term is considered customary English usage, not a descriptive term nutrient content claim, thus *contains 100 mg lycopene* can be used on a food label in the same manner as *100 mg lycopene* can. When a quantitative statement is made about a nutrient for which there is an established daily value, the food must either meet the criteria of an appropriate nutrient content claim or carry a statement disclosing that the food does not comply with an authorized content claim. For example, *low sodium* is defined as less than 140 mg sodium per RACC; therefore, a quantitative statement referencing 200 mg of sodium would need to be qualified such as: *200 mg sodium. Not low in sodium.*

Nutrient content claims include both explicit claims and implied claims. Implied nutrient content claims describe a food or ingredient in a manner that suggests that a nutrient is either present in a certain amount, or is absent. For example, a food labeled *high in oat bran* implies that the food is high in dietary fiber and therefore the food should contain at least 20% of the daily value per RACC of dietary fiber. The term *healthy*, and any derivation of healthy, when used in association with an explicit statement about a nutrient, e.g., *Healthy low sodium*, is considered to be an implied nutrient content claim not only characterizing the levels of the indicated nutrient, but also characterizing the amounts of total fat, saturated fat, cholesterol, and sodium in the food as all being at "healthy" levels. FDA has set criteria to ensure that *healthy* is used consistently and in a meaningful way. There are five criteria to qualify as a healthy food. For most packaged foods these criteria are:

1. meet the criteria for a *low total fat* claim;
2. meet the criteria for a *low saturated fat* claim;
3. have a cholesterol content less than 60 mg per RACC, or per 50 g of food if the RACC is under 30 g or 2 tablespoons;
4. have a sodium content less than 480 mg per RACC, or per 50 g of food if the RACC is under 30 g or 2 tablespoons; and
5. contain at least 10% of the daily value per RACC for one or more of vitamin A, vitamin C, calcium, iron, protein, or dietary fiber.

The fifth criterion (known as the jelly bean rule) is included to preclude foods of little intrinsic nutritional value (such as candy) from being characterized as healthy. There are some variations in these criteria depending upon whether the labeled food is a fruit or vegetable, a grain product, a seafood or game meat product, or a meal-type product. The rules for labeling a food with a "healthy" nutrient content claim are explained in the regulation 21 CFR 101.65(d).

A summary of the rules for the use of nutrient content claims can be found in *A Food Labeling Guide* (Food and Drug Administration, 2013); examples of authorized nutrient content claims are provided in Appendices A and B of this guide.

REFERENCES

Bureau of Consumer Protection, Federal Trade Commission, 2001. Dietary supplements: an advertising guide for industry. April 2001. https://www.ftc.gov/tips-advice/business-center/guidance/dietary-supplements-advertising-guide-industry

Dietary Supplement Health and Education Act of 1994, 1994. Public Law 103–417. 25 October 1994.

Letter to Jonathan Emord (Emord & Associates) from Christine Lewis (FDA, CFSAN), 2001. Re: petition for health claim: antioxidants and cancer (Docket No. 91N-0101). 4 May 2001.

Food and Drug Administration, 1998. Guidance for industry: notification of a health claim or a nutrient content claim based on an authoritative statement of a scientific body, June 1998. http://www.fda.gov/Food/

Food and Drug Administration, 2003a. Consumer health information for better nutrition initiative. Task Force Final Report. July 2003. http://www.fda.gov/Food/IngredientsPackagingLabeling/LabelingNutrition/ucm096010.htm

Food and Drug Administration, 2003b. Guidance for industry and FDA: interim evidence-based review system for scientific data. 10 July 2003. http://www.fda.gov/Food/GuidanceRegulation/GuidanceDocumentsRegulatoryInformation/LabelingNutrition/ucm053832.htm

Food and Drug Administration, 2004. Guidance for industry: substantiation for dietary supplement claims made under section 403(r)(6) of the Federal Food, Drug, and Cosmetic Act, November 2004. http://www.fda.gov/Food/GuidanceRegulation/GuidanceDocumentsRegulatoryInformation/DietarySupplements/ucm073200.htm

Food and Drug Administration, 2009. Guidance for industry: evidence-based review system for the scientific evaluation of health claims—final. January 2009. http://www.fda.gov/Food/GuidanceRegulation/GuidanceDocumentsRegulatoryInformation/LabelingNutrition/ucm073332.htm.

Food and Drug Administration, 2013. Guidance for industry: a food labeling guide. Revised January 2013. http://www.fda.gov/Food/GuidanceRegulation/GuidanceDocumentsRegulatoryInformation/LabelingNutrition/ucm2006828.htm

Food and Drug Administration Modernization Act of 1997, 1997. Public Law 105–115. 21 November 1997.

Nutrition Labeling and Education Act of 1990, 1990. Public Law 101–535. 8 November 1990.

CHAPTER

Nutraceutical and functional food regulations in the European Union 17

Nicola Luigi Bragazzi*, Mariano Martini, Teresa Concetta Saporita[†], Daniele Nucci[‡], Vincenza Gianfredi[§], Francesco Maddalo[¶], Annamaria Di Capua[¶], Flavio Tovani[††], Lorenzo Marensi[†]**

**Department of Health Sciences (DISSAL), School of Public Health, University of Genoa, Genoa, Italy; **Department of Health Sciences (DISSAL), Section of Bioethics, University of Genoa, Genoa, Italy; [†]Hygiene and Public Health Unit, Local Health Unit 3 of Genoa, Genoa, Italy; [‡]Department of Experimental Medicine, Experimental Centre for Health Promotion and Education, University of Perugia, Perugia, Italy; [§]Department of Experimental Medicine, School of Specialization in Hygiene and Preventive Medicine, University of Perugia, Perugia, Italy; [¶]Operative Unit of Hygiene and Public Health, Local Health Service, La Spezia, Italy; [††]Sant'Anna School of Advanced Studies, Pisa, Italy*

Legislation always runs behind reality.
Gulati and Berry Ottaway (2006)

17.1 INTRODUCTION

Since the early 1990s, there has been an exponential growth in the commercial release of nutraceuticals, with a market and an industry expected to reach US$250 billion by 2018, as well as an increased interest from the academic world and the scientific community (Coppens et al., 2006). From the linguistic point of view, the term "nutraceutical" represents a portmanteau of the words "nutrition" or "nutrient" (a nourishing food or food component) and "pharmaceutical" (a medical drug or product). It was coined in 1989 by Stephen L. DeFelice, founder and chairman of the Foundation of Innovation Medicine, who defined a nutraceutical as "a food or parts of food that provide medical or health benefits, including the prevention and/or treatment of disease" (Gulati and Berry Ottaway, 2006).

Despite its increasingly popularity, the term "nutraceutical" harbors some degrees of ambiguity, in that, very often, the label "nutraceutical" has been misused or used in a misleading way to describe a vast array of products ranging from isolated ingredients, nutrients, active principles, and food components, to food/foodstuff,

Developing New Functional Food and Nutraceutical Products. http://dx.doi.org/10.1016/B978-0-12-802780-6.00017-1

dietary supplements, processed food, functional food, enriched/fortified food, and herbal products/botanicals, among others. These products are commercialized as mere food, but with the alleged hype of being able to treat or prevent disorders. This confounding terminology, not being clear-cut, makes it very difficult to determine what is and what is not a nutraceutical.

Despite the fact that nutraceuticals have a millenary history and food has always been considered as a medicine, from the Indians to Hippocrates (Let food be thy medicine and medicine be thy food), the advancements and achievements in the field of biotechnology and food engineering are paving the way to unprecedented possibilities, and therefore there is an urgent need for legislation (Coppens et al., 2006).

17.2 THE EU'S LEGISLATION AND POLICY ON FOOD

If the first concrete application of the European Union (EU) traces back to 1951, we have to wait more than 20 years before the first legislative measures and regulatory provisions specifically devoted to foods.

This long, gradual process, so far still incomplete, reflects the vast debate on the EU, its mission, and its objectives, in a period of great historical changes, ranging from the dramatic period immediately after the Second World War to the recent economic-financial crisis (Table 17.1) (Baumgartner and Uebe, 2014).

The EU's legislation and policy on food are complex and based on different principles:

- the principle of liberalization of trade [article 26(2) and article 114(1) of the Treaty on the functioning of the EU]: that is to say, the elimination of any form of trade restriction and the free circulation of foods/foodstuff (the so-called "common market for food");

Table 17.1 Main Steps in the Process of Formation of the European Union

1951	European Coal and Steel Community (ECSC)
1957	The Treaty of Rome (known also as the EEC Treaty)
1967	The Merger Treaty
1985	The Schengen Agreement
1986	The Single European Act
1990	The Schengen Convention
1993	The Maastricht Treaty
1997	The Amsterdam Treaty
1999–2002	Establishment of the monetary union
2001	The Treaty of Nice
2005	Prüm Convention
2007	The Lisbon Treaty
2012	Treaty Establishing the European Stability Mechanism European Fiscal Compact
2014	The Single Resolution Fund Agreement

- the principle of food safety governance: the market alone can lead to distortions, being potentially plagued by negative externalities and information asymmetry, therefore the regulatory authorities serve in correcting these biases; on the other hand, a too highly regulated market experiences and suffers from a detrimental effect where, for example, competitiveness and innovation are concerned (the so-called "Delaware effect");
- the principle of comitology or committee procedure (article 290 and article 291 of the Treaty on the functioning of the EU): a process by which EU law is modified, adjusted, amended, supplemented, and implemented by "comitology committees" chaired by the European Commission;
- the principle of law harmonization (known also as Home Country principle or the *Cassis de Dijon* principle): that is to say the implementation of legislative measures and regulatory provisions at the level of the entire EU and not at the level of single countries, with a precise collective design and concept rather than a mere collection of local laws;
- the principle of an integrated approach: the so-called "from farm-to-table," "from stable-to-table," or "from farm-to-fork" principle, that is to say the monitoring and management of the entire food production chain, from the preparation, manipulation, and processing to its final delivery to the consumer;
- the principle of food safety (known also as the "no harm" principle), within a public health perspective [article 168(4) (b) of the Treaty on the functioning of the EU]; moreover, food safety is addressed together with animal health and protection, environmental protection, and feed production safety (Baumgartner and Uebe, 2014);
- the principle of consumer protection [article 169(1) and article 169(2) of the Treaty on the functioning of the EU];
- the principle of food traceability or accountability or transparency: every step of the food production chain must be recorded and linked in such a way that it is always possible to trace back to the documentation of every single procedure, as stated in Regulation No. 931/2011;
- the principle of responsibility of food manufacturers;
- the principle of science-based decision making [article 1(2) of the General Food Law]: a rigorous scientific assessment is mandatory before a decision can be reached and the decision makers are advised and supported by scientists in their decisions;
- the precautionary principle, as stated in the 2000 Communication of the European Commission on the Precautionary Principle.

Scandals such as *Listeria* in cheese, *Salmonella* in eggs and in chocolate, mad cow disease (bovine spongiform encephalopathy, a serious fatal neurodegenerative disease due to a proteinaceous infectious agent, termed a prion), the "citrus pulp pellets" case of dioxin contamination (Malisch and Kotz, 2014), the use of hormones in meat preparation and production, and the 2011 Germany *E. coli* O104:H4 outbreak, have contributed to such a strict legislation and policy, leading to the 1997 Green Paper on Food Law, and to the revised 2000 White Paper on Food Safety.

The first Directive on general product safety, namely the Council Directive 92/59/EEC of 29 June 1992, and its subsequent modifications contained in the Directive 2001/95/EC, had the aim of ensuring high safety levels and standards of the foods (considered as goods) freely circulating within the European internal market.

The subsequent framework regulation (EC) No 178/2002 (known also as the General Food Law) makes an explicit mention of food and contributes to the development of the architecture of the organisms and bodies with the responsibilities of ensuring food safety. The General Food Law establishes, indeed, the "Standing Committee on the Food Chain and Animal Health" (SCFCAH), known also as the "Standing Committee on Plants, Animals, Food and Feed" (PAFF), which addresses several competencies originally ascribed to the Standing Veterinary Committee, the Standing Committee for Foodstuffs, and the Standing Committee for Feeding stuffs (Pisanello, 2014).

SCFCAH/PAFF is organized into the following eight sections (article 58 of the General Food Law):

- general food law section;
- biological safety of the food chain section;
- toxicological safety of the food chain section;
- controls and import conditions section;
- animal nutrition section;
- genetically modified food and feed and environmental risk section;
- animal health and animal welfare section; and
- phytopharmaceuticals section.

According to article 53 of the General Food Law, SCFCAH/PAFF is also in charge of adopting adequate emergency measures for food and feed of Community origin or imported from a third country as well as implementing reporting systems (such as the Rapid Alert System for Food and Feed, or RASFF).

Established in 1979, RASFF enables the sharing of early information and notification of warning concerning food safety risk among the 28 national food safety authorities, the EU Commission, the European Food Safety Authority (EFSA), ESA, Norway, Liechtenstein, Iceland, and Switzerland. In this way, a timely, effective, collectively coordinated response can be ensured. Further, RASFF collaborates with the International Food Safety Authorities Network (INFOSAN) of the World Health Organization (WHO).

Apart from the PAFF, there are five other committees in the food and feed area:

- Community Plant Variety Rights (SCPVR);
- Zootechnics (SCZ);
- Regulatory Committee 2001/18/EC;
- Regulatory Committee 2009/41/EC;
- Biocidal products.

Another important organism is the Food and Veterinary Office (FVO), subordinate to the Directorate General for Health and Consumers of the European

Commission. FVO is in charge of assessing the application and the compliance of the EU Member States to EU standards related to food safety, animal health, and phytosanitation, conducting audits, investigations, inspections, and any other related activities.

The precautionary principle helps the decision makers in adopting proper decisions since these rely on the latest scientific evidence available. The precautionary principle substantiates/articulates in:

- risk evaluation: it comprises risk assessment or risk analysis, risk management, and risk communication;
- risk perception, that is to say the public perception of a threat, or said in other words the feeling of being at risk. Risk perception, therefore, depends on different variables, such as, the media coverage or the perceived control over a situation;
- impact assessment: the implementation of certain legislative measures, both ingressive or less ingressive (from issuing prohibitions, restrictions, imposing authorization procedures or any other mandatory conditions, warnings, or instructions to giving recommendations and suggestions), could have social and economic detrimental implications, therefore, the Legislator tries to find a balance between the need of safety and the costs due to the application of the regulatory provisions;
- the ALARA principle or "as low as reasonably achieved," recently replaced by the ALAAMCD principle or "as low as analytical methods can detect").

Risk analysis or risk assessment (articles 6–10 of the General Food Law) is the responsibility of an independent food authority, namely the EFSA, with its four bodies (a management board; an executive director with its staff; an advisory forum; and finally, a scientific apparatus, made up of at least a scientific committee and 10 scientific expert panels), whose establishment in 2002 constituted another important step further for guaranteeing food safety. Different models for assessing risk exist, such as those elaborated by the French Food Safety Agency (AFFSA), the Danish Institute for Food and Veterinary Research, the German BfR (German Federal Institute for Risk Assessment) or other national agencies, the International Life Sciences Institute (ILSI), the European Responsible Nutrition Alliance (ERNA), and the European Federation of Associations of Health Product Manufacturers (EHPM), taking into account, for example, the risk characterization of the vitamins and nutrients and European intake data, based on the population safety index (PSI) of each individual nutrient (Coppens et al., 2006).

Risk management is the responsibility of the decision makers of the EU, namely the Commission, the Parliament, and the Council.

Risk communication consists in properly informing and offering consultation to the public on food risks, eventually interacting with laypeople. New information and communication technologies (ICTs) offer new possibilities for these tasks. The responsibility of risk communication is shared between EFSA and the European Commission (Coppens et al., 2006).

The consumer is protected also by proper labeling, which must contain the following information:

- name of the food;
- list of ingredients;
- allergens and ingredients causing intolerances, if any;
- quantitative ingredient declaration (QID);
- net quantity of the food;
- "use-by" date;
- any special storage conditions and/or conditions of use;
- name or business name and address of the food business operator;
- country of origin or place of provenance;
- instructions for use in case it would be difficult to properly use the food/foodstuff without them;
- alcoholic strength by volume;
- a nutrition declaration.

17.3 FOOD DEFINITION ACCORDING TO THE EU

Regulation (EC) No 178/2002 defines "food" (or "foodstuff") as "any substance or product, whether processed, partially processed or unprocessed, intended to be, or reasonably expected to be ingested by humans."

Article 2 describes what constitutes a food ingredient.

The definition of food/foodstuff explicitly includes also beverages, such as drinks and water (see also Article 6 of Directive 98/83/EC, the Directive 80/778/EEC, and the Directive 98/83/EC), chewing gum, and any substance intentionally incorporated into the food during the different steps and procedures of manufacture, preparation, treatment, and processing.

This definition explicitly excludes (Coppens et al., 2006):

- feed (an edible substance that represents part of animals' diet, endowed with energical or nutritional properties, generally recognized as safe);
- live animals;
- plants prior to harvesting;
- medicinal products (as defined by the Council Directive 65/65/EEC, the Council Directive 92/73/EEC, and article 2.2 of the Directive 2004/27/EC amending a previous Directive, namely the Directive 2001/83/EC);
- cosmetics (as defined by the Council Directive 76/768/EEC);
- tobacco and tobacco products (as defined by the Council Directive 89/622/EEC);
- narcotic or psychotropic substances (as defined by the United Nations Single Convention on Narcotic Drugs, 1961, and the United Nations Convention on Psychotropic Substances, 1971);
- residues and contaminants.

17.4 DEFINITION OF FOOD IMPROVEMENT AGENTS ACCORDING TO THE EU

Food additives, food enzymes, extract solvents, and food flavorings are also known as "food improvement agents."

17.4.1 FOOD ADDITIVES

Regulation EC 1333/2008 emending the Directive 89/107/EEC defines food additives as "any substance not normally consumed as a food in itself and not normally used as a characteristic ingredient of food, whether or not it has nutritive value." In other words, they are substances that are intentionally added to foods in order to perform and exert certain technological functions, numbered in 26 functions, such as:

- to color, that is to say to change, add, or restore colors (food-additives with this main function are known as food colorants);
- to enhance the sweet taste of low-sugar or other dietetic foods, for example foods for diabetics (food-additives with this main function are known as sweeteners);
- to increase the acid taste of food (food-additives with this main function are known as acids);
- to modify the pH of foods (food-additives with this main function are known as acidity regulators);
- to facilitate the action of flavorings (food-additives with this main function are known as flavor enhancers);
- to preserve, that is to say to prolong the shelf-life of foods by protecting them against microorganisms or other insults (food-additives with this main function are known as preservatives);
- to contribute/facilitate the processes of dissolution, dilution, dispersion, or any other physical modification of food-additives (as well as of food supplements, flavorings, food enzymes, or any other foodstuff), without altering their function, that is to say without exerting any technological effect themselves (food-additives with this main function are known as carriers);
- to reduce the adhesive properties of food particles and components (food-additives with this main function are known as anticaking agents);
- to protect against oxidation, fat rancidity, and color change, prolonging the shelf-life of foods (food-additives with this main function are known as antioxidants);
- to create an emulsion or an emulsion-like structure (food-additives with this main function are known as emulsifiers);
- to disperse proteins within fat or other food components in a homogeneous way (food-additives with this main function are known as emulsifying salts);
- to create a rigid, firm, stable structure (food-additives with this main function are known as firming agents and can be used together with gelling agents);

- to prevent superficial oxidation and microbe spreading (food-additives with this main function are known as packaging gases);
- to improve the baking quality (food-additives with this main function are known as flour treatment agents);
- to prevent excessive drying or moisture loss/humidify (food-additives with this main function are known as humectants);
- edible starches chemically and/or physically modified/processed (food-additives of this kind are known as modified starches);
- to prevent or at least to reduce foaming effects (food-additives with this main function are known as antifoaming agents);
- to increase the volumetric properties of foods, without contributing to their energetic value (food-additives with this main function are known as bulking agents);
- to produce foaming effects (food-additives with this main function are known as foaming agents);
- to create a protective exterior shell (food-additives with this main function are known as glazing agents/lubricants);
- to increase the volumetric capacities of a dough or batter (raising agents);
- to increase the thickness of a foodstuff (food-additives with this main function are known as thickeners);
- to create a gel or a gel-like structure (food-additives with this main function are known as gelling agents);
- to expel a food/foodstuff from a container (food-additives with this main function are known as propellants);
- to chelate metal ions (food-additives with this main function are known as sequestrants);
- to stabilize food during its production, packaging, and storage (food-additives with this main function are known as stabilizers).

The EU does not consider food additives (Pisanello, 2014):

- foods in which monosaccharides, disaccharides, or oligosaccharides have been added for their sweetening properties;
- dried or concentrated foods in which flavorings have been incorporated during the manufacturing procedure for their aromatic, sapid, or nutritive properties and for their secondary coloring effect;
- substances used in covering or coating materials, usually not consumed as foods;
- products containing pectin and derived from dried apple pomace or peel of citrus fruits or quinces, or from a mixture of them;
- chewing gum bases;
- white or yellow dextrin, roasted or dextrinated starch, starch modified by acid or alkali treatment, bleached starch, physically modified starch, and starch treated by amylolitic enzymes;
- ammonium chloride;

- blood plasma, edible gelatin, protein hydrolysates and their salts, milk protein, and gluten;
- amino acids and their salts other than glutamic acid, glycine, cysteine, and cystine and their salts without technological function;
- caseinates and casein;
- inulin.

17.4.2 FOOD ENZYMES

Food enzymes can be extracted from plants, animals, or microorganisms (or part of plants, animals, or microorganisms) and include at least one product obtained via a fermentation process. They have specific biochemical and technological actions.

Regulation (EC) No 1332/2008 on food enzymes—the so called "framework regulation"—includes:

- enzymes intended for human consumption, for example, for nutritional or digestive purposes;
- food enzymes used in the production of food additives under Regulation EC 1333/2008 and in the production of processing aids.

The regulation excludes:

- microbial cultures traditionally used in the production of food (cheese, wine), which may incidentally produce enzymes but are not specifically used to produce them;
- food enzyme preparations: mixtures of one or more food enzyme(s) "in which substances such as food additives and/or other food ingredients are incorporated to facilitate their storage, sale, standardization, dilution or dissolution" [as per article 3(2) (b) of the Reg. (EC) No 1332/2008].

Other pieces of EU legislation relevant to food enzymes are the following:

- Regulation (EC) No 1331/2008—establishes and disciplines the common authorization procedure for food additives, food enzymes, and food flavorings;
- Regulation (EU) No 234/2011 [amended by Commission Implementing Regulation (EU) No 562/2012]—implements the common authorization procedure;
- Regulation (EU) No 1169/2011 concerns food information for consumers;
- Enzyme Regulation—Articles 10 to 13 concern specific labeling requirements for food enzymes and food enzyme preparations.

17.4.3 FLAVORINGS

Flavorings are defined by Directive 88/388/EEC and Commission Directive 91/71/EEC, subsequently amended by the Regulation (EC) No 1334/2008 on flavorings and certain food ingredients with flavoring properties for use in/on foods.

EU regulation distinguishes the following kinds of flavorings:

- flavoring substances;
- flavoring preparations;
- thermal process flavorings;
- smoke flavorings (including primary smoke condensates, primary tar fractions, primary products, and derived smoke flavorings);
- flavor precursors;
- other flavorings.

17.4.4 EXTRACTION SOLVENTS

Extraction solvents are defined by Directive 2009/32/EC as "solvents used in an extraction procedure during the processing of raw materials, of foodstuffs, or of components or ingredients of these products and which are removed but which may result in the unintentional, but technically unavoidable, presence of residues or derivatives in the foodstuff or food ingredient."

The definition excludes:

- extraction solvents or any other foodstuffs intended to be exported and released from the EU;
- extraction solvents used in the production of food additives, vitamins, and other nutritional additives, unless required.

17.4.5 PROCESSING AIDS

"Every intentionally added substance in the processing of foods without the clear 'foodstuff' status" is not still addressed by the Regulation (EC) No 1333/2008 (Pisanello, 2014).

17.5 FOOD/DIETARY SUPPLEMENT DEFINITION ACCORDING TO THE EU

Food supplements are concentrated sources of nutrients or other substances with a nutritional or physiological effect, whose purpose is to supplement the normal diet, used alone or in combination. Food supplements are marketed "in dose" form, for example as capsules, pills, pastilles, tablets, or liquids in measured doses, or released by drop-dispensing bottles, ampoules, or sachets of powder, among others. Supplements may be used to correct nutritional deficiencies (Coppens et al., 2006).

Food/dietary supplements are defined by article 2a of Directive 2002/46/EC, amended by the Commission Directive 2006/37/EC, the Commission Regulation (EC) 1170/2009, the Commission Regulation (EU) No 1161/2011, the Commission Regulation (EU) No 119/2014, and the Commission Regulation (EU) 2015/414.

The list of food/dietary supplements initially included only vitamins and minerals, but recently has been revised and includes also:

- amino acids;
- enzymes;
- pre- and probiotics: probiotics is a term coined in 1907 by Elie Metchnikoff and defined as "live microorganisms which when administered in adequate amounts confer a health benefit on the host" (Ebner et al., 2014);
- essential fatty acids;
- fibers, botanicals, and botanical extracts;
- miscellaneous bioactive substances.

17.6 ENRICHED/FORTIFIED FOODS DEFINITION ACCORDING TO THE EU

Enriched/fortified foods are defined by Regulation (EC) 1925/2006 [amended by Regulation (EC) No 108/2008 of the European Parliament and of the Council of 15 January 2008 and by Commission Regulation (EC) No 1170/2009 of 30 November 2009] as "foods to which nutrients or ingredients have been added to add or emphasize particular nutritional characteristics."

This directive does not consider as fortified foods (Pisanello, 2014):

- foods enriched with mandatory fortification of flour and margarine;
- foods which contain trace quantities of vitamins or minerals in alcoholic drinks as authenticity markers;
- foods in which vitamins and minerals have been used as food supplements (as defined by Directive 2002/46/EC);
- foodstuffs for particular nutritional uses or purposes (as defined by Directive 2009/39/EC);
- foods in which vitamins and minerals have been added for additive purposes (as defined by Regulation (EC) No 1333/2008).

17.7 DIETETIC FOODS/FOODS FOR PARTICULAR NUTRITIONAL USE (PARNUTS) DEFINITION ACCORDING TO THE EU

The Directive 2009/39/EC amending the Directive 89/398/EEC defines dietetic foods or foods for particular nutritional use (PARNUTS) as foods, "which, owing to their special composition or manufacturing process are clearly distinguishable from foodstuffs for normal consumption, which are suitable for their claimed nutritional purposes and which are marketed in such a way as to indicate such suitability."

Currently, this Directive is still under force until the New Regulation on Food for Specific Groups, namely the Regulation (EU) No 609/2013 of the European Parliament and the Council on food intended for infants and young children, food for special medical purposes, and total diet replacement for weight control (Food for Specific Groups), adopted on 12 Jun. 2013, will apply (from 20 Jul. 2016).

It should be also emphasized that the label "dietetic foods" is now considered inadequate and as such will be replaced with the new label "food for specific groups."

Examples include:

- infant and follow-on formulae such as growing-up milks, baby foods for infants, and foods for young children (such as processed-cereal based food);
- slimming foods, that is to say foods for people undertaking energy-restricted diets to lose weight;
- foods for special medical uses, such as, foods for diabetics or for patients suffering from other metabolic diseases;
- sports foods;
- food for people with gluten intolerance [gluten-free food or food with very low gluten concentration, as defined by Regulation (EU) No 1169/2011].

17.8 NOVEL FOOD DEFINITION ACCORDING TO THE EU

Novel foods are defined by the Novel Food Regulation (EC) No 258/97 [Council and Commission Regulation (EC) No 1852/2001] as foods or food ingredients, which have not been used to a significant degree for human consumption in the EU before 15 May 1997, when the first Regulation on novel food came into force. This EU law has been recently updated and amended by the Regulation (EU) No 1169/2011 of the European Parliament and of the Council and by the Regulation (EU) 2015/2283 of the European Parliament and of the Council of 25 November 2015 on novel foods.

"Novel Food" can be newly developed, innovative food or food produced using new technologies and production processes (such as, high-pressure pasteurized fruit preparations) as well as food traditionally eaten outside of the EU (exotic products from third countries, such as chia seeds, Noni juice, and Ngali nuts), newly produced nutrients (synthetic zeaxanthin), or extracts from existing food (rapeseed protein) (Constable et al., 2007).

The EU has issued and regularly updates a list of novel foods, namely the "Novel Food Catalogue," which includes all products of animal and plant origin and other substances subject to the Novel Food Regulation, based on information provided by the EU Member States.

The ILSI Europe Novel Foods Task Force has suggested the working concept of "history of safe use" to facilitate the risk assessment of novel foods (Constable et al., 2007).

17.9 NUTRACEUTICAL DEFINITION ACCORDING TO THE EU

As far as nutraceuticals are concerned, due to the increasing number of foods or nutraceuticals reporting nutritional and health claims, the EU has adopted the Regulation (EU) No 1924/2006 on the use of nutritional and health claims made on foods. Nutritional claims are defined as "any claim which states, suggests or implies that a food has particular beneficial nutritional properties due to the presence, absence, increased or reduced levels of a particular nutrient or other substance," while a health claim is "any claim that states, suggests or implies that a relationship exists between a food category, a food or one of its constituents and health."

However, this Regulation does not properly cover products such as botanicals products.

17.10 GENETICALLY MODIFIED ORGANISMS (GMOs)

The term "genetically modified" food describes food containing, consisting of, or produced from "genetically modified organisms" (GMOs). A GMO is defined as "an organism, such as plants or animals, in which the genetic material has been altered in a way that does not occur naturally by mating and/or natural recombination to give them a new property" [Article 2(2) Directive 2001/18/EC, amended by the subsequent Directive (EU) 2015/412]. GMOs are produced in such a way that the new food/foodstuff develops an increased resistance to environmental pressures and competitive/stressful conditions such as diseases, droughts, insects, or herbicides. GMOs may be endowed with an improved quality, such as an increased nutritional value or yield.

Directive 2009/41/EC is specifically devoted to the use of genetically modified microorganisms.

The different requirements, namely the procedures for developing GMOs, their proper usage, the authorizations that are mandatory to obtain, and eventual restrictions or prohibitions of cultivation of GMOs, are contained in Regulation No 1829/2003. Articles 3 and 4 state that GMOs may be placed on the market only for food use, and only after they have been authorized by the EFSA and the European Commission.

Regulation (EC) 1946/2003 regulates transboundary movements of GMOs.

Regulation (EC) 1830/2003 concerns the traceability and the labeling of GMOs and the traceability of food and feed products produced from GMOs.

17.11 CONCLUSIONS

The architecture of the bodies and organisms deputed to ensuring food safety is rather complex, reflecting the complexity of the matter itself. However, this architecture is still far from complete and efforts are required to address further fundamental questions with an important impact on societal life. It is, indeed, true that "legislation always runs behind reality" (Gulati and Berry Ottaway, 2006) and that gene engineering and nanobiotechnologies are achieving unprecedented results. New emerging topics, such as nanofoods (that is to say, foods obtained with the aid of nanotechnologies), need to be properly addressed by the legislator, if possible adopting an anticipatory governance framework and foresight.

REFERENCES

Baumgartner, T., Uebe, W., 2014. European Union. In: Kirchsteiger-Meier, E., Baumgartner, T. (Eds.), Global Food Legislation: An Overview. John Wiley & Sons, Hoboken, NJ.

Constable, A., Jonas, D., Cockburn, A., Davi, A., Edwards, G., Hepburn, P., Herouet-Guicheney, C., Knowles, M., Moseley, B., Oberdörfer, R, Samuels, F., 2007. History of safe use as applied to the safety assessment of novel foods and foods derived from genetically modified organisms. Food Chem. Toxicol. 45 (12), 2513–2525.

Coppens, P., da Silva, M.F., Pettman, S., 2006. European regulations on nutraceuticals, dietary supplements and functional foods: a framework based on safety. Toxicology 221 (1), 59–74.

Ebner, S., Smug, L.N., Kneifel, W., Salminen, S.J., Sanders, M.E., 2014. Probiotics in dietary guidelines and clinical recommendations outside the European Union. World J. Gastroenterol. 20 (43), 16095–16100.

Gulati, O.P., Berry Ottaway, P., 2006. Legislation relating to nutraceuticals in the European Union with a particular focus on botanical-sourced products. Toxicology 221 (1), 75–87.

Malisch, R., Kotz, A., 2014. Dioxins and PCBs in feed and food—review from European perspective. Sci. Total Environ. 491–492, 2–10.

Pisanello, D., 2014. Chemistry of Foods: EU Legal and Regulatory Approaches. Springer, Dordrecht.

CHAPTER

18 Health foods and foods with health claims in the Asian subcontinent: countries of the Association of Southeast Asian Nations

Jerzy Zawistowski
Food, Nutrition and Health, Faculty of Land and Food Systems, University of British Columbia, Vancouver, BC, Canada

18.1 INTRODUCTION

The Asian subcontinent, which is the subject of this chapter, includes countries of South-East Asia (SEA). There are mainland countries such as Malaysia, Thailand, Vietnam, Laos, and Myanmar, and countries located on many islands: Singapore, Indonesia, the Philippines, Brunei, and East Timor-Leste. All of them with the exception of East Timor-Leste belong to the Association of Southeast Asian Nations (ASEAN). This alliance promotes economic and political cooperation between its members. One objective, and also challenge, of the ASEAN is to harmonize health food, health claims, and nutrition labeling regulations between member countries (ASEAN, 2014). Some countries of ASEAN such as Indonesia, Malaysia, and Singapore have already well-established law and regulations for health foods and approved health claims. Others, although they approve foods with health benefits, have limited regulations.

SEA is an Asian subcontinent with a rich history of healthy eating. People of this region recognize that food not only satisfies hunger and satiety but also provides health benefits. It is a common belief that foods may be used for prevention and treatment of specific diseases. This practice is still ongoing, particularly in regions with limited access to health care providers (Van Esterik, 2008). Herbs, spices, teas, fruits, and vegetables have a long tradition as medical therapy. A medley of chilies, red ginger, sesame oil, cotton seed, leaves, and bark is used to formulate various concoctions to heal many diseases (Van Esterik, 2008). Philippine ampalaya (*Momordia charantia*) or bitter gourd, which can be also found in Indonesia and Vietnam, is

Developing New Functional Food and Nutraceutical Products. http://dx.doi.org/10.1016/B978-0-12-802780-6.00018-3

well recognized for its hypoglycemic properties (Murray, 2006). Maunggay, also known as horseradish or drumstick tree (*Moringa oleifera*), is cultivated throughout the SEA region for its culinary, nutritional, and medicinal values. Flowers, seeds, leaves, and green pods are used to cook somlor corko, Cambodian soup, and Thai curry. The edible parts of this tree provide significant amounts of vitamins B, C, K, β-carotene, and minerals. Such essential nutrients are important to combat malnutrition for people in the region who rely on food enrichment. This plant is also used in traditional medicine for its antiinflammatory properties (Leone et al., 2015). In Indonesia, herbs, roots, bark, flowers, seeds, leaves, and fruits along with honey and royal jelly are used to formulate traditional herbal medicine: "jamu." Apparently, jamu is good to enhance health, increase appetite, and treat coughs and ulcers (Van Esterik, 2008). Jamu is particularly popular in Java, from where it originated. Currently, jamu is commercialized as a natural health product in the form of capsules, tablets, and caplets and brings a significant revenue to the Indonesian herbal medicinal industry (Yulisman, 2014).

Although most food components with disease-prevention attributes were lacking scientific support, currently some bioactives have been the subject of research projects. In Malaysia, research and development is conducted, by public universities and scientific institutions, on the characterization and identification of functional components of local plants (FAO, 2004). Scientific recognition of plant bioactives is of great significance, since it can be used to justify formulation of health products. Furthermore, it provides a basis to better regulations and rationale in support of health claims on foods.

Health claims on food labels are an important instrument to communicate health benefits to consumers. They offer the possibility for informed choices. They regulate products and they enhance the marketing of products. Scientifically proven efficacy and safety of foods are a common interest for consumers, regulators, and industry. Foods with validated health benefits contribute to a healthy lifestyle, may maintain a consumer's well-being, and help reduce the risk of chronic diseases. This in turn may reduce the cost of health care. There are also benefits to the food industry such as providing higher profit margins from selling such foods at a premium.

Health foods encompass an array of foods that not only provide direct health benefits by reducing the risk of chronic disease in relatively short periods of time, but also foods enriched with vitamins and minerals, as well as foods that improve the well-being of consumers. Foods may carry a health claim label as long as they are approved by the governing regulatory agency of a region. There are also variations regarding health claims in the degree of scientific supports. Some of these claims may be simple and indicate only the presence of nutrients or nutraceuticals, while some others carry a powerful statement linking consumption of a food and its bioactive with the prevention of a specific disease.

Foods with health claims are considered in many countries as functional foods. There are many definitions of functional foods (Zawistowski, 2011), however, only in a few countries do such definitions have legal meaning. Indonesia is one such country, and Indonesian legislation defines functional foods as "processed foods that contain one or more functional components (proven to be harmless) with a certain physiological function that is proven by scientific studies to be beneficial for health" (HK, 2005). In SEA, however, there is a disparity between regulations and law, and between labeling and allowed health claims, which causes perplexity among consumers and food manufacturers (ILSI, 2015). Some countries adopted and follow the guidelines of Codex Alimentarius, some others accommodate European Union (EU) and US regulations or the food and drug regulations of their own countries. The term "functional foods" has different meanings in different countries. In Singapore and Indonesia functional foods are considered foods, in Malaysia it may be an interface between a food and a drug, while in the Philippines dietary supplements are included in this term. Differences exist within the approval process for allowed health claims. Disease risk reduction claims are allowed in some countries (e.g., Philippines and Indonesia), prohibited in others (e.g., Malaysia), and still under review in some other countries (Zawistowski, 2014).

The purpose of this chapter is to review a regulatory system for functional foods in countries of ASEAN. Diverse acts, regulations, and guidelines that may govern these types of foods, as well as labeling and possible health claims are discussed. In this chapter, the meaning of health foods, foods with health claims, and functional foods is the same. This is particularly true since the International Life Sciences Institute (ILSI) defines functional foods as foods with health claims (Aggett et al., 2005).

18.2 INDONESIA

Herbal and traditional foods with natural bioactive constituents have been available in Indonesia for a long time. These foods, which may provide health benefits to consumers, have the potential to be developed as functional health foods. However, only recently has there been a lot of attention to the functionality of foods that are endemic to this country. A good example of this is the research that has been carried out in Indonesia and Malaysia on crude palm oil. A new study shows that a diet rich in crude palm oil, a trans-fat free product often procured in Malaysia and Indonesia, may reduce blood triacylglycerol levels (Ladeia et al., 2008). This increased attention to health foods is the result of the popularity of healthy lifestyles among consumers in the country. This in turn results in companies launching healthier foods to keep up with the market. Between 2010 and 2013, the annual average growth of food consumption was 7.7%, while the compound annual growth rate for the sale of functional foods was forecast to be as high as 19.5% for 2017 (KITA, 2013). With a population of 250 million, Indonesia is considered a good potential market for functional foods (KITA, 2013).

18.2.1 FUNCTIONAL FOODS AND REGULATORY FRAMEWORK

Indonesia is the only country in SEA, and probably in the world with the exception of Japan, where functional foods have a legal meaning. Functional foods are defined as processed foods that contain one or more functional components (proven to be harmless) with a certain physiological function and proven by scientific studies to be beneficial for health (HK, 2005). These are foods with added health benefits and contain minerals, vitamins, and other nutrients, and/or may be in line with the food fortification program that was launched in 1996 to improve the nutritional status of the community (Bogor Agricultural University, 2003).

In 2005, the Food and Drug Supervisory Agency of the Republic of Indonesia issued provisions on functional foods (HK, 2005). The regulation lists all nutrients and bioactives that are permitted to be used in functional foods. It also specifies allowable health claims and conditions for their uses. Functional foods must comply with the same regulations that are applied for conventional foods (PP 28, 2004). This includes safety, quality, and nutritional standards as well as food packaging and laboratory testing requirements, regardless of whether products are manufactured nationally or imported from outside of Indonesia. The production of functional foods must be conducted in facilities under the Hazard Analysis and Critical Control Points (HACCP) system. Food standards and rules are described in the governmental regulations on food safety, quality, and nutrition (PP 28, 2004; Bill, 2012); regulations on food labeling and advertisement (PP 69, 1999); and the Food Act (Act 7, 1996).

Marketing and sale of foods in Indonesia are governed by the National Agency for Drug and Food Control (BPOM—Badan Pengawas Obat dan Makanan). BPOM controls both domestic and import products and requires premarket approval. If a product meets all its requirements, then it is given a registration number and is approved for marketing. Products with added ingredients from animal sources must be certified in accordance with the Islamic purity law, "Halal." Food products must also have appropriate labelling as specified in Food Law 18/2012, Chapter 8, part one (Bill, 2012). Furthermore, in order to obtain approval for functional foods, producers or importers must submit a dossier with clinical data to substantiate health claims. If a product is intended to carry a nutrient function claim it must be supported by at least one randomized clinical trial (RCT). For supporting a disease risk reduction claim at least three RCT studies are required. Studies must be well designed and executed and published in peer-reviewed journals (Chan, 2011). BPOM registration approval for functional foods (food with claims) and foods with added herbs for medical purposes takes up to 120 days (GAIN, 2015).

18.2.2 HEALTH CLAIMS AND NUTRITION LABELING

Indonesia, like most Asian countries, uses the term "health claim" as defined by Codex Alimentarius (Codex, 2004). Furthermore, the regulations promulgated on Jan. 27, 2005 have approved 14 classes of functional food components that can be used in foods with health claims (HK 00.05.52.0685, 2005). The approved nutrients and nutraceuticals include vitamins; minerals; amino acids; dietary fiber; prebiotics and probiotics; soy isoflavones; phytosterols and phytostanols; and tea polyphenols (Table 18.1).

Table 18.1 Classes of Food Components Approved for Uses in Functional Foods in Indonesia

Class	Food Component
I	Vitamins: A, B1, B2, B3, B5, B6, B12, C, D, E, folic acid, and carotenoids
II	Minerals: Na, K, Zn, Cu, I, Mg, Mn, Se, Cr, V, and B
III	Alcoholic sugar
IV	Unsaturated fat: omega-3 fatty acids, DHA, and ARA
V	Peptides and certain proteins: whey protein, and lactoferin
VI	Amino acids: lysin
VII	Food fiber: insoluble fiber, psillium, oat fiber, and inulin
VIII	Prebiotics
IX	Probiotics: lactobacillus spp. Except lactobacillus bulgaricus, and biofidobacterium
X	Choline, lecithin, inositol
XI	Carnitine and squalene
XII	Soy isoflavones: daidzein, daidzin, genistein, and genistin
XIII	Phytosterols, phytostanols
XIV	Tea polyphenols

Modified from HK (2005).

There are three groups of label claims permitted on foods: nutrient content claims, nutrient function claims, and health claims (HK, 2005). The latter is the disease risk reduction claims that describe the link between a food ingredient and the reduced risk of a chronic disease by altering its major risk factor or factors.

The following health claims were adopted from the US Nutrition Labeling and Education Act of 1990 by the BPOM:

1. calcium and osteoporosis
2. folic acid and neural tube defects
3. sodium and hypertension
4. dietary sugar alcohol and dental caries
5. fiber-containing seeds, fruits, and vegetables and cancer
6. soluble fiber-containing seeds, fruits, and vegetables and coronary heart disease
7. phytosterols and phytostanols and coronary heart disease

In addition, Indonesian regulations permit these additional disease risk reduction claims:

1. whey protein and coronary heart disease
2. green tea extract containing 48–72 g of polyphenols and coronary heart disease

35 nutrition function claims on foods that contain selected vitamins (class I), minerals (class II), omega-3 fatty acids (class IV), dietary fiber (class VII), probiotics (class IX), cholin (class X), squalene (class XI), soy isoflavones (class XII), and tea polyphenols (class XIV) are permitted in Indonesia (HK, 2005).

The following claims are deemed to be acceptable by Indonesian regulations:

- "as antioxidant, vitamin E may help reduce possible risk of coronary heart disease";
- "calcium plays a role in the formation of bones and in maintaining solidity of bones and teeth";
- "the giving of DHA and ARA [docosahexaenoic acid and arachidonic acid] during pregnancy, babyhood, and childhood (up to 2 years old) will optimize brain development and learning competence"; and
- "food fiber (psyllium, oats, and inulin) may help strengthen digestion system."

Some of the claims in this category resemble structure/function claims, for example "calcium and maintaining solid bones and teeth," while others resemble claims of risk reduction and prevention of diseases: soya isoflavones "may reduce the risk of atherosclerosis and coronary heart disease" or "iron may help prevent and overcome anemia as a result of iron deficiency."

Since malnutrition and nutrient deficiency are real problems in Indonesia (Dijkhuizen et al., 2013), the fortification of foods with specific nutrients is a common procedure (Isabelle et al., 2011). Food fortification is governed by Indonesian law and regulations, which also allows nutrient content and comparative claims for energy, protein, fat, and fatty acids content, and enrichment with vitamins and minerals (Department of Health, 1999). Word descriptors such as "Enriched Fortification," "Extra," "Plus," "More, "Added," "Contain," "Provide," "Is a Good Source, "High," "Rich of," and "Is a Very Good Source" are permitted in relation to nutrients used for food enrichment as long as both food and nutrients meet specified requirements. For instance, no claim that food "is a very good source" of folic acid is permitted unless the food pertained to contains at least 20% of the recommended nutrient reference value of this vitamin per serving. In the case of a claim "enriched" with dietary fiber, the pertaining food must contain at least 10% by weight of the recommended value (25 g/day) or 10% more than similar foods for comparative claims such as "more" or "extra."

Specific requirements for nutritional labeling are outlined in a compilation of food regulations prepared by the Department of Health Indonesia (Department of Health, 1994, 1999; PP 69, 1999). Foods with health claims that contain specific nutrients, including energy, protein, and carbohydrate content, levels of vitamins and minerals, as well as fortified and enriched foods with specific nutrients must declare nutrition information as required by the national legislations. Currently, nutrition labeling is mandatory for all prepackaged food products, even in the absence of health claims (EUFIC, 2013). The basic nutrition information panel must include total energy, total fat, protein, total carbohydrate, and sodium (Sihombing, 2010). Additional nutritional information is allowed but it is dependent on the specific food.

The Indonesian Consumer Protection Act contains a provision against misleading claims on foods such as drug claims. Functional foods may carry health claims on their labels and in advertisements only when they are approved by the Head of the Food and Drug Supervisory Board on the recommendations of the Peer Reviewer

Team (HK, 2005). While using a specific claim it may be required to also mention associated warnings and additional information. A health claim stating: "low sodium food may help reduce risk of hypertension" must be accompanied with the statement: "people with hypertension must consult a doctor." Nutrient content claims on magnesium must disclose the statement "not recommended for persons with kidney problems and coronary heart disease, except with doctor's recommendations." Regulations also prohibit some claims related to nutrients and bioactives. For example, a claim stating: "vitamin C plays a role in the formation of collagenic tissue" is allowed, while the claim: "vitamin C increases stamina, and prevents cancer" is prohibited. The claim that links dietary fiber and cancer should not be referred to colon cancer, while a claim of squalene as an antioxidant should not state: "may prevent cancer."

18.3 MALAYSIA

After Singapore and Brunei, Malaysia is the third richest country in SEA with well-established gross domestic product (GDP) growth. Although the agriculture sector is only a small part of the Malaysian economy, the Malaysian government has identified the food processing industry as a priority sector for industrial development (Ahmed, 2012). With a diversity of national ethnic groups, Malaysia's food industry produces a large range of traditional and innovative food products. Consumer awareness toward the health benefits of foods, which has roots in the history of Malays, Indians, and Chinese, has created a demand for health foods. Diet-related chronic diseases such as coronary heart disease are on the rise (Tee et al., 2002). This contributes to why consumers are now paying greater attention to a healthy diet. The food industry has also increased efforts to improve the nutritional quality of foods, including increasing the concentration of desirable nutrients in foods and decreasing the concentration of undesirable nutrients in them.

18.3.1 FUNCTIONAL FOODS AND REGULATORY FRAMEWORK

New scientific developments in food science offer a range of possibilities to increase the health-promoting properties of specific food products. Examples of such foods are functional foods and nutraceuticals. Functional foods are not well defined and regulated in Malaysia. This category may include foods that are enriched with vitamins and minerals as well as other food components, such as plant sterols and dietary fiber. It is worthwhile to note that Malaysian regulations include ingredients such as vitamins, minerals, amino acids, natural substances of plant/animal origin, enzymes, and substances with nutritional/physiological functions within a dietary supplement category in the form of pills, capsules, and other medicinal formats. In Malaysia if a product is not clearly a food or a drug it is classified by the Committee for the Classification of Food-Drug Interface Products (Jawatankuasa Pengkelasan Keluaran Food-Drug Interphase) as a food–drug interface (FDI) based on its composition. Accordingly, if a product contains 80% or more food ingredients and 20% or less bioactive ingredients with pharmacological and/or therapeutic properties, such a product is regulated by the Food Safety and Quality Division. However, if the product contains more than 20% bioactives and less than 80% food-based ingredients, it is regulated by the National Pharmaceutical Control Bureau (NPCB). The sale of FDI, which resembles health food, is governed by either the Drug Act or the Food Act. In addition, NPCB is responsible for the approval of any products in "pure' forms such as vitamins, minerals, amino acids, fatty acids, fiber, and enzymes as well as products which are not traditionally used as foods or beverages with health benefits, such as alfalfa, spirulina, royal jelly, noni juice, rooibos tea, and other herbal products (MHM, 2007).

18.3.2 HEALTH CLAIMS AND NUTRITION LABELING

Similar to other Asian countries, Malaysia uses the term "health claims" as defined by Codex Alimentarius (Codex 2004). The claims follow conditions stipulated in the Codex guidelines (Tee et al., 2002) with a few exceptions that accommodate local requirements. The general principles of the manufacturing, safety, and sale of foods were promulgated in the Malaysian Food Act (1983) and Food Regulations of 1985. Nutrition labeling and claims provision were enacted in 2003 and followed by a number of amendments in 2005 and 2007 (MHM, 2010). These regulations cover requirements for food declaration, compliance and penalties, nutrient content, mandatory and voluntary nutrition labeling, allowed claims, and prohibited claims. The five groups of claims on foods that are permitted are:

- nutrient content claims—describing the level of a nutrient in foods,
- nutrition comparative claims—comparing the nutrient levels and/or energy value between two or more similar foods,
- nutrient function claims—as defined by Codex Alimentarius (Codex, 2004),
- other function claims —as defined by Codex Alimentarius (Codex, 2004),
- claims for enrichment and fortification— regarding specific vitamins and minerals that can be used to fortify, enrich, vitaminize, and supplement foods that meet the conditions described in Food Regulations 1985, Twelfth Schedule, Regulation 26 (Food Regulations, 1985).

Nutrient and other function claims reflect the Codex Alimentarius definition and describe the relationship between a food or food constituent and health. For this reason, these kinds of claims, if used on foods, identify the meaning of functional foods. Currently, Malaysian regulations permit 31 function claims (MHM, 2010). The claims related to vitamins, minerals, omega-3 and omega-6 fatty acids, as well as proteins resemble the Canadian function and nutrient function claims (CFIA, 2016). However, in contrary to the Canadian regulations, Malaysian nutrient function claims are allowed to be labeled on foods that are enriched with bioactives such as probiotics, prebiotics, dietary fiber, lutein, as well as ingredients with hypocholesterolemic properties. It is worthwhile to note that the fat blend, which was patented by the

Brandeis University in the US (Sundram et al., 1996), was granted a function claim under the Malaysian regulations. Functional claims can only be made provided the food meets the criteria for "source of" and other conditions specified by regulations (Tee, 2007b; MHM, 2010). Examples of the function claims for foods containing discussed bioactives are:

- lutein may protect the eye
- *Bifidobacterium lactis* helps improve a beneficial intestinal microflora, may also help to reduce the incidence of diarrhea
- prebiotic claims include:
    - inulin, fructooligosaccharide (FOS), galactooligosaccharide (GOS), and polydextrose help growth of biofidobacteria (it is bifidogenic and maintains a good intestinal function)
- regularity claims include:
    - high amylose maize resistant starch helps improve bowel function
    - resistant dextrin/maltodextrin helps to promote regular bowel movement
- oat soluble fiber (β-glucan) helps lower the rise of blood glucose
- mixture of oligofructose and inulin helps to increase calcium absorption and increase bone mineral density
- hypocholesterolemic claims include:
    - plant sterol or plant stanol and plant sterol ester help lower or reduce cholesterol
    - soy protein helps to reduce cholesterol
    - β-glucan from oat or barley helps lower or reduce cholesterol
    - patented cooking oil blend helps to increase HDL [high-density lipoprotein] cholesterol and improve HDL/LDL cholesterol ratio.

Foods enriched with dietary fiber, soy protein, and plant sterols for lowering blood cholesterol are constituents of truly functional foods in Europe and USA. Table 18.2 lists specific criteria for foods to carry hypocholesterolemic function claims. These claims are considered disease risk reduction claims in various world regulatory jurisdictions. In Malaysia they are in the category of function claims, since disease risk reduction claims are not permitted.

Malaysian Regulations of the Food Act prohibit the description of any foods which include the words "compounded," "medicated," "tonic," "health," or other descriptions that are primarily reserved for reference to drugs. The Act also has provisions that prohibit label claims, "as to the suitability of a food for use in the prevention, alleviation, treatment or cure of a disease, disorder, or particular physical condition unless they are permitted in this regulation," and "which could give rise to doubt the safety of a similar food, or which could arouse or exploit fear in the consumer" (Tee, 2000).

Although nutrition labeling is voluntary for the majority of foods, the law requires mandatory nutrition labels on foods with nutrition and function claims. This also includes special purpose foods. Basic nutrition information includes the amount of protein, carbohydrate, fat, and energy expressed per 100 g or mL. The amount

Table 18.2 Conditions for the Malaysian Nutrient Function Claims for Hypocholesterolemic Ingredients

Component	Minimum Amount Required	Other Conditions
β-Glucan	1.75 g per serving	1. Source of β-glucan shall be from oat and barley 2. Food with β-glucan shall also contain total dietary fiber for not less than amount required to claim as "source": a. 3 g per 100 g (solids) b. 1.5 g per 100 mL (liquids) 3. Following statement is required on the label: a. "Amount recommended for cholesterol lowering effect is 3g/day"
Plant sterol/ stanol	0.4 g per serving on free basis	1. Types of plant sterol or plant stanol permitted: "plant sterol/plant stanol, phytosterols/phytostanol, sitosterol, campesterol, stigmasterol or other related plant stanol" 2. Types of plant sterol esters permitted: "campesterol ester, stigmasterol ester and β-sitosterol ester" 3. Maximum amount in daily serving for product added with plant sterol/plant stanol/plant sterol ester in a free basis is not more than 3 g plant sterol/plant stanol per day 4. Declaration of the total amount of plant sterol/plant stanol contained in the products shall be expressed in metric units per 100 g or per 100 mL or per package if the package contains only a single portion and per serving as quantified on the label 5. Only the terms "plant sterols," "plant stanol," or "plant sterol ester" shall be used in declaring the presence of such components 6. There shall be written on the label of food making such claim a statement: a. "Not recommended for pregnant and lactating women, and children under the age of five years" b. "Persons on cholesterol-lowering medication shall seek medical advice before consuming this product" c. "That the product is consumed as part of a balanced and varied diet and shall include regular consumption of fruits and vegetables to help maintain the carotenoid level"; and d. "With added plant sterols/plant stanol/plant sterol ester" in not less than 10 point lettering.

Soy protein	5 g per serving	Following statement is required on the label: "Amount recommended to give the lowering effect on the blood cholesterol is 25 g per day"
Patented cooking oil blend	Ratio of fatty acid profiles for saturated/monounsaturated/polyunsaturated fatty acids must be 1:1:1	Patented cooking oil blend refers to US patent number 5,578,334 and 5,843,497

Modified from MHM (2010).

of any other nutrient for which a nutrition claim is made should also be included (Tee, 2007a). Examples of such nutrients are dietary fiber, omega-3 fatty acids, cholesterol, vitamins, and minerals. The proposed format for nutrition labeling follows the Codex Alimentarius guidelines (FAO-WHO, 1999). Vitamins and minerals can be only declared if their Nutrient Reference Value (NRV) is listed in the regulations and the amounts must be at least 5% of the NRV per serving (MHM, 2010).

18.4 PHILIPPINES

In the Philippines, one can find an assortment of natural health products. For example, bitter gourd or ampalaya is a vegetable that may lower blood glucose. Maunggay (horseradish tree) and kamote (sweet potato) are good sources of β-carotene: an antioxidant that scavenges oxidative radicals. These types of botanicals are potential bioactives for health foods. At least 10,000 plants have been documented and 2000 species are being subjected to research. Without being labeled as a functional food, some of these foods are recognized to have medicinal functions or health benefits beyond their basic nutritional value. Currently, there are several ongoing research projects in the Philippines studying the efficacy and application of bioactives in functional foods. Scientific validation of medicinal plants for health care is also being conducted. These multidisciplinary projects combine pharmacological and toxicological assessments, formulations, and manufacturing research (Mallillian and Bautista-Batallones, 2004).

18.4.1 FUNCTIONAL FOODS AND HEALTH CLAIMS

With the exception of Indonesia, the Philippines, as with other SEA countries, does not have a legal definition of functional foods. The country follows the Codex Alimentarius definition of functional foods (Codex, 2004). However, there is confusion associated with this term, since in the Philippines it encompasses not only foods with health claims but also fortified foods. Food fortification with vitamins and minerals is an important initiative of governmental policy and is regulated under the Philippine Food Fortification Act of 2000 (Republic Act, 2000). The government has implemented this act to make the fortification of certain staple foods mandatory in order to help combat the problem of malnutrition in the country.

Vitamin A deficiency is endemic to the Philippines. Therefore, under current regulations, micronutrient supplementation and food fortification with vitamin A is mandatory. This applies also to fortification with iron and iodine. There are well-established fortification programs regarding processed food products (Sangkap Pinoy Seal voluntary program) with vitamins and minerals, and staple foods (Diamond Seal mandatory program) such as rice, wheat flour, sugar, cooking oil, and salt with vitamin A and iron (Republic Act, 2000; Mallillian and Bautista-Batallones, 2004). Furthermore, the term functional foods is used to include dietary supplements in medicinal format. A dietary supplement is defined as follows:

> *A processed food product intended to supplement the diet bears or contains one or more of the following dietary ingredients: vitamin, mineral, amino acid, herb, or other dietary substance of botanical, animal, artificial or natural origin to increase the total daily in amounts conforming to the latest Philippine Recommended Energy and Nutrient Intake (RENI) or international daily requirements. It is usually in the form of capsules, tablets, liquid, gels, powders or pills and is not represented for use as a conventional food or as the sole item of a meal or diet or a replacement for drugs and medicines. (Food Safety Act, 2013)*

In 2009, the Philippines promulgated a new law: "the Food and Drug Administration Act of 2009." Based on this act, the government created the Food and Drug Administration (FDA) within the Department of Health (DOH). The FDA has amalgamated two existing DOH agencies: the Philippines Bureau of Foods and Drugs (BFAD) and the Bureau of Health Devices and Technology (BHDT). Subsequently, the FDA assumes BFAD functions over foods and is responsible for governing foods, including those with health claims (de Perio Peralta and Credo Matienzo, 2010). The Act also defines "health products" as "products that may have an effect on health which require regulations as determined by the FDA." This definition also includes health foods.

In 2004, the DOH issued a draft Administrative Order on "Rules and Regulations Governing Nutrition and Health Claims of Prepackaged Food Products Distributed in the Philippines" (Mallillian and Bautista-Batallones, 2004; Austrade, 2007). Afterward, in 2014, the rules and regulations were amended by the DOH (Administrative Order, 2014). The Philippines has adopted the Codex guidelines for use on nutrition and health claims and Codex general guidelines on claims (Codex, 2004). The Codex regulations were implemented in connection with the existing national laws on labeling and advertisement (Consumer Act, 1992; Administrative Order, 2014). Claims not covered by the aforementioned regulations are subject to evaluation based on submitted substantiation (Administrative Order, 2014). Applications for new health claims must be supported by references. For claims substantiating the risk reduction of disease, they must be supported by clinical data from well-designed human intervention studies, and published in peer-reviewed journals (Chan, 2011).

In the Philippines, nutrient content claims, nutrient comparative claims, nutrient function and other function claims, and reduction of disease risk claims are permitted (Chan, 2011).

Nutrient content claims containing the phraseology: "high," "rich," "good source," or "low" are permitted; however, the criteria for making these claims are different from those provided in the Codex guidelines. Approved claims relating to energy, protein, carbohydrates, fat and its components, fiber, sodium, vitamins, and minerals must have an established NRV. For example, "contains low sodium" means that food contains 120 mg or less sodium per 100 g. The NRV is based on the local reference, RENI, and it is used to give recommendations for energy and 21 nutrients (Barba and Cabrera, 2008; FNRI, 2014).

Health claims must be supported by scientific data, which shows a cause and effect relationship established between the consumption of a food or its constituent and the claimed effect. The Philippines does not have a positive list with health claims, however this country authorizes disease risk reduction claims according to the Codex Alimentarius (Tee, 2009). Currently, only two disease risk reduction claims are allowed in the Philippines: "calcium and osteoporosis risk reduction," and "food low in fat and some types of cancer risk reduction" (Zawistowski, 2011). Interestingly, food enriched with plant sterols, which is linked to reducing the risk of heart disease, is approved under the Philippines Food Fortification Act of 2000. However, the eligible claim on foods "this product contains plant sterols that help lower cholesterol" is a nutrient function claim (Zawistowski and Jones, 2015).

According to the Philippines DOH, probiotics are food supplements. The regulators approved the following probiotics species, which are beneficial to human health (BFAD, 2004):

- lactobacilli
- biofidobacteria
- nonpathogenic strains of *Streptococcus*
- *Saccharomyces boulardi*
- *Bacillus clausii*

Furthermore, the following probiotic claims on food labels and for use in advertisement are permitted:

- enhancement of intestinal ecology
- helping improve lactose malabsorption
- improving digestion
- aid to the enhancement of natural resistance to intestinal infections

It is worthwhile to note that probiotics can be also classified as drugs based on specific criteria. The key criterion is the nature of claims made on a product. For example, if a claim relates to altering disease or immune modulatory action, probiotics are classified as drugs with a therapeutic claim. Additional criteria used for classification are formats, such as food or medicinal forms and concentration per dosage unit. In the support of drug claims, a submission dossier to the FDA must contain scientific evidence based on human clinical studies (BFAD, 2004). The classification between drug and food supplement categories also applies to vitamins and mineral products. Drug claims on foods such as "therapeutic," "mitigation," and "treatment of diseases" are strictly prohibited.

18.4.2 NUTRITION LABELING

For a long time, nutrition labeling has been required on foods with nutrition and health claims but since Oct. 2014 it has been mandatory for all foods. Foods must bear the nutrition fact label, which includes calories, total fat, saturated fat, trans fat, cholesterol, sodium, total carbohydrates, dietary fiber, sugar, and protein (Alcaraz, 2015).

18.5 SINGAPORE

Singapore is one of the most developed countries in the world. With high standards of living and one of the world's highest life expectancies, it is no surprise that Singaporeans look for foods that are not only nutritious but also provide additional health benefits. Food products with health claims (functional foods) that state a benefit to well-being or risk reduction of chronic disease are popular on the market. This is particularly true due to an aging population with disposable income that doesn't hesitate to pay a premium price for these kinds of foods.

18.5.1 HEALTH FOODS AND REGULATORY FRAMEWORK

Health foods are regulated by the Agri-Food and Veterinary Authority of Singapore (AVA) or Health Sciences Authority. AVA is the regulatory agency responsible for safeguarding the import and sale of foods on the Singaporean market. In 2009, AVA together with the Health Promotion Board (HPB) initiated an approval of health claims for use on foods. In order to evaluate new health claims, the government of Singapore established the Advisory Committee on the Evaluation of Health Claims. The committee is composed of scientific experts from the government, universities, and consumer and industry associations. It was challenged by AVA to establish a framework for evaluation of health claims based on Codex Alimentarius recommendations. The Singaporean authorities published a guide to food labeling and advertisements in 2010, which was further amended between 2011 and 2015. With references to the Sale of Food Act and the Food Regulations (AVA, 2005), the guide aims to provide information on food labeling as well as health claims (AVA, 2015).

18.5.2 HEALTH CLAIMS AND NUTRITIONAL LABELING

The health claims are defined according to the Codex Alimentarius (Codex, 2004) which specifies "any representation that states, suggests or implies that a relationship exists between a food or a constituent of that food and health." Furthermore, claims are divided into three categories:

1. nutrient function claim—a nutrition claim that describes the physiological role of the nutrient in growth, development, and normal functions of the body.
2. other function claim—this claim concerns specific beneficial effects of the consumption of foods or their constituents, in the context of the total diet, on normal functions or biological activities of the body. Such a claim relates to a positive contribution to health or to the improvement of a function or to modifying or preserving health.
3. reduction of disease risk claim—a claim relating the consumption of a food or food constituent, in the context of the total diet, to the reduced risk of developing a disease or health-related condition. Risk reduction means significantly altering a major risk factor(s) for a disease or health-related condition. Diseases have multiple risk factors and altering one of those risk

factors may or may not have a beneficial effect. In Singapore, this category is named as nutrient-specific diet-related health claims.

Nutrient function claims are allowed on food products if foods meet the specific criteria (AVA, 2015). These include:

- the claim is about essential nutrients that have established recommended intakes and/or are of nutritional importance.
- there is sufficient generally accepted scientific evidence to prove the suggested function or role of the nutrient as claimed.
- the claim enables the public to understand the information provided and its significance to their overall daily diet.
- the particular nutrient mentioned is present in an amount that either meets the requirements of the Food Regulations, that is, 1/6 of daily allowances for vitamins and minerals per reference quantity, or the requirements of the nutrient claim guidelines established by the HPB. The product carrying the claim should also be labelled in accordance with the requirements of the Food Regulations for use of nutrition claims.
- the claim does not state or imply that the nutrient is for prevention or treatment of a disease.

Nutrient function claims are structure/function claims as exemplified by:

- calcium helps build strong bones and teeth (structure)
- fiber helps the digestive system (function).

AVA has published a list of acceptable nutrient function claims for foods (Table 18.3). This list includes claims for protein, lactose, dietary fiber, vitamins, minerals, and other food constituents such as collagen, probiotics, and prebiotics. Interestingly, there is more than one claim for some nutrients. For example, there are four folate claims on foods for pregnant women. Singapore authorities have approved five nutrient function claims on foods for infants and children below six years old. These claims link the roles of choline, DHA and ARA, nucleotides, taurine, and zinc to the mental and physical development of children (AVA, 2015).

Other function claims may only be used on foods when specifically approved by AVA as per an exact wording. These claims include bioactives such as collagen, inulin, probiotics, prebiotics, phytosterols, phytostanols, and their esters. Table 18.4 lists other function claims allowed under the Singapore regulations.

A health claim on foods with phytosterols/phytostanols represents a disease risk reduction claim. The Singaporean regulatory authorities approved this claim on phytosterols-containing vegetable fat or oil containing not more than 20% of saturated fat, margarine and fat spreads containing not more than 27% of saturated fat, and any other foods containing not more than 3%, and 1.5% of total fat in solid and liquid foods, respectively. Interestingly, the eligible products must bear a health claim that is similar to the permitted sterol claim in the EU on nutrition and health claims made on foods (EC, 2006). Food categories and labeling requirements are similar to the

Table 18.3 Acceptable Nutrient Function Claims in Singapore

Component	Health Claims
Claims specific for adults	
Protein	• Protein provides the essential amino acids needed to aid in the building and maintenance of body tissues • Protein helps in tissue building and growth
Lactose	• Low lactose content allows easier digestion/eases digestion for people who are lactose intolerant
Dietary fiber	• Aids the digestive system
Calcium	• Calcium helps build/to support development of strong bones and teeth • Calcium contributes to normal energy metabolism • Calcium is necessary for normal nerve and muscle function • Calcium is necessary for normal blood coagulation
Iodine	• Iodine is essential for the synthesis of thyroid hormones by the thyroid gland • Iodine is necessary for normal energy metabolism • Iodine contributes to normal cognitive function • Iodine contributes to the maintenance of normal skin
Iron	• Iron is an important component of red blood cells which carry oxygen to all parts of the body to help the body's production of energy • Iron is needed to produce haemoglobin, the protein in red blood cells that carries oxygen to tissues • Iron is needed to produce myoglobin, the protein that helps supply oxygen to muscle • Iron contributes to normal cognitive function • Iron contributes to normal cognitive development • Iron contributes to normal energy production • Iron contributes to the reduction of tiredness and fatigue • Iron is necessary for normal immune system function • Iron is necessary for normal cell division
Magnesium	• Magnesium helps in the absorption and retention of calcium • Magnesium contributes to energy metabolism and the maintenance of bone and teeth
Phosphorus	• Phosphorus contributes to bone development • Phosphorus contributes to normal energy metabolism • Phosphorus contributes to the maintenance of normal teeth
Zinc	• Essential for growth

(Continued)

Table 18.3 Acceptable Nutrient Function Claims in Singapore (*cont.*)

Component	Health Claims
Vitamin A	• Vitamin A is essential for the functioning of the eye • Vitamin A helps to maintain normal skin and mucous membrane • Vitamin A contributes to the normal function of the immune system
Vitamin B1 (thiamin)	• Vitamin B1 helps to release energy from proteins, fats, and carbohydrates • Thiamin contributes to normal functioning of the nervous system • Thiamin contributes to the normal function of the heart
Vitamin B2 (riboflavin)	• Vitamin B2 helps to release energy from proteins, fats, and carbohydrates • Riboflavin contributes to the reduction of tiredness and fatigue • Riboflavin contributes to the maintenance of normal skin • Riboflavin contributes to the maintenance of normal red blood cells • Riboflavin contributes to the maintenance of normal vision • Riboflavin contributes to normal functioning of the nervous system • Riboflavin contributes to the protection of cells from oxidative stress
Vitamin B3 (niacin)	• Vitamin B3 helps to release energy from proteins, fats, and carbohydrates • Niacin contributes to the reduction of tiredness and fatigue • Niacin contributes to the maintenance of normal skin • Niacin contributes to normal functioning of the nervous system
Vitamin B6(pyridoxine)	• Vitamin B6 helps to release energy from proteins, fats, and carbohydrates • Pyridoxine contributes to the reduction of tiredness and fatigue • Pyridoxine contributes to the maintenance of normal skin • Pyridoxine contributes to normal functioning of the nervous system • Vitamin B6 contributes to normal homocysteine metabolism • Vitamin B6 contributes to the regulation of hormonal activity

Table 18.3 Acceptable Nutrient Function Claims in Singapore (*cont.*)

Component	Health Claims
Vitamin B12 (cyanocobalamin)	• Vitamin B12 is necessary for fat, carbohydrate, and protein metabolism • Vitamin B12 is needed for/helps in the formation of red blood cells • Vitamin B12 contributes to the reduction of tiredness and fatigue • Vitamin B12 contributes to normal functioning of the nervous system • Vitamin B12 contributes to the normal function of the immune system • Vitamin B12 contributes to normal homocysteine metabolism
Folate	• Folate contributes to normal immune system function • Folate contributes to the reduction of tiredness and fatigue • Folate contributes to normal amino acid synthesis • Folate contributes to normal homocysteine metabolism
Folate (foods for pregnant women)	• Folate helps support fetus growth and overall development • Folate plays a role in the formation of red blood cells • Folate, taken before and during early pregnancy, helps in the mental/normal and overall development of fetus • Folic acid is essential/important for growth and division of cells
Vitamin C	• Vitamin C enhances absorption of iron from nonmeat products • Vitamin C contributes to normal collagen formation for the normal function of blood vessels • Vitamin C contributes to normal collagen formation for the normal function of bones • Vitamin C contributes to normal collagen formation for the normal function of cartilage • Vitamin C contributes to normal collagen formation for the normal function of gums • Vitamin C contributes to normal collagen formation for the normal function of skin • Vitamin C contributes to normal collagen formation for the normal function of teeth • Vitamin C contributes to the reduction of tiredness and fatigue • Vitamin C contributes to the normal function of the immune system • Vitamin C contributes to normal functioning of the nervous system • Vitamin C contributes to the protection of cells from oxidative stress

(*Continued*)

Table 18.3 Acceptable Nutrient Function Claims in Singapore (*cont.*)

Component	Health Claims
Vitamin D	• Vitamin D helps support calcium absorption and improves bone strength • Vitamin D helps the body utilize calcium and phosphorus • Vitamin D contributes to normal blood calcium levels • Vitamin D contributes to the maintenance of normal muscle function • Vitamin D contributes to the maintenance of normal teeth • Vitamin D contributes to the normal function of the immune system
Vitamin E	• Vitamin E is an antioxidant that helps protect cells in the body • Antioxidants like carotenes and Vitamin E help to protect cells from free radicals that may have escaped the natural processes of our body system
Vitamins K and D	• Work synergistically on bone metabolism to improve bone strength/build strong bones
Claims specific for infants and children up to 3 years old	
Choline	Helps support overall mental functioning
DHA and ARA	Are important building blocks for development of the brain and eyes in infant. (only for food for children up to 3 years of age)
Nucleotides	Are essential to normal cell function and replication, which are important for the overall growth and development of infant support body's natural defenses (only for infant formula targeting infants less than 1 year of age)
Taurine	Helps to support overall mental and physical development
Zinc	Helps in physical development
Prebiotic blend (GOS and long chain FOS, zinc, and iron)	Support the child's natural defenses

Modified from AVA (2015).

EU novel foods regulations (EC 333, 2004) while a health claim related to the effect of sterols on blood cholesterol, and the health claim wording are in line with the EU approved claim (EC 983, 2009; EC 384, 2010).

A label must contain the following mandatory statements:

- the product is intended exclusively for people who want to lower their blood cholesterol level;
- the product may not be nutritionally appropriate for pregnant and breastfeeding women and children under the age of five years;

- advice must be given that the product should be used as part of a balanced and varied diet, including regular consumption of fruit and vegetables to help maintain carotenoid levels;
- consumption of more than 3 g/day of added phytosterols does not provide any additional benefit in lowering blood cholesterol levels;
- consumption of at least 2 g of phytosterols/phytostanols per day has been shown to lower blood cholesterol levels;
- a definition of a portion of the food (in g or mL) with a statement of the amount of phytosterols that each portion contains (referred to as a serving).

Table 18.4 Acceptable Other Function Claims in Singapore

Component	Health Claims
Collagen	• Collagen is a protein in connective tissues found in skin, bones, and muscles
Inulin	• Inulin helps in calcium absorption
Probiotics	• Helps to maintain a healthy digestive system • Helps in digestion • Helps to maintain a desirable balance of beneficial bacteria in the digestive system • Helps to suppress/fight against harmful bacteria in the digestive system, thereby helping to maintain a healthy digestive system
Prebiotics	• Prebiotic promotes the growth of good *Bifidus* bacteria to help maintain a healthy digestive system • Inulin helps support growth of beneficial bacteria/good intestinal flora in the gut • Oligofructose stimulates the bifidobacteria, resulting in a significant increase of the beneficial bifidobacteria in the intestinal tract. At the same time, the presence of less desirable bacteria is significantly reduced • Inulin helps increase intestinal bifidobacteria and helps maintain a good intestinal environment
Phytosterols/phytostanols and their esters	• Plant sterols/stanols have been shown to lower/reduce blood cholesterol • High blood cholesterol is a risk factor in the development of coronary heart disease

Modified from AVA (2015).

The nutrient-specific diet-related health claims allowed on foods include five claims related to osteoporosis, heart disease, and cancer (AVA, 2015). All five claims are similar to the US FDA claims, which are approved under the Nutrition Labeling and Education Act (NLEA) (CFR, 2002). Table 18.5 shows this category of claims, principal ingredients, and conditions for using them on food products.

Table 18.5 Approved Nutrient-Specific Diet-Related Health Claims on Foods in Singapore

Principal Ingredient/ Disease	Health Claim	Conditions of Use
Calcium and vitamin D–osteoporosis	"A healthy diet with adequate calcium and vitamin D, with regular exercise, helps to achieve strong bones and may reduce the risk of osteoporosis. (*Name of food*) is a good source of/high in/enriched in/ fortified with calcium'	• The reference quantity of the product must contain at least 50% of calcium RDA (800 mg) • The product should also either be low in fat (≤3 g fat per 100 g or ≤1.5 g fat per 100 mL) or fat free (≤0.15 g fat per 100 g or 100 mL)
Sodium–stroke and heart disease	"A healthy diet low in sodium may reduce the risk of high blood pressure, a risk factor for stroke and heart disease. (*Name of food*) is sodium free/ low in/ very low in/ reduced in sodium'	• Salt/sodium free (≤5 mg sodium per 100 g) • Very low in salt/ sodium (≤40 mg per 100 g) • Low in sodium (≤120 mg per 100 g) • Reduced sodium (≤15% of sodium RDA (2 g) • The product should contain no added salt
Saturated fat and trans fat–heart disease	"A healthy diet low in saturated fat and trans fat, may reduce the risk of heart disease. (*Name of food*) is free of/ low in saturated fats, trans fats'	• Low in saturated fat (≤1.5 g saturated fat per 100 g, and ≤10% of kilocalories from saturated fat) • Free of saturated fat (≤0.5 g saturated fat per 100 g, and ≤1% of the total fat is trans fat, and is free of trans fat (<0.5 g per 100 g) • Product must be at least low in sugar (≤5 g per 100 g or ≤2.5 g per 100 mL), unsweetened, and has no added sugar • Contains ≤100 mg per 100 g cholesterol (1/3 of cholesterol daily allowance) • Contains ≤25% of sodium RDA (2 g)
Dietary fiber–heart disease	"A healthy diet rich in whole grains, fruits, and vegetables that contain dietary fiber, may reduce the risk of heart disease. (*Name of food*) is low/ free of fat and high in dietary fiber"	• Eligible food groups: whole grains, fruit, vegetables, or fiber-fortified foods • Low in fat or fat free • High in dietary fiber (≥3 g per 100 kcal or ≥6 g per 100 g or 100 mL) • Soluble fiber ≥25% of dietary fiber

Table 18.5 Approved Nutrient-Specific Diet-Related Health Claims on Foods in Singapore (*cont.*)

Principal Ingredient/ Disease	Health Claim	Conditions of Use
Dietary fiber–cancer	"A healthy diet rich in fiber containing foods such as whole grains, fruits, and vegetables may reduce the risk of some types of cancers. (*Name of food*) is free/ low in fat and high in dietary fiber"	• Eligible food groups: whole grains, fruit, vegetables, or fiber-fortified foods • Low in fat or fat free • High in dietary fiber (≥3 g per 100 kcal or ≥6 g per 100 g or 100 mL) • Soluble fiber ≥25% of dietary fiber

Modified from AVA (2015).

Mandatory nutrition labeling is required for all foods with health claims. The placement of a nutrition information panel on the label is governed by food regulations (AVA, 2015). The panel must include the serving size as well as a list of a core of four nutrients (energy, protein, fat, and carbohydrate) per serving size and per 100 g or 100 mL. Declaration of other nutrients is mandatory when such nutrients are the subject of a claim (AVA, 2015; MHS, 1998).

Making false, deceptive, or misleading claims that suggest therapeutic or prophylactic actions toward improving human health, or that suggest a food will prevent, alleviate, or cure any disease or conditions affecting the human body is strictly prohibited as per the Sale of Food Act (AVA, 2005).

18.6 THAILAND

Food regulation in Thailand has a 100-year tradition. The very first regulatory act regarding food safety was issued in 1908, while the first food-labeling requirement appeared in 1941 (Rimpeekool et al., 2015). In 1998, the Thai FDA introduced the first nutrition labeling laws and regulations based on the Codex guidelines (Codex, 2004). Since that time, foods with nutrition and health claims have been required to have mandatory nutrition labels (Rimpeekool et al., 2015). However, there is no official regulatory framework for making health claims in Thailand (MPI, 2015). Functional foods, however, are recognized by Thai consumers and are available on the market under different formats. Fermented milk and yogurt with probiotics, energy drinks, and fortified foods with vitamins and minerals are all good examples of functional foods. Methakornkulnan and coworkers conducted a recent survey involving 650 consumers in Bangkok. The results indicate that the functional food market is growing and such foods are a very attractive proposition to Thai consumers. Factors accounting for the increased popularity of this kind of food are: health benefits, particularly "like medicine effects"; safety; and innovation (Methakornkulnan et al., 2013). Foods with "medicine effects" are an essential part of Thai culture and have a long history of consumption. However, only recently have some of these foods been subject to research in an attempt to establish the lacking scientific justification for their health benefits (Charoenpong and Nitihamyong, 2004).

18.6.1 FUNCTIONAL FOODS AND REGULATORY FRAMEWORK

The Thai FDA categorizes functional foods with those that are necessary for special dietary uses. Regulators define functional foods as similar in appearance to conventional foods, but consumed as part of a normal diet with the purpose of attaining claimed health benefits including the possibility of reducing the risk of chronic diseases.

Since Thailand does not have specific regulation for functional foods, they face problem in differentiating these products between foods and drugs in order to decide whether a product is classified under the Drug Act or the Food Act. To classify a functional food as a drug or a food depends mainly on the type and concentration of active ingredients in the product and the claims made. To classify a product as a food, three features: safety, quality, and efficacy, must be used together for the assessment (Charoenpong and Nitihamyong, 2004). If a functional food is classified as a food, it must obey food regulations and must be manufactured according to food safety and quality standards (Charoenpong and Nitihamyong, 2004).

18.6.2 NUTRITION CLAIMS AND NUTRITION LABELING

Thailand has established nutrition claims as relating to energy, protein, fat, carbohydrate, vitamins, and minerals. These claims are based on Codex guidelines and to some extent on US FDA food regulations (MPH, 1998; Kongchuntuk, 2000). The Ministry of Public Health issued a notification No. 182 B.E. 2541, 1998, clause 2, which defines foods with nutrition claims as foods "which indicate nutrients information in nutrition labelling relate to type or quantity of energy or nutrients" (MPH, 1998). There are three categories of nutrition claims in the Thai regulations:

- nutrient content claims
- comparative claims
- nutrient function claims

Examples of nutrient content claims are: "source of calcium," "high in fiber," and "low in fat." Regulations prohibit making the claims: "free" or "low" if a food is naturally free of or low in that nutrient without being processed, adjusted, or formulated to improve nutrient content so as to reflect the claimed conditions.

Examples of comparative claims are: "less than or fewer," "more than," "reduced," "lite, light," and "added, fortified, enriched." To make these claims the claimed food is compared with the reference food. Two categories of reference foods are allowed in comparative claims:

- producer own products
- similar or the same kind of products, which are present on the national market.

The regulations prohibit using a reference food that already has a nutrient claim such as "low" or "less."

Examples of nutrient function claims are:

- calcium is an important component of bones and teeth
- calcium assists in the building process of bones and strong teeth
- folate is an important component of red cell formation
- vitamin B1 and vitamin B12 assist in function of the nervous system

To make a nutrient function claim, the following conditions must be applied:

1. the nutrient should have a Thai Recommended Daily Intake (Thai RDI) for ages of 6 years and up as described in the regulations (MPH, 1998, Appendix 3).
2. the nutrient content of the claimed product must be in the level of the reference serving size and the serving size as displayed on the label shall be classified as a "source' of that nutrient (MPH, 1998, Appendix 2).
3. reliable scientific data must be used in support of the claim.
4. a statement that the claimed food is able to prevent or cure diseases is strictly prohibited.

Currently, there is a positive list authorizing 29 nutrients for nutrition claims (Parinyasiri, 2015). Regulators may allow additional nutrition claims on a case-by-case basis (Chan, 2011). The Thai FDA may approve new claims upon review of a proper submission. Such a dossier should contain scientific substantiation including (Parinyasiri, 2015):

- systematic review, meta-analysis, or
- scientific opinion of a department, organization, or expert scientific bodies, or
- a full version of a well-designed human intervention study.

If a claim on food is approved in other countries but not in Thailand, the importer of such a food must apply to the Thai FDA for approval (MPI, 2015). A good example is the FDA approval of plant sterols under the notification for foods for special dietary uses. Foods with plant sterols are authorized to bear the nutrient function claim "may help lower cholesterol" (Zawistowski and Jones, 2015). Currently, the Thai FDA is considering extending the positive list of nutrient function claims based on industry applications with high-quality supportive evidence. Applications for health claim approval are reviewed by the expert committee and may be endorsed or rejected. Duration time for reviewing is about 1–1½ years (Parinyasiri, 2015).

Other function claims and disease risk reduction claims are not permitted under current food regulations; however, the health authorities are in the process of drafting a document on health claims. The regulation of health foods in support of functional foods will be based on two Codex Guidelines: Guidelines for Use of Nutrition and Health Claims (CAC/GL 23-1997, Rev.1-2004), and Recommendations on the Scientific Substantiation of Health Claims (Step 5/8, 2008) (Parinyasiri, 2015).

Thai food regulations require mandatory nutrition labeling for foods with health claims, regardless of whether a food product bears a nutrient content or nutrient function claim or some other health claim. In addition mandatory nutrition labels must be declared on the following categories of foods (MPH, 1998):

- foods with a comparative claim;
- fortified/enriched foods;
- foods with claims of specific benefits or functions to the body or specific ingredients;
- foods for specific target groups, for instance schoolchildren and the elderly;
- other foods prescribed by the FDA Office.

There are two categories of nutrition labeling (Hamano, 2014):

- full format, which include four core nutrients: energy, carbohydrate, protein, and fat as well as saturated fat, cholesterol, dietary fiber, sugar, sodium, vitamin A, vitamins B1 and B2, calcium, iron, and the claimed nutrient.
- simplified format, which include energy, carbohydrate, protein, and fat as well as sugar, sodium, and the claimed nutrient.

Nutritional labeling may be applied to other foods not mentioned earlier, under a condition that the stipulated format and regulations are observed. The regulations provide examples of the full format and brief format, which is similar to that of the US FDA (Tee et al., 2002).

18.7 BRUNEI DARUSSALAM

Brunei Darussalam is a small country situated in the northwestern part of the Island of Borneo surrounded by the South China Sea. The two larger countries adjacent to Brunei are Malaysia and Indonesia. These occupy the remainder of Borneo Island.

Brunei became a sovereign state in 1984, and since then, the country has experienced significant economic growth based on its natural resources of petroleum and natural gas. Brunei is ranked the fifth richest nation out of 182 countries according to Forbes (Too and Begawan, 2012). With rapid economic growth and development, the government of Brunei promotes education and health. Brunei's Ministry of Health (MOH) is responsible for the oversight and regulation of the health care industry within Brunei.

The MOH has established and developed a modern health care system, which is available for Bruneians; however, like in many other countries, Traditional/Complementary and Alternative Medicine (T/CAM) has its roots. Brunei's tradition and culture go back to the 14th century (Chua, 2010). 80 percent of its area is covered with tropical rain forest with a unique flora made up of over 5000 species of plants and 2000 species of trees, which serve as a diverse bio-resource for natural health care. According to a study performed by the Institute of Health Sciences at the University Brunei, 59% of Bruneians use traditional medicine in some form, considering this to be a safe practice due to long-established traditions and knowledge of natural sources (Nurolaini et al., 2014).

18.7.1 ALTERNATIVE MEDICINE AND REGULATORY FRAMEWORK

On May 26, 2008, the MOH Brunei Darussalam established a separate T/CAM unit under the Department of Medical Services. This emphasizes the important role of T/CAM in the modern holistic health care system of Brunei. Traditional medicine continues to contribute positively in promoting wellness to the population of Brunei. This strategy is in line with global trends and with the position of the World Health Organization (WHO), which encourages all member states to support T/CAM while at the same time encouraging continued evaluation, and formulation of national policies and practices adequate to specific national health care systems. Nowadays the medical practitioners and healers in Brunei are allowed to practice their traditional alternative medicines within the legal framework and guidelines endorsed by the Brunei MOH. With government support, traditional medicine in Brunei, like in other SEA countries, plays an important role and is complementary to the modern health care system (Chua, 2010).

18.7.2 HEALTH SUPPLEMENTS AS FUNCTIONAL FOODS

In 2000, the MOH promulgated the Public Health (Food) Act (Chapter 182), to make provisions to regulate public health in respect to food. The act was revised on Feb. 1, 2001. The Act defines food as "any substance manufactured, processed, sold or represented for human consumption, chewing substances, and any ingredient thereof" and it also governs food labeling (Food Act, 2000). Although the definition encompasses many food products, the Act does not specifically refer to health foods. However, the guidance document (GD24, 2010) issued in May 2010 defines health supplements as "any product that is used to supplement a diet and to maintain, enhance and improve the healthy function of human body." A supplement, which is related to functional foods, contains one or more or a combination of:

1. vitamins, minerals, amino acids, fatty acids, enzymes, probiotics, and other bioactive substances;
2. substances derived from natural sources, including animal, mineral, and botanical materials in the forms of extracts, isolates, concentrate, and metabolite;
3. synthetic sources of ingredients mentioned in (1) and (2) may only be used where the safety of these has been proven.

Supplements are presented in dosage forms such as capsules, tablets, powder, and liquids. It is worthwhile to note that the definition and format of Brunei's health supplements are similar to Canadian natural health products.

Brunei's MOH is responsible for issuing regulations and guidelines for food, pharmaceuticals, traditional medicine, and health supplements. Acts and guidelines include (Chua, 2010):

- The Medicine Order, enforced by Medicines Control Authority, MOH
- Poisons Act 1956 (Cap 114), which regulates the importation, possession, manufacture, compounding, storage, transport, and sale of poisons, enforced by the Pharmacy Enforcement Unit

- The Misuse of Drugs Act (Chapter 27), enforced by the Narcotics Control Bureau, Prime Minister's Department
- The Misuse of Drugs Regulations which regulate the manufacture, possession, supply, import, and export of controlled drugs, jointly enforced by the Narcotics Control Bureau, Prime Minister's Department and Pharmaceutical Services, MOH
- Guidelines for dealing with traditional medicines, homeopathic medicines, and health supplements, enforced by the Pharmacy Enforcement Unit, Pharmaceutical Services, MOH.

The MOH also oversees all regulatory activities related to manufacturing; release and distribution; and the postmarketing follow-up of pharmaceuticals, traditional medicines, and health supplements. Through participation in networking, cooperation, and collaboration with all SEA members as well as internationally, the MOH creates opportunities for Brunei to strengthen collaboration between developed and developing countries.

18.7.3 NUTRITION CLAIMS AND NUTRITION LABELING

Brunei is among the SEA countries, including Laos, Philippines, and Thailand, that do not permit disease risk reduction and other function health claims (Kasapila and Shaarani, 2011). However, nutrient content claims are allowed. Following the Codex guidelines, nutrition claims in respect to energy, protein, carbohydrate, fat, and other nutrients are permitted (Chan, 2011). Brunei has a regulatory system for the approval of new nutrition claims. It is done on a case-by-case basis through formal applications to the Director General of Health Services (Chan, 2011). According to a more recent report, imported and local foods are permitted to carry nutrition claims provided they exhibit a label giving nutrition information (ILSI, 2013). Nutrition claims stating: "source of" and "enriched with" for vitamins and minerals are permitted on foods as long as these foods contain the associated nutrients equal to a specific percentage of the daily allowance value (ILSI, 2013). Furthermore, the MOH is planning to harmonize regulations and bring them in line with international standards.

The Brunei MOH sets forth regulations on food labeling in their Public Health (Food) Act (Food Act, 2000). In this country nutrition labeling is voluntary unless nutritional claims are made on food products. In such a case, the labeling is based on Codex recommendations. Energy and nutrients are expressed per serving sizes of 100 g or 100 mL of food (Kasapila and Shaarani, 2011). Local reference values must be used in the declaration of micronutrients (Keat, 2015).

18.8 CAMBODIA

Based on 2013 data, the Kingdom of Cambodia has a population of approximately 15.4 million. The country's primary industry is agriculture. About 80% of inhabitants live in rural areas and earn their livelihood from agriculture. The agricultural sector is composed of crop production (53.8%), fisheries (27.3%), livestock (12.8%), and forestry (6.1%) (FAO, 2011). Food production is dominated by the "cottage industry" represented by thousands of household small and medium-sized enterprises (SMEs). Since the modern processed food sector is still in its infancy, it is not surprising that the production of functional foods is so far negligible. Yet, there is a great potential for manufacturing these kind of foods. This can be rationalized by observing the abundance of fish, seafood, and fruit. In 2003, agriculture accounted for 40% of GDP (Soeun, 2004).

18.8.1 FUNCTIONAL FOODS AND NATURAL HEALTH PRODUCTS

Indigenous Cambodian fruit has gained increased attention from research institutions as a prospective source of functional food ingredients (Sokneang, 2013).

Noni fruits (*Morinda citrifolia*), also known as Indian mulberry, are indigenous to SEA, Australasia, Hawaii, and Tahiti. Traditionally, people of these regions have been using noni fruits for healing and prevention of various diseases such as cancer, arthritis, and hypertension. In the late 1990s, the processing of noni fruits for functional foods was born. Currently, noni juice and powdered noni fruit are manufactured in Tahiti, Japan, India, China, Europe, and the United States. Furthermore, noni juice is approved under Article 3(1) of the novel food regulations in the EU (EC 258, 1997), which gives it a green light for marketing in Europe. It should be emphasized that the EU authorities have not evaluated or validated possible health claims associated with noni products.

Noni fruits and their health benefits are well known in Cambodia. It is no surprise that in 2003, a Cambodian company named Capa Ltd. established processing for noni fruits, and decided to join other noni world enterprises. The company produces health food products including noni juice, vinegar, wine, powder, and tablets. Varieties of other fruits are processed by SMEs in the country (Sokneang, 2013).

Southeast Asia, including Cambodia, hosts a number of indigenous fruit. Noni, Jamun (*Syzygium cuminii*), tamarind (*Tamarindus indica*), Aonla (*Phyllanthus emblica*), banana, and mango can offer great nutritional value and health benefits. However, research projects are conducted on Cambodian fruits within limited capacities due to lack of proper financial support. In addition, there isn't a regulatory framework or national policy to govern the true health benefits of foods and discriminate them from misleading claims (Sokneang, 2013).

18.8.2 REGULATIONS OF TRADITIONAL MEDICINE

In Cambodia, like in other SEA countries, natural herbal products are the most widely used traditional medicines. According to a WHO survey conducted in 2005 the national policy on Traditional Medicines/Complementary and Alternative Medicines (TM/CAM) was issued in 1996. Regulations of herbal medicines, on the other hand, were introduced in 1998 (WHO, 2005a). However, there is currently no national program for governing these products. In addition, there isn't a national pharmacopoeia with a list of herbal products. There are, however, some monographs such as "Cambodia's Medicinal Plants" and "Your Medicines in Your Garden," which provide some information about plants and their pharmacological effects. Although the Ministry of Health has accepted these publications, their legal status is not clear (WHO, 2005a). The Cambodian Department of Drugs and Food regulates licenses for manufacturing and registration of sales of herbal medicine, but there is no requirement or follow-up of postmarketing or monitoring adverse effects. Herbal drugs are sold as over-the-counter (OTC) medications in pharmacies and special outlets without many restrictions (WHO, 2005a).

18.8.3 REGULATORY FRAMEWORK AND ADOPTION OF FOOD STANDARDS

There are no established laws and regulations to govern functional foods or health foods in Cambodia. However, the existing legal framework for conventional foods may apply to these categories.

The Cambodian authorities commonly issue Prakas for regulatory purposes. Prakas are ministerial or interministerial proclamations (decisions) that are signed by the Minister. They must conform to the Cambodian Constitution and to the law.

Prakas 868 is based on an agreement between multiple agencies and interministerial units regarding the implementation of and institutional arrangements on food safety management (Prakas 868, 2010). The document defines food as "any substance or product, which is intended to be ingested by humans including drink, chewing gum and any substance for manufacture, preparation or treatment." This definition excludes pharmaceuticals. In principle, this is similar to the definition of food as per Codex Alimentarius, the EU, and Canada. It is not clear, however, if health foods and natural health products will fit this definition. It is particularly confusing since Cambodian law defines traditional medicine as "nutrients used to prevent or cure human and animal diseases, including: a product that is mixed or produced by a single or several types of nutrients and is originated from plant, animal or mineral" (Sub-Decree No. 23, 1998). The latter describes a natural health product.

The objective of Prakas 868 is to improve the implementation of food safety and protect consumer health. It gives mandates to six ministries to coordinate activities regarding food safety. Safety, quality, food standards, and current practices are also addressed in the following laws and regulations (Sideth, 2014):

- law on the management of quality and safety of products and services (Jun. 26, 2000),
- law on management of pesticides and fertilizers (Jun. 14, 2012),
- law on Cambodia Standards (Jun. 25, 2007),
- law on Fishery (May 21, 2006),
- sub-decree on Food Hygiene for human consumption (Jun. 12, 2003).

Government policy is to harmonize the Cambodian legal framework with international and regional standards as set up by other countries of the ASEAN to which

Cambodia belongs. The process of harmonizing Cambodia's national standards with ASEAN standards is a stepwise process and is critical for the improvement of food regulations in this country (Theng, 2014).

To enhance the competency of food inspections and law enforcement, line ministries agree with adopting Codex General/guidelines as national standards, but the adoption procedure is still unclear.

In July 2015, the Commerce Ministry released a draft of the law on food safety. This is Cambodia's first law of this kind, that establishes a Food Safety Authority charged with protecting consumer health and ensuring the safety of foods. The authorities are responsible for the regulation of foods, making certain that products meet international standards. With the introduction of the new law, there is an opportunity for developing more regulations under the Food Safety Authority (Retka and Chhengpor, 2015). One may hope that the new Authority establishes a regulatory framework for governing health foods and food supplements.

18.8.4 LABELING AND HEALTH CLAIMS

Cambodia, like most SEA countries, has followed Codex guidelines in developing food labeling regulations. The labeling of foods is regulated by Prakas on Cambodian standard CS 001-2000: labeling of food products (Prakas CS 001, 2000). The law requires basic mandatory labeling of all prepackaged foods. However, there is no provision for nutritional labeling or use of any health-related claims. It is noteworthy that the aforementioned regulation prohibits a label claim stating that foods are "the medicaments for curing illness except the mineral water only" (Prakas CS 001, 2000). A couple of publications have mentioned, in general terms, that nutrition labeling in Cambodia is voluntary unless nutritional claims (Kasapila and Shaarani, 2011) or health claims are made on products as defined in the Guidelines for Use of Nutrition and Health Claims (CAC/GL 23-1997) (Hamano, 2014).

A survey was recently conducted by ILSI among 10 SEA countries including Cambodia (Isabelle et al., 2011). The survey regards information on food regulations, and has published regulatory documents and references. Although primarily focused on food enrichment, the survey contains questions related to health claims permitted in the surveyed countries: nutrient content claims, nutrient comparative claims, function claims, disease risk reduction claims, and conditions for their uses. The survey shows that Cambodia has a mandatory food fortification program aimed at the iodization of salt. However, the survey also reveals that there are no regulations regarding health claims in this country (Isabelle et al., 2011).

18.9 VIETNAM

Vietnam is located on the eastern part of the Indochinese peninsula and borders China to the north, Cambodia, Laos, and the Gulf of Thailand to the west, and the Gulf of Tonkin as a part of the South China Sea to the east. The country's geography and climate play an important role in the Vietnamese diet (Ferro, 1999). Long coastlines and many inland rivers provide fish and shellfish. Wetlands that are critical for rice fields provide rice. Aquatic foods and rice are staples in the Vietnamese diet and account for healthy eating.

Similar to other countries in the SEA region, Vietnam has a long-standing tradition of natural health products that accounts for practicing medicine. It utilizes various foods, herbs, and minerals as well as acupuncture. Vietnamese medical practices are based in Chinese Traditional Medicine (TCM) due to several thousands of years of Chinese influence (Dung, 2001). Thuoc Bac (Northern Medicine) is based on TCM and Yin–Yang interactions. Ginseng is the most used plant for maintaining health and curing diseases (Lopez, 2016).

18.9.1 HERBAL HEALTH PRODUCTS

Agricultural areas were always a significant part of the country. Currently, 67% of Vietnam's population is engaged in agricultural activities (World Bank, 2015). Therefore it is no surprise that rural populations are self-sufficient and rely on herbal treatment of their most common illnesses. The curative properties of most herbal and other health products are only supported by anecdotal data. However, there are 1573 registered herbal remedies in Vietnam and 267 herbal medicines, which are acknowledged as national essential medicines (WHO, 2005b). In addition, new remedies must be supported by records of clinical trials. Implementation of these requirements is ensured by the registration system. Herbal health products are governed by national law and regulations on herbal medicines, which were promulgated in 1989 separately from regulations regarding conventional pharmaceuticals. The Department of Traditional Medicine, under the Ministry of Health, has administrative responsibilities over these products. Herbal medicines are sold as prescription or OTC products. By law, medical, health, and nutrient content claims may be made (WHO, 2005b).

Furthermore, the Vietnamese are very keen on purchasing health foods. According to the Vietnam Supplement Food Association, 56% of people in Hanoi and 48% in Ho Chi Minh City consume functional food (VietNamNet, 2014).

18.9.2 FUNCTIONAL FOODS AND REGULATORY FRAMEWORK

The Vietnamese MOH is the governmental ministry jurisdictionally responsible for development, promulgation, and implementation of national strategies on food and food safety. Together with other Ministries: the Ministry of Agriculture and Rural Development and the Ministry of Industry and Trade, the MOH is responsible for regulating prepackaged and processed foods (GAIN, 2013). In conjunction with other ministries and the prime minister's office, the Ministry is responsible for creating and promulgating long-term health policy programs such as the "National Strategy on Nutrition for the 2001–2010 period" and the "National Policy on Injury Prevention 2002–2010" (MOH, 2016).

In 2014, the MOH issued regulations on management of functional foods (Circular, 2014). The term functional foods includes food supplements, health protection foods, foods for special medical purposes, and foods for special diets. All subgroups are defined as follows (Circular, 2014):

- food supplement (supplemented food) is a common food with micronutrient supplementation such as vitamins, minerals, amino acids, fatty acids, enzymes, probiotics, prebiotics, and other active substances conducive to health.
- health protection food (health supplement, food supplement, or dietary supplement) is a product in the form of capsules, pills, tablets, powder, liquid, and other forms containing one or a mixture of the following compounds:
  - vitamins, minerals, amino acids, fatty acids, enzymes, probiotics, and other bioactives;
  - natural bioactive ingredients originating from animal, mineral, and vegetable in the form of extracts, isolates, and concentrates.
- food for special medical purposes (nutritious medical foods, medical foods) is food that can be consumed orally or with catheters. It should be used only to adjust the patient's diet and can only be administered under the supervision of medical staff.
- food for special diets (dietary food for special uses) is food for dieters, the elderly, and other special subjects based on the recommendation by the Commission of International Food Standards (Codex, 2004). This is formulated food to meet the requirements of specific diets under physical or medical conditions and specific disorders of the user. The composition of these foods must be distinctly different to the composition of ordinary foods of the same nature, if any.

Functional foods, including imported foods, must comply with food safety regulations and law promulgated in the circulars by the MOH: Articles 6 and 7 of Decree No. 38/2012/ND-CP of Apr. 25, 2012 on the food safety law and the articles 4, 5, 7, and 9 of Circular No. 19/2012/TT-BYT of Nov. 9, 2012 on food safety regulations (Decree, 2012; Circular, 2012).

Functional foods that support functions of the human body, increase resistance, and reduce the risk of disease, must have proven efficacy supported by clinical studies published in recognized scientific journals (Circular, 2004, 2014). Clinical data must link food composition to its effect on functionality (Circular, 2004). The main active ingredient of the food must be well characterized (Circular, 2014). The food must be periodically tested for conformity and comply with the provision of Circular No. 19/2012/TT-BYT (Circular, 2012). Promotion of functional foods must be adherent to the Vietnamese law on advertising. A disclaimer "this product is not a medicine and is not intent to replace ineffective medicine" must be stated in the promotional material (Circular, 2014).

With the introduction of functional food regulations, the market for this food category is booming in Vietnam. There is an increasingly high demand for products claiming weight maintenance, improvement of cardiovascular health, and improvement of beauty. The products of international companies such as Amway, NuSkin, and Unicity account for a significant share of the Vietnamese market. It is predicted the food market will grow 20%, while the food supplements market will grow by 9–10% in a couple of years from now (VietNamNet, 2014).

18.9.3 HEALTH CLAIMS AND NUTRITIONAL LABELING

Functional foods like other prepackaged foods must comply with food labeling regulations promulgated by the Vietnamese authority (Joint Circular, 2014). Functional foods must bear the statement "functional foods," and must not imply that they can replace medicine (Russin and Vecchi, 2014).

In Vietnamese legislation terms, nutrition and health claims are based on Codex International Food Standards (Codex, 2004). The following label claims are allowable in Vietnam (Chan, 2011; Tee, 2015):

- nutrient content and comparative claims as related to the presence of vitamins, minerals, and other nutrients.
- nutrient function claims as related to several bioactives including dietary fiber, nondigestible oligosaccharides, phytosterols, and polyunsaturated fatty acids. Although there is no positive list of permitted claims, industry or distributors may apply for approval to the MOH for additional claims.
- disease risk reduction claims are not yet permitted.

Functional foods (food supplements) with added vitamins, minerals, amino acids, fatty acids, enzymes, probiotics, prebiotics, or other bioactive substances must refer to the recommended nutrient intakes (RNI) for each group as specified in Appendix 1 of Circular 43/2014 (Circular, 2014). Health claims are only permitted if the content of nutrients in the food is 10% RNI or more, and this is proven by scientific evidence (Circular, 2014).

Food for special medical purposes including food for pregnant women, infant formulas, and intravenous feeding should display a dosage applicable for each user group in accordance with the nutritional demand and physician's instructions (Anh, 2012). A disclosure "use only under supervision of medical staff" should be displayed on the label (Circular, 2014).

Currently, nutrition labeling on foods in Vietnam is voluntary and follows Codex recommendations (Ngoc, 2015). In 2014, the regulatory authority published guidelines for permitted nutrient content claims and comparative claims (Joint Circular, 2014). This is a positive list that includes: energy, fat and saturated fat, cholesterol, sugar, sodium, protein, vitamins and minerals, and fiber (Joint Circular, 2014). Conditions for permitted nutrition claims are shown in Table 18.6.

18.10 LAOS

The official name of Laos is Lao People's Democratic Republic (Lao PDR). Laos is a landlocked country in SEA occupying the northwest part of the Indochinese peninsula. It borders Vietnam to the east, Thailand to the west, China and Myanmar to the north, and Cambodia to the south.

Table 18.6 Conditions for Nutrient Content and Comparative Nutrition Claims in Vietnam

Component	Claim	Conditions
Energy	Low calories	4 kcal (170kj) in 100 g (solid) or 20 kcal (80kj) in 100 mL (liquid)
	No calories	4 kcal in 100 mL (liquid)
Fat	Low fat	3 g in 100 g (solid) or 1.5g in 100 mL (liquid)
	No fat	0.5 g in 100 g (solid) or 100 mL (liquid)
Saturated fat	Low saturated fat	1.5 g in 100 g (solid) or 0.75 g in 100 mL (liquid) and providing 10% of the energy from saturated fat
	No saturated fat	0.1 g in 100 g (solid) or 0.1 g in 100 mL (liquid)
Cholesterol	No cholesterol	5 mg in 100 g (solid) or 5 mg in 100 mL (liquid). Smaller than 1.5 g of saturated fat in 100 g (solid) or 0.75 g of saturated fat in 100 mL (liquid) and providing 10% of the energy from saturated fat
Sugar	No sugar	0.5 g in 100 g (solid) or 0.5 g in 100 mL (liquid)
Sodium	Low sodium	120 mg in 100 g
	Very low sodium	40 mg in 100 g
	No sodium	5 mg in 100 g
Protein	As supplements	Providing 10% of nutrient reference value in 100 g (solid) or 5% of nutrient reference value in 100 mL (liquid) Providing 5% of nutrient reference value for 100 kcal (12% nutrient reference value for 1 MJ) or providing 10% of nutrient reference value for a serving
	High protein	Twice as much as the value of protein supplements
Vitamins and minerals	As supplements	Providing 5% of nutrient reference value in 100 g (solid) or 7.5% of nutrient reference value in 100 mL (liquid) Providing 5% of nutrient reference value for 100 kcal (12% nutrient reference value for 1 MJ) or providing 15% of nutrient reference value for a single serving
	High vitamins and minerals	Twice as much as the value of vitamin and mineral supplements
Fiber	As supplements	3 g in 100 g or 1.5 g in 100 kcal, or providing 10% of daily nutrient reference value for a single serving
	High fiber	6 g for 100 g or 3 g for 100 kcal, or providing 20% of daily nutrient reference value for a single serving

Adapted from Appendix II of Joint Circular (2014).

The major crop of Laos is rice, which is cultivated on rain-fed and irrigated paddy areas. Coffee and maize are other significant crops. In addition, vegetables, tropical fruits, peanuts, roots, and tubers are also locally produced. Forest covers over 60% of Laos's territories, and provides wild foods and herbs. Despite an abundance of agricultural resources, there is a limited production of manufactured food. Laos, therefore, relies on importing processed food (FAO-WHO, 2002).

Since the introduction of economic reforms in 1986, Laos has undergone socioeconomic change, which is responsible for significantly improving the nation's health status.

18.10.1 FOOD AND HEALTH REGULATORY FRAMEWORK

The health sector is governed by the MOH, which is responsible for the implementation and administration of policy and law promulgated by the Law on Health Care of 2005 (WHO, 2009). In 2009, MOH introduced the National Food Safety Policy (NFSP) for the control and management of quality and safety of foods based on science and a risk approach to protect consumer health and ensure fair trade. The policy was adopted by the Prime Minister, at the time, and involves other relevant ministries and departments. The emphasis of said policy is on strengthening food safety management systems; adaptation and implementation of food standards that are in agreement with Codex Alimentarius, WHO, and the ASEAN policy; and training regulators and food inspectors (NFSP, 2009).

In Laos, the MOH is the key ministry with legislative power over various aspects of food. Duties include policy making, improving regulations and measures related to foods as disseminated in the Law on Food of 2004, which was amended in 2013 (National Assembly, 2004). The food law defines food as "any substance that is consumed or drunk by human beings in the form of food that is fresh, cooked, raw or processed, except drugs." It also defines three groups of foods (National Assembly, 2004):

- *carbohydrates*: food that provides energy, such as: rice, animal and plant oils, flour, sugar, and others that create energy;
- *proteins*: a food group that helps to build up body growth and immunity to germs, such as: meat, fish, milk, eggs, and various types of beans;
- *vitamins and minerals*: that is necessary for physical and mental development and that helps to increase physical immunity against germs, such as: vegetables, fruits, vitamins, iodine, and other minerals.

The main focus of the Law on Food is food safety and food inspection, which aims to control the quality and safety standards, and protect public health. Pursuant to the Food Law, there are a number of legislative documents related to food regulations as follows:

- ministerial regulations on the basic principles in the application of sanitary and technical measure for the food safety management No. 518/MoH (MOH, 2009a);
- decision on food inspection No. 297/MOH (MOH, 2012);
- regulation on the control, on production, export-import safe food No. 856/MoH (MOH, 2006);
- regulation on labeling of prepackaged food No. 159/MoH (MOH, 2009b);

- provisions on: the control of imported-exported foods No. 035/FDA, quality control of domestically food products No. 048/FDA, quality control of domestically circulated foodstuff No. 105/FDA (FAO-WHO, 2002);
- decree on universal salt iodination No. 42/PM (FAO-WHO, 2002).

18.10.2 HEALTH CLAIMS AND NUTRITION LABELING

Laos as with most of the ASEAN countries follows Codex guidelines in developing national regulations. Nutrition claims such as nutrient content claims, nutrient comparative claims, and nutrient function claims may be allowed as long as they meet "the minimum criteria set by Codex Guidelines for the Use of Nutrition and Health Claims" (MOH, 2009b). Laotian regulations define claims in general as: "any information which states, suggests or implies that a food has particular qualities relating to its origin, nutritional properties, nature, processing, composition or any other quality." There are, however, no specific directives on governing health claims, functional foods, or health foods (MOH, 2009b). If nutrition claims are made on foods, nutrient information is mandatory and should include a declaration of energy, protein, fat, and carbohydrate content. Information on energy must be expressed in kcal per 100 g or 100 mL, while all three types of nutrients should be stated in g per 100 g or 100 mL of food product. Vitamins and minerals should be declared in metric units and/or as a percentage of the NRV per 100 g or per 100 mL. Food claims referring to prevention, alleviation, treatment, or cure of a disease, disorder, or other physiological condition are prohibited (MOH, 2009b).

A noteworthy fact is that in 2000, the Laotian government promulgated the national policy on traditional medicine (TM) as a part of its drug policy. TM is derived from natural sources including medicinal plants, trees, animals, and minerals and its composition resembles natural health products and health supplements. In Laos, both TM and health supplements are regulated by the Law on drugs and medical products (National Assembly, 2011).

18.11 MYANMAR

The Union of Myanmar (Burma) is the largest country by territorial area in mainland SEA. With a population of over 52 million people, Myanmar is the fourth largest country of the ASEAN. It is considered a "least developed country" on the human development index by the United Nations (UNDP, 2014). The country borders Bangladesh to the west, India to the north-east, China to the north-east, Laos to the east, Thailand to the south-east, and the Andaman Sea to the south. Myanmar is a rural country. About 63% of the working-age population is employed in the agricultural sector. The country's economy is based on agriculture and natural resources (FAO, 2011). The major agricultural crops are rice, pulses, beans, sesame, groundnuts, sugarcane, lumber, and fish. Agricultural commodities contribute approximately 23% of export revenues (FAO, 2011).

18.11.1 FOOD AND HEALTH REGULATORY FRAMEWORK

The Myanmar MOH is responsible for policy making and administration of health affairs, including public health, health care services, occupational health, disease control, medical research, and drugs and foods. There are seven departments within the Ministry:

- Department of Health
- Department of Health Planning
- Department of Medical Sciences
- Departments of Medical Research for Lower Myanmar, and Upper Myanmar
- Department of Traditional Medicine
- Department of Food and Drug Administration

The Department of FDA is a national agency, which was established in 1995 under the provision of the National Drug Law of 1992. The agency oversees safety and quality of food, drugs, medical devices, and cosmetics. The FDA has two divisions: the Food Division and Drug Division, and each of them have enforcement and laboratory units. Food Division responsibilities include regulatory affairs, food safety and food safety management systems, food safety training, research and development, premarket and postmarket surveillance of domestic and imported foods, food inspection and analysis, as well as manufacturing plant certification.

In 1992, the Myanmar Food and Drug Board Authority was created under the Ministry of Health. The Authority is empowered by the Ministry to develop policies, guidelines, and procedures related to various aspects of foods as enacted by the National Food Law (Council, 1997). The Food Law was promulgated in 1997 with its aims to protect public health, and systematically control and regulate the production, import, export, storage, distribution, and sale of foods (Council, 1997).

18.11.2 FOOD CLAIMS AND NUTRITIONAL LABELING

In Myanmar, there are no regulations for functional foods. Although the National Food Law has provisions for food labeling, there are no specifics on nutrition labeling. However, there are plans to modernize the current law according to international standards such as Codex Alimentarius. This includes nutritional labeling (Kyaw, 2015).

In addition, there is no law and regulation related to nutrition and health claims on foods (Isabelle et al., 2011). Mandatory salt ionization is the only provision made regarding fortification in Myanmar (Isabelle et al., 2011).

In 1996, the State Law and Order Restoration Council enacted the Traditional Drug Law. The aims of the law are to promote traditional and herbal medicine, control the safety, quality, and efficacy of products, control manufacturing of products, and oversee their registration. Most traditional medicines are natural health products, based primarily on botanical plants. This is a separate category from regular drugs. Safety and efficacy assessments of these products are required by law and enforced through registration, inspection, and laboratory analysis (Council, 1996). In 2000, the Department of Traditional Medicine compiled and published a monograph of medicinal plants of Myanmar (DTM, 2000).

ACKNOWLEDGMENTS

My sincere gratitude and thanks to my wife Ursula and son Robert for the continuous support and helpful suggestions during writing this chapter.

REFERENCES

Act 7, 1996. Act of the Republic of Indonesia No. 7 of 1996 on food. The President of the Republic of Indonesia, Jakarta, Indonesia.

Administrative Order, 2014. Revised rules and regulations governing the labeling of prepackaged food products further amending certain provisions of Administrative Order No. 88-B s. 1984 or the "Rules and regulations governing the labeling of prepackaged food products distributed in the Philippines," and for other purposes. Administrative Order No. 2014-0030 dated September 08, 2014, Department of Health, Office of the Secretary, Republic of Philippines, Manila, Philippines.

Aggett, P.I., Antoine, J.M., Asp, N.G., Bellisle, F., Contor, L., Cummings, J.H., Howlett, J., Müller, D.J.G., Persin, C., Pijls, L.T.J., Rechkemmerl, G., Tuijtelaars, S., Verhagen, H., 2005. Passclaim process for the assessment of scientific support for claims on foods. Consensus on criteria. Eur. J. Nutr., Supplement 1 44:I/3-I/30.

Ahmed, E.M., 2012. Malaysia's food manufacturing industries productivity determinants. Mod. Econ. 3, 444–453.

Alcaraz, H., 2015. FDA Philippines update. 9th Seminar on Nutrition Labeling, Claims and Communication Strategies, August 4-5, 2015, ILSI SEA Region, Manila, Philippines.

Anh, B.H.M., 2012. Nutrition labeling and claims Vietnam. Seminar on Nutrition Labeling and Claims Regulations. August 6-8, 2012, ILSI SEA Region, Bangkok, Thailand.

ASEAN, 2014. White paper on harmonisation in the food sector prepared for senior economic officials meeting, ASEAN Asian Food and Beverage Alliance.

Austrade, 2007. Wellbeing Philippines, health biotechnology and wellbeing. Australian Trade Commission.

AVA, 2005. Sale of Food Act. Chapter 283. Previous versions of 1973, 1985, 2002. Amended in 2015 as Sale of Food Act (Chapter 283) Food (Amendment) Regulations 2015 by the Minister for National Development. Agri-Food and Veterinary Services, Singapore.

AVA, 2015. A guide to food labelling and advertisements. Agri-Food & Veterinary Authority, First published February 2010. Amendments October 2011, October 2013, July 2014, January 2015 and July 2015, Singapore.

Barba, C.V.C., Cabrera, M.I.Z., 2008. Recommended energy and nutrient intakes for Filipinos 2002. Asia Pac. J. Clin. Nutr. 17, 399–404.

BFAD, 2004. Guidelines on probiotics. Circular No. 16 s., Department of Health Bureau of Food and Drugs, Republic of the Philippines, Manila, Philippines.

Bill, 2012. Bill of the Republic of Indonesia concerning food. House of Representatives of the Republic of Indonesia, October 18, 2012, Jakarta, Indonesia.

Bogor Agricultural University, 2003. National Fortification Commission, Directorate General of Public Health of MOH, Deputy of Food Safety and Hazardous Substances of NADFC, Directorate General of Chemical, Agro-Forestry Based Industry of MOIT, Wheat Flour Producers Association of Indonesia, Cooking Oil Industries Association of Indonesia, Infant Food Producers Association of Indonesia, Iodized Salt Producers Association of Indonesia. Country Investment Plan for Food Fortification in Indonesia, Report, Jakarta, Indonesia.

CFIA, 2016. Health claims. http://www.inspection.gc.ca/food/labelling/food-labelling-for-industry/health-claims/eng/1392834838383/1392834887794

CFR, 2002. Health Claims. Code of Federal Regulations, Food and Drugs, 21, Vol. 2, Chapter I, Part 101, Food Labeling, §101.71–101.83, 124–149.

Chan, P., 2011. Regulations and guidelines relating to functional foods in South-East Asia, April 30, 2011, ILSI SEA Region Conference.

Charoenpong, C., Nitihamyong, A., 2004. Thailand. In: Report of the Regional Expert Consultation of the Asia-Pacific Network for Food and Nutrition of Functional Foods and their Implications in the Daily Diet; Bangkok, Thailand, 16–19 November 2004. RAP Publication 2004/33 FAO, Rome, Italy, pp. 18–19.

Chua K.H., 2010. Traditional medicine in State of Brunei Darussalam. Report of Department of Medical Services, Ministry of Health. http://herbalnet.healthrepository.org/bitstream/123456789/2288/2/01-p.1-12.pdf

Circular, 2004. Guidelines for the management of functional food products. Circular No. 08/2004/TT-BYT dated August 23, 2004, The Ministry of Health, Hanoi, Vietnam.

Circular, 2012. The Ministry of Health guiding the regulation conformity announcement and announcement on conformity with regulation on food safety. Circular No. 19/2012/TT-BYT dated November 9, 2012, The Ministry of Health, Hanoi, Vietnam.

Circular, 2014. Regulations on management of functional foods. Circular No. 43/2014/TT-BYT dated November 24, 2014, The Ministry of Health, Hanoi, Vietnam.

Codex, 2004. Guidelines for the Use of Nutrition and Health Claims. CAC/GL 23-1997, Adopted at 22nd Session in 1997, Revised in 2004, Amended in 2001, 2008, 2009, 2010 and 2011. Annex adopted 2009. Codex Alimentarius Commission.

Consumer Act, 1992. Consumer Act of the Philippines (Republic Act No. 7394) dated April 13, 1992, Republic of the Philippines, Congress of the Philippines, Manila, Philippines.

Council, 1996. The Traditional Drug Law. The State Law and Order Restoration Council No. 7/96, July 25, 1996, Myanmar.

Council, 1997. The National Food Law. The State Law and Order Restoration Council No. 5/97, March 3, 1997, Myanmar.

Decree, 2012. The Government detailing the implementation of a number of articles of the law on food safety. Decree No. 38/2012/ND-CP dated April 25, 2012, The Government of the Socialist Republic of Vietnam, Hanoi, Vietnam.

Department of Health, 1994. Compilation of food regulations. Directorate General of Food and Drug Control, Department of Health Indonesia, Jakarta, Indonesia.

Department of Health, 1999. Government regulation No. 69 regarding food labelling and advertisement. Directorate General of Food and Drug Control, Department of Health Indonesia, Jakarta, Indonesia.

de Perio Peralta, A., Credo Matienzo, M.C., 2010. The new Food and Drug Administration (FDA) of the Republic of the Philippines. First WHO Global Forum on Medical Devices, September 9-11, 2010, Bangkok, Thailand.

Dijkhuizen, M.A., Wieringa, F.T., Soekarjo, D., Van, K.T., Laillou, A., 2013. Legal framework for food fortification: examples from Vietnam and Indonesia. Food Nutr. Bull. 34, S112–S123.

DTM, 2000. Medicinal plants of Myanmar. Monograph. Ministry of Health, Department of Traditional Medicine, Myanmar.

Dung, T.N., 2001. Tue Tinh: founder of Vietnamese traditional medicine. J. Altern. Complem. Med. 7, 401–403.

EC 258, 1997. Regulation of the European Parliament and of the Council EC No. 258/97 of January 27, 1997 concerning novel foods and novel food ingredients. Official Journal of the European Communities L 43/1.

EC 333, 2004. Commission Regulations EC No 333/2004 of 31 March 2004 authorising the placing on the market of yellow fat spreads, salad dressings, milk type products, fermented milk type products, soya drinks and cheese type products with added phytosterols/ phytostanols as novel foods or novel food ingredients under Regulation (EC) No 258/97 of the European Parliament and of the Council. Official Journal of European Union L 105/40.

EC 1924, 2006. Regulation (EC) No 1924/2006 of the European Parliament and of the Council of 20 December, 2006 on nutrition and health claims made on foods. Official Journal of European Union L 404/9.

EC 983, 2009. Commission Regulations EC No 983/2009 of 21 October, 2009 on the authorisation and refusal of authorisation of certain health claims made on food and referring to the reduction of disease risk and to children's development and health. Official Journal of European Union L 277/3.

EC 384, 2010. Commission Regulation EU No 384/2010 of May 5, 2010 on the authorisation and refusal of authorisation of certain health claims made on foods and referring to the reduction of disease risk and to children's development and health. Official Journal of European Union L 113/6.

EUFIC, 2013. Global update on nutrition labelling. European Food Information Council. February 2013, Brussels, Belgium.

FAO, 2004. Report of the Regional Expert Consultation of the Asia-Pacific Network for Food and Nutrition on Functional Foods and their Implications in the Daily Diet. RAP publication 2004/33, Food and Agriculture Organization for the United Nations Regional Office for Asia and the Pacific. November 16-19, 2004, Bangkok, Thailand.

FAO, 2011. Report of the twenty-seventh session of the Asia and Pacific plant protection commission. FAO RAP Publication 2011/24. August 15-19, 2013, Manila, Philippines.

FAO-WHO, 1999. Food Labelling: Complete Texts (Revised 1999). Joint FAO/WHO Food Standards Programme. FAO/WHO, Rome, Italy.

FAO-WHO, 2002. Global Forum of Food Safety Regulators. January 28-30, 2002, Marrakech, Morocco.

Ferro, J., 1999. Vietnamese Foods and Culture. Rourke Press, Vero Beach, FL.

FNRI, 2014. Annual report. Food and Nutrition Research Institute. Department of Science and Technology, Philippines.

Food Act, 1983. Act 28 Food Act 1983 with amendment – Act A1266/2006. Laws of Malaysia.

Food Act, 2000. Public Health (Food) Act. Chapter 182. S 17/98 Amended by S 72/2000. Revised edition. Laws of Brunei.

Food Regulations, 1985. The Food Regulations 1985, Ministry of Health Malaysia.

Food Safety Act, 2013. The implementing rules and regulations of Republic Act no. 10611, "an act to strengthen the food safety regulatory system in the country to protect consumer health and facilitate market access of local foods and food products, and for other purposes" otherwise known as the "Food Safety Act of 2013". February 20, 2015. Republic of the Philippines Department of Agriculture and Department of Health. Joint DA-DOH Administrative Order no. 2015-007.

GAIN, 2013. Vietnam. Food safety law and guiding decree released. Global Agricultural Information Network (GAIN). Gain Report No. VM3032. December 6, 2013. USDA Foreign Agricultural Services.

GAIN, 2015. Indonesia. Food and agricultural import regulations and standards—narrative. Global Agricultural Information Network (GAIN). Gain Report No. ID1547. December 24, 2015. USDA Foreign Agricultural Services.

GD24, 2010. Brunei Darussalam guidelines for manufacturing and handling of halal medicinal products, traditional medicines and health supplements. Guidance Document. The Religious Council Negara Brunei Darussalam.

Hamano, H., 2014. Workshop and roundtable discussion on food. ILSI Meeting. Yangon, Myanmar, March 04, 2014.

HK, 2005. Basic provisions on functional food supervision. Regulation of the Head of the Food and Drug Supervisory Agency of the Republic of Indonesia, Board No. HK 00.05.52.0685 dated January 27, 2005, Indonesia.

ILSI, 2013. 8th seminar on nutrition labeling, claims and communication strategies. Report of International Life Sciences Institute Southeast Asia Region. November 26-27, 2013, Jakarta, Indonesia.

ILSI, 2015. 9th seminar on nutrition labeling, claims and communication strategies. Report of International Life Sciences Institute Southeast Asia Region. August 4-5, 2015, Manila, Philippines.

Isabelle, M., Chan, P., Wijaya, S.Y., 2011. Report on regulatory status of micronutrient fortification in Southeast Asia. International Life Sciences Institute South Asia Region Report Series. ILSI, Singapore.

Joint Circular, 2014. Guiding the goods label for some foods, food additives and aids for processing packaged foods. Joint Circular No. 34/2014/TTLT-BYT-BNNPTNT-BCT. The Ministry of Health, The Ministry of Agriculture and Rural Development, The Ministry of Industry and Trade. October 10, 2014, Hanoi, Vietnam.

Kasapila, W., Shaarani, S., 2011. Harmonisation of food labelling regulations in Southeast Asia: benefits, challenges and implications. Asia Pac. J. Clin. Nutr. 20 (1), 1–8.

Keat, N.K., 2015. Fostering harmonization of labelling and claims in ASEAN – industry perspective. 9th seminar on nutrition labeling, claims and communication strategies, August 4-5, 2015, Manila, Philippines.

KITA, 2013. Indonesian market focus. Functional food. Korea International Trade Association. Vol. 8.

Kongchuntuk, H., 2000. Thailand experience in nutrition labelling regulations and education. In: Proceedings of the National Seminar on Nutrition Labelling: Regulations and Education, August 7–8, 2000, Kuala Lumpur, Malaysia, pp. 36–45.

Kyaw, H.N., 2015. Nutrition labeling and claims: regulatory updates from Myanmar. 9th seminar on nutrition labeling, claims and communication strategies, August 4–5, 2015, Manila, Philippines.

Ladeia, A.M., Costa-Matos, E., Barata-Passos, R., Guimaraes, A.C., 2008. A palm oil rich diet may reduce serum lipids in healthy young individuals. Nutrition 24, 11–15.

Leone, A., Spada, A., Battezzati, A., Schiraldi, A., Aristil, J., Bertoli, S., 2015. Cultivation, genetic, ethnopharmacology, phytochemistry and pharmacology of Moringa oleifera leaves: an overview. Int. J. Mol. Sci. 16, 12791–12835.

Lopez, L., 2016. Vietnamese traditional medicine. http://www.vietspring.org/custom/tradmedicine.html

Mallillian, A.C., Bautista-Batallones, C., 2004. Philippines. In: Report of the Regional Expert Consultation of the Asia-Pacific Network for Food and Nutrition of Functional Foods and their Implications in the Daily Diet, Bangkok, Thailand, 16–19 November 2004. RAP Publication 2004/33 FAO, Rome, Italy, pp. 15–17.

Methakornkulnan, P., Hsieh, Y.J., Huang, L.Y., 2013. Consumer market for functional foods in Thailand. In: Official Conference Proceedings of the Asian Conference on Psychology & the Behavioral Sciences. Osaka, Japan.

MHM, 2007. Guide to classification of food-drug interface products. Standards section food safety and quality division. Ministry of Health Malaysia. http://fsq.moh.gov.my/v5/images/filepicker_users/5ec35272cb-78/Penerbitan/risalah/GuidetoFDIproducts.pdf

MHM, 2010. Guide to nutrition labelling and claims dated of December 2010, Ministry of Health Malaysia.

MHS, 1998. Nutrition labelling: a handbook on the nutrition information panel. Department of Health, Ministry of Health Singapore.

MOH, 2006. Regulation on the control, on production, export-import safe food No. 856/MoH. Ministry of Health, May 12, 2006, Vientiane Capital, Lao PDR.

MOH, 2009a. Ministerial regulations on the basic principles in the application of sanitary and technical measure for the food safety management No. 518/MoH. Ministry of Health, 18 March 2009, Vientiane Capital, Lao PDR.

MOH, 2009b. Regulation on labeling of prepackaged food No. 159/MoH. Ministry of Health, March 18, 2009, Vientiane Capital, Lao PDR.

MOH, 2012. Decision on food inspection No. 297/MOH. Ministry of Health, February 24, 2012, Vientiane Capital, Lao PDR.

MOH, 2016. Strategy, orientation, planning and development plan. Ministry of Health Portal. http://moh.gov.vn/province/pages/ChienLuocDinhHuong.aspx

MPH, 1998. Nutrition labeling. Notification of Ministry of Public Health No. 182 B.E. 2541. The Government Gazette Volume 115, Part 47 (Ngor), dated June 11, 1998, Bangkok, Thailand.

MPI, 2015. Global regulatory environment of health claims on foods. MPI technical paper no. 2015/16. Prepared by Food Science and Risk Assessment Group, July 2015. Ministry for Primary Industries, Wellington, New Zealand.

Murray, F., 2006. Ampalaya. Nature's Remedy for Type 1 and Type 2 Diabetes. Basic Health Publications, Inc., Laguna Beach, CA.

National Assembly, 2004. Law on food. National Assembly, May 15, 2004, Vientiane Capital, Lao PDR.

National Assembly, 2011. Law on drugs and medical products, amended. National Assembly, December 21, 2011, Vientiane Capital, Lao PDR.

NFSP, 2009. National Food Safety Policy. Ministry of Health, No. 020/MOH, dated January 13, 2009. Adopted by Prime Minister, No. 028/PMO, dated February 3, 2009, Vientiane Capital, Lao PDR.

Ngoc, T.T., 2015. Nutrition labeling and claims regulations in Vietnam. 9th seminar on nutrition labeling, claims and communication strategies, August 4–5, 2015, ILSI SEA Region, Manila, Philippines.

Nurolaini, K., Wint, Z., Jaini, N.H., Jaludin, N.H., Hamid, M.A., Linda, Y.Y.L., 2014. Prevalence of the use of traditional medicine in Brunei Darussalam. Homeopathy Ayurvedic Med. 3, 155–160.

Parinyasiri, T., 2015. Updates on nutrition labeling and claims regulations in Thailand. 9th seminar on nutrition labeling, claims and communication strategies, August 4–5, 2015, ILSI SEA Region, Manila, Philippines.

PP 69, 1999. Government regulation of the Republic of Indonesia Number 69/1999 on food labelling and advertisement. President of the Republic of Indonesia, Jakarta, Indonesia.

PP 28, 2004. Government regulation of the Republic of Indonesia Number 28/2004 on food safety, quality and nutrition. President of the Republic of Indonesia, Jakarta, Indonesia.

Prakas CS 001, 2000. Prakas on Cambodian standard CS 001-2000 labelling of food product. Regulation No. 1045 dated December 28, 2000. Ministry of Industry, Mines and Energy. Kingdom of Cambodia.

Prakas 868, 2010. Inter-ministerial prakas on the implementation and industrial arrangements of food safety based on the farm to table approach. Prakas No. UATH.BRK 868. Deputy Prime Minister, Minister of Economy and Finance, Senior Minister, Minister of Commerce, Minister of Industry, Mines and Energy, Minister of Agriculture, Forestry and Fisheries, Minister of Health, Minister of Tourism. Kingdom of Cambodia.

Rimpeekool, W., Seubsman, S., Banwell, C., Kirk, M., Yiengprugsawn, V., Sleigh, A., 2015. Food and nutrition labelling in Thailand: a long march from subsistence producers to international traders. Food Policy 56, 59–66.

Republic Act No. 8976, 2000. An Act establishing the Philippine Food Fortification Program and for other purposes. The implementing rules and regulations of Republic Act No. 8976, Department of Health, Office of the Secretary, Republic of the Philippines, Manila, Philippines.

Retka, J., Chhengpor, A., 2015. Government releases draft law on food safety. July 16, 2015, The Cambodia Daily.

Russin, J., Vecchi, S.E., 2014. Food Law in Vietnam. Russin & Vecchi, LLC, Bangkok, Moscow, New York.

Sideth, S., 2014. Cambodia. Updates on food safety regulatory frameworks. Workshop and round table discussion on food safety and standards. March 4-5, 2014, Yangon, Myanmar.

Sihombing, T.H., 2010. Nutrition labelling and claims in Indonesia. Seminar and workshop on nutrition labeling, claims and communication strategies for the consumers, September 20-21, 2010, Kuala Lumpur, Malaysia.

Sokneang, I., 2013. Overview of fruit processing in Cambodia: current status and future prospects. Workshop: valorisation of traditional processing of indigenous and underutilised fruits. Institute of Technology of Cambodia, January 14–16, 2013, Phnom Peng, Cambodia.

Soeun, M., 2004. Food safety overview in Cambodia. Country paper for Global Food Safety Forum, November 18-19, 2004, Beijing, China.

Sub-Decree No. 23, 1998. Sub-Decree No. 23 on the production, import-export and trade of traditional medicines for the health sector date April 28, 1998, Ministry of Health. Royal Government of Cambodia.

Sundram, K., Perlman, D., Hayes, K.C., 1996. Increasing the HDL level and the HDL/LDL ratio in human serum with fat blends. November 26, 1996, US Patent 5,578,334.

Tee, E.S., 2000. Proposed requirements for nutrition labelling, Malaysia: part II. In: Proceedings of the National Seminar on Nutrition Labelling: Regulations and Education, 7–8 August 2000. Kuala Lumpur, Malaysia, vol. 1, pp. 71–80.

Tee, E.S., Tamin, S., Ilyas, R., Ramos, A., Tan, W.-L., Lai, D.K.-S., Kongchuntuk, H., 2002. Current status of nutritional labelling and claims in the South-East Asian region: are we in harmony? Asia Pac. J. Clin. Nutr. 11, S80–S86.

Tee, E.S., 2007a. Claims and scientific substantiation: efforts in harmonizing in Asia. Conference on Functional Foods, May 9–11, 2007, Malta.

Tee, E.S., 2007b. Labelling guideline, February 14, 2007, Ministry of Health Malaysia.

Tee, E.S., 2009. Nutrition labelling and health claims: Codex guidelines. 8th international food data conference on functional foods, October 1–3, 2009, Bangkok, Thailand.

Tee, E.S., 2015. Health claims and scientific substantiation in SEA – process and key learnings. 9th seminar on nutrition labeling, claims and communication strategies, August 4–5, 2015, ILSI SEA Region, Manila, Philippines.

Theng, D., 2014. Current situation of national standard setting in Cambodia. National training workshop on strengthening food standard setting and participation in Codex activities in Cambodia, 19–21 February 2014, Phnom Penh, Cambodia.

Too, D., Begawan, B.S., 2012. Forbes ranks Brunei fifth richest nation. February 25, 2012, Jakarta Post.

UNDP, 2014. Human development report 2014. Sustaining human progress: reducing vulnerabilities and building resilience. United Nations Development Programme, New York, NY.

Van Esterik, P., 2008. Food Culture in Southeast Asia. Greenwood Press, Westport, CT.

VietNamNet, 2014. Foreigners earn big money from Vietnamese functional food market. VietNamNet Bridge, May 5, 2014. http://english.vietnamnet.vn/fms/business/101305/foreigners-earn-big-money-from-vietnamese-functional-food-market.html

WHO, 2005a. Cambodia. National Policy of Traditional Medicine and Regulation of Herbal Medicines. Report of a WHO Global Survey. Word Health Organization Press, Geneva, pp. 128–129.

WHO, 2005b. Vietnam. National Policy of Traditional Medicine and Regulation of Herbal Medicines. Report of a WHO Global Survey. Word Health Organization Press, Geneva, pp. 138–139.

WHO, 2009. Lao PDR – Country Cooperation Strategy at a Glance. Briefs of the World Health Organization Press, Geneva.

World Bank, 2015. Vietnam. World development indicators. The World Bank data. http://data.worldbank.org/country/vietnam

Yulisman, L., 2014. Government to strengthen "jamu" quality through identification, November 18, 2014, The Jakarta Post.

Zawistowski, J., 2011. Legislation of functional foods in Asia. In: Saarela, M. (Ed.), Functional Foods: Concept to Profit second ed. Woodhead Publishing Ltd, Oxford, Cambridge, pp. 73–108.

Zawistowski, J., 2014. Regulation of functional foods in selected Asian countries in the Pacific Rim. In: Bagchi, D. (Ed.), Nutraceuticals and Functional Foods Regulations in the United States and Around the World, second ed. Elsevier Press, Amsterdam, pp. 419–479.

Zawistowski, J., Jones, P., 2015. Regulatory aspects related to plant sterol and stanol supplemented foods. J. AOAC Int. 98, 1–7.

CHAPTER

Nutraceutical innovation and health claims in China 19

Chun Hu
Nutrilite Health Institute, Buena Park, CA, United States

19.1 INTRODUCTION

Even though functional foods or health foods emerged in China from the early 1990s, it was in 1995 that the legal status of health foods was officially established under China's Food Hygiene Law, followed by the announcement and implementation of a series of provisions, regulations, and protocols. When the first Chinese health food was approved in late 1996, the Chinese health food industry started its journey into an ever-bigger segment in the food industry and market value has expanded dramatically from the early days. It was estimated that annual sales of health foods in China were RMB 20 billion yuan, and reached RMB 100 billion in 2010 (Li and Li, 2011), and market size was estimated to be RMB 260 billion yuan by the China Health Care Association in 2012 (Hu, 2014). The China health foods (including nutrient supplements) market is estimated as worth more than RMB 600 billion yuan in 2014 (HKTDC, 2014). China's health food market including dietary supplements has become one of the largest markets in the world, and the aspirational goal for the nutrition and health food industry is RMB 1000 billion yuan by the end of 2015 according to China's "12th-Five Year Development Plan for Food Industry."

Since health foods include both food with specific functions and nutrient supplements in China, the focus of this chapter is limited to the data for products with health function claims. The purpose of this chapter is to focus on reviewing previous regulatory change and analyzing China's State Food and Drug Administration (CFDA)-approved domestic health foods (excluding dietary supplements) between 2003 and Oct. 2014, and providing a prospect on current active research in China regarding the health foods area.

19.2 HISTORY AND EVOLUTION OF CHINESE HEALTH (FUNCTIONAL) FOODS

The estimated goal for nutrition and health food production value is RMB 1000 billion yuan at the end of 2015, according to the National Development and Reform Committee and "The 12th-five year development plan for food industry." The focus

Developing New Functional Food and Nutraceutical Products. http://dx.doi.org/10.1016/B978-0-12-802780-6.00019-5

of this period is to develop new food resources, to identify the relationship between structure and function for those substances with biological activities as well as their efficacy, and the life quality and technology of food and health food. The development of health foods with Chinese characteristics and new functions was encouraged in a combination of traditional health care theory and unique resources and technology in leveraging the rich history of traditional Chinese medicine and food therapy. China's Ministry of Science and Technology launched a strategic research program to support the research and development of functional foods. For instance, during the 11th 5-year period (2006–2011), the focused funding areas were critical technology, functional product development, and strategy research categories to support universities, government research institutes, and industry working collaboratively to explore basic and application research in the functional food area. The funding areas cover the validation of functional assessment techniques for health foods, detection techniques for functional components and antiadulteration in the functional foods, stability technology for functional components for the application, and highly efficient separation and preparation. This supportive R&D program also covered functional product development and industrialization, including categories such as lowering blood lipids or lowering blood pressure or lowering blood sugar, antioxidants, weight loss, and memory improvement, as well as traditional functional foods. This program also supported the strategy of research for a functional food resource analysis and development strategy.

Health (functional) foods are defined as foods with a specific health functional claim or supplementation of vitamins/minerals, which are suitable to be consumed by a particular population, with a function to modify body function, but are not intended to treat disease, and shall not cause any acute, subacute, or chronic harm. The concept was developed in China in the mid-1990s when the Ministry of Health (MOH) was in charge of administrative approval of health foods, and such authority was transferred to the CFDA at the end of 2003. Since then the CFDA has evaluated and approved health food applications, and proposed the amendment of the allowed functional claims, respective assessment protocols, as well as restrictions, while the MOH (which was renamed as the National Health and Family Planning Commission, NHFPC, on Mar. 17, 2013) has shifted its focus to the safety assessment and approval of novel food materials and new food additives, as well as revision and publication of food safety national standards. The permitted functional claims have evolved from 12 claims in 1996 to 27 functional claims in 2003 (in addition to a nutrient supplement procedure added in 2005), to the most recent CFDA new proposal for consolidation of 18 claims in 2012 (Table 19.1). However, CFDA's new proposal was still under discussion at the time when this chapter was drafted, and it is unclear when the new revised claims will be finalized and effective.

Correspondent to the functional claim and administrative authority, a series of assessment protocols were proposed for public comments in 2011, and a few were finalized and became effective in May 2012. These include assessment protocols for "antioxidant function," "to assist protect gastro mucosa injury," " to assist lowering blood sugar," "vision fatigue relief," "to improve nutritious anemia," "to assist

Table 19.1 Evolution of Chinese Health Food Claims

Year	Claims
1996	*12 functional claims*: modulating immunity, antiaging, improving memory, promoting growth, antifatigue, weight loss, improving hypoxic endurance, antiirradiation, antimutagenesis, inhibiting tumor, modulating blood lipids, and improving sexual function
1997	*22 functional claims*: modulating immunity, antiaging, improving memory, promoting growth, antifatigue, weight loss, hypoxic endurance, antiirradiation, antimutagenesis, inhibiting tumor, modulating blood lipids, improve sexual function, modulating blood sugar, improving gastrointestinal function (with specific function), improving sleep, improving nutritious anemia, protecting liver from chemical-induced injury, promote milk secretion, cosmetics (specific function), promoting lead excretion, clearing throat, modulating blood pressure, improving osteoporosis
2000	*22 functional claims*: modulating immunity, modulating blood lipids, modulating blood sugar, antiaging, improving memory, promoting lead excretion, clearing throat, modulating blood pressure, improving sleep, promoting milk secretion, antimutagenesis, antifatigue, hypoxic endurance, antiirradiation, weight loss, promoting growth, improving osteoporosis, improving anemia, assisting to protect liver from chemical-induced injury, cosmetic (acne removal/chloasma removal/improve skin moisture and oil content), improving gastrointestinal function (modulate gut flora/promote digestion/bowel movement/assist to protect gastric mucosa)
2003	*27 functional claims*: to improve immunity, to assist lowering blood lipids, to assist lowering blood sugar, antioxidant, to assist to improve memory, to relieve visual fatigue, to promote lead excretion, to clear throat, to assist to lower blood pressure, to improve sleep, to relieve physical fatigue, to improve hypoxia endurance, to assist to protect from irradiation damage, weight loss, to promote growth, to improve bone density, to promote milk secretion, to improve anemia, to assist to protect liver from chemical-induced injury, to remove acne, to remove chloasma, to improve skin moisture content, to improve skin oil content, to modulate colon flora, to promote digestion, to promote bowel movement, to assist to protect gastric mucosa injury (nutrient supplement, when guidance for nutrient supplements became available in 2005)
2012 proposal but has not been finalized	*Proposed for 18 functional claims*: to help improving immunity, to help lowering blood lipids, to help lowering blood sugar, to help improving sleep, antioxidant, to help relieving physical fatigue, to help reducing body fat, to help improving bone density, to help improving nutritious anemia, to help improving memory, to clear throat, to help improving hypoxia endurance, to help protecting from alcohol-induced liver injury, to help lead excretion, to help milk secretion, to help relieving visual fatigue, to help improving gastrointestinal function (bowel movement, modify colon flora, promote digestion, and assist to protect gastric mucosa), to help improving facial skin health (acne removal, chloasma removal, and improve skin moisture content)

(Hu, 2014).

blood lipids lowering," "to promote lead excretion," "weight loss," and "to clear throat" functions. These new protocols require both animal testing and human trials except for vision fatigue which requires human testing only (CFDA, 2012a); the other 18 protocols still followed the 2003 working protocol at the time when this chapter was drafted. Depending on health function, some protocols require assessment on both animal and human models (antioxidant, blood lipids lowering, blood sugar lowering, memory improvement, lead excretion, clear throat, blood pressure improvement, promote breast milk secretion, weight loss, improve growth, improve nutritious anemia, modulate intestinal flora, promote digestion, anticonstipation, protect gastric mucosa injury) whilst others only require either animal tests (immunity, sleep improvement, antiphysical fatigue, hypoxic endurance improvement, radiation protection, bone density improvement, protect liver from chemical-induced injury) or human studies (vision fatigue, acne removal, chloasma removal, skin moisture content improvement, and skin oil content improvement).

The new protocols have changed the model used, biomarkers, and criteria for effectiveness assessment. Comparing antioxidant assessment protocols between 2003 and 2012, the bromobenzene and radiation induced animal models were eliminated in the new protocol, and an alcohol-induced oxidative damage model was introduced while the original D-galactose-induced oxidation model remained unchanged. In the lipid peroxidation measurement, serum 8-isoprostane and carbonyl content determinations were added as new requirements, and both blood and tissue reduced form glutathione measurement was added. In human trials, 18–65 years old is the age of subjects in this protocol, and the serum 8-isoprostane test is needed. In both the animal model and human model, the criteria of effectiveness assessment became stricter in the new protocol. Table 19.2 illustrates a brief comparison between the 2003 and 2012 antioxidant protocols, indicating that the new protocol requires more testing of biomarkers and stricter criteria for judgment of effectiveness.

In addition to the current 27 function claims and nutrient supplements, innovation on new function exploration has always been encouraged by government policy. The working guidance on health food registration issued in 2005 allowed the applicant to pursue novel functions (CFDA, 2005a). The applicants are expected to design and execute independent animal and human studies, and the government inspection agency assessed the research and development report submitted by the applicant. At the end of 2012, a detailed guidance was proposed by CFDA for public comment, which clearly defined the required criteria for novel function claims (CFDA, 2012b). In this proposal, a novel function claim is defined as "to regulate body function, improve body health, or reduce the risk of disease without any implication of prevention or treatment of any disease." The name and evaluation protocol of the novel function must be completely different from current allowed function claims; the scientific evaluation protocol and assessment criteria shall be provided by the applicant, with sufficient amount of scientific and generally acceptable evidence to support them; and the novel function claim shall comply with relevant law and regulations. In return, the novel function claim is granted exclusively to the successful applicant for a certain period of time and no other competitor's application for the same function

Table 19.2 Comparison Between 2003 Antioxidant Protocol and 2012 Antioxidant Protocol (CFDA, 2011)

	2003 Antioxidant Protocol	2012 Antioxidant Protocol
Animal test	Aged animal Rats, >12 months old Mice, 8–10 months old Induced oxidative damage: – D-Galactose injection – Bromobenzene induced – Radiation induced Experiment duration is 30 days, can be extended to 45 days if necessary Measurements include: – Lipid peroxidation products: malondialdehyde or lipofasci – Enzyme activity: superoxide dismutase and glutathione peroxidase Criterial for effectiveness assessment: Positive *if* – Either lipid peroxidation product result is positive, and – Either antioxidant enzyme activity result is positive	Aged animal Rats, >10 months old Mice, >8 months old Induced oxidative damage: – D-Galactose injection (mice) – Acute alcohol induced (mice and rats) Experiment duration is 30 days, and can be extended to 60 days if necessary Measurements include: – Lipid peroxidation products: malondialdehyde or 8-isoprostane – Protein oxidation product: carbonyl – Enzyme activity: superoxide dismutase and glutathione peroxidase – Nonenzymatic antioxidant: glutathione (reduced form) Criterial for effective assessment: Positive *if* – Three of the following measurements are positive: • Lipid peroxidation products • Protein carbonyl • Antioxidant enzyme activity • Glutathione (reduced form)
Human trial	Human subjects are between 45 and 65 years old, with at least 50 subjects in each group; Experiment duration is 3 months and can be extended to 6 months if necessary; Criterial for effectiveness assessment: product is positive *if any of the following measurements is positive*: – Lipid peroxidation products – Superoxide dismutase – Glutathione peroxidase	Human subjects are between 18 and 65 years old, with at least 50 subjects in each group; Experiment duration is 3 months and can be extended to 6 months if necessary; Criterial for effectiveness assessment: product is positive *if any two of the following measurements are positive*: – Lipid peroxidation products – Superoxide dismutase – Glutathione peroxidase

will be considered within this protected phase. Even though the concept of "novel functional claim" has been available for a long time, no product has been granted novel functional claim status yet, suggesting that challenges still exist in realizing the opportunity of novel function development.

19.3 THE USE OF ALLOWED INGREDIENTS IN APPROVED HEALTH FOODS

In addition to general food ingredients, food additives, and novel food ingredients, product developers rely on the long-standing ingredients lists issued by MOH in 2002, that is, "ingredients as both food and medicine" and "ingredients only permitted in the health foods," to cover 201 ingredients from botanical and animal sources; however, neither the species nor the material parts were specified in either document. CFDA issued compiled technical specifications of raw materials and excipients that are commonly used in health food registration (CFDA, 2014) in early 2014, which consolidated the technical specifications of those commonly used food ingredients, food additives, vitamins and minerals used in nutrient supplements, nutrient fortifiers in health food, "ingredients as both food and medicine" and "ingredients only permitted in the health foods," approved novel food ingredients, and other commonly used raw materials in health foods. This working document for the first time specified the genera, species, and parts for those listed in "ingredients as both food and medicine" and "ingredients only permitted in the health foods" and provided the respective quality standard to each food item. It was an update from the original MOH documents, and several ingredients were either removed or combined in this revision. For example, the pagoda tree (*Sophora Japonica* L) flower and bud were listed in the 2002 MOH list as two items (Hu, 2014), and now is combined as one item in the new CFDA document. All health food ingredients compiled in this document must comply with relevant food safety standards for the myotoxin, contaminants, and pesticide residues. Any contents with implication for China Pharmacopeia must comply with the specifications defined in China Pharmacopeia.

At the time this chapter was drafted, the national standards on botanical extracts were limited. In addition to 47 botanical extract standards listed in China Pharmacopeia, only 7 new botanical standards were implemented in 2013 by the Ministry of Commerce, of which 5 ingredients are allowed in functional foods (including panax ginseng extract, ginkgo leaves extract, *gotu kola* extract, bilberry extract, and milk thistle extract) (Yi and Chao, 2014), while other botanical ingredients have not been standardized yet.

The National Food Safety Standard of Health Food (GB 16740-2014) was revised in 2014 to replace the previous version published in 1997. Health foods are those that are legally approved with specific health functions or are intended to supplement vitamins and minerals, that is, foods that are suitable for consumption by a specific population with the function to regulate body function but without intention to cure disease, nor cause any acute, subacute, or chronic harm. This new national food safety standard redefined the scope, definition, technical specification, and labeling requirements for health foods. In general, all ingredients including both active and functional ingredients as well as excipients must comply with food safety standards and/or relevant provisions. The new national standard also defines the sensory characteristics which shall comply with the following specifications. Sensory characteristics, limits of heavy metals, microbial, myotoxin, and pesticides were specified in GB 16740-2014 or referred to other relevant National Food Safety Standards and provisions, such as GB 2762-2012 (contaminant level in food), GB 2761-2011 (myotoxin limit levels in food), and GB2763-2014 (upper limit of pesticide residue in food).

CFDA started to administrate the approval of health foods in Nov. 2003; about 7089 health food licenses (excluding nutrient supplements) have been approved and granted to domestic manufacturers to the end of Oct. 2014 (data source: CFDA health food online database). Among the current 27 valid health functional claims, immune function represents the most approved category, accounting for 41.2% of total approvals, followed by 15.1% for antiphysical fatigue function, 9.4% for blood lipids lowering function, 5.7% for bone density improvement function, and 5.1% for blood sugar regulation function (Fig. 19.1). Other claims such as "sleep improvement," "protect liver from chemical-induced injury," "anticonstipation," "weight loss," "antioxidant," "chloasma removal," "hypoxia endurance improvement," and "memory improvement" represented 2–5% of approvals respectively between 2003 and 2014. Other function claims were less than 2% in total approved products, and it is interesting to note that there has been no approval for the "skin oil content improvement" claim, suggesting that there was less consumers' need for this, therefore CFDA has proposed to remove this claim in the 2012 function claim consultation proposal. CFDA allows products to carry more than one functional claim (provided that the validation of each intended function assessment is approved). The number of approvals each year has varied; in recent years, CFDA approved 497, 625, and 840 health

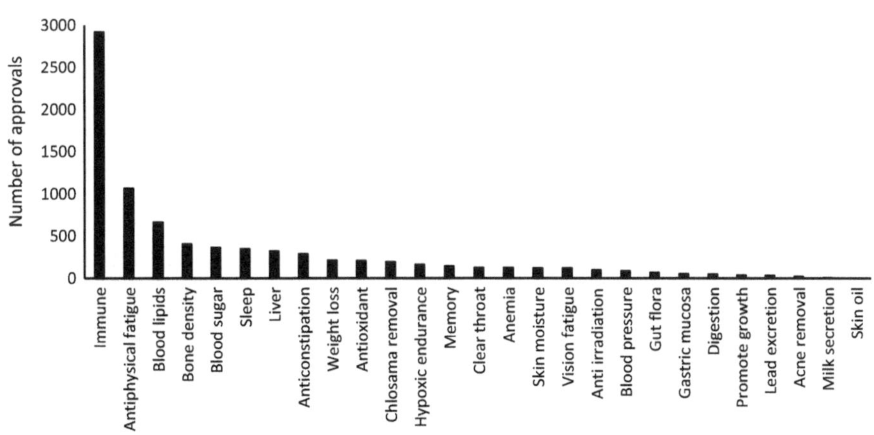

FIGURE 19.1 Distribution of the Functional Claims in CFDA-Approved Health Products Between November 2003 and October 2014 (Excluding Nutrient Supplements)

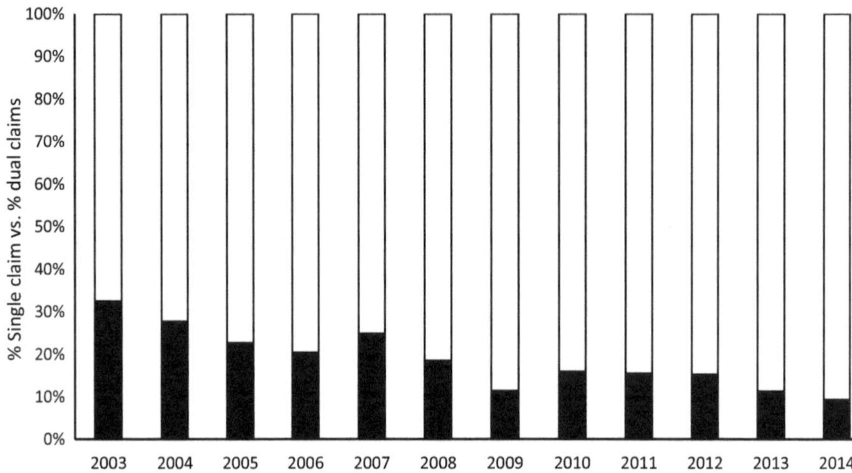

FIGURE 19.2 Health Products With one Claim or two Claims Approved by CFDA Between November 2003 and October 2014 (Excluding Nutrient Supplements)

■, Two claims; □, one claim.

products (excluding nutrient supplements) in 2012, 2013, and the first 10 months of 2014, respectively. During the first few years, 20–25% of products were approved with more than one functional claim, however, the current trend indicates that fewer products have carried more than one claim in recent years; Fig. 19.2 shows that less than 10% of approved health foods have carried two claims in recent years, indicating that the multiple-claim product is on a declining trend.

Health food delivery forms are not limited to typical food forms; pharmaceutical oral delivery forms are allowed and expected to include sublingual tablets and spray (CFDA, 2005b). Fig. 19.3 illustrates that pharmaceutical oral delivery formats are the primary forms all-time for CFDA-approved health foods, representing about 2/3–3/4 of total approvals. Of all approved health foods (2003–2014), 41% of products were found in hard-shell capsules, 17% as softgels, 14% as tablets, 13% in liquid, 5% in granule or powder drink forms, and 4% of products were delivered as tea. Looking into delivery forms year on year, softgels has maintained around a 20% share since 2007. In the approved softgel health foods, 30–40% were filled with traditional Chinese medicine materials or extracts with immune, antiphysical fatigue, and blood lipid lowering functions (Sa, 2013). Hard-shell capsules continue to be the primary delivery form, though the percentage has dropped from 44% in 2004 to 37% during the first 10 months of 2014; tablets remain consistently between 13% and 18%, with their functions being focused on immune and bone density functions (Jiang and Dong, 2014).

Fig. 19.4 illustrates the 10 most used ingredients in the approved domestic health foods (excluding nutrient supplements) between 2003 and 2014. Their appearances were greater than 5% of the total 7089 products. They were classified into three categories, that is, "ingredients as both food and medicine," "ingredients only allowed

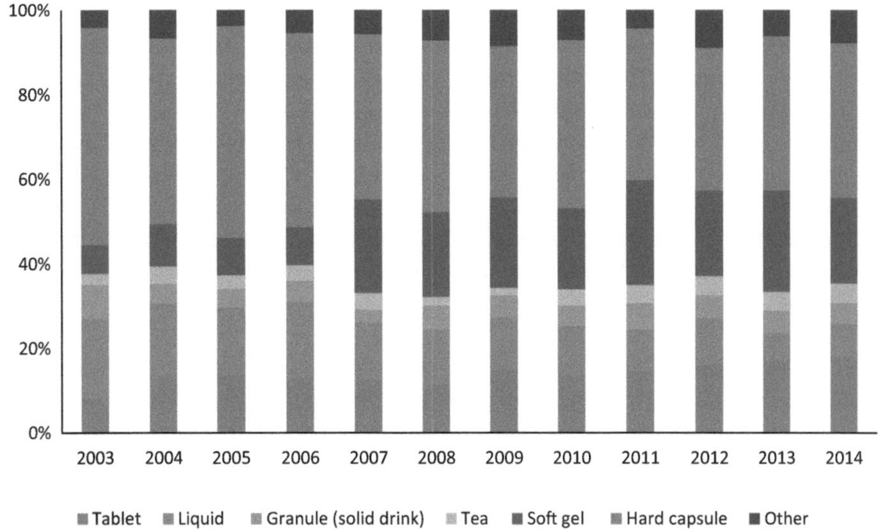

FIGURE 19.3 Delivery Forms of CFDA-Approved Health Foods Between November 2003 and October 2014 (Excluding Nutrient Supplements)

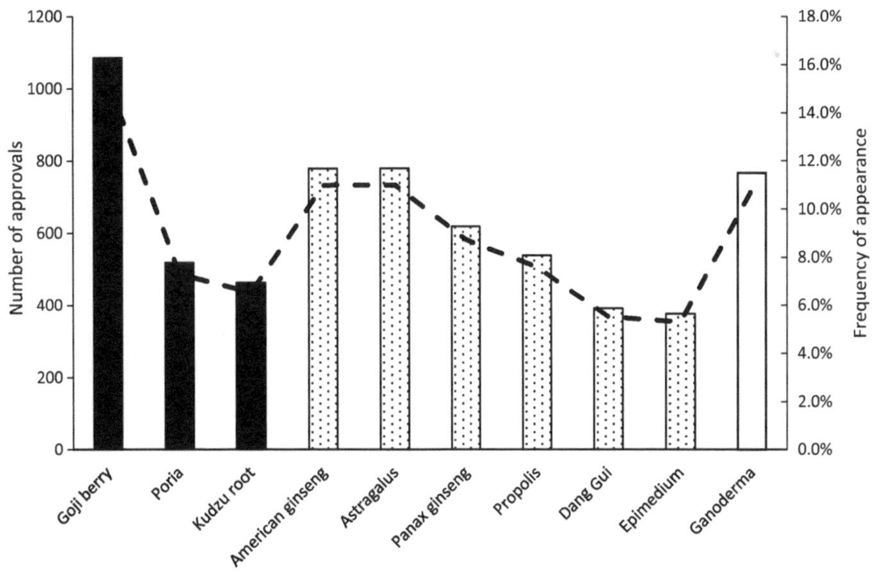

FIGURE 19.4 10 Most Used Ingredients in Approved Chinese Health Foods Between November 2003 and October 2014 (Excluding Nutrient Supplements)

Bar, number of approval; *dashed line*, frequency of appearance; ■, ingredient as both food and medicine; ░, ingredient only allowed in health food; □, other allowed ingredients in health food.

in health food," and others which include "a series of fungus, probiotic strains, novel food ingredients" approved by MOH. Among these 10 most used ingredients, goji berry or wolfberry (*Lycium barbarum* L.), poria (*Poria cocos* (Schw) wolf), and kudzu root (*Pueraria lobate* (willd.) Ohwi, *P. thomsonii.* Benth.) are listed in the list of "ingredients for both food and medicinal use." American ginseng (*Panax quinquefolium* L.), astragalus (*Astragalus membranaceus* (Fisch.) Bge.), ginseng or panax ginseng (*Panax ginseng* C. A. Mey), propolis (*Apis mellifera* L.), and Dang Gui (Chinese angelica) (*Angelica sinensis* (Oliv.) Diels) are listed in the list of "ingredients only allowed in health food," while ganoderma [*Ganoderma lucidum* (Leyss.ex Fr.) Karst., *G. sinense* Zhao Xu et Zhang] is one listed under fungi strains allowed in health foods (CFDA, 2005b, 2014).

Goji berry or wolfberry (*Lycium barbarum* L.) has a long history of application in traditional Chinese medicine and day-to-day food therapy practice, and it is believed to nourish the liver and kidneys with the function of replenishing essence and brightening vision (Pharmacopoeia, 2010). The health benefits of goji berries on aging, neuroprotection, antifatigue, endurance improvement, glucose metabolism, antioxidant activity, and immune regulation have been documented (Amagase and Farnsworth, 2011). Goji berry was listed by MOH in 2002 as one of the ingredients allowed in both food and medicine. Of the 1085 health products containing goji berry, 585 products were granted for immune function, 433 for antiphysical fatigue function, 43 for vision fatigue relief function, 50 for blood lipid regulatory function, and 55 for preventing chemical-induced liver injury function, which suggested that this well-recognized super-fruit aligns well with historical beliefs and its known mechanisms of action. The percentage of products containing goji berry seems to be on a declining trend in recent years (Fig. 19.5). For example, goji berry accounted for 15.5, 22.9, and 20.7% of approved health foods in 2004, 2005 and 2006 respectively, but declined to 9.7, 10.1, and 11.5% in 2012, 2013, and the first 10 months of 2014, respectively, suggesting that low-level repetition (Su et al., 2002) with goji berry product development is less attractive for product developers, and innovation on resource, key component separation, and efficacy demonstration have become keys for success in functional food development.

Recently a wild variety of goji berry, one less explored, *Lycium ruthenicum* Murr, has been studied extensively in China. *L. ruthenicum* Murr is a goji berry variety naturally distributed in northwest China and produces dark black fruits with strong resistance to harsh environments. Its anthocyanin profile was characterized: petunidin (2-(3,4-dihydroxyl-5-methoxyphenyl)-3,5,7-trihydroxychromenylium) glycoside was the primary anthocyanin with some being acylated, and it exhibits in vitro antioxidant activities (Yan et al., 2014; Zheng et al., 2011); on the other hand, cyanidin-3-glucoside (a common anthocyanin in other small berries) was not found in this black goji berry fruit (Li et al., 2013). The phytochemical fingerprint profiles of *L. ruthenicum* Murr were similar in the samples collected from 26 different production areas, and total phenolics, total flavonoids, and total anthocyanins vary dramatically among samples (Yan et al., 2014). Polysaccharide from *L. ruthenicum* suppressed the nitric oxide production, iNOS expression, and toll-like receptor 4 in lipopolysaccharide-induced mouse macrophage RAW264.7 cells through downregulation of the

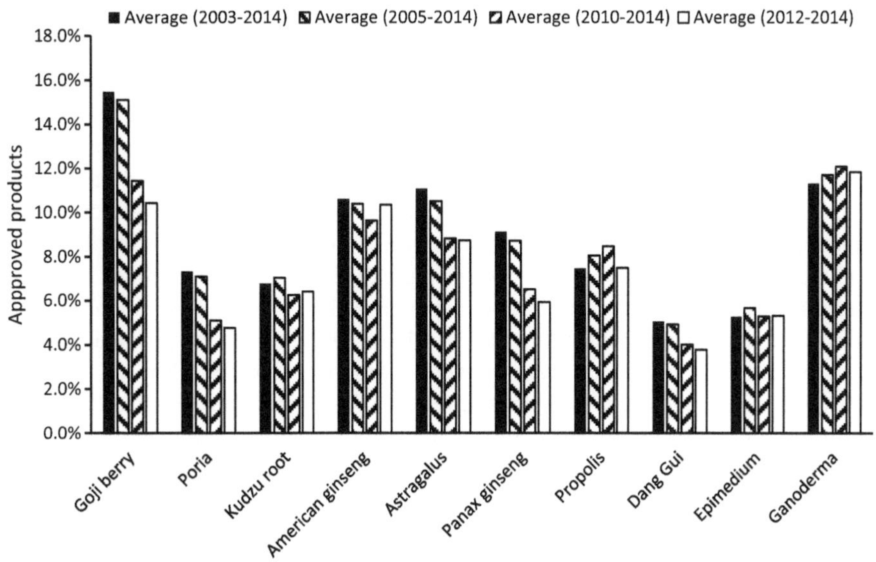

FIGURE 19.5 Average Appearance of the 10 Most Used Botanical Ingredients in Chinese Health Foods (Excluding Nutrient Supplements) Between November 2003 and October 2014

NF-κB singling pathway (Peng et al., 2014). *L. ruthenicum* polysaccharide reduced fasting blood sugar level in alloxan-induced diabetic mice after 1 month feeding. The hepatic antioxidant enzyme activities were improved, and lipid peroxidation product levels in both the blood and liver were lower (Wang et al., 2009). Following China's antioxidant protocol, a *L. ruthenicum* Murr buccal tablet significantly improved oxidative defense in the mice model (Li et al., 2014). Polysaccharide isolated from *L. ruthenicum* Murr exhibited an antiphysical fatigue function with improved blood chemistry analysis and longer forced swimming time (Ni et al., 2013).

In traditional Chinese medicine, poria possesses the function "to promote urination, leach out dampness, nourish spleen, and calm spirit" (Pharmacopoeia, 2010). Active phytochemicals in poria include triterpenes and polysaccharide to improve immune function (Sun, 2014). Nearly 40% of poria-containing products were granted an immune function claim, followed by 15.1% with an antiphysical fatigue claim. Kudzu root is considered "to expel pathogenic factor in muscles to abate heat, to promote body fluid production and quench thirst, to lift *yang* and relieve diarrhea, and to detoxify alcohol" (Pharmacopoeia, 2010). Isoflavone, 4′,7-hydroxyl-isoflavone-8-β-D-glucoside (puerarin), identified from kudzu root supports the traditional application in diabetes management (Wong et al., 2011). Puerarin significantly reduced serum AST, ALT level, improved liver homogenate malondialdehyde, and relieved liver tissue degeneration and necrosis in the CCl4-induced acute rat liver injury model (Zhao and Wei, 2005). The benefit of puerarin was also revealed in the rat model by shortening drunkenness time (Zhu et al., 2014). The most used application of kudzu root in health foods is for "preventing liver from chemical-induced injury" (31.5%) followed by "lowering blood sugar" (19.7%).

American ginseng is considered to supplement *qi* and nourish *yin*, clear heat, and produce body fluids (Pharmacopoeia, 2010). The functions of American ginseng-containing products were primarily related to immune function (50%) and antiphysical fatigue function (41.3%). Panax ginseng "invigorates renal *qi*, strengthens spleen and lung, promotes body fluids production and replenishes blood, calms mind, and promotes intelligence" (Pharmacopoeia, 2010). The claims of approved panax ginseng products equally split into immune function (42.1%) and antiphysical fatigue function (41.1%). Astragalus "replenishes *qi* and invigorates *yang*, strengthens the body, promotes diuresis and relieves enema, expels pus, and regenerates flesh" in traditional Chinese medicine (Pharmacopoeia, 2010). The leading approvals for astragalus were for immune function (44.1%), followed by antiphysical fatigue function (21.5%) and blood sugar lowering function (15.8%). Propolis "replenishes weakness, melts turbid lipids, and relieves thirst" (Pharmacopoeia, 2010); the majority of propolis approvals were for immune function (67.1%) and blood sugar lowering function (18.8%). *Dang Gui* "nourishes and promotes blood circulation, regulates menstruation and relieves pain, moisturizes the intestine, and stimulates bowel movement" (Pharmacopoeia, 2010). Though organic acids, phthalide, and polysaccharides were found as primary active components, polysaccharides from *Dang Gui* were found to be beneficial for immunity, antioxidant, and hematopoietic activities (Chen et al., 2013). About 30% of *Dang Gui* products carry an immune function claim, and 20% a chloasma removal claim. Epimedium "invigorates kidney and strengthens *yang*, strengthens sinew and bone, eliminates wind and dampness" (Pharmacopoeia, 2010). More than half (57.4%) of claims associated with epimedium are for antiphysical fatigue, 26.6% for immune function, and 20.2% for improving bone density. Evidence suggested that prenyl flavonoids in epimedium exhibit hormone regulation, antiosteoporosis, immunological regulation function, antioxidant, and antiaging activities (Ma et al., 2011). Flavonoids' metabolites and their aglycones support the antiosteoporosis, and of these, baohuoside I (3,5,7-trihydoxy-4′-methoxyl-8′-prenylflavone-3-O-α-L-rhamnopranoxide) was thought be more beneficial than other highly glycosylated epimedium flavonoids due to its balance between solubility and permeability in relation to antiosteoporosis function (Jiang et al., 2014). Ganoderma "supplements *qi* and calms spirit, relieves cough, and clams panting" (Pharmacopoeia, 2010); a traditional water decoction of ganoderma was found to regulate immune function (Qin and Qian, 2003). Three-quarters of this ingredient made an immune function claim, 11.2% on antiphysical fatigue, and 10.7% claimed preventing the liver from chemical-induced injury.

19.4 RECENT NOTABLE INGREDIENT INNOVATION AND APPLICATIONS IN HEALTH FOODS

Oligosaccharides have drawn a good amount of attention in the last two decades. With oligosaccharide health and functional benefits being revealed by scientists worldwide, research on production and application of oligosaccharides also took off, and various oligosaccharides have been manufactured in China from the 1990s, particularly the most studied and utilized isomaltose-oligosaccharide, xylose-oligosaccharide, fructooligosaccharide, and soy oligosaccharide (including stachyose and raffinose). These products are derived from food material or food waste. Oligosaccharides are not digested and absorbed by humans, but can be fermented by gastrointestinal microbes. The primary health functions of oligosaccharides are gastrointestinal benefits to improve intestinal microflora, improve fecal volume, relieve constipation symptoms, and improve immune function (Yang et al., 2004; Yao et al., 2013). Using a constipated mice model induced by oral administration of compound diphenoxylate, the anticonstipation effect of fructooligosaccharide was proved by improving the gastric emptying rate and fecal volume (Chu et al., 2010). Xylose oligosaccharide, regardless of the degree of oligomerization, produced either by microwave or by enzymatic synthesis, was found to improve oxidative defense, blood lipids status, and gut flora balance in a mice model stressed by a high-fat diet (Ma et al., 2013). In addition to chicory and Jerusalem artichoke being known as good sources of fructooligosaccharide, the traditional Chinese dietary herb burdock root contains 3.6% fructooligosaccharide (Yang et al., 2004). The structure of burdock oligosaccharide was identified as a 12:1 composition of D-fructose:D-glucose, of which 12 fructose residuals linked through β (2 $\rightarrow$1) glucoside with one glucose residue linked through α (1$\rightarrow$2) glucoside at the end of a linear structure (Hao et al., 2005), and it may possess immune regulatory function, blood sugar and lipid regulations, as well as gastrointestinal benefits (Cao et al., 2012). Some of these oligosaccharides gained novel food status in recent years. Xylose-oligosaccharide and galactose-oligosaccharide were approved as novel foods in 2008 by MOH, and stachyose was approved in 2010 as a novel food ingredient. Subsequently, the National Food Safety Standard for Nutrient Fortifier (GB 14880-2012) stipulates the amounts and food types where galactose-oligosaccharide (derived from lactose), fructooligosaccharide (derived from chicory), and raffinose (derived from beet) can be used as nutrient fortifiers in baby formula and baby cereal supplementary food. They can be used individually or in combination but the total amounts shall not exceed 64.5 g/kg.

As of Oct. 2014, CFDA has approved 334 health foods containing oligosaccharide as health foods, with claims varying from immune function (150 approvals) and anticonstipation (113 approvals) to intestinal flora improvement (43 approvals). Over 112 applications makes fructooligosaccharide the leading oligosaccharide used in health foods, followed by xylose-oligosaccharide (72 approved products), isomaltose-oligosaccharide (55 approved products), stachyose (30 approved products), and soy oligosaccharide (25 approved products). Among all health foods, less than 1% of products (or 66 out of all 7089 products) carry an "improve intestinal flora" claim, of which 65% or 43 products are formulated with oligosaccharide, suggesting the prebiotic function of oligosaccharides is accepted in this category. The delivery forms include chewable, powder drink, and powder supplementary food suitable for not only infants but also the general population, with very low known adverse effects (Xiao et al., 2010; Yang et al., 2004). Besides the health benefits of oligosaccharide, one study showed that isomaltose-oligosaccharide could also be used in coextrusion to encapsulate ascorbic acid to generate a glassy solid solution to provide a unique microencapsulation method for thermally sensitive ingredients and potential application to consumers with digestive sugar sensitivity (Chang et al., 2011).

Cistanche is a parasitic higher plant distributed in deserts, primarily parasitizing on roots of *Haloxylon ammodendron* and *H. persicum*. Cistanche has a long history of medicinal and food use in China; commonly available cultivars in China include *Cistanche deserticola* Y.C. Ma, *C. salsa* (C.A. Mey) G. Beck, *C. salsa var. albiflora* P.F. et Z.C. Lou, *C. sinensis* G. Berk, and *C. tubulosa* (Schenk) Wight, distributed in northwest regions in China, in which the Xinjiang Uyghur Autonomous Region becomes a leading production area for *C. tubulosa* and the Inner Mongolia Autonomous Region for *C. deserticola*. Both *C. deserticola* Y.C. Ma and *C. tubulosa* (Schenk) Wight are listed in the current Chinese Pharmacopeia (Pharmacopoeia, 2010) as authentic species. Cistanche has been used for a long time "to nourish kidney and strengthen *Yang*." The primary phytochemicals identified from cistanche include phenylethanoid glycoside, iridoid and glycoside, lignan and glycoside, phenolics, terpenoids, alkaloids, and phytosterols (Lei et al., 2003; Song et al., 2010; Tu et al., 1997). With 16 phenylethanoid glycosides being identified, echinacoside and acteoside are used as marker compounds for cistanche material in the Chinese Pharmacopeia. The abundance and variety of phenylethanoid glycosides vary by the cistanche species and production areas (Lei et al., 2003; Tu et al., 1997). Seven phenylethanoid glycosides were identified from *C. deserticola*, while six phenylethanoid glycosides were found in *C. salsa* and five from *C. tubulosa*, suggesting the variety of phenylethanoid glycosides could be used to differentiate different cistanche species (Wang et al., 2000). *C. tubulosa* water extract (concentrate to dry, 1:6 ratio to *cistanche* raw material) was tested for toxicity, and it was found that the oral acute maximum tolerated dose for mice was greater than 20 g/kg bw, without mutagenicity in the bone marrow micronucleus assay, sperm malformation test, and Ames test, and no obvious adverse effect was found in rats in a 90-day feeding test, suggesting *C. tubulosa* is well tolerated and safe (Zhang et al., 2014a). Similarly, using 2500–7500 mg/kg bw (equivalent to 100–300 times the human daily recommended dose), no toxic effect was observed in SD rats (Jiang et al., 2013). Phenylethanoid glycoside extracted from *C. salsa* protected dopaminergic neurons against dopamine neurotoxicity induced by 1-methyl-4-phenyl-1, 2, 3, 6-tetradropyridine in C57 mice (Geng et al., 2004). Nie et al. (2012) evaluated the antiphysical fatigue effect of tablets formulated with *C. tubulosa*, eleuthero, and American ginseng. In this study, the mice were randomized in different dose treatment groups with oral gavage of 0.167, 0.333, and 1.000 g/kg bw (a 1 g tablet was equivalent to 2.4 g *C. tubulosa*) once per day for 30 days. The high-dose group was found to have a significantly longer weight-bearing swimming time, higher liver glycogen storage, and lower blood urea nitrogen after exercise, indicating that the *C. tubulosa* compound tablet exhibited an antiphysical fatigue function according to the antiphysical fatigue function assessment protocol (Ni et al., 2012).

For neuroprotection, phenylethanoid glycoside from *C. salsa* prevented 1-methyl-4-phenylpridinium ion-induced apoptosis in cerebella granule neurons by inhibiting caspase-3 and caspase-8 activities (Tian and Pu, 2005). Using a standardized *C. tubulosa* water extract (containing 25.4% phenylethanoid glycoside, 3.8% acteoside, and 4.1% isoacteoside), Wu et al. found this extract ameliorated the cognitive dysfunction in an Alzheimer's disease-like rats model through blocking amyloid deposition, reversing cholinergic and hippocampal dopaminergic neuronal function (Wu et al., 2014). Wang et al. found that phenylethanoid glycoside derived from *C. salsa* oral administration improved the ultrastructure of the hippocampus of D-galactose-induced brain aging mice with antioxidant activity being attributed to the mechanism underlying this beneficial effect (Wang et al., 2005). Phenylethanoid glycoside from *C. deserticola* improved mice learning and memory abilities in aged mice induced by D-galactose; the phenylethanoid improved the serum and brain SOD activities and reduced the hepatic and serum malondialdehyde level (Xuan and Liu, 2008).

Oral administration of *C. tubulosa* extract (equivalent to acteoside amounts of 24.2, 72.6, and 120.9 mg/kg bw) to male *db/db* mice for 45 days showed that cistanche extract suppressed the fasting blood glucose and postprandial blood sugar levels, and exhibited improved insulin resistance and dyslipidemia in *db/db* mice (Xiong et al., 2013). Antioxidant activities of cistanche were evaluated in various models. Six structural-related phenylethanoid glycosides isolated from *C. salsa* were evaluated for their potentials in DPPH radical scavenging, and the results illustrated that the activities of these compounds were as follows: 2′-acetylacteoside > acteoside ≥ tubuloside B ≥ isoacteoside > echinacoside > cistanoside A, indicating that the number of the phenolic hydroxyl group, steric hindrance, and the positions of hydroxyl substitution structural characteristics are all important factors in determining the activity effectiveness (Yang et al., 2009). Wister rats were subjected to oral gavage with *C. deserticola* extract (daily dose equivalent to 6.01–17.90 g/kg bw) for 4 weeks; malondialdehyde in heart mitochondria was significantly lowered in the treatment compared to the control group after rats completed exhaustive swimming whilst cistanche-treated groups also led to improvement in the activities of heart antioxidant enzymes (SOD, GSH-Px), suggesting that cistanche feeding improved the exercise endurance and antioxidant defense in responding to the oxidative damage incurred to the myocardial mitochondria during exercise (Zhou et al., 2012). In addition, acteoside from *C. tubulosa* improved the telomerase activity in the heart and brain of an aging mice model induced by D-galactose, and lowered the malondialdehyde accumulation tissues, suggesting this might be related to an antiaging property of this herb. The acteoside feeding also resulted in an enhancement of immunity of these aging mice (Zhang et al., 2008).

Growing in harsh environments, cistanche has long gained its reputation as a rare and valuable traditional Chinese medicine ingredient. Studies have been conducted in China to explore innovative procedures to produce cistanche, to enhance the processability and efficiency in utilizing this material. Gao et al. compared drying efficiency between different drying methods (tray drying, belt drying, spray drying, and microwave drying) and capsule-filling characteristics; although there was no significant impact on the echinacoside yield by these drying methods, however, only spray drying provided good solid dose processability with both good powder flowability and compressibility, and the spray-drying condition was optimized (Gao et al., 2015). Since phenylethanoid glycoside has poor water solubility, Sheng et al. extracted phenylethanoid glycoside from cistanche scrap, followed by β-cyclodextrin encapsulation, and found the benzene ring of the echinacoside entered into the cavity of β-cyclodextrin during the formation of the echinacoside-β-cyclodextrin inclusion complex; this preparation provided a unique method in preparing cistanche extract with improved water solubility and thermal stability with a potential in functional food application (Sheng et al., 2013). Ren et al. used macroporous resin with ethanol of mobile phase to purify and enrich the phenylethanoid glycoside from *C. tubulosa* extract (Ren et al., 2005). With the resource from the wild and cultivation being limited, researchers in China explored the opportunity of producing cistanche in tissue culture. For example, Li et al. (2006) compared vitamin C, citric acid, and sodium sulfite in preventing subcultures of *C. deserticola* cells; by inhibiting polyphenol peroxidase activity, they optimized the growth condition with a 25.4% (dry weight) increment of cell growth. Similarly, glucose as carbon source was found to be beneficial for the production of echinacoside and acteoside in a *C. salsa* callus culture condition, and the authors found that the phenylethanoid glycoside composition from the *C. salsa* callus culture was similar to those from wild craft, though the concentrations from the callus culture were higher than wild crafted samples, suggesting that callus culture may provide a viable path for producing cistanche phenylethanoid (Guo et al., 2004).

There were 19 approved health foods containing *cistanche* herb, of which three products were clearly formulated from *C. tubulosa* from the respective ingredients lists. The formulas often contain other traditional Chinese medicinal ingredients, such as American ginseng, rhodiola, astragalus, panax ginseng, goji berry, eleuthero, and epimedium. Echinacoside and acteoside were used as reference compounds in some products. Most of these products make immunity claims, followed by anti-physical fatigue claims, and memory improvement claims.

Application of novel food ingredients is another approach to drive functional food innovation in China. Maca (*Lepidium meyenii*) is another new emerging ingredient in China in the last decade. Maca originated in high elevations in the Andes with a long history of traditional use in South America (Feng et al., 2009; Wang et al., 2007). The biological activities of this Andes herb include fertility improvement, sexual performance, antiproliferation, and improvement in vitality and stress tolerance (Wang et al., 2007). A small double-blind placebo-controlled study illustrated that 12 weeks of maca supplementation showed a small but significant improvement in subjective perception of general and sexual well-being in adult patients with mild erectile dysfunction (Zenico et al., 2009). This herb was introduced to China, primarily in southwest Yunnan province in early 2000. The mountain region in Yunnan province is considered to have maca growth conditions similar to those in the Andes, and cultivation testing began in 2002 at 2800–3200 meters altitude above sea level. Analysis showed that cultivated Yunnan maca contains 16.7% protein, 21.5% crude fiber, and 12.2% ash with similar mineral and vitamins to maca root from Peru (Feng et al., 2009), hence the nutrition profile of Yunnan maca root indicates the Yunnan cultivated maca is of nutritional value and business value. Phytosterol contents from maca root cultivated in Yunnan were similar to those from Peru, with white maca containing the highest amount of total phytosterol, though brassicasterol and ergosterol were low to nondetectable in the Yunnan samples (Gan et al., 2013). Besides the rich content of protein and carbohydrates, maca contains other phytochemicals with biological functions; these include macaenemacarmide, glucosinolates, alkaloids, and phytosterols (Wang et al., 2007). The maca plant growth pattern and harvest pattern were outlined in a recent report (Shang et al., 2013). The field study and promotion of the Maca production have led to a scale-up of maca production in the mountain regions of Yunnan province, and maca production was reported to exceed 2400 tons in 2012, accounting for more than 95% of national production in China.

Maca gained its novel food status in May 2011, which allows cultivated maca root as a food ingredient after being processed by slicing, drying, pulverization, and pasteurization. The novel food approves daily consumption of maca of no more than 25 g (quality spec, protein ≥10%, dietary fiber ≥10%, and moisture content ≤10%), and maca products are not recommended for infants, women in lactation and pregnancy, and nonsuitable populations and daily allowances are also required to be declared on the product label. This novel food approval paved the path for maca as an ingredient in food development. The spray-drying process was optimized with consideration of feed concentration, inlet air temperature, and hot airflow rate to produce maca fine powder with maximization of the active fraction in maca and physical properties (Zhang et al., 2014b). By comparing different extraction methods (ethanol extraction, supercritical extraction, and ultrasonic assistant extraction), it was found that a 95% ethanol, 50% ethanol, and water sequential extract resulted in the highest yield of maca root, identified with macaene and macamide (Tang et al., 2015). Following the Chinese health food assessment protocol, Zhang et al. fed mice at a daily dose of 250–1000 mg/kg bw (equivalent to 5–20 times the recommended human dose, that is, 10 mg/kg bw human) for 30 days to assess immune functions and physical fatigue function. The authors reported that maca powder improved immune function by improving the response on a battery of immune response measurements; and in a separate animal experiment, physical fatigue improvement was demonstrated with prolonged burdened swimming time, higher hepatic glycogen reservation, and lower lactic acid after swimming, providing the evidence to qualify this maca powder for immune function support and physical fatigue prevention (Zhang et al., 2014c; Li et al., 2014).

Upon the approval of maca as a novel food ingredient, 12 maca-containing products were approved during the first 10 months of 2014, representing 1.3% of total approved health foods within the same period of time. All of these products claimed "to relieve physical fatigue" with dosage form varying from hard-shell capsule, tablet, and powder to maca liquor. Total protein content was used as the quantification mark in these products (10 out of 12), instead of using secondary metabolic phytochemicals for ingredient markers. Some other ingredients used to formulate with maca include American ginseng, goji berry, astragalus, epimedium, and rhodiola.

REFERENCES

Amagase, H., Farnsworth, N.R., 2011. A review of botanical characteristics, phytochemistry, clinical relevance in efficacy and safety of Lycium barbarum fruit (Goji). Food Res. Int. 44, 1702–1717.

Cao, X., Cao, J., Chen, K., 2012. Advances in pharmacological effects of Arctium lappa L. Chinese J. Mod. Appl. Pharmacol. 29, 975–979.

CFDA, 2005a. Decree 19, 2005, Interim administrative measures of health food registration, issued April 30, 2005, Beijing, China. http://www.sda.gov.cn/WS01/CL0053/24516_1.html

CFDA, 2005b. Notification regarding interim provision of nutrient supplement application and evaluation etc eight provision (CFDA registration division No. 202, 2005), issued May 20, 2005. Beijing, China. http://www.sda.gov.cn/WS01/CL0055/10396_9.html

CFDA, 2011. Letter to seek public comments on revision of function evaluation of blood sugar lowering and other protocols (CFDA health food and cosmetic division letter No 325, 2011, issue August 2, 2011), Beijing, China. http://www.sda.gov.cn/WS01/CL0780/64478.html

CFDA, 2012a. CFDA announcement 2012, No. 17, Notification on issuing nine health function evaluation protocols including antioxidant function. Beijing, China. http://www.sda.gov.cn/WS01/CL0847/71257.html

CFDA, 2012b. Letter to seek public comments again on the provision of novel function application and evaluation of health food registration (consulting version). CFDA health food and cosmetic division letter No 490, 2012, issued November 5, 2012, Beijing, China. http://www.sda.gov.cn/WS01/CL0780/75896.html

CFDA, 2014. CFDA Notification 2014-No. 60 1st guidance on compendium of technical specifications of raw material and excipient used for health food registration. April 14, 2014. Beijing, China. http://www.sfda.gov.cn/WS01/CL1136/98317.html

Chang, D., Zhang, X., Xia, S., Jia, C., 2011. Encapsulation of ascorbic acid in glassy isomalto-oligosaccharides by extrusion. J. Food Sci. Biotech. 30, 185–190.

Chen, X.-P., Li, W., Xiao, X.-F., Zhang, L.-L., Liu, C.-X., 2013. Phytochemical and pharmacological studies on Radix Angelica sinensis. Chinese J. Nat. Med. 11, 577–587.

Chu, W., Fu, X., Fan, Z., Li, S., 2010. Study on effect of fructooligosaccharide of bowel movement on the constipation model mice. Chinese Remed. Clin. 10, 1135–1136.

Feng, Y., He, Z., Xu, L., Zhang, Z., Shi, L., Chen, X., 2009. Nutritive elements analysis and evaluation of maca (*Lepidum meyenii*) cultivated in Yunnan. Forest Res. 22, 696–700.

Gan, J., Feng, Y., Zhang, H., He, Z., Zheng, H., Li, X., 2013. Analysis on composition and content of sterols in three color types of maca, *Lepidium meyenii*. Forest Res. 26, 129–133.

Gao, J., Xu, Z., Luo, Y., Xu, L., Bi, Y., Wang, Z., Xiao, W., 2015. Studies on drying process of extract of *Cistanche tubulosa*. Mod. Chinese Med. 17, 387–391.

Geng, X., Song, L., Pu, X., Tu, P., 2004. Neuroprotective effects of phenylethanoid glycosides from *Cistanches salsa* against 1-methyl-4-phenyl-1,2,3,6-tetrahydropyridine (MPTP)-induced dopaminergic toxicity in C57 mice. Biol. Pharm. Bull. 27, 797–801.

Guo, Z., Yu, J., Liu, R., Ju, Y., Xiao, Q., 2004. Studies on culture of *Cistanche salsa* callus and synthesis of phenylethanoid glycosides. Chinese Trad. Herbal Drugs 35, 204–207.

Hao, L., Chen, L., Zhong, N., Chen, K., Li, G., 2005. Separation, purification and structure of burdock oligosaccharide. Chem. J. Chinese U. 26, 1242–1247.

HKTDC, 2014. China's health food market. http://china-trade-research.hktdc.com/business-news/article/China-Consumer-Market/China-s-health-food-market/ccm/en/1/1X000000/1X002L54.htm

Hu, C., 2014. Historical change of raw materials and claims of health food regulations in China. In: Bagchi, D. (Ed.), Nutraceutical and Functional Food Regulations in the United States and Around the World. second ed. Academic Press, San Diego, CA, pp. 363–388, (Chapter 21).

Jiang, J., Cui, L., Sun, E., Li, J., Cheng, X., Ding, S., Feng, L., Jia, X., 2014. Material basis for anti-osteoporosis efficacy of Epimedium flavonoids based on their in vivo metabolism. Chinese Trad. Herbal Drugs 45, 721–729.

Jiang, Y., Dong, N., 2014. Registration, review and approval of tablets functional food in China. Chinese J. Exp. Trad. Med. Formul. 20, 225–227.

Jiang, Z., Xue, W., Liu, K., Jin, W., Wang, R., Lu, R., Xie, H., 2013. Subchronic toxicity of *Cistanche tubulosa* extract on SD rats. J. Prev. Med. Info. (Chinese), 29.

Lei, L., Song, Z., Tu, P., 2003. Advances in research of chemical constituents in plants of cistanche Hoffing et Link. Chinese Trad. Herbal Drugs 35, 473–475.

Li, J., Li, C., Qiu, Z., Shen, H., Liu, H., 2006. Control of browning in cell subculture of *Cistanche deserticola*. Chinese J. Process Eng. 6, 450–453.

Li, J., Li, D., 2011. Current status of Chinese legal system and standard system for health food. Chinese J. Food Sci. 32, 318–323.

Li, L., Li, J., Ding, C., 2013. Determination of anthocyanins in fruits of *Lycium ruthenicum* Murr. by HPLC. Chinese J. Food Sci. 34, 164–166.

Li, Y., Li, R., Yang, M., Sun, Z., Shan, Y., 2014. Study on antioxidant effect of *Lycium ruthenicum* buccal tablet in vivo. Chinese J. Exp. Trad. Med. Formul. 20, 190–193.

Ma, H., He, X., Yang, Y., Li, M., Hao, D., Jia, Z., 2011. The genus Epimedium: an ethnopharmacological and phytochemical review. J. Ethnopharmacol. 134, 519–541.

Ma, S., Le, G., Shi, Y., Shun, J., Jia, L., Wang, H., 2013. Effects of xylooligosaccharides on antioxidant capacity, blood lipid and microorganism of the intestinal tract in high fat mice. Science, available from: http://www.paper.edu.cn/html/releasepaper/2013/03/203/

Ni, H., Zhang, J., Qing, D., Jia, X., 2012. Study on the anti-fatigue effect of compound *Cistanche tubulosa* tablet. J. Xinjiang Medical U. 35, 687–689.

Ni, W., Gao, T., Wang, H., Du, Y., Li, J., Li, C., Wei, L., Bi, H., 2013. Anti-fatigue activity of polysaccharides from the fruits of four Tibetan plateau indigenous medicinal plants. J. Ethnopharmacol. 150, 529–535.

Peng, Q., Liu, H., Shi, S., Li, M., 2014. *Lycium ruthenicum* polysaccharide attenuates inflammation through inhibiting TLR4/NF-κB signaling pathway. Int. J. Biol. Macromol. 67, 330–335.

Pharmacopoeia, C.O.C., 2010. Pharmacopoeia of People's Republic of China, 2010 vol. 1. Chinese Medical Science Press, Beijing, China.

Qin, K., Qian, Y., 2003. Advance in garnoderma immune pharmacology. Pharmacol. J. Chinese PLA 19, 360–362.

Ren, L., Gu, X., Tao, G., Tang, J., 2005. Study on absorption and isolation of the macroporous resin for the phenylethanoid glycosides from *Cistanche tubulosa* (Schenk) R Wight. Chinese J. Food Sci. 26, 227–231.

Sa, Y., 2013. Approval and common questions on the registration assessment of softgel health food containing traditional Chinese medicine. Chinese J. Info. Trad. Chinese Med. 20, 4–6.

Shang, R., Wang, B., Xu, L., 2013. Study on the growth dynamic of medicine-food dural used maca (*Lepidum meyenii*). Chinese Agr. Sci. Bull. 29, 134–138.

Sheng, Z., Hu, X., Xu, X., 2013. Preparation and identification of the cistanche phenylethanoid glycosides-β-cyclodextrin inclusion complex. http://www.paper.edu.cn/html/releasepaper/2013/04/7/

Song, H., Jia, X., Xie, H., 2010. Research progress on chemical composition of *Cistanche tubulosa*. Xinjiang Trad. Chinese Med. 28, 76–78.

Su, Y., He, H., Sun, Z., 2002. China goji berry resource and its current status and prospective in food industry. Chinese J. Food Sci. 23, 292–294.

Sun, Y., 2014. Biological activities and potential health benefits of polysaccharides from Poria cocos and their derivatives. Int. J. Biol. Macromol. 68, 131–134.

Tang, Y., He, W., Liao, X., Yang, B., 2015. Comparison of extraction methods of active ingredients of maca. Sci. Technol. Food Ind. (Chinese) 36, 189–193.

Tian, X.F., Pu, X.P., 2005. Phenylethanoid glycosides from *Cistanches salsa* inhibit apoptosis induced by 1-methyl-4-phenylpyridinium ion in neurons. J. Ethnopharmacol. 97, 59–63.

Tu, P., Wang, B., Deyama, T., Zhang, Z., Lou, Z., 1997. Analysis of phenylethanoid glycosides of herba cistanche by RP-HPLC. Acta Pharmacol. Sinic. 32, 294–300.

Wang, J., Chen, X., Zhang, W., 2009. Study on hypoglycemic function of polysaccharides from lycrium ruthenium Murr. fruit and its mechanism. Chinese J. Food Sci. 30, 244–248.

Wang, X., Wang, X., Wang, X., Jiang, X., 2005. Influence of glycosides of cistanche on the ultrastructure of hippocampus of the D-galactose induced brain aging model mice. Chinese J. Behav. Med. Sci. 14, 966–967.

Wang, Y., Luo, G., Hu, Y., Zhu, Y., Zhang, S., Hu, J., Liu, L., Wang, H., 2000. Analysis of phenylethanoid glycosides in the extract of herba cistanchis by LC/ESI-MS/MS. Acta Pharmacol. Sinic. 35, 839–842.

Wang, Y., Wang, Y., McNeil, B., Harvey, L.M., 2007. Maca: an Andean crop with multi-pharmacological functions. Food Res. Int. 40, 783–792.

Wong, K.H., Li, G.Q., Li, K.M., Razmovski-Naumovski, V., Chan, K., 2011. Kudzu root: traditional uses and potential medicinal benefits in diabetes and cardiovascular diseases. J. Ethnopharmacol. 134, 584–607.

Wu, C.R., Lin, H.C., Su, M.H., 2014. Reversal by aqueous extracts of *Cistanche tubulosa* from behavioral deficits in Alzheimer's disease-like rat model: relevance for amyloid deposition and central neurotransmitter function. BMC Complement. Altern. Med. 14, 202.

Xiao, L., Ma, J., Qin, S., 2010. Safety research on functional oligosaccharides. Sci. Technol. Food Ind. (Chinese), 24–26.

Xiong, W.T., Gu, L., Wang, C., Sun, H.X., Liu, X., 2013. Anti-hyperglycemic and hypolipidemic effects of *Cistanche tubulosa* in type 2 diabetic db/db mice. J. Ethnopharmacol. 150, 935–945.

Xuan, G.D., Liu, C.Q., 2008. Research on the effect of phenylethanoid glycosides (PEG) of the *Cistanche deserticola* on anti-aging in aged mice induced by D-galactose. J. Chinese Medicin. Mater. 31, 1385–1388.

Yan, Y., Dai, G., Ran, L., Luo, Q., Li, X., Qin, K., Wu, P., Zeng, X., Cao, Y., 2014. The polyphenols composition of *Lycium ruthenicum* Murr. from different places. Sci. Agric. Sinic. 47, 4540–4550.

Yang, J.H., Hu, J.P., Rena, K., Du, N.S., 2009. Structure-activity relationships of phenylethanoid glycosides in plants of *Cistanche salsa* on antioxidative activity. J. Chinese Medicin. Mater. 32, 1067–1069.

Yang, Z., Pu, Y., He, R., 2004. Recent advances on biological effects and safety of fructooligosaccharide. Life Sci. Res. 8, 122–126.

Yao, X., Zhang, L., Yang, Y., 2013. Research progress on 4 oligosaccharides and L-arabrose in intestinal health. J. Hyg. Res. 41, 341–345.

Yi, S., Chao, Y., 2014. Analysis of approved health food containing Chinese materia medica and advice for supervision and management. Chinese Trad. Herbal Drugs 45, 1353–1536.

Zenico, T., Cicero, A.F., Valmorri, L., Mercuriali, M., Bercovich, E., 2009. Subjective effects of *Lepidium meyenii* (Maca) extract on well-being and sexual performances in patients with mild erectile dysfunction: a randomised, double-blind clinical trial. Andrologia 41, 95–99.

Zhang, H., Li, B., Gao, Y., Qin, H., 2014c. Experimental study of maca powder on immunity enhancement and anti-fatigue effect in mice. Mod. Prevent. Med. 41, 1096–1099.

Zhang, H., Weng, X., Chen, L., Li, X., 2008. Effect of *Cistanche tubulosa* (Scheuk) Whight acteoside on telomerase activity and immunity of aging mice. Chinese J. Pharmacol. Toxicol. 22, 270–273.

Zhang, J., Qing, D., Sun, Y., Ni, H., Jia, X., 2014a. Study of safety of water extract of cistanche tubulosa. Chinese J. Trad. Med. Technol. 21, 413–414.

Zhang, W., Zheng, H., Li, K., Zhang, H., Xu, J., Zhang, L., 2014b. Optimization of spray drying process for the preparation of maca fine powder. Chinese J. Food Sci. 35, 34–38.

Zhao, C., Wei, L., 2005. Protective effects of puerarin on CCl4-induced acute hepatic injury in rats. Acta Acad. Medic. Milit. Tert. 27, 625–628.

Zheng, J., Ding, C., Wang, L., Li, G., Shi, J., Li, H., Wang, H., Suo, Y., 2011. Anthocyanins composition and antioxidant activity of wild *Lycium ruthenicum* Murr. from Qinghai-Tibet Plateau. Food Chem. 126, 859–865.

Zhou, H., Cao, J., Lin, Q., 2012. Effect of *Cistanche herba* on the swimming ability and oxidation resistance of mitochondrial in rats. Chiense J. Exp. Trad. Med. Formul. 18, 229–234.

Zhu, Z., Xue, J., Liu, X., Meng, M., Yang, Y., Zhang, J., Guo, R., 2014. Puerarin and isoflavones from Kudzu root (*pueraria lobata ohwi*) prevent acute drunkenness and relieve alcoholism in mice. Chinese J. Food Sci. 35, 247–251.

CHAPTER

Functional foods in line with young consumers: challenges in the marketplace in Slovenia

20

Tina Vukasović

DOBA Faculty, Prešernova, Maribor, Slovenia; International School for Social and Business Studies, Mariborska cesta, Celje, Slovenia; Faculty of Mathematics, Natural Sciences and Information Technologies, University of Primorska, Glagoljaška, Koper, Slovenia

20.1 INTRODUCTION

Development of the world food market has, in recent years, been marked by rapid, unexpected, and complex changes. The world food industry is operating in an explicitly dynamic environment which demands constant adjustments and responses. These intensive processes increase the competitive advantage of operating activities on the global market and are the result of numerous changes in the period of the new economy (Vukasović, 2009). In this saturated market environment, distribution channels, marketing activities, diversification strategies, and food quality are increasingly important. In addition, consumers have become more concerned about the nutrition, health, and quality of food they eat (Gil et al., 2000).

The Slovenian agrofood sector plays an important socioeconomic role but it has been affected recently by many crises. Food quality and safety became the most important concern of consumers. Therefore, new safety and quality rules and standards have been designed by the public and private sectors and they are widely adopted as brands and labels. Functional foods are foods that provide the body with not only the necessary nutrients, but because of their composition also an additional beneficial contribution to health. The consumer usually identifies it by giving notice to various messages on the specific characteristics of food: nutrition and health claims. They were allowed on the Slovenian market after 2006, when the European Union adopted a Regulation which unifies the use of claims across the entire territory of the Community.

Research related to consumer attitudes and preferences for functional foods is very sparse, especially in Slovenia. Results of previous studies (Nashar and Ivanova, 2015; Weststrate et al., 2002; Brečić et al., 2014; Schnettler et al., 2015; Markovina et al., 2011) confirmed that consumers have positive attitudes toward

Developing New Functional Food and Nutraceutical Products. http://dx.doi.org/10.1016/B978-0-12-802780-6.00020-1

functional foods. Because of the opportunity to make statements on food labels related to the health benefits of functional foods, it is not surprising that major companies are interested in developing such foods for the health and wellness market.

To the best of our knowledge, there has been no latest research pertaining to consumer attitudes toward functional foods in Slovenia. The goal of this chapter is to explore the Slovenian young consumers' perception of functional foods, to investigate underlying attitudes and their willingness to buy functional foods in the future. Consumer survey was conducted using a self-administered questionnaire on a sample of 150 young consumers aged between 14 and 35 years. Research questions focused on knowledge about functional foods, attitudes, and buying behavior regarding functional foods. Slovenian young consumers express a positive attitude toward functional foods. The majority of the respondents purchase functional foods, usually in hypermarkets and supermarkets. Apart from taste, the most important functional foods attribute is price/quality ratio. Three factors that explain young consumers' attitudes toward functional foods are health awareness and confidence, lack of trust for functional foods, and food's price and quality. The results of this research could be used in planning further development of the functional foods market for young consumers. Overall positive attitudes toward functional foods should be reinforced and young consumers can be influenced through targeted advertising. This chapter gives the latest account of young consumers' perception of functional foods in Slovenia. The results of this research can be used to plan further marketing activities.

20.2 FUNCTIONAL FOODS CONCEPT AND DEFINITION

Functional foods cannot be a single well-defined/well-characterizable entity. Indeed, a wide variety of food products are, or will in the future be, characterized as functional foods with a variety of components, some of them classified as nutrients, affecting a variety of body functions relevant to either a state of well-being and health and/or the reduction in risk of a disease. Thus no simple, universally accepted definition of functional foods exists. Especially in Europe, where even the common term "dietary fiber" has no consensual definition, it would be unrealistic to try to produce such a definition for something as new and diverse as functional foods. Functional foods has thus to be understood as a concept. Moreover, if it is function driven rather than product driven, the concept is likely to be more universal and not too much influenced by local characteristics or cultural traditions (Diplock et al., 1999).

Japan is the birthplace of the term "functional foods." Moreover, that country has been at the forefront of the development of functional foods since the early 1980s when systematic and large-scale research programs were launched and funded by the Japanese government on systematic analysis and development of food functions, analysis of physiological regulation of function by food, and analysis of functional foods and molecular design. As a result of a long decision-making process to establish a category of foods for potential enhancing benefits as part of a national effort to reduce the escalating cost of health care, the concept of foods for specified health use (FOSHU) was established in 1991. These foods, which are intended to be used to improve people's health and for which specific health effects (claims) are allowed to be displayed, are included as one of the categories of foods described in the Nutrition Improvement Law as foods for special dietary use. According to the Japanese Ministry of Health and Welfare, FOSHU are (Roberfroid, 2001, pp. 11, 12):

- foods that are expected to have a specific health effect due to relevant constituents, or foods from which allergens have been removed, and
- foods where the effect of such an addition or removal has been scientifically evaluated, and permission has been granted to make claims regarding the specific beneficial effects on health expected from their consumption.

Functional foods has as many definitions as the number of authors referring to it. These definitions go from simple statements such as:

- foods that may provide health benefits beyond basic nutrition,
- foods or food products marketed with the message of the benefit to health (Riemersma, 1996),
- everyday food transformed into a potential lifesaver by the addition of a magical ingredient (Coghlan, 1996)

To very elaborate definitions such as:

- food and drink products derived from naturally occurring substances consumed as part of the daily diet and possessing particular physiological benefits when ingested (Hillian, 1995)
- food derived from naturally occurring substances that can and should be consumed as part of the daily diet and that serve to regulate or otherwise affect a particular body process when ingested (Smith et al., 1996)
- food similar in appearance to conventional food, which is consumed as part of a usual diet and has demonstrated physiological benefit and/or reduces the risk of chronic disease beyond basic nutritional functions (Health Canada, 1997)
- food that encompasses potentially helpful products including any modified food or food ingredient that may provide a health benefit beyond that of the traditional nutrient it contains (Food and Nutrition Board, 1994)
- food similar in appearance to conventional food that is intended to be consumed as part of a normal diet, but has been modified to subserve physiological roles beyond the provision of simple nutrient requirements (Roberfroid, 2001, p. 13).

Whatever definition is chosen, "functional foods" appears as a quite unique concept that deserves a category of its own, a category different from nutraceutical, f(ph)armafood, medifood, designer food, or vitafood, and a category that does not include dietary supplements. It is also a concept that belongs to nutrition and not to pharmacology. Functional foods are and must be foods, not drugs, as they have no therapeutic effects. Moreover their role regarding disease will, in most cases, be in reducing the risk of disease rather than preventing it (Roberfroid, 2001, pp. 13, 14).

20.3 FUNCTIONAL FOOD: EUROPEAN PERSPECTIVES

The unique features of functional foods are (Diplock et al., 1999):

- being a conventional or everyday food
- to be consumed as part of the normal/usual diet
- composed of naturally occurring (as opposed to synthetic) components perhaps in unnatural concentration or present in foods that would not normally supply them
- having a positive effect on target function(s) beyond nutritive value/basic nutrition
- may enhance well-being and health and/or reduce the risk of disease or provide health benefits so as to improve the quality of life including physical, psychological, and behavioral performances
- have authorized and scientifically based claims.

It is in that general context that the European Commission's Concerted Action on Functional Food Science in Europe (FUFOSE), which actively involved a large number of the most prominent European experts in nutrition and related sciences, has been coordinated by the International Life Science Institute—ILSI Europe. It developed in early 1996 to reach a European Consensus on "Scientific Concepts of Functional Foods" which was published in 1999. It should be emphasized that a functional food will not necessarily be functional for all members of the population, and that matching individual biochemical needs with selected food component intakes may become a key task as we progress in our understanding of the interactions between genes and diet (Kok, 1999).

From a practical point of view, a functional food can be (Roberfroid, 2001, pp. 15, 16):

- a natural food
- a food to which a component has been added
- a food from which a component has been removed
- a food where the nature of one or more components has been modified
- a food in which the bioavailability of one or more components has been modified, or
- any combination of these possibilities

20.4 THE CONSUMER SELF-CARE PHENOMENON

Numerous surveys conducted over the last decade have indicated that increasing numbers of consumers are taking greater responsibility for their own health and well-being, and that they are increasingly turning to their diet to enable them to do so. The tendency for consumers to view the "kitchen cabinet as the medicine cabinet" was initially identified as a leading trend in the food industry in 1994 (Sloan, 1994). This "self-care" phenomenon remains a leading consumer trend today (Sloan, 2002). The 10th annual consumer trend report from the Food Marketing Institute and Prevention Magazine found that 76% of consumers strongly or mostly agree that eating healthfully is a better way to manage illness than medication (Food Marketing Institute and Prevention Magazine, 2001).

The aging demographics of the 21st century will continue to fuel this self-care phenomenon. Those over the age of 50 years will increase by 48% compared with 16% for the 13–24-year-old age group over the next decade. Of greater importance, however, is the growth in the number of individuals >65 years old. By 2035, ~70 million people will be in this age bracket (Dychtwald, 1999). With a continued increase in the overall age of the population, chronic diseases of aging such as heart disease, cancer, osteoporosis, Alzheimer's disease, and age-related macular degeneration, among others, are inevitable, imposing an enormous stress on the cost of health care. Consumer interest in self-care and dissatisfaction with the current health care system will continue to be a leading factor motivating consumer food purchasing decisions (Hasler, 2002).

20.5 THE FUTURE OF FUNCTIONAL FOODS

Extensive research is currently directed toward increasing our understanding of "functional foods." Academic, government, and private research institutes around the globe are devoting substantial efforts to identifying how functional foods and food ingredients might help prevent chronic disease or optimize health, thereby reducing healthcare costs and improving the quality of life for many consumers. An emerging discipline that will have a profound effect on future functional foods research and development efforts is *nutrigenomics*, which investigates the interaction between diet and development of diseases based on an individual's genetic profile (Fogg-Johnson and Meroli, 2000). Interest in nutrigenomics was greatly augmented by the recent announcement that a rough draft of the complete sequence of the human genome had become available. In Feb. 2001, the complete sequence of the human genome was announced by Ventor and colleagues (The Celera Genomics Sequencing Team, 2001). This technological breakthrough could eventually make it feasible to tailor a diet for an individual's specific genetic profile. Nutrigenomics will have a profound effect on future disease prevention efforts including the future of the functional foods industry. Another technology that will greatly influence the future of functional foods is biotechnology (Gura, 1999). Recent examples of biotechnology-derived crops which have tremendous potential to improve the health of millions worldwide include golden rice and iron-enriched rice (Institute of Food Technologists, 2000). These grains are genetically engineered to provide enhanced levels of iron and β-carotene which could, in turn, help prevent iron deficiency anemia and vitamin A deficiency-related blindness worldwide. In the future, other foods enhanced with other nutritive or nonnutritive substances may even help to prevent chronic diseases such as heart disease, osteoporosis, or cancer (Falk et al., 2002). The acceptance of biotechnology by consumers (currently a major issue in Europe) will be important if the potential of this powerful methodology is to be realized (Hasler, 2002).

Because of the opportunity to make statements on food labels related to the health benefits of functional foods, it is not surprising that major companies are interested in developing such foods for the health and wellness market (Hasler, 2002). The market for foods positioned for their health benefits will continue to be strong for the next several decades given the consumer interest in self-care, aging demographics, and increasing health care costs.

20.6 MARKERS: A KEY TO DEVELOPMENT OF FUNCTIONAL FOODS

The development of functional foods will, in most cases, rely on measurements of "markers." These markers need to be identified and validated for their predictive value of potential benefits to a target function or the risk of a particular disease. Markers of correlated events are "indicators" whereas markers representing an event directly involved in the process are "factors" (Diplock et al., 1999). When related to the risk of disease, indicators and even factors might, in some instances, be equivalent to "surrogate markers" defined as a biological observation, result, or index that predicts the development of chronic disease (Keystone, 1997).

The more that is known about the mechanisms leading to health outcomes, the more refined will be the identification of the markers and their appreciation. The markers should be feasible, valid, reproducible, sensitive, and specific. They can be biochemical, physiological, behavioral, or psychological in nature. However, dynamic responses might be as useful as, or more useful than, static or single point measurements. In many cases, a battery of markers might be needed in order to create a decision tree from multiple tests (Fig. 20.1).

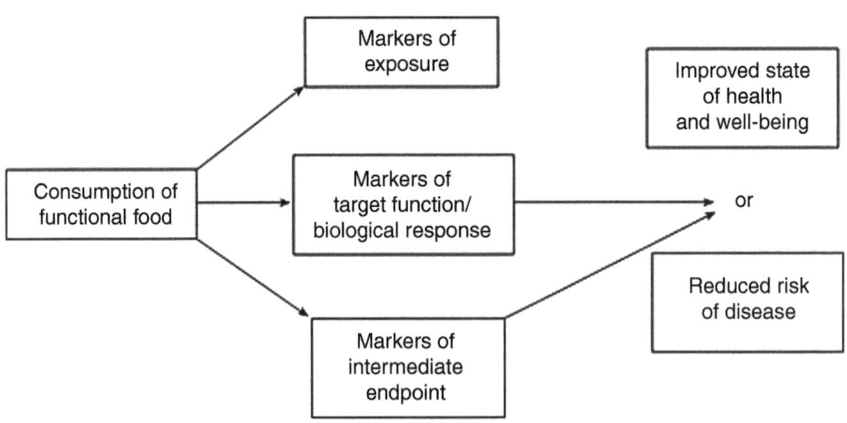

FIGURE 20.1 Markers for Functional Foods Development

Roberfroid (2001, p. 18).

These markers, most of which still need to be identified and validated, will relate to (Roberfroid, 2001, p. 18):

- exposure to the food component under study by measuring serum, fecal, urine, or tissue level of the food component itself or its metabolite(s), or the concentration of an endogenous molecule that is directly influenced by the consumption of the food component
- target function(s) or biological response such as change in serum or other body fluids of the concentration of a metabolite, a specific protein, an enzyme, a hormone, etc.

These first two markers are either indicators or factors (Roberfroid, 2001, p. 18):

- An appropriate endpoint of an improved state of well-being and health and/or reduction of a disease risk. Such a marker is likely to be a factor rather than an indicator.
- Individual susceptibility or genetic polymorphism controlling the metabolism and/or the effect of the food component under study.

To further develop these markers, a state-of-the-art literature review will be necessary to identify, define, and characterize potential markers. Furthermore, the basic scientific knowledge underpinning these markers will be evaluated. The next step will include assessment of their relevance to physiological function, to well-being and health, and eventually to disease risk. A validation will then be necessary both for the methodology and for biological relevance. Finally, classification as indicator or factor will be made and potential dietary modulations demonstrated. New techniques such as those used by molecular and cellular biologists will be useful in identifying target groups who could benefit from the consumption of specific functional foods (Roberfroid, 2001, p. 19).

20.7 RESEARCH MATERIAL AND METHODS

20.7.1 METHOD AND SAMPLE

In the next part of the chapter the key research characteristics are summarized. Details of the research methods can be found in Vukasović (2013, 2014). Consumer survey was conducted using a self-administered questionnaire on a sample of 150 young consumers aged between 14 and 35 years. Research questions focused on knowledge about functional foods, attitudes, and buying behavior regarding functional foods. The sample consisted of 85 women and 65 men. The most respondent of consumers were in the group between 25 and 35 years; 85% of respondents currently living in cities visit rural areas regularly, at least once a month. The respondents are interested in functional foods, suggesting that the survey respondents provide an interesting study group for this issue.

20.7.2 QUESTIONNAIRE

The introduction letter, reminder, and draft questionnaire were developed for the purpose of the research. The guidelines were used in order to give the questionnaire a good look and feel, and to ensure that respondents could progress quickly through it. All questionnaires carried a stamped number in order to be able to add factual consumers' data. The survey examined various factors and determinants that influence the attitudes, perceptions, knowledge, and consumption of functional foods in Slovenia. The data were collected using a structured questionnaire administered on the field survey in Apr. 2015 with the use of closed (questions with multiple choice answers) and open questions (questions that were not given alternatives as answers). Several attributes of functional products were measured by means of the five-point Likert scale anchors of 1 = low or bad degree of the characteristic, and 5 = high or good degree of the characteristic.

The survey concept is based on prior literature. Magnusson et al. (2001) learned that the most important purchase criteria for food are: taste, health, long shelf-life, and quality aspects of the product. A survey conducted in European Community (EC) countries (Lennernas et al., 1997) has shown that quality/freshness, taste, healthy diet, price, family preferences, and habits are the most important food-choice factors. Wandel and Bugge (1997) showed that taste, freshness, appearance, and nutritional value are significant in choice of fruits and vegetables. Verbeke (2001) demonstrated that the top five important attributes for fresh meat in Belgium are freshness, quality, taste, healthiness, and free of hormones (Radman, 2005). Vukasović (2011) demonstrated that important attributes for fresh meat in West Balkans Countries (WBC) are taste and healthiness and the most important attributes for traditional products in WBC are special taste, consistent quality, nutritional value, healthy, and safe products (Vukasović, 2014).

20.7.3 DATA ANALYSES

Analysis of the quantitative data provided by the mentioned questionnaire used the Statistical Package for Social Sciences (SPSS 17.0). The data obtained from the survey were analyzed with univariate analysis in order to check distributions of frequencies and to detect possible errors occurring during the research and/or data entering. Chi-square, ANOVA, and correlation analysis were performed to examine the differences in buying behavior and attitudes among respondents. The level of comparison was set at 0.05. Prior to hypotheses testing, factor analysis was conducted to determine the independent variables of the study. Ambiguous items were eliminated from the survey by varimax rotation. The data collected from the surveys were put through a validity assessment (KMO = 0.862; Bartlett's test sign = 0.000), thus revealing that the sample of the study was appropriate for factor analysis and that there is a strong relationship between the variables. The reliability analysis of the research instrument yielded a Cronbach's alpha value of 0.88 and a significance level of 0.000.

20.8 RESULTS AND DISCUSSION

20.8.1 LEVEL OF RECOGNITION OF THE TERM "FUNCTIONAL FOODS"

The data of quantitative research have shown that knowledge and recognition are not high for functional foods. About 43% of young consumers are familiar with the concept of functional foods. Slovenian young consumers define functional foods as tasty foods, quality foods, healthy foods, natural foods, and safe foods. The conclusion that can be drawn from these facts is that better knowledge could be achieved through constant education of young consumers. This means that there is a need in Slovenia to continuously inform the young consumer more about the meaning of functional foods and the meaning of the labeling put on products. It is in the best interest of all subjects participating in the processes of production, research, and consumption of food to lift the quality of nutrition in Slovenia to a higher level. Information on functional foods is an important factor, representing the only instrument that young consumers have to differentiate the attributes of functional foods from others, and to form positive attitudes and quality perceptions toward these foods. The knowledge on functional foods is determined by sociodemographic variables (education level, income, values, lifestyle), but it is also affected by information provided by public administration, mass media, environmental associations, and shopping sites. Similar results were found in the research by Verbeke (2005) where about 46.5% of consumers are familiar with the concept of functional foods. Also similar results were found in the research by Stojanović and Dragutinović-Mitrović (2012) for consumers in Serbia who cannot verify health claims either before or after purchase/consumption, since the property rights on information are extremely weak.

20.8.2 CONSUMERS' FUNCTIONAL FOODS BUYING BEHAVIOR

In the next step of the survey all young consumers who are not purchasing functional foods were excluded from further evaluation. The reasons for not purchasing functional foods were evaluated. Particular reasons why the young consumers were denying the functional foods were the higher price (88%) and the unavailability of the purchase of functional foods offered (82% of respondents pointed this out).

The next question the researcher asked respondents about functional foods was whether they believed differences between nonfunctional and functional foods truly existed. The great majority of young respondents believed there were differences (91%); only 9% doubted that such differences existed. Only young consumers that claimed to believe in the differences were considered for further analysis.

In the case of the frequency and location of functional foods purchase, most young consumers purchase functional foods in hypermarkets (86%) and supermarkets (81%) (Fig. 20.2). Similar results were found in the research by Markovina et al. (2011) for Croatian young consumers. We assessed whether the gender variable has any effect on the purchase place of functional foods. T-test was performed and Levene's test results indicated equal variances between both male and female young consumer groups ($F = 0.979$; $p = 0.312$). No significant differences could be detected ($p = 0.225 > 0.05$).

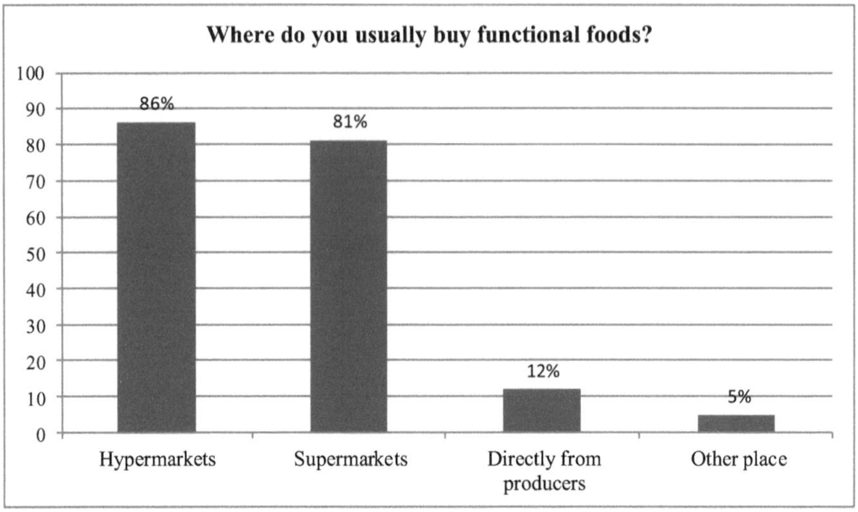

FIGURE 20.2 Usual Purchase Places for Functional Foods

About 35% of respondents decided to buy functional foods "rarely." Further, 48% buy such products "often" or "very often," so they are regular buyers, and 18% of respondents bought functional foods very rarely.

The data for analyzing the assortment of functional foods purchasing were collected by means of open-ended questions. The market for milk and milk products is seen as being the most valuable category of functional foods in the future (91%). The second most frequent purchases are meat and meat products which represent 48%.

Among the purchase motives, tasty foods, quality foods, healthy foods, and safe foods were indicated as the most important reasons of purchasing functional foods (Table 20.1). Similar results were found in the research by Markovina et al. (2011) and Tomić et al. (2014).

20.8.3 PERCEPTION OF FUNCTIONAL FOODS

Since the beginning of the 1980s some trends in food consumption have become evident in WBC: increased vegetables, fruit, cheese, meat, and fish consumption, on the one hand, and decreased sugar consumption on the other. Reasons for these trends which are usually mentioned by experts are increased health concerns and preferences for diet as well as lower demand for calorie intake (Vukasović, 2009). Those trends were also found in our quantitative research. In this part of the research, using the method of semantic differential, in assembly of the structured questionnaire, the standpoints and the perceptions of young consumers toward functional foods have been researched. Characteristics of functional foods, prepared in advance, were in the questionnaire in order to help the young consumers express their points of view toward the mentioned products. Respondents expressed their opinions by marking

Table 20.1 Motives for Functional Foods Choices

Motives for Functional Foods Choices	Mean
Tasty foods	4.83
Quality foods	4.76
Healthy foods	4.65
Safety	4.51
Diet foods	4.03
Nutritional value	3.89
Provides welfare	3.86
Easy for use	3.59

Table 20.2 Functional Foods Perception

Perception	Mean	SD
Healthy	4.96	1.15
Tasty	4.75	1.18
Safety	4.73	1.19
Consistent quality	4.68	1.21
Diet food	4.63	1.22
Provides welfare	4.57	1.14
Natural	4.45	1.29
Easy for use	4.34	1.33
Ethical	4.23	1.27

an answer in the proper spot. They were evaluating the characteristics that refer to functional foods. The average evaluations of the recounted characteristics were calculated based on the acquired evaluations and by that an image of the views of young consumers toward functional foods was determined. This research made in Slovenia shows that young consumers perceive functional foods as healthy, tasty, safe, of consistent quality, diet food, provides welfare, natural, easy for use, and ethical. Most perceptions of functional foods were positive (Table 20.2).

20.8.4 PRICE WILLINGNESS AND FUTURE BUYING INTENTIONS

Respondents were asked to indicate if they were willing to pay a higher price for functional foods compared to nonfunctional foods, and how much extra they were willing to pay. 8.5% of respondents definitely would not pay more money and 12% did not know whether they would pay premium prices for functional foods. Most of the young consumers (57.3%) willing to pay a premium price would pay 11–20% extra, 21% would pay 21–30% extra, 12.1% would pay 31–50% extra, and 9.6% of respondents were willing to pay 51–100% extra for functional foods.

Future buying intention was measured on a five-point scale with one meaning "no future purchase" of functional foods and five meaning "very high probability" of future purchase. About 53% of the respondents intended to purchase functional foods in the future and only 5% claimed that probably or surely they will not purchase functional foods.

In these times of congested markets, companies are increasingly aware of the importance of familiarity with the characteristics of consumers, their motives for the purchase of certain products, and their standpoints on traded objects. Companies pay more attention to market and consumer researches, which are one of the more important factors that determine success on the market (Vukasović, 2009).

20.9 FUTURE TRENDS FOR FUNCTIONAL FOODS MARKET

By reference to the conclusions of the FUFOSE concerted action, future trends will be as follows (Diplock et al., 1999):

- Components in foods have the potential to modulate target functions in the body so as to enhance these functions and/or contribute toward reducing the risk of disease, and functional foods science will contribute to human health in the future provided that evidence is supported by sound scientific, mostly human, data.
- Nutritionists and food scientists have the possibility through the development of functional foods to offer beneficial opportunities related to well-being and health and reduction of risk of disease. Such a new approach in nutrition is strongly dependent upon the identification, characterization, measurement, and validation of relevant markers as defined previously. The design of such studies still needs to be carefully analyzed and specifically developed by reference to, but differently from, classical clinical studies that have been elaborated to help in developing drugs, not food products.
- Major target functions in the body that are or can be modulated by specific food products will have to be identified or characterized. The basic science to understand these functions and how they relate to well-being and health or a particular pathological process needs to be developed so as to give the necessary scientific base to develop new functional foods products.
- Progress in food regulation, which is the means to guarantee the validity of the claims as well as the safety of the food, will have to be made.

In Slovenia the market for functional foods identified the following trends:

- in the interests of consumers: healthy eating and the potential impact of certain foods or ingredients on health and well-being,
- synergy between science and the development of foods for the benefit of the consumer,
- the target consumer of functional foods: a person who is very busy, with a little free time, but most of the day spent away from home, eating alone, with family

meetings in the late afternoon, a homemade breakfast, and one evening meal at the afternoon,

- characteristics of functional foods: easy for use and quick to prepare, positive impact on health, sophistication, diversity of flavors, natural and diet foods, provides welfare, ethical, and suitable for anytime and anywhere,
- from strict frameworks of medicine and pharmacy to the new dimensions of efforts to maintain and improve health.

Optimized nutrition is a major challenge for nutritional science in the 21st century. The development of functional foods is part of this challenge but elaboration of claims should remain basically a scientific challenge, and not primarily a marketing one. The proper scientific validation of functional claims is critical to the success of functional foods, both for the benefit of human health and of the food industry (Roberfroid, 2001, p. 25).

20.10 CONCLUSIONS

This study provides interesting insights for future developments of the Slovenian functional foods market. The definition reveals what young consumers bear in mind when conceptualizing functional foods. In general, an important task for the producers will be to increase consumers' knowledge of what functional foods are and how to differentiate them in the marketplace. Along with knowledgeable and educated young consumers, consumption could be raised to another level. These results are particularly useful for product marketing and future product development in the functional foods sector, as they provide an indication as to how it can better reach and satisfy existing young consumers of functional foods, and possibly, adapt functional foods for other consumer groups in order to attract new segments and a new generation of consumers. This paper gives the latest account of young consumers' perception of functional foods in Slovenia. The results of this research can be used to plan further marketing activities. A limitation of this study is the sample including only young consumers. Further research is needed to analyze other consumer groups, and to determine which segments are most appropriate for functional food products marketing. Research results indicate a need to increase consumer familiarity with functional foods and improvements in some of its characteristics. Overall positive attitudes toward functional foods should be reinforced and young consumers can be influenced through targeted advertising, and also point-of-sale promotional activities.

REFERENCES

Brečić, R., Gorton, M., Barjolle, D., 2014. Understanding variations in the consumption of functional foods—evidence from Croatia. Brit. Food J. 116 (4), 662–675.

Coghlan, A., 1996. A plateful of medicine. New Sci. 2054, 12–13.

Diplock, A.T., Aggett, P.J., Ashwell, M., Bornet, F., Fern, F.B., Roberfroid, M.B., 1999. Scientific concepts of functional foods in Europe: consensus document. Brit. J. Nutr. 81 (1), 1 S1–S28.

Dychtwald, K., 1999. Age power. Age Power: How the 21st Century Will be Ruled by the New Old. Jeremy P. Tarcher/Putnam, New York, NY, pp. 1–30.

Falk, M.C., Chassy, B.M., Harlander, S.K., Hoban, 4th, T.J., McGloughlin, M.N., Akhlaghi, A.R., 2002. Food biotechnology: benefits and concerns. J. Nutr. 132, 1384–1390.

Fogg-Johnson, N., Meroli, A., 2000. Nutrigenomics: the next wave in nutrition research. Nutraceut. World 3, 86–95.

Food and Nutrition Board, Institute of Medicine, National Academy of Sciences, 1994. In: Thomas, P.R., Earl, R. (Eds.), Opportunities in the Nutrition and Food Sciences. National Academy Press, Washington, DC.

Food Marketing Institute and Prevention Magazine, 2001. The top 10 functional food trends: the next generation. Shopping for Health. Reaching Out to the Whole Health Consumer. Food Marketing Institute, Washington, DC.

Gil, J.M., Gracia, A., Sanchez, M., 2000. Market segmentation and willingness to pay for organic products in Spain. Int. Food Agribus. Manag. Rev. 3 (2), 207–226.

Gura, T., 1999. New genes boost rice nutrients. Science 285, 994–995.

Hasler, C.M., 2002. Functional foods: benefits, concerns and challenges—a position paper from the American Council on Science and Health. J. Nutr. 132 (12), 3772–3781.

Health Canada, 1997. Policy Options Analysis: Nutraceuticals/Functional Foods. Health Canada, Health Protection Branch, Therapeutic Products Programme and Food Directorate, Ottawa.

Hillian, M., 1995. Functional foods: current and future market developments. Food Technol. Internat. Europe, 8, 25–31.

Institute of Food Technologists, 2000. IFT expert report on biotechnology and foods. Food Technol. 54, 61–80.

Keystone, 1997. The Keystone national policy dialogue on food nutrition and health: executive summary. J. Nutraceut. Funct. Med. Foods 1, 11–32.

Kok, F.J., 1999. Functional foods: relevance of genetic susceptibility.Proceedings of Forum on Functional Food, Council of Europe Publications, Strasbourg, pp. 217–229.

Lennernas, M., Fjellstrom, C., Becker, W., Giachetti, I., Schmitt, A., Remaut de Winter, A.M., Kearney, M., 1997. Influences on food choice perceived to be important by nationally representative samples of adults in the European Union. Eur. J. Clin. Nutr. 51 (2), S8–S15.

Magnusson, M.K., Arvola, A., Koivisto Hursti, U.K., Aberg, L., Sjoden, P.O., 2001. Attitudes towards organic foods among Swedish consumers. Brit. Food J. 103 (3), 209–226.

Markovina, J., Čačić, J., Gajdoš Kljusurić, J., Kovačić, D., 2011. Young consumers' perception of functional foods in Croatia. Brit. Food J. 113 (1), 7–16.

Nashar, M., Ivanova, D., 2015. Functional foods: the new concept emerging from the traditional nutrition. Scripta Sci. Pharmaceut. 2 (1), 15–28.

Radman, M., 2005. Consumer consumption and perception of organic products in Croatia. Brit. Food J. 107 (4), 263–273.

Riemersma, R.A., 1996. A fat little earner. Lancet 347, 775–776.

Roberfroid, M., 2001. In: Gibson, G.R., Williams, C.M. (Eds.), Functional Food: Concept to Product. Woodhead Publishing, Cambridge, UK.

Schnettler, B., Horacio, M., Lobos, G., Sepulveda, J., Orellana, L., Mora, M., Grunet, K., 2015. Willingness to purchase functional foods according to their benefits: consumer profiles in Southern Chile. Brit. Food J. 117 (5), 1453–1473.

Sloan, A.E., 1994. Top 10 trends to watch and work on: the more things change, the more they stay the same. Food Technol. 48, 89–100.

Sloan, A.E., 2002. The top 10 functional food trends: the next generation. Food Technol. 56, 32–58.

Smith, B.L., Marcotte, M., Harman, G.A., 1996. Comparative Analysis of The Regulatory Framework Affecting Functional Food Development and Commercialization in Canada, Japan, the European Union and the United States of America. Intersector Alliance Inc., Ottawa.

Stojanović, Ž., Dragutinović-Mitrović, R., 2012. The Serbian functional food market: does regulation make a difference? Econ. Ann. LVII (193), 53–69.

The Celera Genomics Sequencing Team, 2001. The sequence of the human genome. Science 291, 1304–1351.

Tomić, M., Cerjak, M., Rupčić, I., 2014. Functional foods and the young. J. Food Prod. Market. 20 (5), 441–451.

Verbeke, W., 2001. Beliefs, attitude and behaviour towards fresh meat revisited after the Belgian dioxin crisis. Food Qual. Prefer. 12 (8), 489–498.

Verbeke, W., 2005. Consumer acceptance of functional foods. Socio-demographic, cognitive and attitudinal determinants. Food Qual. Prefer. 16 (1), 45–57.

Vukasović, T., 2009. Consumer perception of poultry meat and the importance of country of origin in a purchase making process. World Poultry Sci. J. 65 (1), 65–74.

Vukasović, T., 2011. The importance of national chicken meat origin in Central and South-Eastern Europe. World Poultry Sci. J. 67 (2), 237–242.

Vukasović, T., 2013. Attitude towards organic meat: an empirical investigation on West Balkans Countries (WBC) consumers. World Poultry Sci. J. 69 (3), 527–539.

Vukasović, T., 2014. Consumer preference for traditional products in West Balkans countries. J. Int. Food Agribus. Market. 26 (3), 173–188.

Wandel, M., Bugge, A., 1997. Environmental concern in consumer evaluation of food quality. Food Qual. Prefer. 8 (10), 19–26.

Weststrate, J.A., van Poppel, G., Verschuren, P.M., 2002. Functional foods, trends and future. Brit. J. Nutr. 88 (2), S233–S235.

CHAPTER

Medical foods and US regulations

21

Claudia Lewis, Michelle Jackson, Kristen Klesh, Matthew Poliner

Venable LLP, Washington, DC, United States

21.1 WHAT EXACTLY IS A "MEDICAL FOOD"?[a]

A "medical food" is a narrow category of food that is defined under the Federal Food, Drug, and Cosmetic Act (FDCA). Specifically, a medical food is "a food which is formulated to be consumed or administered enterally *under the supervision of a physician* and which is intended for the *specific dietary management* of a disease or condition for which *distinctive nutritional requirements*, based on recognized scientific principles, are established by medical evaluation" [emphasis added].[b] Companies may lawfully make disease-related claims in association with medical foods, which is a key distinction between medical and conventional foods. Claims about diseases are generally prohibited for conventional foods, with the limited exception of approved health claims,[c] qualified health claims,[d] and classical nutrient claims.[e]

Unfortunately, the US Food and Drug Administration (FDA), the federal agency that regulates medical foods, has not provided an abundance of rules and regulations regarding "medical foods," and the legislative history is also minimal. In 2013, however, FDA published a draft guidance document entitled, *Draft Guidance for Industry: Frequently Asked Questions about Medical Foods; Second Edition*, which explains that, in the agency's view, medical foods are foods that are "specially formulated and processed (as opposed to a naturally occurring foodstuff used in a natural state) for a patient who is seriously ill or who requires use of the product as a major component of a disease or condition's specific dietary management."[f]

[a]The FDCA defines "food" as "(1) articles used for food or drink for man or other animals, (2) chewing gum, and (3) articles used for components of any such article." 21 USC § 321(f).

[b]See 21 USC § 360ee(b)(3).

[c]See 21 USC § 343(r)(3); 21 CFR § 101.72–101.83.

[d]See FDA, Qualified Health Claims: Letters of Enforcement Discretion, available at <http://www.fda.gov/Food/IngredientsPackagingLabeling/LabelingNutrition/ucm072756.htm>.

[e]21 USC § 343(r)(6); 21 CFR § 101.93(g)(2).

[f]FDA, Draft Guidance for Industry: Frequently Asked Questions about Medical Foods; Second Edition (August 2013).

Developing New Functional Food and Nutraceutical Products. http://dx.doi.org/10.1016/B978-0-12-802780-6.00021-3

Importantly, medical foods have a distinct regulatory framework from other types of foods, including "foods for special dietary use." As discussed below, foods for special dietary use supply special dietary needs for patients prevented from eating normal foods. But, unlike medical foods, foods for special dietary use are "not specifically tailored for use as the nutritional component of the patient's treatment." Examples of foods for special dietary use include certain hypoallergenic foods, certain infant foods, and food products useful in reducing or maintaining weight. To better understand how FDA regulates medical foods as compared to other food categories, it is helpful to review the legislative and regulatory history of medical foods.

21.2 A BRIEF HISTORY OF MEDICAL FOOD REGULATION

Because of their role in mitigating serious symptoms of diseases, what we now consider medical foods were regulated until 1972 as prescription drugs under section 201(g)(1)(B) of the FDCA.[g] Indeed, prior to 1972, manufacturers of medical foods were subject to burdensome requirements, such as submitting Investigational New Drug applications (INDs), conducting clinical trials, and obtaining approval of New Drug Applications (NDAs) before they could bring new medical food products to market. Given the time-consuming and cost-restrictive nature of these requirements, such requirements severely curtailed medical food innovation.

In 1972, as the result of the agency's desire to foster innovation in the development of medical foods and to ensure product availability at reasonable costs, the agency reassessed its position on medical foods. Safety concerns, however, prompted the agency to continue differentiating medical foods from general use foods. By way of example, FDA raised concerns that Lofenalac, an infant product designed for use in the dietary management of a rare genetic condition known as phenylketonuria (PKU), would pose a hazard to healthy infants since it would be nutritionally inadequate for such infants. As a way to help address the agency's safety concerns, FDA reclassified medical foods provided *enterally* (i.e., ingested via the digestive tract) as "foods for special dietary use," but the agency continued to designate medical foods provided parenterally (injected into the body) as drugs subject to FDA's Drug Efficacy Study (DESI) program.[h]

[g]Advanced Notice of Proposed Rulemaking, Regulation of Medical Foods, 61 Fed. Reg. 60661, 60662 (November 29, 1996).

[h]FDA recognized foods for "special dietary use" as early as 1941. Per regulation, FDA stated that the term "special dietary uses" as applied to food for man, meant, among other things, "uses for supplying particular dietary needs which exist by reason of a physical, physiological, pathological or other condition, including but not limited to the conditions of disease, convalescence, pregnancy, lactation, allergic hypersensitivity to food, underweight, and overweight." Amendment to the General Regulations, Regulations for the Enforcement of the Federal Food, Drug, and Cosmetic Act, 6 Fed. Reg. 5921 (November 22, 1941). This part of the regulation remains unchanged in the Code of Federal Regulations. See 21 CFR § 105.3(a)(1) (March 15, 1977).

One year later, when the agency promulgated rules requiring nutrition labeling for certain foods, FDA exempted specific types of foods for special dietary use from the nutrition labeling requirements.[i] As a result, two categories of food for special dietary use were exempted from the nutrition labeling required of other food: (1) any food represented for use as the sole item of the diet and (2) foods represented for use solely under medical supervision in the dietary management of specific diseases and disorders (i.e., what we now consider to be medical foods). FDA indicated that it was exempting two categories because nutrition labeling for foods intended for consumption by the general population was not appropriate for these food products. The term "medical food" was finally defined by statute in the Orphan Drug Amendments of 1988.[j]

In 1990, Congress passed the Nutrition Labeling and Education Act of 1990 (the "NLEA"), which affirmed FDA's authority to mandate nutrition labeling on most foods and also clarified the agency's role in regulating nutrient content claims and health claims on food labels. Additionally, and importantly, it incorporated the definition of medical foods contained in the Orphan Drug Amendments of 1988 into section 403(q)(5)(A)(iv) of the FDCA and *exempted* medical foods from the nutrition labeling, health claim, and nutrient content claim requirements applicable to foods generally.[k] On Jan. 6, 1993, FDA published the final rule on mandatory nutrition labeling. The final rule incorporated the statutory definition of medical foods into the agency's regulations at section 101.9(j)(8) and exempted medical foods from the nutrition labeling requirements.[l]

21.3 "MEDICAL FOODS" VERSUS OTHER FOOD CATEGORIES REGULATED BY FDA

Following the formal statutory definition of medical foods and FDA's decision to exempt this product category from certain food labeling and claim regulations, FDA grew increasingly concerned about the rapid increase in the variety and number of products marketed as medical foods, safety problems associated with the manufacture and quality control of these products, and the potential proliferation of fraudulent claims not supported by sound science.[m]

Accordingly, in 1996, FDA issued an Advanced Notice of Proposed Rulemaking (ANPR) on the "Regulation of Medical Foods."[n] The agency also sought to clarify

[i] 38 Fed. Reg. 2124, 2126 (January 19, 1973).
[j] A "medical food" was defined as "a food which is formulated to be consumed or administered enterally under the supervision of a physician and which is intended for the specific dietary management of a disease or condition for which distinctive nutritional requirements, based on recognized scientific principles, are established by medical evaluation." 21 USC § 360ee(b)(3). The amendments also introduced a subcategory called "orphan medical foods" to be used in the management of "…any disease or condition that occurs so infrequently in the United States that there is no reasonable expectation that a medical food for such a disease or condition will be developed without assistance." Idem.
[k] Nutrition Labeling and Education Act, P.L. 101-535 (1990), 21 USC § 343(q)(5)(A)(iv).
[l] 58 Fed. Reg. 2079, 2151 (January 6, 1993).
[m] See 61 Fed. Reg. 60661–60662 (November 29, 1996).
[n] 61 Fed. Reg. 60661 (November 29, 1996).

the distinct differences between medical foods and foods for special dietary use. Although this ANPR was withdrawn in 2004,[o] due to the sparse regulatory history available to shed light on the agency's views on medical foods, the document remains a go-to source for industry to understand the agency's views of medical food regulation. Indeed, FDA has indicated that its opinions on medical foods remain the same as expressed in that ANPR. The ANPR references several criteria to define a medical food but, in general, there are three broad conditions that a product must satisfy in order to qualify as a "medical food" under the FDCA.

Each of these conditions is discussed in detail next.

Condition 1: "Distinctive Nutritional Requirements": FDA acknowledged in the 1996 ANPR that products in the marketplace purporting to be medical foods were not consistent with the statutory definition of a medical food and that the product category had come to include foods that the agency felt would more appropriately be categorized as "foods for special dietary use." As such, FDA sought to distinguish these two types of foods; the agency started by attempting to clarify what was meant by "distinctive nutritional requirements" in the FDCA's definition of "medical food." Under the FDCA, distinctive nutritional requirements must be based on recognized scientific principles and be established by medical evaluation. As FDA noted in the ANPR, however, the law does not define "distinctive nutritional requirements." The agency proposed two possible interpretations of the phrase: (1) the "Physiological Interpretation" and (2) the "Alternative Interpretation."

In its discussion of the Physiological Interpretation, FDA acknowledged that "distinctive nutritional requirements" could be understood as referring to the body's need for specific amounts of nutrients to maintain homeostasis and sustain life. Based on this interpretation, medical foods are

> *foods that are formulated to aid in the dietary management of a specific disease or health-related condition that causes distinctive nutritional requirements that are different from the nutritional requirements of healthy people. Foods for special dietary use, on the other hand, are foods that are specially formulated to meet a special dietary need, such as a food allergy or difficulty in swallowing, but that provide nutrients intended to meet ordinary nutritional requirements. The special dietary needs addressed by these foods do not reflect a nutritional problem per se; that is, the physiological requirements for nutrients necessary to maintain life or homeostasis addressed by foods for special dietary use are the same as those of normal, healthy persons. These foods are formulated in such a way that only the ingredients or physical form of the diet is different.*[p]

---

[o]69 Fed. Reg. 68834 (November 26, 2004).

[p]61 Fed. Reg. at 60667. The agency provided an example of a person possessing a special dietary need for a food that is in liquid form due to problems swallowing, noting that this special dietary need does not change his or her physiologic nutrient requirements. Along the same lines, a person allergic to gluten may need foods specially formulated, but the food would still provide the same amount of amino acids as needed by the general population because the quantitative and qualitative amount of overall protein required by the body is similar in both healthy and protein-sensitive individuals. Idem.

In its discussion of the Alternative Interpretation, FDA acknowledged that "distinctive nutritional requirement" could be construed to encompass physical and physiological limitations in a person's ability to ingest or digest conventional foods, as well as distinctive physiological nutrient requirements.[q] FDA recognized that this interpretation is consistent with a report of the Life Sciences Research Office of the Federation of American Societies for Experimental Biology (LSRO/FASEB), which stated in its 1990 *Guidelines for the Scientific Review of Enteral Food Products for Special Medical Purposes* that medical foods are for "patients with limited or impaired capacity to ingest, digest, absorb, or metabolize ordinary foodstuffs or certain nutrients contained therein, or (who) have other specialized medically determined nutrient requirements."[r] FDA indicated that the Alternative Interpretation of "distinctive nutritional requirement" would include

> *foods intended for persons not able to ingest foods in certain physical forms (e.g., solid food), foods intended for persons who need a concentrated form of nutrition because of reduced appetite as a result of disease or convalescence, or foods intended for persons who may have other physical limitations on the amount or composition of food that they can consume. Although these types of conditions do not necessarily result in nutrient needs different from those of healthy persons, they represent a situation where it may be necessary that the food be formulated and manufactured within very narrow tolerances to ensure that the food provides most or all of the essential nutrients, as the person for whom the food is intended may not be able to eat a variety of foods to ensure that they meet their nutritional requirements.*[s]

Because the ANPR was a proposal and not a final rule, FDA did not indicate whether it viewed the "Physiological Interpretation" or the "Alternative Interpretation" as more appropriate. Rather, the agency requested comments on which of the two definitions would better protect and promote the public's health.

Condition 2: "Formulated to be Consumed or Administered Enterally under the Supervision of a Physician": FDA recognized the second element required of medical foods as the statutory requirement that a medical food be "formulated to be consumed or administered enterally under the supervision of a physician." Importantly, users of medical foods must be receiving short-term or long-term "active and ongoing" medical supervision (e.g., in a healthcare facility or as an outpatient).[t] Contrary to foods for special dietary use, FDA views medical foods as an integral component of the clinical management of the patient's disease. Again, FDA contrasted foods for special dietary use from medical foods, noting that unlike foods for special dietary

---

[q]Idem.

[r]J.M. Talbot, *Guidelines for the Scientific Review of Enteral Food Products for Special Medical Purposes*, Life Sciences Research Office, Federation of American Societies for Experimental Biology, Bethesda, MD (1990).

[s]61 Fed. Reg. at 60668.

[t]Idem. (citing 56 Fed. Reg. 60377).

use, medical foods are not meant to be merely recommended by a physician as a "part of an overall diet designed to reduce the risk of a disease or medical condition, to lose or maintain weight, or to ensure the consumption of a healthy diet."[u]

Condition 3: "Specific Dietary Management": Finally, the 1996 ANPR recognized the third fundamental statutory element required of a medical food, namely, that it be intended for the "specific dietary management" of a disease or condition. According to FDA, the term "'specific dietary management' evidences that Congress intended [medical] foods to be an integral part of the clinical treatment of patients."[v] FDA further acknowledged, as consistent with this intent, the LSRO/FASEB Panel's conclusion that the objective of incorporating the use of medical foods into patient management was, in part, to "ameliorate clinical manifestations of the disease," "favorably influence the disease process," and "positively influence morbidity and mortality (patient outcomes)."

21.4 CONSIDERATIONS FOR BRINGING A MEDICAL FOOD TO MARKET IN THE UNITED STATES

Once companies have determined that their product meets the requirements set forth above (i.e., meets the definition of a medical food), there are a number of other items that companies must consider. These topics include formulation restrictions, premarket and administrative requirements, manufacturing requirements, claim substantiation requirements, labeling requirements, and FDA's enforcement history.

21.4.1 PREMARKET AND ADMINISTRATIVE REQUIREMENTS

While both medical foods and drugs are permitted to make claims that the product helps mitigate disease, there are significant differences between the premarket approval requirements for the two categories. Unlike drugs, medical foods need not undergo premarket FDA review or approval. And currently, clinical trials evaluating medical foods are not subject to the IND process, although FDA policy on INDs for medical foods is unsettled.[w] This makes the medical food designation appealing to investors looking to send their companies' products on a quicker, cheaper route to market.

[u]Idem.

[v]Idem. at 60668.

[w]In September 2013, FDA issued a Guidance document entitled, *Guidance for Clinical Investigators, Sponsors, and IRBs: Investigational New Drug Applications (INDs) – Determining Whether Human Research Studies Can Be Conducted Without an IND*, that stated "a clinical investigation intended only to evaluate the nutritional effects of a food (including medical foods) would not require an IND, but an investigation intended to evaluate other effects of a food on the structure or function of the body would." However, on October 30, 2015, FDA stayed this section of the guidance document (80 Fed. Reg. 66907).

As a subcategory of foods, however, facilities that manufacture, process, pack, or hold medical foods must still register with FDA as a food facility.[x] In addition, companies importing medical foods into the United States must provide prior notice to FDA, as required of all other types of food.[y]

21.4.2 FORMULATION CONSIDERATIONS

It is important for companies to keep in mind while formulating medical foods that, similar to all other foods, all ingredients in medical foods must be approved food additives (used in compliance with FDA regulations at 21 CFR Parts 172–173 and 179), subject to a prior sanction, or generally recognized as safe (GRAS).[z] In addition, as with all other types of foods, companies are prohibited from using in medical foods any ingredient that was approved as a drug under 21 USC § 355, licensed as a biological product under 42 USC § 262, or for which substantial clinical investigations (as a drug or biological product) have been instituted and for which the existence of such investigations has been made public, unless the ingredient was first used in food or otherwise meets one of the limited exceptions set forth in 21 USC § 331(ll).[aa]

21.4.3 LABELING REQUIREMENTS

While medical foods are exempt from FDA's nutrient content claim and health claim regulations, medical food labels must comply with all other FDA regulations governing food labeling. For example, FDA regulations and the FDCA require the following items to appear on the label of a medical food: statement of identity, net quantity of contents, ingredient declaration, allergen labeling, and the name and place of business of the manufacturer, packer, or distributor of the product.[bb]

21.4.4 COMPLIANCE WITH FDA'S GOOD MANUFACTURING PRACTICES AND OTHER MANUFACTURING RESTRICTIONS

As of the date of publication of this chapter, medical foods are subject to FDA's current good manufacturing practice (cGMP) regulations for food set forth at 21 CFR Part 110. However, companies should take note that FDA finalized new regulations in late 2015 that set forth cGMPs, hazard analysis, and risk-based preventive controls for human food that will apply to medical food companies once they become effective.[cc] The agency has phased-in compliance dates for these new regulations beginning in 2016. In 1996, FDA published its Medical Foods Compliance Program

[x]21 CFR § 1.225.
[y]21 CFR § 1.276.
[z]FDA, *Draft Guidance for Industry: Frequently Asked Questions about Medical Foods*; Second Edition (August 2013).
[aa]21 USC § 331(ll).
[bb]21 CFR § 101.3(a); 21 CFR § 101.105(a); 21 CFR § 101.4; 21 USC § 343(w); 21 CFR § 101.5.
[cc]21 CFR Part 117.

for domestic and imported products as part of the agency's Compliance Program Guidance Manual.[dd] It is not yet clear how the agency will change this program, if at all, in response to the new regulations governing cGMPs, hazard analysis, and risk-based preventive controls. However, as of the date of publication of this chapter, the manual discusses FDA inspections of medical food manufacturers and indicates that FDA will look to the following areas when inspecting medical food companies:[ee] (1) suitability of buildings and facilities, including sanitation; (2) receipt, storage, and testing of raw materials, including packaging; (3) warehousing and distribution procedures; (4) production and processing procedures and controls, including the production and processing procedures used to control potential microbiological contamination; and (5) laboratory quality control specifications and procedures for raw material, in-process, and finished product testing (including the firms' criteria for acceptance/rejection of test results). Moreover, the Compliance Manual discusses identity testing to: (1) ensure nutrient uniformity and stability; (2) ensure nutrient content and quantity as represented on the labeling; (3) identify potential microbiological and environmental contaminants; (4) ensure packaging and seal integrity; (5) production master and batch records; and (6) records/procedures relating to equipment cleaning, laboratory controls, complaints, and returned products.[ff]

In addition, companies should be aware of additional FDA regulations that are applicable to medical foods, generally. Indeed, medical foods are subject to the FDA regulations governing thermally processed low-acid foods packaged in hermetically sealed containers,[gg] regulations governing acidified foods,[hh] and FDA's emergency permit control regulations.[ii]

21.4.5 CLAIM SUBSTANTIATION REQUIREMENTS

At the federal level, both the FDA and the US Federal Trade Commission (FTC) regulate the substantiation for claims made for medical foods. Both agencies prohibit "false and misleading" claims,[jj] and the standard utilized by the agencies is largely consistent. Whether a claim is false and misleading is a fact-specific inquiry. According to the FTC, "Most deception involves written or oral misrepresentations, or omissions of material information. Deception may also occur in other forms of conduct associated with a sales transaction."[kk] Advertising is considered deceptive if it is "likely to mislead reasonable consumers under the circumstances."[ll]

[dd]FDA, Compliance Program Guidance Manual, Medical Foods Compliance Program § 7321.002 (1996).
[ee]Id.
[ff]Id.
[gg]21 CFR Part 113.
[hh]21 CFR Part 114.
[ii]21 CFR Part 108.
[jj]21 USC § 343(a); 15 USC §§ 45, 52, 55.
[kk]See *FTC Policy Statement on Deception*, <https://www.ftc.gov/system/files/documents/public_statements/410531/831014deceptionstmt.pdf>.
[ll]Idem.

In short, the FDA and FTC require that health-related claims be substantiated by "competent and reliable scientific evidence." "Competent and reliable scientific evidence" is defined by the FTC to mean "tests, analyses, research, studies, or other evidence based on the expertise of professionals in the relevant area, that have been conducted and evaluated in an objective manner by persons qualified to do so, using procedures generally accepted in the profession to yield accurate and reliable results."[mm] The FDA and most states use a similar standard.[nn] The FTC notes that "competent and reliable scientific evidence" is a "flexible" standard and that there is "no fixed formula for the number or type of studies required." However, in practice, the regulators will typically expect at least one well-designed human clinical trial to substantiate any health-related claims made for a product.

21.4.6 FDA ENFORCEMENT

To the best of our knowledge, FDA issued its first Warning Letter addressing medical foods in 2001.[oo] Since then, 15 more Warning Letters have been issued, the most recent of which was on Jan. 15, 2016.[pp] In a recent Warning Letter from Dec. 2013, FDA provided some insight into its views on the limitations of the medical food category.[qq] As in its other letters, FDA stated that it considers the statutory definition of medical food to "narrowly constrain" the types of products that fit within this category. Accordingly, FDA told the recipient that

> *a medical food must be intended for a patient who has a limited or impaired capacity to ingest, digest, absorb, or metabolize ordinary foodstuffs or certain nutrients, or who has other special medically determined nutrient requirements, the dietary management of which cannot be achieved by the modification of the normal diet alone and a medical food must provide nutritional support specifically modified for the management of the unique nutrient needs that result from the specific disease or condition.*

In addition, FDA issued a Warning Letter in Aug. 2013 in which the agency indicated that it does not consider the following as diseases or conditions to be eligible

---

[mm] *United States v. Bayer Corp.*, No. CV 07-01(JLL), 2015 WL 5822595, at *3 (D.N.J. September 24, 2015); *see also* FTC, *Dietary Supplements: An Advertising Guide for Industry* (April 2001).

[nn] FDA, *Guidance for Industry: Substantiation for Dietary Supplement Claims Made Under Section 403(r)(6) of the Federal Food, Drug, and Cosmetic Act* (December 2008); see also *State v. Lee*, 2004 WL 5656237, No. 3AN-01-10170 (February 4, 2014) (Alaska judgment citing FTC's standard as the definition of "competent and reliable scientific evidence").

[oo] FDA, Warning Letter, Bristol-Myers Squibb Company (August 29, 2001), <http://www.fda.gov/iceci/enforcementactions/warningletters/2001/ucm178451.htm> (accessed 13.09.13).

[pp] FDA, Warning Letter, Nutri-Dyn Midwest, Inc. (January 15, 2016), <http://www.fda.gov/ICECI/EnforcementActions/WarningLetters/2016/ucm482627.htm> (accessed 01.02.16). However, it is important to note that this letter did not provide a discussion of the agency's specific objections to the medical food claim.

[qq] FDA, Warning Letter, NVN Therapeutics (December 26, 2013), <http://www.fda.gov/ICECI/EnforcementActions/WarningLetters/2013/ucm381353.htm> (accessed 14.12.15).

for treatment and/or maintenance with medical foods: chronic fatigue syndrome, fibromyalgia, leaky gut syndrome, metabolic syndrome, cardiovascular disease, inflammatory bowel disease and/or conditions, Type 2 diabetes, atopic disorders such as eczema, rhinitis, and allergy-responsive asthma, bariatric patients pre- and post-operatively, and peripheral artery disease.[rr]

FDA has also identified other conditions in Warning Letters over the years that the agency considers to be inappropriate for treatment with medical foods, including the following: inflammatory conditions, migraines,[ss] immune system deficiencies, Alzheimer's disease,[tt] arthritis, colitis, constipation, lactose intolerance, diarrhea,[uu] chronic illnesses, failure to thrive, pre/postsurgery conditions,[vv] and vitamin deficiency throughout pregnancy, postnatal, and the lactating periods.[ww] The agency provided one or more of the following reasons for its objections to these conditions: (1) there was inadequate evidence that the disease had distinct nutritional requirements, (2) there was inadequate evidence that the particular food at issue would meet the distinct nutritional requirements, and/or (3) there was no available evidence illustrating that the nutrient levels cannot be achieved by diet modification alone.

Apart from the Warning Letters, FDA has not actively enforced the medical food regulations.

21.5 CONCLUSIONS

The medical food category presents exciting opportunities for companies to make disease-related claims for their products without being subject to the rigors of the drug approval process. However, given the very narrow interpretation of the category by the FDA, companies should be sure to conduct a thorough analysis to confirm that the product in development fits squarely within the agency's constraints on the category. In addition, companies should be aware of the claim substantiation requirements associated with any claims they intend to make for the product. Finally, companies should be careful to comply with all other FDA requirements related to medical foods, including manufacturing, importation, registration, and labeling requirements.

[rr]FDA, Warning Letter, Metagenics (August 13, 2013), <http://www.fda.gov/iceci/enforcementactions/warningletters/2013/ucm367142.htm> (accessed 02.02.16). This is the second Warning Letter Metagenics has received, and the same four products, plus an additional eight, are at issue.

[ss]FDA, Warning Letter, Metagenics (October 1, 2003), <http://www.fda.gov/ICECI/EnforcementActions/WarningLetters/2003/ucm147751.htm> (accessed 13.09.13).

[tt]FDA, Warning Letter, Neuroscience, Inc. (December 19, 2011), <http://www.fda.gov/ICECI/EnforcementActions/WarningLetters/2011/ucm284391.htm> (accessed 02.02.16).

[uu]FDA, Warning Letter, Ganeden Biotech Inc. (December 8, 2006), <http://www.fda.gov/ICECI/EnforcementActions/WarningLetters/2006/ucm076208.htm> (accessed 02.02.16).

[vv]FDA, Warning Letter, Nestle Healthcare Nutrition (December 3, 2009), <http://www.fda.gov/ICECI/EnforcementActions/WarningLetters/2009/ucm194121.htm> (accessed 02.02.16).

[ww]FDA, Warning Letter, Pan American Laboratories, Inc. (November 20, 2009), <http://www.fda.gov/ICECI/EnforcementActions/WarningLetters/2009/ucm191841.htm> (accessed 02.02.16).

CHAPTER

22 Regulatory approvals, intellectual property, branding and trademark in nutraceuticals and functional foods

Leighton K. Chong, Lawrence J. Udell, Bernard W. Downs
Udell Associates, Castro Valley, CA, United States

22.1 INTRODUCTION

Intellectual property (IP) can be a relatively mundane topic. However, the information presented in this chapter opens up a variety of creative issues that not only demonstrate how to acquire and position IP (and IP relationships), but how to utilize that IP properly. This information will be practical, innovative, and useful in bridging the gap between the creation and development of IP and successfully marketing it using powerful and protective branding strategies that better insulate the long-term return on investment (ROI). Having been involved in the big, wide world of IP creation, management, and marketing for more than 70 years, the authors offer a unique perspective based on a wealth of successful experiences. How do you transfer to paper diversified experiences in working with inventors and entrepreneurs in a multitude of industries?

Let us first explore and understand the values of IP and its relationship to the industry of nutraceuticals. The knowledge of this very special business started many years ago with the inception of "alternative" and complementary medicine, even before it was discussed openly and generally accepted.

It is important to understand that IP in this industry does not only encompass what is normally recognized as patents and trademarks. The business of harnessing and commercializing the power of nature relies on the ability to transform nature's creations into proprietary compounds marketed in the forms of liquids, powders, capsules, tablets, and topicals. The resulting IP is comprised of formulas, methods, and procedures for combining those derivatives in a unique manner into products that have the *novel* ability and characteristics positively to affect the user. Novel product technologies are derived from the soil and sea and every imaginable plant, insect, reptile, mammal, marine creature, etc. For example: (1) Plants are the source of a

Developing New Functional Food and Nutraceutical Products. http://dx.doi.org/10.1016/B978-0-12-802780-6.00022-5

plethora of phytonutrients; (2) various proteins, with therapeutic and analytical benefits, are being isolated from insects; (3) reptiles are a source for a number of novel substances (i.e., antivenom and immune complexes, etc.); (4) mammals contribute a host of therapeutic agents from proteins to undenatured collagen (Bagchi et al., 2002); (5) algae, seaweed, and mussels, and fish and krill oils, etc. offer a host of therapeutic options; (6) glandular extracts provide therapeutic naturopathic interventions; and everything in between. The scientists who are creating these potentially remarkable products are a special group imbued with a lust for finding the secrets that will reduce needless pain and suffering and increase the opportunities to improve health, possibly thrusting themselves into fame and fortune in the process.

However, their motivation for discovery is not totally aligned with recognition, but truly enhanced by reaching out to touch the future and the lives of potentially millions of people they will never know. All of the results and motivation are enhanced by the knowledge that their creations will (hopefully) be protected by the existing IP laws that exist in the majority of countries throughout the world along with powerful branding and marketing strategies.

This is the primary focus of this chapter. After spending untold millions of dollars in research and development (R&D), the benefits to the researchers and their companies are weighed in the potential value of their inventions and their subsequent transformation into commercial products.

22.2 NUTRACEUTICALS, PATENT RIGHTS, AND BIOPROSPECTING

As defined in Wikipedia, "nutraceutical" is a word phrase combination of "nutrition" and "pharmaceutical" and refers to foods, and more generally supplements, intended to have a therapeutically beneficial effect on human health. It can also refer to individual chemicals which are present in common foods (and therefore may be delivered in a nondrug form). Many such nutraceuticals are phytonutrients, which are plant compounds possessing both health protecting and promoting properties. Nutraceuticals are often used in nutrient premixes or nutrient systems in the food and pharmaceutical industries. Nutraceuticals that are in or from foods are also sometimes called "functional foods," but are probably more accurately identified as components of functional foods. For example, red wine (containing resveratrol) is an antioxidant (Quincozes-Santos et al., 2007) and an anticholesteremic (Pal et al., 2004), broccoli (containing sulforaphane) is a cancer preventive (Dinkova-Kostova et al., 2007), and soy and clover (containing isoflavonoids) are nutraceuticals used to improve arterial health in women (Clarkson, 2002), inhibit undesirable blood vessel formation in tumorous tissues (antiangiogenic) (Fotsis et al., 1993), and possess antitumor effects (Barnes et al., 1996; Wietrzyk, 2007).

Except for newly discovered foods, which must be an extremely rare occurrence given the course of human history and foods, all nutraceuticals by definition involve food or food nutrients that are already known to be used by the world's populations and cultures. Any "discovery" in nutraceuticals then would consist of identifying, isolating, and making a composition containing the active ingredient or nutrient in an already known food substance found to have a *previously unknown* beneficial effect on human health or formulated with other ingredients in an effective but previously unknown manner and/or for a previously unknown benefit or purpose.

Commercial entities commonly seek to obtain IP rights in discoveries and utilize strong branding strategies in order to reap the benefits of their investment of time and resources in research and to incentivize discoverers. Under the IP regimes existing in the world today, the primary form of IP protection for discoveries is to obtain a patent, that is, to obtain a government patent grant for the exclusive right to a discovery. The patent has a limited term, typically 20 years from the first application. But, in seeking to obtain a patent for identifying, isolating, and making a composition of an active ingredient, from an already known food substance, the inventor is only entitled to patent claim coverage of only that contribution found by the inventor to be both "new" and "nonobvious" over all prior public knowledge about the already known food substance.

Thus, attempts by food and pharmaceutical companies to secure patents on nutraceuticals will invariably raise tensions between a company's claim for exclusive patent rights in making a new and nonobvious discovery and the common heritage of the public to continue to access what was already known about foods. The more well known an existing food substance and its benefits are, the less likely is a company's claim to exclusive rights to a nutraceutical made from the food's active ingredient to be broadly defined, granted, and/or upheld. Conversely, if a food substance and its benefits are not considered to be widely known, or perhaps are hitherto undocumented, the more likely it will be that a company can obtain broad patent rights in them. Knowledge about a food substance and its benefits are often undocumented when it is known only to a primitive or indigenous people and transmitted only through folklore or sacred practices.

The practice of companies conducting research in remote areas for naturally occurring biological substances that may be found to have benefits undocumented or unknown to the world at large is called "bioprospecting." The ethics, fairness, and potential harm from current practices in biotechnology research are being increasingly questioned as third world governments, indigenous peoples, and ethnic populations worldwide have developed a growing awareness and interest in controlling the use of and sharing the benefits of bioprospecting research.

An increasingly complex national and international policy framework arising out of the 1992 World Convention on Biodiversity (CBD) seeks to have companies conduct bioprospecting research under formal agreements entered into with the sovereign governments, indigenous peoples, and local communities who control or occupy the lands on which such research is to be conducted (Convention on Biodiversity [CBD], 2016). Article 15 of the Convention addresses the terms and conditions for access to genetic resources and traditional knowledge, informed consent, and benefit-sharing. It recognizes the sovereignty of states over their natural resources and provides that access to these resources shall be subject to the prior informed consent of the contracting party (signatory country) providing such resources. It also provides that access shall be based on mutually agreed terms in order to ensure the sharing of benefits arising from the commercial or other utilization of these genetic resources with the contracting party providing such resources.

In 1999, work was begun to operationalize the mandates of the CBD which resulted in the issuance in Apr. 2002 by the CBD Secretariat of the "Bonn Guidelines on Access to Genetic Resources and Benefit Sharing," which were adopted unanimously by CBD treaty members (BGAGRBS, 2002). The Bonn Guidelines were adopted to assist contracting parties, governments, providers, users, and other stakeholders in developing overall access and benefit-sharing strategies and in identifying the steps involved in the process of obtaining access to genetic resources and benefit-sharing. More specifically, the Guidelines are intended to help them when establishing legislative, administrative, or policy measures on access and benefit-sharing and/or when negotiating contractual arrangements for access and benefit-sharing.

While a signatory to the CBD, the USA has in fact not yet ratified the treaty. The US State Department has formed a coordinating agency as the US National Focal Point under the CBD Clearing-House Mechanism (Convention on Biological Diversity [CBD], 2016). For the time being, the US National Focal Point has deferred responsibility to the various national agencies having jurisdiction of regions or lands under federal control in which bioprospecting may be permitted, such as national park lands, federal conservation trust lands, Indian reservations, and so on. If the United States ratifies the CBD Treaty, the US National Focal Point and designated national authorities would have a duty to assist participating parties in bioprospecting negotiations. Furthermore, these authorities would need to advise and assist state agencies (as regional authorities) in the development of policies regulating bioprospecting in compliance with the CBD framework in an effort to ensure the equitable sharing of benefits with all involved parties from the commercialization of these resources.

The Bonn Guidelines place great emphasis on the obligation for research users to seek the prior informed consent of bioaccess providers. They also identify the basic requirements for mutually agreeable terms and define the main roles and responsibilities of, and stress the importance of, the substantive involvement of all stakeholders. They also cover other elements such as incentives, accountability, means for verification, and dispute settlement. Finally, they enumerate suggested elements for inclusion in material transfer agreements and provide a suggested checklist for monetary and nonmonetary benefit sharing.

The focal point of the ongoing arbitration of rights in bioprospecting has centered on intellectual property rights (IPR) and, specifically, patent rights in discoveries derived from bioprospecting. The Conference of Parties (COP) organized for ongoing negotiations of Contracting Parties under CBD requested the World Intellectual Property Organization (WIPO), an international agency dealing with world intellectual property issues under the auspices of the United Nations, to conduct studies on the interface of IPR protocols with bioprospecting mandates. WIPO formed the Intergovernmental Committee (IGC) on Genetic Resources, Traditional Knowledge and Folklore in 2000 to conduct, among other subjects, studies of the interface of bioprospecting mandates and IPR rights (World Intellectual Property Organization [WIPO], 2016). At least 17 countries are active participants on the WIPO/IGC Committee and many nongovernmental organizations and activist groups have observer status to the Committee's deliberations.

Patents are government-sanctioned grants of exclusive rights to inventors (and their companies) to exploit commercially their new discoveries, mainly through deriving profits from the sale of products based on those discoveries. When discoveries are based on genetic resources or traditional knowledge obtained from bioprospecting, issues arise as to whether the researchers had proper permission to access biological materials in the field, including obtaining the informed consent of the providing party affected, whether they derived their inventions from already known or traditionally known substances or healing practices, whether their use of the biological materials and information was proper, and whether there was an equitable sharing of benefits with the local authorities who permitted access for such research.

The WIPO/IGC studies have raised the issue whether the patent systems of participating countries should be changed to implement CBD mandates by requiring disclosure of bioprospecting agreements in patent applications (World Intellectual Property Organization [WIPO], 2016). A majority of IGC members favor imposing an affirmative duty on research companies to state that their research was conducted pursuant to and in legal compliance with a valid bioprospecting agreement, subject to sanctions including patent invalidity if the requirement is not met. More conservative IGC members (such as Japan and the USA) have expressed concern that such a substantive bioprospecting disclosure requirement could have unintended adverse consequences on patent systems and the objectives of rewarding innovation and investment in research.

A WIPO draft statement on a bioprospecting disclosure requirement was recently transmitted to the CBD COP in May 2006 (World Intellectual Property Organization [WIPO], 2016).

The biotechnology industry in the USA has generally recognized the implications of the developing international framework on bioprospecting under the CBD, and the significant milestone reached by unanimous approval of the Bonn Guidelines for member countries, users, and providers. Although the Bonn Guidelines are not yet legally binding, biotechnology companies have had to weigh the possible losses that might be incurred through research foregone and benefits relinquished and the considerable costs of compliance, against the risks of incurring local protests, possible legal sanctions, loss of public goodwill, and invalidation of patents. Many are concluding that compliance with international mandates on bioprospecting is inevitable, or at least would be good business practice to achieve some degree of compliance with mandates that are likely to become widely adopted.

As an example, the Biotechnology Industry Organization (BIO), representing more than 1100 biotechnology companies, academic institutions, state biotechnology centers, and related organizations in 50 US states and 31 other nations, recently adopted its recommended Guidelines on bioprospecting to its members (Guidelines for BIO Members Engaging in Bioprospecting, 2005). The BIO Guidelines identify "best practices" that should be followed by companies engaging in bioprospecting.

BIO advises its members that the Guidelines provide only a "road map" but that they "have extensive discretion to shape their conduct to meet whatever requirements countries impose with respect to bioprospecting activities." A member company is not required to follow the Guidelines and those Guidelines are not enforceable against a member company. However, it is conceivable that companies that do not engage in conduct consistent with the Guidelines might be subject to criticism for not following "best practices." The biotechnology industry thus acknowledges that the proper conduct of bioprospecting activities requires the use of formal bioprospecting agreements with the relevant national and local authorities.

US Patent Laws require that an applicant disclose, in any patent application, information known to the applicant that is material to the patentability of the patent claims submitted. The relevant Patent Office Rules, 37 Code of Federal Regulations, state in the pertinent part, as follows:

§ 1.56 Duty to disclose information material to patentability

a. … Each individual associated with the filing and prosecution of a patent application has a duty of candor and good faith in dealing with the Office, which includes a duty to disclose to the Office all information known to that individual to be material to patentability as defined in this section.

The duty to disclose information exists with respect to each pending claim until the claim is cancelled or withdrawn from consideration, or the application becomes abandoned.

The duty to disclose all information known to be material to patentability is deemed to be satisfied if all information known to be material to patentability of any claim issued in a patent was cited by the Office or submitted to the Office in the manner prescribed by §§ 1.97(b)–(d) and 1.98. However, no patent will be granted on an application in connection with which fraud on the Office was practiced or attempted or the duty of disclosure was violated through bad faith or intentional misconduct.

The Office encourages applicants to carefully examine:

b. The closest information over which individuals associated with the filing or prosecution of a patent application believe any pending claim patentably defines, to make sure that any material information contained therein is disclosed to the Office.

The applicant's duty of disclosure under existing US Patent Office rules would require the disclosure of relevant information as to prior uses, traditional knowledge, and/or source materials where relevant to a particular patent application. Likewise, the written description and enablement requirements of the US Patent Laws, Section 112, would require US applicants to identify relevant source materials on which claims to invention are made. However, it is important to keep in mind that biotech discoveries are often made following many steps or levels of extraction, modification, refinement, or synthesis away from a naturally occurring material or traditional remedy or practice product, such that the importance of the initial material investigated may no longer be relevant to the patentability of the invention. Nevertheless, a patent applicant may deem it safer to disclose source materials initially investigated in the patent application or during patent prosecution rather than take the risk of having to litigate the issue later.

22.3 BRANDING, AND A HYPOTHETICAL CASE SCENARIO

22.3.1 NOW THAT WE HAVE PATENT PROTECTION, WE CAN MARKET OUR NATURAL CREATIONS WITH THE CONFIDENCE THAT THEY ARE COMPLETELY PROTECTED ... RIGHT? (NOT QUITE ... MORE IS NEEDED!)

Great disappointment and potentially huge legal defense expenses await the inventors of patent-protected nutraceutical products and/or technologies if they attempt to take their creations to market in a big way solely on the protective strength of the patents, without establishing powerful identity, branding, and trademark strategies. The key factor here is "in a big way." Just ask the original IP-based suppliers of products like Cat's claw, Ginkgo, St John's wort, *Garcinia cambogia, Hoodia, Citrus aurantium*, grape seed extracts, chromium nicotinate, noni, mangosteen, and countless others. The evidence-based technology and patent protection are only half the story when it comes to propelling and insulating the market success of a patented nutraceutical product. To complete the equation required for long-term market success, strong branding strategies are a must.

The principle seems to be that the greater the market impact and success an invention/creation has, the greater the lure for copycat knockoff suppliers to bring competing products to market. "But I have a patent and am protected," you might say. The patent provides a basis to defend the invention, not to prevent it from being infringed, although the hope is that patent protection should discourage infringement. But, to savvy knockoff artists, a commercially successful patented product can offer an enormous opportunity for copycat windfall profits, especially if the patent holder is a financially fragile start-up venture … an easy mark.

How so? The following is an (oversimplified) example. Suppose you discovered a botanical species (call it "*Onlyess importantium*") that was a powerful antioxidant and had a profound effect on turning off chronic inflammation. For the purpose of this illustration, *O. importantium* has already been sold as a botanical detoxifier since before 1994 (no New Dietary Ingredient approvals by the Food and Drug Administration (FDA) are required). So the species has been commercially available, but for other applications. Further investigation finds that one previously inconspicuous component in *O. importantium* made the most significant contribution to achieving the greatest therapeutic effect on inflammation; call it an "inflammanoid." So, you developed a solvent-free extraction method to achieve a 50% concentration of the inflammanoid (and inflammanoid glycosides), which naturally occurs at about 6% of the total composition. When tested, this product had an oxygen radical absorbance capacity (ORAC) value off the charts and turned off the inflammation process like a switch, promoting profound and rapid healing (which required conservatively a \$250,000 investment in a clinical study). It is sensational. You go through the process of applying for and receiving patents on the extraction method, the novel composition of the 50% inflammanoid, and for its application as an antioxidant and downregulator of (chronic) inflammation (requiring a significant investment as well). Keep in mind that you are still incurring business and research expenses during the period of research, presentations, and publication (hopefully) of the study, prior to its intended market launch.

Now, you have to tell the world about your invention. So, you create a new raw material product and call it "AntioxInflam." In one go-to-market strategy, you contract with an infomercial company to promote AntioxInflam a proprietary source of inflammanoid. The infomercial costs about \$200,000 to produce and about \$1.5 million per month to run. Plus, you have to invest in manufacturing a few thousand bottles to start and arrange for fulfillment, unless the infomercial company takes care of this. All of this costs money.

The infomercial is a success and the phones at the call center are ringing off the hook. Your inventory requirements at the fulfillment center skyrocket. Assuming that you are in strict compliance with FDA and Federal Trade Commission (FTC) guidelines and are making only supportable structure function claims, you are selling 100,000 bottles a month at \$34.95 each. You are making a boatload of money and helping an enormous number of people. Life is good! However, you have now jumped high onto *everybody's* radar, especially that of opportunistic competitors.

Keep in mind that once you make AntioxInflam and *O. importantium* household names, competitors can enter the market, but don't need to make any claims at all. They also don't need to duplicate your extraction methods. They don't need to infringe on your patents to sell products. They only need to manufacture a reasonable concentrate, extracted the old (noninfringing) way, and supply a very cheap generic/commodity version of *O. importantium* (Oi) to as many branded finished product marketers as possible. Recall that these copycat merchants have *none of the costs* associated with R&D nor creating and building the market. So, eliminate the R&D and market-building costs (that you incurred), factor in the significantly cheaper raw material costs, and even with a product costing a fraction of your original product, these copycat merchants can make a significantly higher profit percentage than you on a much lower suggested manufacturers retail price (SMRP).

Again, they only need to supply other finished product companies with their version of the raw material for which *no claims* need to be made. You already popularized the *O. importantium* and have an exponentially higher breakeven point than them. They don't have to do anything except supply the market. And, the more successful and greater your market impact is, the more likely it will be that copycat knockoff merchants will enter the market and achieve success in proportion to your market success. The objective of course, will be for them to appear as much like your original product as possible, but sell for a much less expensive price. Most consumers will purchase the cheaper knockoff version in hopes of receiving similar benefits. Knockoff artists are experts at creating the illusion of quality without actually fulfilling that value promise. This results in ever-increasing numbers of consumers switching from your product to counterfeit versions, cratering your sales and potential for ROI. They also undermine and destroy the credibility of the product (because "it's not working"). The knockoff counterfeiter just *loves* your infomercial. And, they will enter the market in increasing numbers as long as the opportunity to sell *O. importantium* exists.

But, the market pie can only be sliced into so many pieces before the ROI potential is destroyed for everyone, including the knockoff artists. They will even put each other out of business in vicious price wars. By the way, purchasing agents of major companies are often paid a bonus based on the amount of money saved in lowering product costs; and most purchasing agents (and executives and investors in profit-sensitive public companies) are not scientists with an understanding of analytical chemistry or the differences between different sources of the same raw materials. So, if it looks the same and costs less, chances are the lower-priced product will replace higher-priced versions; and lost consumer confidence is the irrevocable casualty. It has been proposed that this phenomenon played a major role in the nutraceutical market crash of the mid- to late 1990s.

This is the very real market threat that can reduce a wonderfully effective patented natural ingredient to a nearly worthless generic commodity. So, the real dilemma regards how long and to what magnitude will you continue to invest in your market-building promotions once competing low-cost counterfeit market predators have a plethora of alternative products blitzing consumers? In this scenario, the answer is not very long. And, now the activities and costs of defending your patents become an excruciating requirement against diminishing sales and profits. In addition, your ability to monitor the influx and potential of patent infringers diminishes against the increasing magnitude of counterfeit product marketers over time. For all the companies that don't make any claims, there will be some that do. But, as the crowd of competitors gets larger, the ability to identify these violators gets more difficult. And, to further aggravate the situation, your "benefit sharing" partners of the source technologies are becoming increasingly cynical over the sudden reduction in fortunes and don't understand what has happened, because the product category (*all O. importantium* products) *appears* to be gaining in popularity and experiencing explosive growth. Potential suspicion, friction, blame, and retribution by them could inflame things further. So, can this process be averted? Absolutely yes ... but how?

22.3.2 PROTECTING YOUR PATENTS WITH CODED IDENTITIES AND BRANDING ... THE RIGHT WAY

Not to overstate the obvious, but the traditional role of branding is to establish a value promise that is linked to a *unique* "branded" trade name (and supposedly a uniquely superior product or service). This certainly sounds simple enough. However, most people view branding in a very one-dimensional way, limited to a consumer product or service label (i.e., Band-Aid). In the nutraceutical, cosmeceutical, functional food, and beverage categories, branding can incorporate the value promise/reputation of the company (brand), the finished product (brand), the ingredient (brand), and/or any branded valuations/validations (i.e., Good Housekeeping, ANSI, USP, GRAS, Kosher, Organic, GMP, etc.). This strategy validates the value promise with "layers of branding" representing layers of validation.

In this aspect, the most important factor is differentiating the unique identity and therefore the superiority of the product. In regard to where competition is directed (at you), this means beginning with your raw material. So, even the species of *O. importantium* must now somehow be differentiated from all the other sources. Since you have developed a different product, even though you have trade named the finished raw material (AntioxInflam), you still need to identify that YOUR *unique* raw material has different or special properties than any of its generic counterparts or else there will be severe market consequences.

When research is published (in scientific journals, and industry and lay publications), editors and publishers are most often very reluctant to publish branded trade names for fear of appearing to endorse products and losing the credibility of being unbiased. For this reason, they (at least the respected publications) almost unanimously publish only the generic name of an ingredient. Therefore, consider that when your "independent" research is published, if the benefits of AntioxInflam are attributed only to the *O. importantium* species, then your potential competitors will be given the most powerful market weapon to use against you … scientific support for Generic *O. importantium.* And, to any industry or lay publications that revile using branded trade names, attributing the research benefits to just *O. importantium* bolsters the opportunities for market exploitation with counterfeit products by all of your knockoff competitors.

Therefore, you must give your raw material a unique identity in addition to the generic name. One tactic is to assign a product identity or a research code to the material (Ex. XYZ123). Now AntioxInflam is officially a *unique* XYZ123 form of *O. importantium.* By the way, this code must appear in the research publications, at least in the "Materials and Methods" section. If publishers and editors refuse to identify the brand name, they will be very reluctant to delete the product identity research code in association with the generic name, which only you have.

This tactic is also valuable for creating other branding options when and/or if needed. As an example, if one very large company wants to have an exclusive for a period of time in a specific channel, you have the means to create a separate brand name exclusive to that customer, as long as you link the brand to the XYZ123 research code. This enables that different brand to base the product claims on original validated research. And you have successfully created your own competition, which is a far superior option to having a vicious counterfeit competitor. The more you establish your other patent-protected brands of XYZ123 *O. importantium* in the market, the more of a disincentive you can create for knockoff products, as the market can appear to be filling up with competitive products, reducing the knockoff profit potential and ROI. Now your science, IP, significant investments, and the product's ultimate value promise are all better insulated from market piracy.

This strategy also applies to multiple ingredient formulas in finished products. Research and patents on finished products increase their market potential exponentially. Up to this time, the vast majority of research and patent protections in the natural products industry has been established by ingredient manufacturers on single ingredients. Research and patents on finished products amount to only a very small fraction of those done on nutraceutical ingredients. Yet, this offers the greatest revenue potential and ROI.

There are other effective strategies that improve the potential for long-term success in the nutraceutical industry. However, discussions regarding those methods are more the subject of an advanced marketing course on branding and outside the scope of this chapter. Suffice it to say, this information provides a basic knowledge and guide on how to create a successful strategic agenda for developing, validating, and marketing nutraceutical IP from source to consumer.

22.4 CONCLUSIONS

Prior to DSHEA in 1994, the nutritional products industry was characterized more by flash-in-the-pan product popularity that usually lasted at most for about 2 years. These popular products sparked a fad-like following that usually dissipated with the next new product fad. During those years preceding 1994, there was not a very significant incentive to do the kind of research that would reveal a nutritional product's powerful health correcting and/or promoting properties. The regulations prohibited making just about any kind of important and/or informative claims. So the market opportunities were relatively small and the arsenal of IP was comparatively sparse. Since the passage of DSHEA, the opportunities to do exciting research, develop new product technologies, and obtain important IP are almost endless. We believe that this growing industry will find, through extensive R&D, some remarkable and beneficial new materials that will continue to benefit humankind in ways that will amaze the medical and scientific communities. Furthermore, we expect that in the future, systemic biological premises of nutraceutical technologies and IP will be increasingly administered with and/or in place of conventional "reductionist" single loci pharmacological medicine. The next phase of market growth is dependent on valid scientific research for new product technologies, patents, stronger and more effective branding and trademark strategies in product marketing, and international regulatory compliance.

REFERENCES

Bagchi, D., Misner, B., Bagchi, M., Kothari, S.C., Downs, B.W., Preuss, H.G., 2002. Effects of orally administered undenatured type II collagen against arthritic inflammatory diseases: a mechanistic exploration. Int. J. Clin. Pharmacol. Res. 22, 101–110.

Barnes, S., Sfakianos, J., Coward, L., Kirk, M., 1996. Soy isoflavonoids and cancer prevention. Underlying biochemical and pharmacological issues. Adv. Exp. Med. Biol. 401, 87–100.

Bonn Guidelines on Access to Genetic Resources and Benefit Sharing, Secretariat of Convention on Biological Diversity, April 2002. http://www.biodiv.org/doc/publications/cbd-bonn-gdls-en.pdf

CBD, 2016. National Focal Points, see listing at http://www.biodiv.org/convention/default.shtml

Clarkson, T.B., 2002. Soy, soy phytoestrogens and cardiovascular disease. J. Nutr. 132, 566S–569S.

Convention on Biological Diversity (CBD), 2016. National Focal Points, see listing at https://www.cbd.int/doc/lists/nfp-cbd.pdf

Dinkova-Kostova, A.T., Fahey, J.W., Wade, K.L., et al., 2007. Induction of the phase 2 response in mouse and human skin by sulforaphane-containing broccoli sprout extracts. Cancer Epidemiol. Biomarkers Prev. 16, 847–851.

Fotsis, T., Pepper, M., Adlercreutz, H., et al., 1993. Genistein, a dietary-derived inhibitor of in vitro angiogenesis. Proc. Natl. Acad. Sci. USA 90, 2690–2694.

Guidelines for BIO Members Engaging in Bioprospecting, July 2005. http://www.bio.org/ip/international/200507guide.asp

Pal, S., Naissides, M., Mamo, J., 2004. Polyphenolics and fat absorption. Int. J. Obes. Relat. Metab. Disord. 28, 324–326.

Quincozes-Santos, A., Andreazza, A.C., Nardin, P., Funchal, C., Goncalves, C.A., Gottfried, C., 2007. Resveratrol attenuates oxidative-induced DNA damage in C6 glioma cells. Neurotoxicology 28, 886–891.

Wietrzyk, J., 2007. The influence of isoflavonoids on the antitumor activity of vitamin D3. Postepy. Hig. Med. Dosw. (Online) 61, 253–260.

WIPO Technical Study on Disclosure Requirements Concerning Genetic Resources et al., 2016. Documents available at http://www.wipo.int/tk/en/genetic/proposals/index.html#resources

World Intellectual Property Organization (WIPO), 2016. Intergovernmental Committee (IGC), Genetic Resources and Intellectual Property, http://www.wipo.int/tk/en/genetic

CHAPTER

Nutritional supplements in wound care

23

Ryan Dickerson, Scott Chaffee, Amitava Das, Sashwati Roy, Chandan K. Sen

Department of Surgery, Comprehensive Wound Center, Dorothy M. Davis Heart & Lung Research Institute, The Ohio State University Wexner Medical Center, Columbus, OH, United States

23.1 INTRODUCTION

The growing health and economic burdens derived from chronic wounds are difficult to ignore, as nearly 6.5 million people in the United States alone are affected by chronic wounds (Crovetti et al., 2004; Broughton et al., 2006). Such a significant number of chronic wound patients warrants them as a developing important public health concern, and the money spent to treat complications associated with the condition is rising. Recent reports indicate that more than US$25 billion are currently being spent annually on treating patients with chronic wounds, and this figure is expected to rise in upcoming years (Sen et al., 2009). These aforementioned monetary figures are solely representative of what the United States is spending on this problem, and it must be scaled to understand the global implications of chronic wounds on health care as a whole. Chronic wounds will continue to have an enormous impact on both social and economic platforms until we deepen our understanding of the issue and better recognize possible treatments and preventative measures (Sen et al., 2009).

Chronic wounds are an example of a diverse etiological stalling of the normal healing process (Menke et al., 2007). Wound healing is a multifaceted biological process that consists of overlapping phases that aim to restore the proper physiology and barrier function of the tissue (Falanga, 2004; Singer and Clark, 1999; Gilmore, 1991). The three distinct phases of the wound healing cascade include hemostasis/inflammation, proliferation, and maturation (Broughton et al., 2006; Singer and Clark, 1999; Gilmore, 1991). In order for a wound to heal properly, the early inflammatory phase must be resolved in a timely fashion to allow for remodeling by way of the addition of new granulation tissue and arrangement of the new extracellular matrix. Should a wound progress through the phases of wound healing without issue or delay and the wound site be reconstructed, the wound is deemed acute. Typically, acute wounds will heal within a matter of a few days. However, if a wound does not progress through the wound healing cascade in a timely manner, usually 3 or 4 weeks postinjury, then the wound is labeled as chronic (Sen et al., 2009). In the case of chronic wound development, the wound healing progression is usually stopped in the early inflammatory phase and will continue to be delayed in a proinflammatory stage for an extended period of time.

Developing New Functional Food and Nutraceutical Products. http://dx.doi.org/10.1016/B978-0-12-802780-6.00023-7

23.2 NUTRITION STATUS AND MALNUTRITION IN WOUND HEALING

Nutrient-deficient malnutrition, especially over a long period of time, can lead to many severe consequences regarding stamina, body weight, and ability to fight off infections (Stechmiller, 2010). More specifically, malnutrition can significantly inhibit the tensile strength of wounds, and a person may become prone to delays in the wound healing cascade (Irvin, 1978). Children are amongst the most susceptible to malnutrition which could have lifelong effects on both physical and cognitive growth (Sermet-Gaudelus et al., 2000; Ge and Chang, 2001). There are many factors that lead to nutrient deficiencies, which may differ depending on different parts of the world and common diets. Some parts of the world may experience dire nutrient deficiencies due to lack of sanitary water and access to certain nutrient-dense foods (Diaz et al., 2003). Whereas, other groups of people may experience nutrient deficiency from patterns of eating and habits of eating certain high-calorie, low-nutrient density foods (Bandini et al., 1999). Nutritional status is assessed through a variety of methods, but two standard clinical tools used are the short nutritional assessment questionnaire (SNAQ) and the malnutrition universal screening tool (MUST). Despite some sources of nutrient deficiencies, the clinical problems are nonetheless quite severe. A weakened immune system and the inability to heal in a proper manner may lead to serious infections and a system stalled in a proinflammatory environment, leading to chronic issues.

In a normal, healthy individual, the ability to fight infection and experience proper wound healing is common. Chronic wounds are rarely seen in individuals who have proper nutrition and do not have a preexisting disease that may complicate their ability to heal. In situations where people are absent of illness and disease yet are living an unhealthy, physically inactive lifestyle, acute wounds and illnesses are much more likely to become chronic. The opportunity to combat this growing shift in lifestyle and societal norms is available, and nutraceutical intervention may be a vital tool for the management of chronic conditions (Al-Okbi, 2014).

Wound healing impairments and delays may be attributed to many different factors including excess or deficient consumption of macronutrients, vitamins, and minerals: including proteins, carbohydrates, fats, vitamin A, vitamin C, vitamin D, vitamin E, copper, zinc, iron, and magnesium among others (Dryden et al., 2013). With the involvement of this list of factors in many different bodily processes, it is to be expected that they all have varying roles in the resolution of inflammation and wound healing.

Minerals such as copper, zinc, iron, and magnesium all have very distinct roles in the timely process of proper wound healing. Specifically, copper displays angiogenic properties, serving as a cofactor for enzymes necessary for remodeling, and inducing the expression of vascular endothelial growth factor (VEGF) (Xie and Kang, 2009; Sen et al., 2002). Zinc also serves as a cofactor for important wound healing enzymes and transcription factors such as matrix metalloproteinases, promotion of re-epithelialization, and generation of new tissues by promoting cell division (Agren et al., 1991; Agren, 1990). Magnesium is involved in the wound healing process by serving as a critical cofactor for enzymes involved in the synthesis of both protein and collagenase. Additionally, iron is important in oxidative stress, as well as the hydroxylation of the amino acids lysine and proline (Guo and Dipietro, 2010). Recent research has focused on the outcome of iron deficiency on wound healing, and it shows it is harmful as it led the body to be stranded in a proinflammatory environment (Sindrilaru et al., 2011). In humans, deficient iron levels have been linked to chronic inflammatory diseases including rheumatoid arthritis and lupus erythematosus (Wright et al., 2014). Similar to deficient levels of iron, zinc deficiency was also determined to have a negative impact on early inflammatory phases in CD-1 mice (Blain et al., 1998). It is obvious that these minerals all play critical roles in proper wound healing, particularly in regards to preliminary enzymes in the early inflammatory phases. Additionally, vitamins such as A, C, D, and E have varying roles in the promotion of collagen synthesis, production of healthy scar tissue, and also promotion of the timeliness of the transition between the different wound healing phases (MacKay and Miller, 2003).

23.3 PHASES OF WOUND HEALING

A prototypical wound healing cascade requires a smooth transition between the phases and proper communication between inflammatory factors and immune cells. As previously mentioned, the distinct phases of wound healing are the preliminary inflammation stage, followed by the angiogenic stage, and culminating with the maturation/proliferation phase (Fig. 23.1). The purpose of the initial phase of the wound healing response is to halt the spread of infection and stabilize injuries, which

- Chronic wounds result from dysregulation of macrophage function; progress is halted in proinflammatory environment
- Macrophage Phenotypes: M1 macrophages are proinflammatory; M2 macrophages are pro-healing

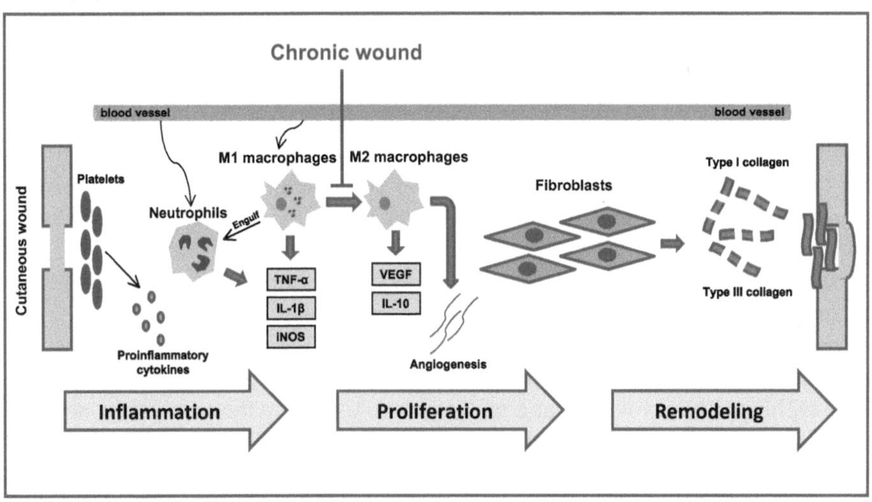

FIGURE 23.1 Derailment of physiological wound healing processes in chronic wounds

allows the body to recognize the problems and recruit the proper inflammatory and wound healing factors to the environment (DiPietro, 1995). Ways in which the body stabilizes and neutralizes problems are by constriction of blood vessels and the development of blood clots. Factors including platelet-derived growth factor (PDGF), TGF-β, and VEGF are released at the site of the clot to help initiate the healing process (Schar et al., 2015).

After the initial hemostatic phase has been completed, the blood vessels reverse course and begin to vasodilate in an effort to bring more blood and wound healing factors to the site of infection or injury (Steed, 1997). Leukocytes are major producers of inflammatory cytokines, which direct much of the inflammatory process (Kolaczkowska and Kubes, 2013). The early inflammatory phase is critical for the expression of proinflammatory cytokines, TNF-α and IL-1 in particular, which work to recruit various sources of immune cells including neutrophils and monocytes/macrophages to the wound site in an effort to resolve the issue (Zhang and An, 2007). Both TNF-α and IL-1 are upregulated in the wound environment itself, which leads to the initial recruitment of polymorphonuclear leukocytes (Wright et al., 2010). Neutrophils are antimicrobial in nature, and their primary role is to clear any harmful microbes that are causing damage to the body (Wilgus et al., 2013). A major role of neutrophils is to produce additional proinflammatory cytokines in the wound site. Because neutrophils are releasing strong antimicrobial properties, it is vital that these cells are cleared in a timely manner in order to prevent prolonged inflammation. Since neutrophils act in such a rapid manner, many of the enzymes and proteases that they release are not specific to a type of pathogen, and without proper removal of these cells, damage may be done to the host tissue (Wilgus et al., 2013).

Macrophages are other major immune cell players that are recruited to the wound site by proinflammatory cytokines. Macrophages have many important roles such as host defense and transition of wound sites from proinflammatory to antiinflammatory environments; however, possibly the most critical role of macrophages in wound healing is the engulfment and clearance of apoptotic cells and neutrophils (Fujiwara and Kobayashi, 2005). Recent research has shown that macrophage dysfunction in particular is a leading cause of impaired wound closure, leading to chronicity. Similar to neutrophils, macrophages also secrete proinflammatory cytokines including TNF-α and IL-1, which continue to drive the inflammatory phase to resolve in preparation for the angiogenic/proliferative phase (Agbanoma et al., 2012; Murray and Wynn, 2011).

In a normal progression of wound healing, the inflammatory phase resolves in a timely fashion and the necessary growth factors have been recruited to the site in preparation of growth and organization of new tissue. This critical phase is initiated by the movement of blood vessels to the wound site. As the cascade progresses, blood vessels will form a new network to supply the new tissue, and collagen will be organized in an effort to return the appropriate barrier properties of the skin (Tonnesen et al., 2000). The proinflammatory role of macrophages has been previously discussed; however, these diverse immune cells also have a job in the later stages of wound healing. Macrophages are capable of shifting from proinflammatory in nature to an antiinflammatory state in which they recruit anti-inflammatory cytokines to the site to trigger the final steps of wound healing (Das et al., 2015). Studies show that any dysregulation of antiinflammatory cytokine and growth factor recruitment doesn't allow the wound to completely heal, which often leads to the production of a chronic wound environment (Khanna et al., 2010).

23.4 FATTY ACIDS AND INFLAMMATION

Despite the importance and necessity of both neutrophils and macrophages in a wound environment, there are many other factors that assist the body throughout the process. For example, prostaglandins are essential to the initial inflammatory response, as well as aiding in the eventual resolution of inflammation (Ricciotti and FitzGerald, 2011). Prostaglandins are lipid compounds that are derived from the polyunsaturated fatty acid (PUFA) arachidonic acid, and these compounds are important for processes such as inflammation and aggregation of platelets (Kuehl and Egan, 1980). The enzyme cyclooxygenase-2 (COX-2) is responsible for the synthesis of prostaglandins (Ricciotti and FitzGerald, 2011). Prostaglandins are also responsible for the release of unsaturated fatty acids, which are notorious for expressing antiinflammatory behaviors. PUFAs are important to the shift between pro-inflammatory environments and ones that are antiinflammatory (Kirkup et al., 2010). Specifically, PUFAs act to limit the production of different types of inflammatory markers and cytokines, along with reactive oxygen species (ROS) and expression of adhesion molecules (Calder, 2013). n-6 and n-3 PUFAs generate different products, which eventually have contrasting effects on inflammation. Eicosanoids, such as prostaglandin E2 (PGE_2), are essential to the progression of the inflammatory phase, and these mediators originate from PUFA (Dennis and Norris, 2015). n-6 PUFAs are from an arachidonic acid origin and are proinflammatory in nature; whereas, n-3 PUFAs are from an eicosapentaenoic acid (EPA) origin and are antiinflammatory (Khanapure et al., 2007; Patterson et al., 2012). Conjugated linoleic acids (CLA) are a class of essential fatty acids with conjugated double bonds in their carbon structure with various geometrical isomers. In a rodent wound healing model, supplementation of CLA resulted in a dose-dependent improvement in wound healing rates (Park et al., 2010). The improved wound healing rate is suggested to be the result of down-regulation of proinflammatory pathway nuclear factor-κB (NFκB). CLA supplementation was also found to downregulate COX-2. Because NFκB is redox sensitive, the antioxidant properties of CLA are thought to block activation of the transcription factor, leading to reduced COX-2 expression and reduced inflammation.

Recently, studies have focused on C-reactive protein (CRP) as an indicator of inflammation. For example, elevated CRP levels are a great indicator of whether or not a person is likely to develop inflammatory conditions such as diabetes mellitus and heart disease in the future (Galland, 2010). Preliminary studies using glycemic index (GI) led investigators to state that consumption of a diet with a high GI is correlated with an overall increase in susceptibility to diabetes, heart disease, and stroke (Galland, 2010).

Also, preliminary evidence suggests that a diet containing increased fiber intake is correlated with decreased CRP concentrations (Ajani et al., 2004). Alternative evidence did find the direct correlation of fiber and CRP; nonetheless, they discovered that increased fiber consumption led to decreased levels of IL-6 and TNF-α, both proinflammatory cytokines. Also, recent evidence regarding fat and prevalence of CRP indicates a positive correlation between saturated fatty acids and CRP expression, with an inverse relationship detected between PUFA and CRP (Santos et al., 2013). Additionally, recent evidence indicates an inverse relationship between levels of EPA and docosahexaenoic acid (DHA) with CRP and IL-6 levels (Bowden et al., 2009).

23.5 DIETARY INTERVENTION AND INFLAMMATION

Certain dietary habits are well established as having the ability to lower general inflammation, and some of these habits include increased consumption of omega-3 fatty acids, reduction of trans-fats intake, and increased consumption of fruits, vegetables, nuts, and whole grains (Giugliano et al., 2006). The benefits of increasing fruit and vegetable intake have been proven by recent research indicating the negative correlation of heart disease, diabetes, and even cancer. Also, the herbal supplement market in the US has been expanding rapidly, despite the products not being evaluated and approved by the Food and Drug Administration (FDA). As far back as 1999, nearly half of adults in the United States admitted using herbal supplements in that current year (Abebe, 2002). Herbal supplements that may offer antiinflammatory properties include ginkgo, ginseng, garlic, and turmeric, to name a few. These herbal supplementations are most commonly thought to act through antioxidant pathways or antiplatelet aggregation (Abebe, 2002).

It is clear that a committed change from an unhealthy, sedentary lifestyle will provide long-term benefits, particularly in terms of developing diseases related to chronic inflammation. Dietary intervention is a preliminary step in providing the body the necessary nutrients to combat infections and allow inflammation to resolve in a timely manner. Conditions including obesity often put the body in a position where it cannot readily combat inflammatory environments, especially not in a timely manner, and these conditions often lead to the production of nonhealing wounds and chronic inflammation.

23.6 NUTRACEUTICALS AND FUNCTIONAL FOODS

Holistic approaches and herbal medicines have been widely used and accepted in Eastern societies, particularly Indian and Chinese cultures. A handful of nutraceuticals have been investigated in regards to their role in inflammatory pathways, and examples of these nutraceuticals are as follows: capsaicin, cinnamaldehyde, and *Boswellia serrata* gum resin (Gupta et al., 1997). *B. serrata* contains boswellic acids (BA), which are known to exhibit antiinflammatory properties (Ammon, 2006). BA acts to inhibit the production of leukotrienes, which are a distinct family of eicosanoid proinflammatory molecules, either by altering their translocation or by a specific interaction with 5-lipoxygenase (Safayhi et al., 1995; Schweizer et al., 2000). Leukotrienes are generated via the lipoxygenase pathway using arachidonic acid as the starting material (Abe and Yoshimoto, 2004). Leukotrienes have an impactful role on inflammation as they are now widely accepted as proinflammatory mediators (Busse and Gaddy, 1991). Evidence indicates that many chronic inflammatory disorders are propagated by the presence of leukotrienes. Finally, recent studies have proven the antiinflammatory effects of BA in patients that suffered from chronic inflammatory diseases (Ammon, 2006, 2003).

Diabetes mellitus is associated with aberrant cytokine expressions that negatively impact the wound healing process (Abdel-Salam, 2014). It has recently been demonstrated that camel whey protein supplementation in a streptozotocin-induced diabetic rat model was able to correct diabetes-associated dysregulated cytokine profiles after dermal wounding (Abdel-Salam, 2014). Diabetic rats exhibited delayed wound healing time compared to nondiabetic animals. Diabetic rats supplemented with whey protein (100 mg/kg body weight) showed improved wound healing rates similar to control rats. In addition, proinflammatory cytokines IL-1β, IL-6, and TNF-α and antiinflammatory cytokines IL-4 and IL-10, which were measured to be deregulated in the diabetic model, showed expression levels comparable to control rats after whey supplementation.

Resveratrol, a polyphenolic compound found in red wine, peanuts, and berries, is another natural product that is being investigated in diabetic wound healing (Bashmakov et al., 2011). Resveratrol has been reported to activate sirtuins (SIRTs), a family of class III histone deacetylases, leading to increased longevity in yeast in *Drosophila*. Activation of SIRT1 correlates with reduced TNF-α production. Antiinflammatory TGF-β1 in the skin is also observed during resveratrol supplementation. By reducing proinflammatory cytokines and raising resolution of inflammation signals, resveratrol may represent a natural compound capable of regulating inflammation.

Curcumin, apigenin, and naringenin are phytophenols that have been shown to have antiinflammatory properties by inhibiting the proinflammatory transcription factor NFκB (Lim et al., 2013). Curcumin, found in high concentrations in the spice turmeric, has been touted in traditional medicine for centuries but has recently been shown to have benefits to resolve inflammation (Goel et al., 2008). The flavonoids naringenin (found in grapefruit) and apigenin (found in chamomile and parsley) have been shown to reduce production of nitric oxide (NO) and PGE2 in a dose-dependent manner by downregulation of inducible nitric oxide synthase (iNOS) and COX-2 (Park et al., 2012). One recent study focused on apigenin's protective role during lipopolysaccharide (LPS)-induced inflammation in endothelial cells (Duarte et al., 2013). The authors found that apigenin reduced endothelial cell apoptosis by decreasing the activity of caspase-3. Additionally, they reported that apigenin reduces mitochondrial ROS production by modulating Complex I of the electron transport chain, restoring mitochondrial function.

It has recently been reported that naringenin supplementation reduces hepatic inflammation in rats by modulating matrix metalloproteinases (MMP) by inhibiting NFκB (Chtourou et al., 2015). In their study, the authors showed evidence that rats treated with naringenin (50 mg/kg body weight, 90 days) exhibited reduced macrophage infiltration and oxidative stress markers. In a dermal setting, the potential effect of naringenin treatment to reduce ultraviolet B (UVB)-induced inflammation has also been reported (Martinez et al., 2015). First, Martinez et al. reported that hairless mice intraperitoneally injected with naringenin prior to UVB exposure exhibited less oxidative stress, recruitment of neutrophils, edema, and many proinflammatory cytokines. The same group has also just recently published evidence suggesting that topical application of naringenin may prove efficacious in the prevention of UVB skin inflammation and resulting damage (Martinez et al., 2016).

23.7 FERMENTED PAPAYA PREPARATION

Fermented papaya preparation (FPP) is a natural supplement produced by yeast fermentation of nongenetically modified organically grown *Carica papaya* Linn from Hawaii, USA. Papaya has long been considered a medicinal fruit in various parts of the world. Our lab has shown that oral supplementation of FPP (0.2 g/kg body weight) to leptin receptor mutant (*db/db*$^{-/-}$) diabetic mice improved wound healing outcomes (Collard and Roy, 2010). FPP-supplemented animals exhibited increased macrophage recruitment to wound sites and increased expression of the angiogenic factor VEGF. In addition to chronic unresolved inflammation, diabetes mellitus is associated with increased prevalence of wound infection resultant from compromised bactericidal respiratory burst production (Daoud et al., 2009). We measured a significant increase in inducible oxidant production by wound macrophages in the FPP-supplemented mice. We also have observed that human peripheral blood mononuclear cells (PBMC) supplemented ex vivo with FPP exhibit a similar increase in inducible respiratory burst oxidant production (Dickerson et al., 2012). Generation of reactive oxygen species (ROS) during respiratory burst is facilitated primarily through the multiprotein enzyme complex NADPH oxidase (Nunes et al., 2013). When stimulated, cytosolic and membrane-bound protein subunits colocalize and assemble on the membrane to catalyze the oxidation of molecular oxygen to superoxide anions at the expense of NADPH. Human PBMC supplemented ex vivo with FPP showed increased expression of Rac2, a Rho-family GTPase and essential activating component of the NADPH oxidase complex. Additionally, FPP-supplemented PBMCs exhibited increased phosphorylation of the p47phox protein subunit component of NADPH oxidase, which is necessary for activation of the complex (Dickerson et al., 2012).

After establishing that supplementation of FPP, a carbohydrate-rich product, did not negatively affect blood glucose levels in type 2 diabetes mellitus (T2DM) patients, we then investigated oral supplementation of FPP to T2DM patients. We observed that T2DM donors' PBMCs exhibited a compromised respiratory burst compared to nonT2DM donor-derived samples and similar to our mouse studies, we observed that oral supplementation of FPP (3 g, 3 times per day) increased inducible respiratory burst superoxide production (Dickerson et al., 2015). While increasing production of ROS in response to stimuli, we did not measure any increase in markers of lipid peroxidation that would be indicative of increased oxidative stress burden. This observation underscores the action of FPP to support the innate immune system to produce ROS in response to injury or infection, but not contribute to raising chronic oxidative stress. Taken together, FPP is a potential nutritional supplement which may be of benefit for T2DM wound healing.

REFERENCES

Abdel-Salam, B.K., 2014. Modulatory effect of whey proteins in some cytokines involved in wound healing in male diabetic albino rats. Inflammation 37 (5), 1616–1622.

Abe, M., Yoshimoto, T., 2004. [Leukotriene-lipoxygenase pathway and drug discovery]. Nihon yakurigaku zasshi Folia pharmacologica Japonica 124 (6), 415–425.

Abebe, W., 2002. Herbal medication: potential for adverse interactions with analgesic drugs. J. Clin. Pharm. Ther. 27 (6), 391–401.

Agbanoma, G., Li, C., Ennis, D., Palfreeman, A.C., Williams, L.M., Brennan, F.M., 2012. Production of TNF-alpha in macrophages activated by T cells, compared with lipopolysaccharide, uses distinct IL-10-dependent regulatory mechanism. J. Immunol. 188 (3), 1307–1317.

Agren, M.S., 1990. Studies on zinc in wound healing. Acta Derm-Venereol. Suppl. 154, 1–36.

Agren, M.S., Chvapil, M., Franzen, L., 1991. Enhancement of re-epithelialization with topical zinc oxide in porcine partial-thickness wounds. J. Surg. Res. 50 (2), 101–105.

Ajani, U.A., Ford, E.S., Mokdad, A.H., 2004. Dietary fiber and C-reactive protein: findings from national health and nutrition examination survey data. J. Nutr. 134 (5), 1181–1185.

Al-Okbi, S.Y., 2014. Nutraceuticals of anti-inflammatory activity as complementary therapy for rheumatoid arthritis. Toxicol. Ind. Health 30 (8), 738–749.

Ammon, H.P., 2003. Boswellic acids (the active agent in incense) for the treatment of chronic inflammatory diseases. Med. Monatsschr. Pharm. 26 (9), 309–315.

Ammon, H.P., 2006. Boswellic acids in chronic inflammatory diseases. Planta Med. 72 (12), 1100–1116.

Bandini, L.G., Vu, D., Must, A., Cyr, H., Goldberg, A., Dietz, W.H., 1999. Comparison of high-calorie, low-nutrient-dense food consumption among obese and non-obese adolescents. Obes. Res. 7 (5), 438–443.

Bashmakov, Y.K., Assaad-Khalil, S., Petyaev, I.M., 2011. Resveratrol may be beneficial in treatment of diabetic foot syndrome. Med. Hypotheses 77 (3), 364–367.

Blain, D., Kubow, S., Chan, H.M., 1998. Zinc pretreatment inhibits isotretinoin teratogenicity and induces embryonic metallothionein in CD-1 mice. J. Nutr. 128 (7), 1239–1246.

Bowden, R.G., Wilson, R.L., Deike, E., Gentile, M., 2009. Fish oil supplementation lowers C-reactive protein levels independent of triglyceride reduction in patients with end-stage renal disease. Nutr. Clin. Pract. 24 (4), 508–512.

Broughton, II G., Janis, J.E., Attinger, C.E., 2006. The basic science of wound healing. Plast. Reconstr. Surg. 117 (7 Suppl.), 12s–34s.

Busse, W.W., Gaddy, J.N., 1991. The role of leukotriene antagonists and inhibitors in the treatment of airway disease. Am. Rev. Respir. Dis. 143 (5 Pt 2), S103–S107.

Calder, P.C., 2013. n-3 fatty acids, inflammation and immunity: new mechanisms to explain old actions. Proc. Nutr. Soc. 72 (3), 326–336.

Chtourou, Y., Fetoui, H., Jemai, R., Ben Slima, A., Makni, M., Gdoura, R., 2015. Naringenin reduces cholesterol-induced hepatic inflammation in rats by modulating matrix metalloproteinases-2, 9 via inhibition of nuclear factor kappaB pathway. Eur. J. Pharmacol. 746, 96–105.

Collard, E., Roy, S., 2010. Improved function of diabetic wound-site macrophages and accelerated wound closure in response to oral supplementation of a fermented papaya preparation. Antioxid. Redox Sign. 13 (5), 599–606.

Crovetti, G., Martinelli, G., Issi, M., Barone, M., Guizzardi, M., Campanati, B., et al., 2004. Platelet gel for healing cutaneous chronic wounds. Transfus. Apher. Sci. 30 (2), 145–151.

Daoud, A.K., Tayyar, M.A., Fouda, I.M., Harfeil, N.A., 2009. Effects of diabetes mellitus vs. in vitro hyperglycemia on select immune cell functions. J. Immunotoxicol. 6 (1), 36–41.

Das, A., Sinha, M., Datta, S., Abas, M., Chaffee, S., Sen, C.K., et al., 2015. Monocyte and macrophage plasticity in tissue repair and regeneration. Am. J. Pathol. 185 (10), 2596–2606.

Dennis, E.A., Norris, P.C., 2015. Eicosanoid storm in infection and inflammation. Nat. Rev. Immunol. 15 (8), 511–523.

Diaz, J.R., de las Cagigas, A., Rodriguez, R., 2003. Micronutrient deficiencies in developing and affluent countries. Eur. J. Clin. Nutr. 57 (Suppl 1), S70–S72.

Dickerson, R., Deshpande, B., Gnyawali, U., Lynch, D., Gordillo, G.M., Schuster, D., et al., 2012. Correction of aberrant NADPH oxidase activity in blood-derived mononuclear cells from type II diabetes mellitus patients by a naturally fermented papaya preparation. Antioxid. Redox Sign. 17 (3), 485–491.

Dickerson, R., Banerjee, J., Rauckhorst, A., Pfeiffer, D.R., Gordillo, G.M., Khanna, S., et al., 2015. Does oral supplementation of a fermented papaya preparation correct respiratory burst function of innate immune cells in type 2 diabetes mellitus patients? Antioxid. Redox Sign. 22 (4), 339–345.

DiPietro, L.A., 1995. Wound healing: the role of the macrophage and other immune cells. Shock 4 (4), 233–240.

Dryden, S.V., Shoemaker, W.G., Kim, J.H., 2013. Wound management and nutrition for optimal wound healing. Atlas Oral Maxillofac. Surg. Clin. N. Am. 21 (1), 37–47.

Duarte, S., Arango, D., Parihar, A., Hamel, P., Yasmeen, R., Doseff, A.I., 2013. Apigenin protects endothelial cells from lipopolysaccharide (LPS)-induced inflammation by decreasing caspase-3 activation and modulating mitochondrial function. Int. J. Mol. Sci. 14 (9), 17664–17679.

Falanga, V., 2004. The chronic wound: impaired healing and solutions in the context of wound bed preparation. Blood Cell. Mol. Dis. 32 (1), 88–94.

Fujiwara, N., Kobayashi, K., 2005. Macrophages in inflammation. Curr. Drug Targets Inflamm. Allergy 4 (3), 281–286.

Galland, L., 2010. Diet and inflammation. Nutr. Clin. Pract. 25 (6), 634–640.

Ge, K.Y., Chang, S.Y., 2001. Definition and measurement of child malnutrition. Biomed. Environ. Sci. 14 (4), 283–291.

Gilmore, M.A., 1991. Phases of wound healing. Dimens. Oncol. Nurs. 5 (3), 32–34.

Giugliano, D., Ceriello, A., Esposito, K., 2006. The effects of diet on inflammation: emphasis on the metabolic syndrome. J. Am. Coll. Cardiol. 48 (4), 677–685.

Goel, A., Kunnumakkara, A.B., Aggarwal, B.B., 2008. Curcumin as "Curecumin": from kitchen to clinic. Biochem. Pharmacol. 75 (4), 787–809.

Guo, S., Dipietro, L.A., 2010. Factors affecting wound healing. J. Dent. Res. 89 (3), 219–229.

Gupta, I., Parihar, A., Malhotra, P., Singh, G.B., Ludtke, R., Safayhi, H., et al., 1997. Effects of Boswellia serrata gum resin in patients with ulcerative colitis. Eur. J. Med. Res. 2 (1), 37–43.

Irvin, T.T., 1978. Effects of malnutrition and hyperalimentation on wound healing. Surg. Gynecol. Obstet. 146 (1), 33–37.

Khanapure, S.P., Garvey, D.S., Janero, D.R., Letts, L.G., 2007. Eicosanoids in inflammation: biosynthesis, pharmacology, and therapeutic frontiers. Curr. Top. Med. Chem. 7 (3), 311–340.

Khanna, S., Biswas, S., Shang, Y., Collard, E., Azad, A., Kauh, C., et al., 2010. Macrophage dysfunction impairs resolution of inflammation in the wounds of diabetic mice. PLoS One 5 (3), e9539.

Kirkup, S.E., Cheng, Z., Elmes, M., Wathes, D.C., Abayasekara, D.R., 2010. Polyunsaturated fatty acids modulate prostaglandin synthesis by ovine amnion cells in vitro. Reproduction 140 (6), 943–951.

Kolaczkowska, E., Kubes, P., 2013. Neutrophil recruitment and function in health and inflammation. Nat. Rev. Immunol. 13 (3), 159–175.

Kuehl Jr., F.A., Egan, R.W., 1980. Prostaglandins, arachidonic acid, and inflammation. Science 210 (4473), 978–984.

Lim, R., Barker, G., Wall, C.A., Lappas, M., 2013. Dietary phytophenols curcumin, naringenin and apigenin reduce infection-induced inflammatory and contractile pathways in human placenta, foetal membranes and myometrium. Mol. Hum. Reprod. 19 (7), 451–462.

MacKay, D., Miller, A.L., 2003. Nutritional support for wound healing. Altern. Med. Rev. 8 (4), 359–377.

Martinez, R.M., Pinho-Ribeiro, F.A., Steffen, V.S., Caviglione, C.V., Vignoli, J.A., Barbosa, D.S., et al., 2015. Naringenin inhibits UVB irradiation-induced inflammation and oxidative stress in the skin of hairless mice. J. Nat. Prod. 78 (7), 1647–1655.

Martinez, R.M., Pinho-Ribeiro, F.A., Steffen, V.S., Silva, T.C., Caviglione, C.V., Bottura, C., et al., 2016. Topical formulation containing naringenin: efficacy against ultraviolet B irradiation-induced skin inflammation and oxidative stress in mice. PLoS One 11 (1), e0146296.

Menke, N.B., Ward, K.R., Witten, T.M., Bonchev, D.G., Diegelmann, R.F., 2007. Impaired wound healing. Clin. Dermatol. 25 (1), 19–25.

Murray, P.J., Wynn, T.A., 2011. Protective and pathogenic functions of macrophage subsets. Nat. Rev. Immunol. 11 (11), 723–737.

Nunes, P., Demaurex, N., Dinauer, M.C., 2013. Regulation of the NADPH oxidase and associated ion fluxes during phagocytosis. Traffic 14 (11), 1118–1131.

Park, N.Y., Valacchi, G., Lim, Y., 2010. Effect of dietary conjugated linoleic acid supplementation on early inflammatory responses during cutaneous wound healing. Mediators Inflamm. 2010.

Park, H.Y., Kim, G.Y., Choi, Y.H., 2012. Naringenin attenuates the release of pro-inflammatory mediators from lipopolysaccharide-stimulated BV2 microglia by inactivating nuclear factor-kappaB and inhibiting mitogen-activated protein kinases. Int. J. Mol. Med. 30 (1), 204–210.

Patterson, E., Wall, R., Fitzgerald, G.F., Ross, R.P., Stanton, C., 2012. Health implications of high dietary omega-6 polyunsaturated fatty acids. J. Nutr. Metab. 2012, 539426.

Ricciotti, E., FitzGerald, G.A., 2011. Prostaglandins and inflammation. Arterioscler. Thromb. Vasc. Biol. 31 (5), 986–1000.

Safayhi, H., Sailer, E.R., Ammon, H.P., 1995. Mechanism of 5-lipoxygenase inhibition by acetyl-11-keto-beta-boswellic acid. Mol. Pharmacol. 47 (6), 1212–1216.
Santos, S., Oliveira, A., Casal, S., Lopes, C., 2013. Saturated fatty acids intake in relation to C-reactive protein, adiponectin, and leptin: a population-based study. Nutrition 29 (6), 892–897.
Schar, M.O., Diaz-Romero, J., Kohl, S., Zumstein, M.A., Nesic, D., 2015. Platelet-rich concentrates differentially release growth factors and induce cell migration in vitro. Clin. Orthop. Relat. R. 473 (5), 1635–1643.
Schweizer, S., von Brocke, A.F., Boden, S.E., Bayer, E., Ammon, H.P., Safayhi, H., 2000. Workup-dependent formation of 5-lipoxygenase inhibitory boswellic acid analogues. J. Nat. Prod. 63 (8), 1058–1061.
Sen, C.K., Khanna, S., Venojarvi, M., Trikha, P., Ellison, E.C., Hunt, T.K., et al., 2002. Copper-induced vascular endothelial growth factor expression and wound healing. Am. J. Physiol. Heart 282 (5), H1821–H1827.
Sen, C.K., Gordillo, G.M., Roy, S., Kirsner, R., Lambert, L., Hunt, T.K., et al., 2009. Human skin wounds: a major and snowballing threat to public health and the economy. Wound Repair Regen. 17 (6), 763–771.
Sermet-Gaudelus, I., Poisson-Salomon, A.S., Colomb, V., Brusset, M.C., Mosser, F., Berrier, F., et al., 2000. Simple pediatric nutritional risk score to identify children at risk of malnutrition. Am. J. Clin. Nutr. 72 (1), 64–70.
Sindrilaru, A., Peters, T., Wieschalka, S., Baican, C., Baican, A., Peter, H., et al., 2011. An unrestrained proinflammatory M1 macrophage population induced by iron impairs wound healing in humans and mice. J. Clin. Invest. 121 (3), 985–997.
Singer, A.J., Clark, R.A., 1999. Cutaneous wound healing. New Engl. J. Med. 341 (10), 738–746.
Stechmiller, J.K., 2010. Understanding the role of nutrition and wound healing. Nutr. Clin. Pract. 25 (1), 61–68.
Steed, D.L., 1997. The role of growth factors in wound healing. Surg. Clin. North Am. 77 (3), 575–586.
Tonnesen, M.G., Feng, X., Clark, R.A., 2000. Angiogenesis in wound healing. J. Investig. Dermatol. Symp. Proc. 5 (1), 40–46.
Wilgus, T.A., Roy, S., McDaniel, J.C., 2013. Neutrophils and wound repair: positive actions and negative reactions. Adv. Wound Care 2 (7), 379–388.
Wright, H.L., Moots, R.J., Bucknall, R.C., Edwards, S.W., 2010. Neutrophil function in inflammation and inflammatory diseases. Rheumatology 49 (9), 1618–1631.
Wright, J.A., Richards, T., Srai, S.K., 2014. The role of iron in the skin and cutaneous wound healing. Front. Pharmacol. 5, 156.
Xie, H., Kang, Y.J., 2009. Role of copper in angiogenesis and its medicinal implications. Curr. Med. Chem. 16 (10), 1304–1314.
Zhang, J.M., An, J., 2007. Cytokines, inflammation, and pain. Int. Anesthesiol. Clin. 45 (2), 27–37.

CHAPTER

24 Development of low cost and eco-friendly feed for various candidate species for the sustainability of commercial aquaculture and reduction of aquatic pollution

Kartik Baruah*, Parisa Norouzitallab*,, Asim Kumar Pal[†]**

**Laboratory of Aquaculture &* Artemia *Reference Center, Department of Animal Production, Faculty of Bioscience Engineering, Ghent University, Belgium; **Laboratory of Immunology and Animal Biotechnology, Department of Animal Production, Faculty of Bioscience Engineering, Ghent University, Belgium; [†]APC Nutrients, Secunderabad, India*

24.1 INTRODUCTION

Global aquaculture fish production has grown steadily in the past few decades. According to the latest available statistics collected globally by the United Nations' Food and Agricultural Organization (FAO), world aquaculture production attained another all-time high of 90.4 million tons in 2012, with a worth of US$144.4 billion (FAO, 2014). According to the FAO, this sharp increase in aquaculture production is predicted to continue further. Although that might sound like good news for the world's food supply, however there is a hidden cost behind some of the farmed fishes on supermarket shelves. Many of these farmed fishes, including the popular salmon, trout, cod, and shrimps, are fed on wild fishes (as fish meal), which are an expensive source of protein. It takes about 2–5 kg of wild fish to produce 1 kg of farmed fish, such as salmon. However, currently wild stocks have already reached their maximum biological limits, so any increase in fishing efforts is unsustainable (Powell, 2003).

The most commonly cultured fishes in the world, the carp and tilapia, are mainly vegetarian, and so can be farmed on a plant-based diet. In Asia, where currently about 90% of the world's aquaculture takes place, these fishes are big business. However, in the Western world, where carp is often considered to be bony and tilapia is a relatively new and unfamiliar product, these fishes still fail to compete with meaty

Developing New Functional Food and Nutraceutical Products. http://dx.doi.org/10.1016/B978-0-12-802780-6.00024-9

carnivorous fishes like salmon, trout, and cod. The production of these farmed carnivorous fishes is lagging behind in comparison to herbivorous fishes, such as tilapia and carp. However, in recent years, because of the increasing market demand, farming of other carnivorous fish, such as cod, halibut, sea-bass, sea bream, and tuna, is markedly increasing. This has caused a huge problem because farmed fish as mentioned earlier are fed on a diet that relies heavily on fish oil and fishmeal as the source of proteins and nutrients. A simple calculation showed that the current haul of fish oil and fishmeal would soon be surpassed by the needs of global aquaculture.

Over the past few decades, various researchers and aquafarmers have begun to think that one way to make aquaculture more sustainable is to change the diets of farmed fishes, that is, "to turn carnivores into vegetarians" (Naylor et al., 2000). There are also commercial reasons to look for alternatives to fishmeal and oil. Currently fishmeal is the major dietary protein source, comprising between 20% and 60% of the aquafeed. Feed makes up three-quarters of fish farmers' total running costs, and the price of fishmeal/oil is escalating with the increasing demand. The need to replace fishmeal in aquafeed is therefore highly warranted and even organizations like the FAO have strongly recommended that. Today the options for vegetarian fish diets include plant ingredients like soybean, corn, rapeseed, sunflower seeds, flax seeds, wheat glutens, sorghum, groundnut, cottonseed, and sesame. These ingredients are the most promising protein and energy sources for aquafeed. In fact, some farmed salmon already get up to 50% of their protein or oils from such ingredients, without their health or growth rates being affected (Refstie et al., 2001). However, attempts to increase the inclusion level of plant-based nutrients above 50% tend to cause serious problems in fish including irritated lower intestines, depressed immune system, and inability to absorb key minerals such as phosphorus, calcium, zinc, and iron (Baruah et al., 2009). This problem is due to the presence of various antinutritional factors (i.e., natural compounds that interfere with the digestibility/absorption of nutrients) in the plant ingredients.

24.2 PHYTIC ACID: THE CHELATOR OF NUTRIENTS

Among the different antinutrients, phytic acid is one found in most of the ingredients commonly used in aquafeed like barley, rice, sorghum, wheat, maize, gram, groundnut, rapeseed, soyabean, cottonseed, and sesame (De Silva and Anderson, 1995). Phytic acid isolated from plants belongs to the group of organic phosphates and is a mixture of calcium–magnesium salt of inositol hexaphosphoric acid, also known as phytin. Salts of phytic acid are also called phytate. It is an abundant plant constituent comprising 1–5% by weight of the legumes, cereals, oil seeds, pollens, and nuts (Vohra and Satyanarayan, 2003). It is an organic form of phosphorus, which is chemically a myo-inositol hexakisdihydrogen-phosphate (IP6). The occurrence of phytate in plant foodstuffs is well documented. It constitutes between 0.7% and 2% of most cereal grains and oilseeds (Adeola and Sands, 2003) and is the primary source of inositol and storage form of phosphorus and other minerals in plant seeds that are used as animal feed ingredients (Powell, 2003). Most foods of plant origin contain 50–80% of their total phosphorus, or even higher in selected varieties, as phytate (Raboy, 1997). On average, the phytate-phosphorus content of cereals (like maize, rice, wheat, sorghum, and barley) and oilseed meals (like groundnut, soybean, cottonseed, and sunflower meals) varies from 51% to 82% of the total phosphorus present in them (Tyagi and Verma, 1998).

24.3 PHYTIC ACID: TOUGH TO INGEST AND DIGEST

The most important antinutritional effect of phytate is that it contains phosphorus that cannot be digested by monogastric animals like pigs, chickens, and fish as their intestinal mucosa do not secrete the enzyme phytase, and hence its bioavailability is very limited (Vielma et al., 1998). In rainbow trout, for instance, only 22% of the phosphorus in the soybean meal was available to the fish (Sugiura et al., 1999). Further data from several studies on both herbivorous and carnivorous fishes have shown that only a certain percentage (e.g., about 20% in salmon) of dietary phosphorus in typical commercial feeds is retained by the fish (Ketola and Harland, 1993). This means that the unutilized phosphorus gets discharged into the aquatic environment and causes aquatic pollution (Sugiura et al., 1999). Phytic acid has also other vital effects on feed constituents. Because of its high density for negatively charged phosphate groups, phytate chelates directly or indirectly with other mineral cations (as cited earlier) in smaller amounts and forms a poorly soluble complex. Zinc availability was greatly reduced in Chinook salmon when sodium phytate was added to their feed and this resulted in the development of cataracts (Richardson et al., 1985). Apart from minerals, phytate also forms complexes with other nutrients like protein and amino acids, vitamins, lipids, and starch (Cao et al., 2007). The interaction between phytic acid and proteins is believed to be an ionic type and is dependent on pH (Cosgrove, 1966). At low pH, phytic acid forms electrostatic linkages with the basic arginine, lysine, and histidine residues resulting in insoluble complexes. As the pH approaches the isoelectric point, the charge on the protein is neutralized and the phytate is no longer bound and becomes soluble. In its soluble state phytate complexes with protein because of the presence of divalent cations. These cations, usually calcium, magnesium, or zinc, act as a bridge between negatively charged protein, carboxyl groups, and the phytate (Anderson, 1985). In vivo studies have shown that phytate–protein complexes are insoluble and are less subjected to attack by proteolytic enzymes than the same protein alone. The reduced solubility of proteins because of such complexing can negatively influence certain functional properties of proteins, which are dependent on their hydration and solubility. Phytate is known to inhibit a number of digestive enzymes, such as pepsin, α-amylase, and trypsin (Ravindran et al., 1995). In protein–phytate interaction, the amino group present on the side chain of the amino acids is one of the functional groups involved in the interaction, thereby decreasing the digestibility of proteins. Even the action of certain enzymes, such as amylase, trypsin, acid phosphatase, and tyrosinase, has been shown to be inhibited by phytic acid and also by inositol pentaphosphate (Harland and Morris, 1995). It was reported that phytate–protein complexes may be formed postfeeding in the gut at pH between 2 and 3 (Jongbloed et al., 1997). At this pH range, it was found that soluble proteins in casein, corn, rice polish, soybean meal, and sunflower meal were substantially precipitated in the presence of phytic acid. In rainbow trout fed purified diets containing phytic acid, it was reported that the fish suffered marked reduction in protein digestibility and about 10% reduction in growth and feed conversion (Spinelli et al., 1983). It was also found that increasing calcium and magnesium content in the diet in the presence of phytic acid did not affect growth and feed conversion. Fish fed diets containing over 1% calcium without phytic acid had a 5% reduction in growth and feed conversion. Therefore, it was concluded that reduced growth was due to reduced protein bioavailability and not due to a reduction in zinc, iron, or copper availability.

Phytic acid affects the digestibility of starch through interaction with the amylase enzyme (Baruah et al., 2007a). It also reduces the solubility of starch by binding to starch and reducing its absorption, which results in lowering glucose utilization. Starch binding occurs because of hydrogen bond formation (Thompson, 1986). The effect of phytic acid on starch digestibility was studied in vitro and negatively correlated with the blood glucose response (glycemic index) in healthy volunteers (Yoon et al., 1983). It was found that the glycemic index was correlated negatively with the phytic acid content of the food tested. In vitro digestion studies involving human saliva at physiological pH and temperature showed that in the presence of sodium phytate, the rate of digestion of raw wheat starch was reduced by 50%. This was reversed by the addition of calcium, which is known to complex with phytic acid. In addition, phytic acid also complexes with lipids, thereby reducing their digestibility (Vohra and Satyanarayan, 2003). A study carried out on carp showed that the fish fed diets with phytic acid contained lower fat compared to those without phytic acid (Usmani and Jafri, 2002).

24.4 PHYTASE: A NOVEL NUTRACEUTICAL TO DIGEST PHYTATE

Using phytase (myo-inositol hexaphosphate hydrolase) is an ideal approach to hydrolyze indigestible phytate in plant ingredients. Phytase is a phosphatase enzyme that sequentially cleaves orthophosphate groups from the inositol ring of phytic acid to yield available free inorganic phosphorus, and a series of lower phosphoric esters (inositol pentaphosphate to inositol monophosphate) as intermediates, thereby decreasing phytates' affinity for different cations (Lei et al., 1993). The reaction ultimately leads to the production of free myoinositol (Harland and Morris, 1995). Based on the type of reactions catalyzed, they are classified as 3-phytase (EC 3.1.3.8) and 6-phytase (EC 3.1.3.26). Phytases are widespread in nature, occurring in animals, plants, and microorganisms. Phytate-degrading enzymes from microorganisms are considered to be 3-phytases, whereas seeds of higher plants are said to contain 6-phytases. Microbial phytases have pH optima in the range of 2–6, while plant

Table 24.1 Commercial Production Information of Microbial Phytase

Company	Country	PhytaseSource	Production Strain	Trademark
AB Enzymes	Germany	*Aspergillus awamori*	*Trichoderma reesei*	Finase
Alko Biotechnology	Finland	*A. oryzae*	*A. oryzae*	SP, TP, SF
Alltech	USA	*A. niger*	*A. niger*	Allzyme phytase
BASF	Germany	*A. niger*	*A. niger*	Natuphos
Biozyme	USA	*A. oryzae*	*A. oryzae*	AMAFERM
DSM	USA	*P. lycil*	*A. oryzae*	Bio-Feed Phytase
Fermic	Mexico	*A. oryzae*	*A. oryzae*	Phyzyme
Finfeeds International	Finland	*A. awamori*	*T. reesei*	Avizyme
Genencor International	USA	*P. simplicissimum*	*Penicillium funiculosum*	ROVABIO
Roal	Finland	*A. awamori*	*T. reesei*	Finase
Novozymes	Denmark	*A. oryzae*	*A. oryzae*	Ronozyme

Adapted from Cao et al. (2007).

phytases tend to have an optimal pH of 5. In recent years, several phytases with different biochemical properties have been characterized and their effectiveness in animal feeding has been demonstrated (Stahl et al., 1999). Several microbial phytases either as a dry powder or as liquid are available commercially. The first commercial phytase products (Natuphos) derived from a genetically modified *Aspergillus niger* strain were introduced onto the market in 1991. After the mid-1990s, several distinct microbial phytase products become commercially available (Table 24.1).

24.5 SIGNIFICANCE OF PHYTASE IN AQUAFEED

24.5.1 IMPACT ON PHOSPHORUS BIOAVAILABILITY

Phosphorus is a key mineral for fishes and prawns. Fish can absorb soluble phosphorus through the skin, fins, and gills; however, the concentration of phosphorus in fresh water and seawater is low (NRC, 1993). Therefore, the phosphorus requirement for fish is dependent on feed. It is also important to mention that phosphorus is a critical pollutant in the aquatic environment. Excessive phosphorus concentration is the most common cause of eutrophication of aquatic systems (Correll, 1999). Phytase supplementation makes the chelated phytate-phosphorus available to fish and reduces its excretion into the environment (Debnath et al., 2005; Baruah et al., 2007a) (Fig. 24.1). About 20–40% of phytate-phosphorus can be released by adding phytase at concentrations of 500 and 1000 U/kg, respectively, in carp fed with a soybean meal

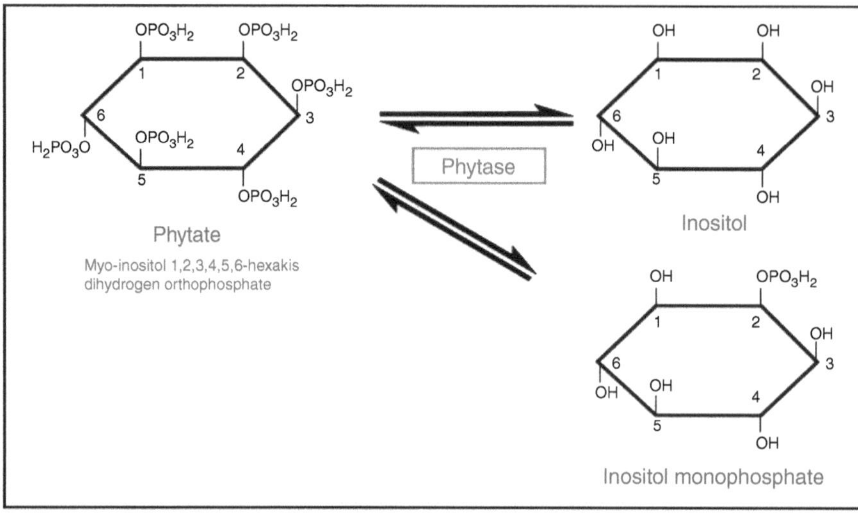

FIGURE 24.1 Mode of action of phytase in releasing bound phosphates

Adapted from Debnath et al. (2005).

based diet. Similar reports have been reported for different fish species (for review, see Debnath et al., 2005; Cao et al., 2007). Robinson et al. (2002) reported that a 250 FTU/kg diet could effectively replace a 0.8% dicalcium phosphate supplement in channel catfish diets without affecting growth, feed efficiency, or bone phosphorus deposition. All these studies demonstrated that phytase was effective in enhancing phosphorus bioavailability considerably, thereby reducing the release of fecal-phosphorus into the environment. These studies indicate that phytase may provide a remarkable platform for sustainable and eco-friendly farming.

24.5.2 IMPACT ON THE BIOAVAILABILITY OF OTHER MINERALS

Phytate–mineral complexes may be formed with various di- and trivalent cations, as well as with proteins (Wise, 1980). For example, calcium-bound phytate increases chelation with trace minerals especially with zinc to form coprecipitates that make the zinc unavailable to animals. Phytase supplementation hydrolyzes phytate and increases the concentration of minerals like magnesium, calcium, manganese, and zinc in plasma, bone, and the whole body (Baruah et al., 2007a,b). Yan et al. (2002) conducted a feeding trial to determine the effects of phytase on utilization of dietary protein and minerals by fingerlings Channel catfish, *Ictalurus punctatus*, fed an all plant protein diet composed of soybean meal, corn, and wheat middlings. After 14 weeks, it was found that fish fed phytase supplements had higher concentration of ash, calcium, phosphorus, and manganese in the bone than fish fed a diet without phytase supplementation. Similar results were reported for salmonids, common carp, stripped bass, Nile tilapia, *P. pangasius*, and *L. rohita* (for a review, see Cao et al., 2007).

However, it is pertinent to note that the absorption and utilization of particular minerals are species specific and also dependent on the feed ingredients used. Further research is warranted in this direction, in different commercially important fishes.

24.5.3 IMPACT ON PROTEIN DIGESTIBILITY

As mentioned earlier, phytate can nonselectively bind to proteins, inhibit activities of enzymes including pepsin, trypsin, and alpha-amylase (Liener, 1994), and decrease protein digestibility. Phytase treatment of plant-based diets can effectively increase phosphorus availability in fishes; however, there is not enough information on protein and amino acid utilization. Better protein economy of plant-based aquafeeds by phytase would increase the interest of the feed industry toward this relatively new nutraceutical. In fish, there are conflicting and inconsistent reports as to the efficacy of phytase in improving protein digestibility. Phytase has been shown to increase (Baruah et al., 2007a), not affect (Yan et al., 2002) or decrease (Teskeredzic et al., 1995) protein digestibility. These variations in the results may be attributed to multiple factors that include variation in phytic acid content in different feedstuffs, species used, inherent characteristics of feed ingredients, and/or probably the presence or absence of a true stomach in different fish species as phytase activity is pH specific. Thus, at present, the mechanisms underlying the protein-associated responses to added phytase remain largely speculative and therefore need further research.

24.5.4 IMPACT ON GROWTH PERFORMANCE

Both herbivorous and carnivorous fishes fed semipurified diets containing various levels of calcium, phosphorus, zinc, and sodium phosphate with a high dietary phytic acid (2.6%) content exhibited depressed growth. But the growth performance was increased when microbial phytase was incorporated in the diets. For instance, in Channel catfish fed with diets containing only plant protein sources, phytase supplementation was shown to increase weight gain and feed consumption by 23.5 – 11.6%, respectively (Jackson et al., 1996). Similar performance was also observed in *Pangasius pangasius* (Debnath et al., 2005), and the African catfish *Clarias gariepinus* (Van Weerd et al., 1999). Besides these, several researchers also found positive effects of phytase treatment on weight gain in other fishes (Channel catfish, Li and Robinson, 1997; Rainbow trout, Rodehutscord and Pfeffer, 1995). Positive effects of dietary phytase on the growth of several fishes could be attributed to better bioavailability of phosphorus and other minerals, and improved protein digestibility.

24.5.5 IMPACT ON FORMULATION OF LOW-COST FEED

Low-cost feed formulation involves feed production at the most cost-effective level in terms of resource, time, energy, and money by minimizing input while maximizing output. For many years, fish nutritionists had looked for ways to utilize cheaper and readily available plant protein sources. Despite having these resources, their inclusion in aquafeed formulation was limited due to the antinutritional activities of phytate. However, with the introduction of phytase, fish nutritionists can formulate and produce a cheaper plant-based aquafeed. This will ultimately lead to profit maximization of aquaculture enterprises.

24.5.6 IMPACT ON AQUACULTURE POLLUTION

The environmental impact of aquaculture activities is increasingly under review and ecologists as well as governments are putting restrictions on this industry. Both marine and freshwater aquafarmers are facing increasing pressure concerning the discharges into the surrounding ecosystems. Aquaculture pollution occurs because of these discharges in general or because of loading of phosphorus into the environment in particular. Phosphorus in the feed ingredients occurs in a number of forms. In ingredients of animal origin, it occurs in inorganic form as well as phosphate complexes of protein, lipid, and carbohydrate. These forms are usable by fish. In contrast, plant feedstuffs contain phosphorus in the form of phytate, which is generally unavailable to fish as mentioned previously. It comes out through the excreta to the water which is enzymatically cleaved by soil and waterborne microorganisms, thereby releasing phosphorus into the water bodies causing eutrophication, which in turn results in oxygen depletion due to excessive algal growth. Dietary microbial phytase supplementation can overcome this problem by making the chelated phosphorus available to fish and lessening its excretion into the environment through the fecal matter. In the catfish *Ictalurus punctatus*, dietary phytase was reported to reduce the excretion of fecal phosphorus by about 60% (Li and Robinson, 1997). Beside this, many studies also suggested the potential environmental benefits of phytase application by reducing the excretion of phosphorus by 30–40% (Omogbenigun et al., 2003). The environmental benefits of using this enzyme are: (1) less mineral supplementation is required and therefore less inorganic phosphate is required in the diet; (2) less organic phosphate (phytic acid) is excreted and thereby less phosphate loads into the environment in intensive aquaculture settings.

24.6 SYNERGISM BETWEEN THE NUTRACEUTICAL PHYTASE AND ACIDIFIERS

Several lines of evidence suggest that microbial phytase can partly render phytate-phosphorus available to animals (Cromwell et al., 1995; Jongbloed et al., 1996). This indicates that part of the phytate-phosphorus is still unavailable to the animal. Previous studies had shown that the optimum activity of microbial phytase occurs at pH of 5.0–5.5 and 2.5 (Simons et al., 1990). Phytase activity changes along the digestive tract, with the most efficient phytate hydrolysis occurring in the stomach (Yi et al., 1996). In terrestrial animals, it was shown that about 43% of dietary phytate hydrolysis by phytase takes place in the stomach and only 7% takes place in the small intestine (Jongbloed et al., 1992). Gut pH influences the bioavailability of dietary minerals by regulating the chelation and complex formation of the element and by altering the transport mechanism of minerals (Baruah et al., 2007b). Studies on rainbow trout showed that acidification of experimental diets with acidifiers like formic acid influence the gut pH and enhance the apparent bioavailability of key minerals like phosphorus, calcium, and magnesium (Vielma and Lall, 1997). A similar result was also observed in rainbow trout and *L. rohita* fed citric acid supplemented diets (Baruah et al., 2007b). Similar to other higher vertebrates, fish also absorb phosphorus as orthophosphates. The digestibility of simple phosphates is relatively high (> 80%) in both rainbow trout and common carp but the complex phosphates are more efficiently absorbed by gastric rainbow trout, which have a true stomach, than by agastric carp, which do not have a true stomach (Ogino et al., 1979).

Microbial phytase was inactive at pH 7 or above (Sugiura et al., 2001). A typical soybean meal based diet for fish has a pH of approximately 6.0 and addition of phytase into the diet further increases the pH which is not suitable for its optimum activity. In rainbow trout, acid production assists in lowering the (dietary) pH but in stomachless fishes, no such mechanism exists and thus supplementation of acidifiers such as citric acid, formic acid, and fumaric acids will be needed to lower diet acidity. Reducing the dietary pH might lower the gut pH and thus increase the efficacy of phytase. This could lead to reduced feed costs and to a greater use of phytase, which would be of importance to the environment.

24.7 DRAWBACKS OF USING EXOGENOUS PHYTASE IN FEED

Phytase shows optimum activity at two pH, that is, alkaline and acidic pH. The pH of the gastrointestinal tract of fish depends on the presence or absence of a stomach. Phytase activity therefore is species specific. The application of phytase as a feed additive is limited due to several factors like cost, inactivation at high temperatures required for pelleting (>80°C), loss of activity during storage, and narrow optimum pH range. The activity of natural phytase in certain plant feedstuffs is high enough to be considered in feed formulation (Paik, 2003). This phytase may not hydrolyze phytate in the gut of fish. Moreover, wide diversity in the feeding habits, and the presence or absence of a functional stomach in fish can markedly influence the dietary requirement of phytase. Further research needs to be carried out to determine the optimum dose of dietary phytase for different commercially important species.

24.8 NEED OF THE HOUR

Beneficial effects of feed enzymes cannot be denied; however, the usefulness of such enzymes will increase when new forms of enzymes with the following properties are available: increased activities under normal conditions, high thermal resistance, cheap to produce, long storage life under ambient conditions, resistant to proteolysis, and activity across a broad pH range. The nature of interactions of phytase with different dietary constituents needs to be established. For instance, whether the enhancement of phosphorus digestibility due to phytase supplementation has any effect on other minerals and nutrients is to be urgently investigated. So far, a limited number of studies has been conducted to determine the effects of enzyme supplementation on various physiological and endocrine parameters like secretion of other enzymes, bile salts, on the immune response, and hormone levels including growth hormone, thyroid hormone, insulin, etc. Comprehensive studies need to be carried out to characterize these effects.

24.9 FLAVOR OF FISH FED A VEGETARIAN DIET

Even if fish can be cultured feeding exclusively on an all-vegetarian diet, an important question that needs to be addressed is whether fish fed such a diet taste the same as carnivorous ones. Fish fed a diet supplemented with only plant-based oils resulted in milder-flavored, less oily fish fillets with a different texture. However, experts disagree on whether consumers will notice. In taste tests, professional food tasters could not tell the difference between salmon raised on vegetarian diets and those raised on animal-based diets (Powell, 2003).

Fish fed plant-based diets also tend to possess low amounts of omega-3 fatty acids in their flesh. However, many fish nutritionists have overcome this problem by simple supplementation of feed additives to the diet. In pilot studies, Samuelsen et al. (2001) added the enzyme lipase to fish diets and showed an increase in the level of omega-3 fatty acids in the fish. Another way to rectify the fat content is to give a fish a vegetarian diet for most of its life, and then follow it with a more "natural", fishier diet just before going to market. By this method, the amount of fish oil used over a fish's life could be cut by about 85% (Powell, 2003).

24.10 CONCLUSIONS

Many scientists believe that the final obstacles to making plant-based aquafeed realistic and affordable will be cleared in the next 5–10 years, allowing fish diets to be close to 100% vegetarian. If carnivorous fish can at least be weaned off a diet of wild fish or can be fed a 100% vegetarian diet, then the aquaculture industry will continue to grow in a sustainable way. The new diet will reduce pollution, fish costs, and even world hunger, as at least some of the trash fish currently fed to farmed fish could instead be used to feed people. If aquaculture as a whole is to succeed, it needs at the very least to be sustainable—not only for the benefit of wild stocks, but ultimately for the industry and consumers too.

REFERENCES

Adeola, O., Sands, J.S., 2003. Does supplemental dietary microbial phytase improve amino acid utilization? A perspective that it does not. J. Anim. Sci. 81, E78–E85.

Anderson, P.A., 1985. Digestibility and Amino Acid Availability in Cereals and Oilseeds. American Association of Cereal Chemists, St. Paul, MN.

Baruah, K., Sahu, N.P., Pal, A.K., Jain, K.K., Debnath, D., Mukherjee, S.C., 2007a. Dietary microbial phytase and citric acid synergistically enhances nutrient digestibility and growth performance of *Labeo rohita* (Hamilton) juveniles at sub-optimal protein level. Aquac. Res. 38, 109–120.

Baruah, K., Pal, A.K., Sahu, N.P., Debnath, D., Yengkokpam, S., Mukherjee, S.C., 2007b. Interactions of dietary microbial phytase, citric acid and crude protein level on mineral utilization by rohu, *Labeo rohita* (Hamilton) juveniles. J. World Aquacult. Soc. 38, 238–249.

Baruah, K., Pal, A.K., Sahu, N.P., Debnath, D., Yengkokpam, S., Norouzitallab, P., Sorgeloos, P., 2009. Dietary crude protein, citric acid and microbial phytase interacts to influence the hemato-immunological parameters of rohu, *Labeo rohita,* juveniles. J. World Aquacult. Soc. 40, 824–831.

Cao, L., Wang, W., Yang, C., Yang, Y., Diana, J., Yakupitiyage, A., Luo, Z., Li, D., 2007. Application of microbial phytase in fish feed. Enzyme. Microb. Tech. 40, 497–507.

Correll, D.L., 1999. Phosphorus: a rate limiting nutrient in surface waters. Poult. Sci. 78, 674–682.

Cosgrove, D.J., 1966. The chemistry and biochemistry of inositol polyphosphates. Rev. Pure Appl. Chem. 16, 297–335.

Cromwell, G.L., Coffey, R.D., Monegue, H.J., Randolph, J.H., 1995. Efficacy of low-activity, microbial phytase in improving the bioavailability of phosphorus in corn-soyabean meal diets for pigs. J. Anim. Sci. 73, 449–456.

De Silva, S.S., Anderson, T.A., 1995. Fish Nutrition in Aquaculture. Chapman and Hall Aquaculture Series 1. Chapman and Hall, London, UK.

Debnath, D., Sahu, N.P., Pal, A.K., Baruah, K., Yengkokpam, S., 2005. Present status and future prospects of phytase in aquafeed—review. Asian-Aust. J. Anim. Sci. 18, 1800–1812.

FAO, 2014. The State of World Fisheries and Aquaculture (SOFIA) 2014. FAO Fisheries Department, Rome, Italy.

Harland, F.B., Morris, E.R., 1995. Phytin: a good or a bad food component. Nutr. Res. 15, 733–754.

Jackson, L.S., Li, M.H., Robinson, E.H., 1996. Use of microbial phytase in channel catfish *Ictalurus punctatus* diets to improve utilization of phytate phosphorus. J. World Aquacult. Soc. 27, 309–313.

Jongbloed, A.W., Mroz, M., Kemme, P.A., 1992. The effect of supplementary Aspergillus niger phytase in diets for pigs on concentration and apparent digestibility of dry matter, total phosphorus, and phytic acid in different sections of the alimentary tract. J. Anim. Sci. 70, 1159–1168.

Jongbloed, A.W., Kemme, A., Mroz, A., 1996. The effect of organic acids in diets for growing pigs on the efficacy of microbial phytase. In: Coelho, M.B., Kornegay, E.T. (Eds.), Phytase in Animal Nutrition and Waste Management. BASF Corporation, Mount Olive, NJ, p. 515.

Jongbloed, A.W., deJonge, L., Kemme, P.A., Mroz, Z., Kies, A.K., 1997. Proceedings of the Sixth BASF Forum on Animal Nutrition, Ludwigshafen, Germany.

Ketola, H.G., Harland, B.F., 1993. Influence of phosphorus in rainbow trout diets on phosphorus discharge in effluent water. T. Am. Fish. Soc. 122, 1120–1126.

Lei, X., Pao, K.K., Miller, E.R., Ullrey, D.E., Yokoyama, M.T., 1993. Supplemental microbial phytase improves bioavailability of dietary zinc to weaning pigs. J. Nutr. 123, 1117–1123.

Li, M.H., Robinson, E.H., 1997. Microbial phytase can replace inorganic phosphorus supplements in channel catfish *Ictalurus punctatus* diets. J. World Aquacult. Soc. 28, 402–406.

Liener, I.E., 1994. Implications of antinutritional components in soybean foods. Crit. Rev. Food Sci. 34, 31–67.

Naylor, R.L., Goldberg, R.J., Primavera, J.H., Kautsky, N., Beveridge, M.C., Clay, J., Folk, C., Lubchenco, J., Mooney, H., Troell, M., 2000. Effect of aquaculture on world fish supplies. Nature 405, 1017–1024.

NRC (National Research Council), 1993. Nutrient Requirements of Fish. National Academy Press, Washington, DC, USA.

Ogino, C., Takeuchi, L., Takeda, H., Watanabe, T., 1979. Availability of dietary phosphorus in carp and rainbow trout. B. Jpn. Soc. Sci. Fish 45, 1527–1532.

Omogbenigun, F.O., Nyachoti, C.M., Slominski, B.A., 2003. The effect of supplementing microbial phytase and organic acids to a corn-soybean based diet fed to early-weaned pigs. J. Anim. Sci. 81, 1806–1813.

Paik, I.K., 2003. Application of phytase, microbial or plant origin, to reduce phosphorus excretion in poultry production. Asian Aust. J. Anim. Sci. 16, 124–135.

Powell, K., 2003. Eat your veg. Nature 24, 378–379.

Raboy, V., 1997. Accumulation and storage of phosphate and minerals. In: Larkins, B.A., Vasil, I.K. (Eds.), Cellular And Molecular Biology of Plant Seed Development. Kluwer Academic, Dordrecht, The Netherlands, pp. 441–477.

Ravindran, V., Bryden, W.L., Kornegay, E.T., 1995. Phytates: occurrence, bioavailability and implications in poultry nutrition. Poult. Avian Biol. Rev. 6, 125–143.

Refstie, S., Storebakken, T., Baeverfjord, G., Roem, A.J., 2001. Long-term protein and lipid growth of Atlantic salmon (*Salmo salar*) fed diets with partial replacement of fish meal by soy protein products at medium or high lipid level. Aquaculture 193, 91–106.

Richardson, N.L., Higgs, D.A., Beames, R.M., McBride, J.R., 1985. Influence of dietary calcium, phosphorus, zinc and sodium phytate level on cataract incidence, growth, and histolopathology in juvenile Chinook salmon (*Oncorhynchus tshawytscha*). J. Nutr. 115, 553–567.

Robinson, E.H., Li, M.H., Manning, B.B., 2002. Comparison of microbial phytase and dicalcium phosphate for growth and bone mineralization of pond-raised channel catfish, *Ictalurus punctatus*. J. Appl. Aquacult. 12, 81–88.

Rodehutscord, M., Pfeffer, E., 1995. Effects of supplemental microbial phytase on phosphorus digestibility and utilization in rainbow trout (Oncorhynchus mykiss). Water Sci. Technol. 31, 143–147.

Samuelsen, T., Isaksen, M., McLean, E., 2001. Influence of dietary recombinant microbial lipase on performance and quality characteristics of rainbow trout, Oncorhynchus mykiss. Aquaculture 194, 161–171.

Simons, P.C.M., Versteegh, H.A.J., Jongbloed, A.W., Kemme, P.A., Slump, P., Bos, K.D., Wolters, W.G.E., Beudeker, R.F., Verschoor, G.J., 1990. Improvement of phosphorus availability by microbial phytase in broilers and pigs. Brit. J. Nutr. 64, 525–540.

Spinelli, J., Houle, C.R., Wekell, J.C., 1983. The effects of phytates on the growth of rainbow trout (Salmo gairdneri) fed purified diets containing varying quantities of calcium and magnesium. Aquaculture 30, 71–83.

Stahl, C.H., Han, Y.M., Roneker, K.R., House, W.A., Lei, X.G., 1999. Phytase improves iron bioavailability for hemoglobin synthesis in young pigs. J. Anim. Sci. 77, 2135–2142.

Sugiura, S.H., Raboy, V., Young, K.A., Dong, F.M., Hardy, R.W., 1999. Availability of phosphorus and trace elements in low-phytate varieties of barley and corn for rainbow trout (*Oncorhynchus mykiss*). Aquaculture 170, 285–296.

Sugiura, S.H., Gabaudan, J., Dong, F.M., Hardy, R.W., 2001. Dietary microbial phytase supplementation and the utilization of phosphorus, trace minerals and protein by rainbow trout (*Oncorhynchus mykiss* Walbaum) fed soybean meal-based diets. Aquac. Res. 32, 583–592.

Teskeredzic, Z., Higgs, D.A., Dosanjh, B.S., McBride, J.R., Hardy, R.W., Beames, R.M., Simell, M., Vaara, T., Bridges, R.B., 1995. Assessment of unphytinized and dephytinized rapeseed protein concentrate as sources of dietary protein for juvenile rainbow trout, Oncorhynchus mykiss. Aquaculture 131, 261–277.

Thompson, L.U., 1986. Phytic acid: a factor influencing starch digestibility and blood glucose response. In: Graf, E. (Ed.), Phytic Acid: Chemistry and Applications. Pilatus Press, Minneapolis, MN, p. 173.

Tyagi, P.K., Verma, S.V.S., 1998. Phytate phosphorus content of some common poultry feedstuffs. Indian J. Poult. Sci. 33, 86–88.

Usmani, N., Jafri, A.K., 2002. Influence of dietary phytic acid on the growth, conversion efficiency, and carcass composition of mrigal Cirrhinus mrigala (Hamilton) fry. J. World Aquacult. Soc. 33, 199–204.

Van Weerd, J.H., Khalaf, K.H.A., Aartsen, F.J., Tijssen, P.A.T., 1999. Balance trials with African catfish *Clarias gariepinus* fed phytase-treated soybean meal-based diets. Aquacult. Nutr. 5, 135–142.

Vielma, J., Lall, S.P., 1997. Dietary formic acid enhances apparent digestibility of minerals in rainbow trout, *Oncorhynchus mykiss* (Walbaum). Aquacult. Nutr. 3, 265–268.

Vielma, J., Lall, S.P., Koskela, J., Schöner, F.J., Mattila, P., 1998. Effects of dietary phytase and cholecalciferol on phosphorus bioavailability in rainbow trout (*Oncorhynchus mykiss*). Aquaculture 163, 309–323.

Vohra, A., Satyanarayan, T., 2003. Phytases: microbial sources, production, purification and potential biotechnological applications. Crit. Rev. Biotechnol. 23, 29–60.

Wise, A., 1980. Dietary factors determining the biological activities of phytate. Nutr. Abs. Rev. 53, 791–806.

Yan, W., Reigh, R.C., Xu, Z., 2002. Effects of fungal phytase on utilization of dietary protein and minerals, and dephosphorylation of phytic acid in the alimentary tract of channel catfish *Ictalurus punctatus* fed an all-plant protein diet. J. World Aquacult. Soc. 33, 10–22.

Yi, Z., Kornegay, E.T., Denbow, D.M., 1996. Supplemental microbial phytase improves zinc utilization in broilers. Poult. Sci. 75, 540–546.

Yoon, J.H., Thompson, L.U., Jenkins, D.J., 1983. The effect of phytic acid on *in vitro* rate of starch digestibility and blood glucose response. Am. J. Clin. Nutr. 38, 835–842.

CHAPTER

Strategies on adverse event reporting: a global framework for nutraceutical industry

25

Kristy Appelhans, Joyce Cao
Herbalife Nutrition, Los Angeles, CA, United States

25.1 INTRODUCTION

Globally, nutraceutical products or analogous categories (e.g., dietary supplements, herbal or botanical medicines, functional foods, alternative and complementary medicines, natural health products, etc.) are being reviewed for safety due to potency and adulteration by many of the regulatory authorities in the respective countries where they are marketed. In general, however, regulatory requirements aimed at substantiating the safety or efficacy of these products prior to marketing, and monitoring product safety postmarket, are highly inconsistent globally. In fact, some markets have no codified requirements for premarket or postmarket product safety measures specifically for nutraceutical products. As a result, both regulators and industry are largely naïve in their understanding of establishing relevant product safety measures for nutraceuticals, implementing processes and procedures associated with pharmacovigilance, monitoring the efficacy of safety measures, and/or properly identifying significant safety signals throughout the product life cycle. Pharmacovigilance is defined as "the science and activities relating to the detection, assessment, understanding and prevention of adverse effects or any other possible drug-related problems" (World Health Organization, 2004). While the term was developed within the context of the pharmaceutical industry, it is typically used to describe similar safety activities conducted within the scope of the nutraceutical industry.

The absence of robust safety regulations globally does not mean that regulators are unaware of the importance of product safety and ongoing surveillance for nutraceuticals. On the contrary, prudent monitoring (by industry as well as regulators) and regulations for surveillance are currently in various stages of development and enforcement worldwide with a more recent push amongst regulators of nutraceuticals to prioritize efforts in pharmacovigilance-related legislation.

By enhancing the global understanding of pharmacovigilance and establishing best practices to properly implement safety surveillance industry-wide, dietary

Developing New Functional Food and Nutraceutical Products. http://dx.doi.org/10.1016/B978-0-12-802780-6.00025-2

supplement companies have a unique opportunity to help regulatory authorities shape regulations regarding these matters and better standardize pharmacovigilance requirements globally. To that end, organizing, harmonizing, and enhancing global monitoring and regulations associated with nutraceuticals is crucial to continuously support consumer safety related to nutraceutical use.

Nutraceutical trade associations serve as an important resource for the industry by acting as a unified voice regarding regulatory matters and may act on this responsibility in several ways including the submission of comments and suggestions in response to draft guidance or regulations and publishing scientific opinion papers to rebut general inaccuracies associated with the industry, discuss/debate regulatory questions, set standards for evaluating the industry or its products, or defend a specific product or ingredient. Major trade organizations include the Council for Responsible Nutrition (US and international divisions), the American Herbal Products Association, the Natural Products Association, and the American Botanical Council.

25.2 NUTRACEUTICAL INDUSTRY BACKLASH FOR PERCEIVED LACK OF REGULATORY OVERSIGHT

Even in the most regulated environments like the United States, Canada, and Australia, scientific and mainstream media often scrutinize the safety of nutraceuticals, which is largely led by the perception that nutraceuticals lack adequate regulatory oversight (Cohen, 2014; Consumer Reports magazine, 2012; Vatistas and Samuels, 2012). Specifically, one of the most frequent criticisms of nutraceutical and dietary supplement regulations is that manufacturers often are not required to adhere to the equivalent of the prescription drug marketing process with premarket clinical studies for safety and efficacy and other requirements held to the rigor of a new drug application (NDA). Upon this premise, it is further surmised that without such premarket processes in place, the products must be unsafe or not properly or thoroughly assessed to determine safety. What escapes most industry skeptics is the fact that although nutraceuticals contain active ingredients, many of these ingredients have long-standing histories of safe use in the diet or for traditional health (supportive and complementary) purposes and unlike conventional drug products, are not typically formulated to prevent, treat, mitigate, or cure disease. For example, teas, vitamins, minerals, essential fatty acids, amino acids, and caffeine are very common substances that are included in today's nutraceutical product formulations and have extensive safety and efficacy data behind their use in global populations for a variety of well-substantiated claims. Where the industry brings concern, and rightfully so, is when nutraceutical product manufacturers (often referred to as the "bad actors") are found to be noncompliant with current manufacturing regulations. Some common examples of noncompliance include: the production of formulas which contain adulterated ingredients, or components which are otherwise poorly characterized. However, these types of issues may be best addressed by first implementing additional regulatory enforcement actions with an intention to identify noncompliant companies. With this approach, compliant companies are less likely to be subjected to excessive and unnecessary scrutiny. Therefore, this strategy may also help to improve upon the general misconception and concern that the nutraceutical industry as a whole is unsafe or unregulated. Furthermore, it may be advantageous to make sure current regulations are being properly enforced and followed before proposing additional regulations in an attempt to control for manufacturing deficiencies.

It should be clarified that even in the absence of codification, nutraceutical companies are often subject to a marketing application or product registration process which typically includes documentation to substantiate the safety, efficacy, and/or quality of the finished good as determined by the regulatory authorities in the respective markets where products are intended to be sold (Health Canada, 2013). In the event that a regulatory authority does not require the manufacturer to provide any or all of this information premarket, compliance with postmarket reporting and/or current good manufacturing practices (cGMP) are typically required to: (1) keep regulators informed of potential quality or safety signals or other significant observations obtained via inspections, etc., and (2) provide relevant guidance and/or establish or improve upon industry regulations for producing a safe product and monitoring the safety of that product throughout its lifecycle (Electronic Code of Federal Regulations, 2015; ICH Harmonised Tripartite, 2000; Instant GMP, 2012; International Alliance of Dietary Food Supplement Associations, 2011; Srinivasan, 2014). However, the level of enforcement that a regulatory authority exercises with regard to postmarket safety surveillance, data mining/trending for signal detection, industry compliance, etc., can be dependent on available internal or staff resources and sometimes the perceived threat to consumer safety according to the scientific and medical community (published and nonpublished opinions and conclusions), media coverage, investigative outcomes and/or public advisories from other regulatory officials, basic risk of the product type, indication, and/or its ingredients, etc. That said, regulatory authorities may place the majority of responsibility on manufacturers to be compliant and to provide pertinent notifications rather than allocating resources to systematically police these activities to ensure industry-wide compliance. These considerations may lead to some potential variability in product safety and quality between different manufacturers depending on the manufacturer's level of due diligence and even the extent of expertise that a company may have available to help make prudent decisions regarding these matters. In contrast, regulators may not trust dietary supplement companies to be transparent regarding product safety data or significant emerging safety trends. If this is the case, approaches for enforcement may include an obligation for manufacturers to report all adverse event cases which they receive (serious and nonserious with or without substantiation for causal association), systematic or even spontaneous pharmacovigilance inspections, and formal inquiries regarding any or all adverse event case reports received directly by the regulatory authority (from the manufacturer or otherwise) (Korean Food and Drug Administration, 2010). The subsequent sections will provide a closer look at key concepts, objectives, and regulations which are associated with establishing pre- and postmarket safety for nutraceuticals while discussing some of the variations as well as harmonization for pharmacovigilance activities worldwide.

25.3 REGULATORY OVERVIEW FOR NUTRACEUTICALS

There are various similarities globally in regard to nutraceutical regulations. For example, many governments require that manufacturers register or obtain licenses for products prior to marketing. During the registration process, it is common for regulators to ask for any known adverse event data associated with the formulation in other countries where it may be marketed, obtained via premarket studies, or other sources of safety data. Additionally, a thorough product review (in the form of a product dossier) is often requested to scientifically substantiate product safety and label claims and to reassure regulators that there is no apparent, potential, and/or significant safety risk associated with the product or any of its ingredients. Another similarity in the regulation of nutraceuticals globally is that these products are often classified in a category that is regulated separately from food products and often with more rigorous requirements pre- and/or postmarket.

25.4 US CODIFICATION FOR MARKETING NUTRACEUTICALS

The Dietary Supplement Health Education Act of 1994 (DSHEA) was created "to amend the Federal Food, Drug, and Cosmetic Act (FD&C Act) to establish standards with respect to nutraceuticals, and for other purposes" (Food and Drug Administration, 1994a). As such, the FD&C Act is referenced herein with discussion of specific provisions for dietary supplements marketed in the US. However, further discussion to follow will briefly address the significance of DSHEA for dietary supplement regulation especially as it relates to the premarket "approval" processes for these products in the US.

A dietary supplement is specifically labeled as such and is not marketed as a conventional food (Food and Drug Administration, 1994b). In addition, "dietary supplement" adheres to the following criteria: "...means a product (other than tobacco) intended to supplement the diet that bears or contains one or more of the following dietary ingredients: (A) a vitamin; (B) a mineral; (C) an herb or other botanical; (D) an amino acid; (E) a dietary substance for use by man to supplement the diet by increasing the total dietary intake; or (F) a concentrate, metabolite, constituent, extract, or combination of any ingredient described in clause (A), (B), (C), (D), or (E)."

New dietary ingredient (NDI) Notifications must be submitted to the US Food and Drug Administration (FDA) for a product prior to marketing if one or more of the ingredients were not marketed in a US product categorized as a dietary supplement prior to Oct. 15, 1994 (preDSHEA) (Food and Drug Administration, 2011). Therefore, verification for the distribution/marketing of products in accordance with this requirement includes the provision of sales records, manufacturing reports, and/or promotional materials that are dated prior to Oct. 15, 1994.

The documentation which is required for NDI Notifications is fairly comprehensive. Products which may be "grandfathered" into the marketplace with preDSHEA distribution have the advantage of relying on FDA for determining product safety (and compliance) through inspections conducted at the manufacturing site, business entity, etc. However, manufacturers which intend to market NDIs must bear the burden of proving that the ingredient(s)/product is safe when used as intended. Therefore, the NDI Notification will include relevant safety studies (e.g., toxicological evidence and human clinical trial data), identity analyses, ingredient specifications including those associated with contaminants (heavy metal, microbial, etc.), and a comprehensive description of manufacturing practices to further establish the safety, purity, and quality of the product/ingredients. NDI Notifications must be submitted 75 days prior to the intended product launch date and manufacturers will be notified of the FDA's observations regarding the notification. FDA's decision is communicated in one of four main types of letters: (1) no objection, (2) submission incomplete, (3) safety-related objection, (4) other objections (e.g., regulatory classification of the NDI inappropriately assigned/not a dietary ingredient).

25.5 GLOBAL CODIFICATION FOR NUTRACEUTICALS

It should be noted that global legislation for nutraceuticals is typically not comparable to the extensive amount of codification that the United States has established for pre- and postmarket activities associated with these products. However, Canada and Australia are two examples of exceptions in the definitions for nutraceuticals (categorized as Natural Health Products in Canada and Listed Medicines in Australia) are very similar to those in the United States. Neither Canada nor Australia has an NDI process, but all products must undergo a registration/approval process prior to marketing in both of these countries; so overall, the process is slightly more rigorous as the majority of US nutraceuticals on the market today fall under the "grandfathered" list and do not require any notification or approval before they are marketed. In addition, Health Canada has established product and ingredient monographs and specification requirements which serve as templates for proposed labeling (including cautionary or warning statements and contraindications as well as acceptable claims), durations and directions of use, and ingredient dosage and/or strength specifications (Health Canada, 2009). Manufacturers are required to be compliant with all monographs and other product specifications (e.g., finished goods specifications described in Health Canada's NHPD Quality of Natural Health Products Guide) prior to gaining approval and product licensure from Health Canada. However, these monographs support expeditious premarket product approval processes because the manufacturer does not need to provide scientific substantiation for products or ingredients which are deemed safe and efficacious within the context of the established monographs.

Although country-specific codification for nutraceuticals is sporadic worldwide, it should be noted that the current trend is to create requirements that establish product and ingredient safety prior to marketing. These discussions are also starting to surface in the US amongst industry experts and regulators who feel that premarket notification processes are a crucial component to ensuring product and consumer safety (Schultz, 2015).

25.6 GLOBAL POSTMARKET SAFETY SURVEILLANCE REQUIREMENTS FOR NUTRACEUTICALS

Similar to the main objective for codified and/or other mandatory premarketing requirements, there is a general goal for most regulators to ensure products are safe after they enter the marketplace. However, there are several differences in the ways that regulators expect to monitor product safety. First and foremost, regulations will vary depending on which party is expected to regulate product safety throughout the product life cycle, whether the manufacturer (i.e., primarily self-regulation), the regulator, or both. In the US, product safety for most "grandfathered" nutraceuticals is heavily self-regulated. As discussed, there are no premarket notifications or other requirements (with the exception of NDI applications) and the market is left to implement a proper level of due diligence for product quality and safety. This includes establishing adequate and compliant manufacturing processes (in accordance with Current Good Manufacturing Practice) and a robust adverse event complaint collection, management, and reporting system (in compliance with mandatory postmarket reporting requirements for serious adverse events) (Electronic Code of Federal Regulations, 2015; Food and Drug Administration, 2007). Additionally, US FDA inspections for dietary supplement manufacturers are performed on a spontaneous basis which is determined by FDA. In contrast, in countries such as South Korea and Canada, regulatory authorities carry much of the monitoring and compliance burden by conducting routine inspections.

Amongst the existing regulations for postmarket surveillance and vigilance activities, some regulatory authorities have in place postmarket notification (reporting) requirements that typically focus on complaints, which potentially pose the greatest risk to consumers if there should be a causal association established between such complaints and product consumption (European Medicines Agency, 2013; Food and Drug Administration, 2007; Health Canada, 2011; Medicines and Healthcare Products Regulatory Agency, 2011; Ministry of Health Malaysia, 2002; Therapeutic Goods Administration, 2011). However, the terms which define such regulations and level of enforcement vary greatly around the world. For example, even the term "adverse event" (AE) presents yet another potential difference in country-specific regulations and definitions. In some cases, an adverse event refers to a spontaneous health-related complaint which the consumer, or another individual reporting on the consumer's behalf, allegedly associates with the consumption of a suspect product(s) (Food and Drug Administration, 1994b). The US FDA's regulation requires that even temporal associations be reported, and the MedWatch 3500A form specifically states that, "Submission of a report does not constitute an admission that medical personnel, user, facility, importer, distributor, manufacturer or product caused or contributed to the event." For example, in this type of case, a manufacturer may be required to report that a consumer alleged consumption of a product(s) was associated with their cholecystectomy without further corroboration (e.g., physiological plausibility, medical opinion, etc.). Conversely, an adverse reaction (AR) or adverse drug reaction (ADR) usually refers to a clinically proven or substantiated causal association between the suspect product and the reported reaction. Mandatory serious adverse drug reaction reporting to the Therapeutic Goods Administration (TGA) in Australia is based on reasonable medical/clinical judgment that an association exists between the reported product(s) and the event reported (Therapeutic Goods Administration, 2011). As an example, a case of severe acute (and transient) tachycardia with reasonable duration to onset following the ingestion of a highly caffeinated substance (i.e., event reportedly occurred within 30 min of ingestion) and spontaneous recovery upon "withdrawal" in the absence of other factors (e.g., extreme stress or anxiety, relevant medical history, concurrent medication use, etc.) is likely attributable to the reported caffeinated product. Therefore, reasonable medical judgment would dictate that the case be reported to TGA.

As suggested earlier, causality is typically defined in two general ways which are "as determined" and "as reported" causality. As determined causality is usually a causal association reported or identified by a qualified health professional (e.g., doctor, pharmacist, nurse) and as reported causality is essentially an allegation which can be reported by the consumer or other individuals who do not necessarily have to be qualified health care professionals. Consumer family members, friends, attorneys, law enforcement, Poison Control Centers, regulatory authorities, and treating health care professionals are common sources for adverse event information.

Additionally, the events which are deemed significant and/or are placed under specific regulatory obligations vary from country to country, and this is usually defined by case "seriousness." However, there is some harmonization globally when it comes to defining serious adverse events (European Medicines Agency, 2013; Food and Drug Administration, 1994b, 2007; Health Canada, 2011; Medicines and Healthcare Products Regulatory Agency, 2011; Ministry of Health Malaysia, 2002; Therapeutic Goods Administration, 2011). The US, Canada, Australia, Malaysia, and EU member states are some jurisdictions which have established similar definitions and criteria for serious adverse events that include death, life-threatening events, an event that requires medical intervention to prevent a life-threatening condition, congenital anomalies/birth defects, disability/permanent damage, hospitalization, and/or other serious or important medical events not fully defined by the foregoing criteria (e.g., bronchospasm or acute and severely elevated liver enzymes or jaundice commonly meet the definition of serious medical events).

The mandated frequency of reporting adverse events also varies between different regulatory authorities. For example, some regulations require "expedited" reporting, which is generally reporting an event upon receipt of, or within a short duration after receiving, the complaint. Others may require aggregate or periodic reporting of adverse events—periodic safety update reports (PSURs). Furthermore, some regulatory authorities may require both periodic and expedited reporting. The formats for reporting adverse event cases also vary by jurisdiction, although many regulatory authorities have adopted documentation styles which capture similar types of information including the name of the suspect agent(s), affected product lot numbers, nature of the event, the age and gender of the complainant/consumer, duration to onset for the reported event, withdrawal and/or recovery data, and an event narrative to describe what reportedly occurred, including any relevant analyses or conclusions which the manufacturer or reporter (in situations where the manufacturer does not initiate the adverse event submission) may have.

Further differentiation for adverse event reporting or notification can be determined by the expectedness of the event—that is, the reported reaction was expected or not expected according to product labeling or other relevant product packaging or adjunct literature. Expectedness can be determined by a number of factors including: regulatory mandate to label for expectedness on certain products based on literature reviews or other authoritative opinions, a company's personal experience with their product, such as clinical testing and/or postmarket observations, or even company-specific approaches to limit product liability (e.g., when specific populations are targeted for product use, ambiguous evidence exists regarding the expectedness of an event, etc.) (Health Canada, 2011).

25.7 OVERVIEW OF GOOD MANUFACTURING PRACTICE REQUIREMENTS, SPECIFICATIONS, AND STANDARDIZATIONS

Like the pharmacovigilance requirements for nutraceuticals, global GMP regulations are equally sporadic in codification status and vary greatly in degrees of implementation and/or the interpretation of due diligence throughout the industry worldwide. For this reason, considerable efforts are being put forth to improve upon the international alignment for these regulations, as well as global compliance. Currently the International Conference on Harmonization (ICH) member countries including the US, Japan, Australia, Canada, Norway, and 27 EU member states are following a harmonized approach to cGMP (Instant GMP, 2012). In further support of harmonization, the Council for Responsible Nutrition and other international industry trade associations, as well as representatives for the trade associations, such as The International Alliance of Dietary/Food Supplement Associations (IADSA), have come together to articulate the principles of GMP and best practices in a global technical document called the Guide for Global Manufacturing Processes for Supplements (International Alliance of Dietary Food Supplement Associations, 2011). The guidance is intended for all manufacturers and allows for both ICH-harmonized and non-harmonized markets to benefit from the general direction which is provided for the "production (manufacturing, quality control, packaging, distribution and storage) of supplements." The Guide is nonprescriptive, and manufacturers who may refer to this document are still expected to be compliant in regard to all applicable local regulatory requirements which correspond with the market where they manufacture or distribute products. To that end, the Guide is significant in that it may provide a useful reference for companies that need to implement or update their GMP processes and procedures. It may even serve as an informative resource for regulators who are codifying GMP requirements in their jurisdictions or developing a framework for codification.

Basic principles of best practices for GMP involve adequate quality control systems which ensure that manufacturing procedures are clearly defined in documentation (including standard operating procedures and employee training completions) and are executed in accordance with the general precautions for safety and sanitation. It should also be noted that regulators often require supplement manufacturers to apply food GMP or at a minimum, general hygiene standards for the manufacturing of these substances where supplement-specific GMP legislation is absent. Some of the key components in either case are as follows:

1. Standard parameters for hygienic practices including, but not limited to, facility cleanliness, hygiene, pest control, and proper personal protective equipment for employees (e.g., gloves, bibs, shoe and hair covers, and goggles).
2. Maintain adequate product storage conditions for temperature and humidity.
3. Raw materials and finished goods must be held to valid and verifiable specifications including those that apply to ingredient purity, identity, and quantity. Source of raw materials should be assessed as to country of origin and the potential risk they pose with cross-contamination and substandard material.
4. Must be able to demonstrate ingredient traceability.

Further to the basic sanitary provisions described earlier, US dietary supplement cGMP regulations dictate that the manufacturer's inability to properly characterize ingredients on a product label (with validated analytical methods) and/or an observation that the labeled ingredients are not an accurate representation of what is inside the product container, qualifies the product as misbranded and presumed as adulterated. Therefore, the obligation is for manufacturers to have appropriate analytical measures in place for ingredient identification and quantification to meet label specifications for each product. In fact, US cGMP for dietary supplements explicitly express (amongst other requirements) that an identity specification for each component in a product is required, and that specifications are to be verified/validated through at least one testing procedure before using or distributing the component in a product for public consumption (Electronic Code of Federal Regulations, 2015). For common ingredients characterized in the United States Pharmacopoeia (USP) and/or USP Dietary Supplement Compendium (Food Chemicals Codex), identification, strength, purity, packaging/labeling, and dosage are conveniently standardized for industry use (Srinivasan, 2014). However, standardization for some botanical ingredient types or other novel components is still ongoing as industry's innovation continues.

25.8 WHERE QUALITY MEETS PHARMACOVIGILANCE

Quality control and GMP are significant components to any pharmacovigilance system, as adverse events can be associated with a variety of factors including: deviations from established product specifications, failures to maintain adequate manufacturing conditions, or, in some instances, a combination of the two. In fact, cGMP compliance for US regulations includes the manufacturer's ability to show adequate investigative processes for complaints, both product-related (e.g., organoleptic, packaging, and foreign matter) and health-related (adverse events) (Electronic Code of Federal Regulations, 2015). It is with good reason that this type of provision is part of the GMP requirements as typically any necessary corrective and preventive action (CAPA) cannot take place without a comprehensive understanding of the issue, how it occurred, and, most importantly, when it occurred. A company's ability to avoid public exposure to a safety or quality-related incident and to mitigate potential risks to the end-user is highly dependent on the sensitivity of both the quality and pharmacovigilance acute signal detection programs (identifying a relevant issue/deviation acutely). Unlike product quality complaints, which are usually limited to a single product per event, signal detection for adverse events can be very challenging for complex (multiingredient) substances and also in instances where multiple product use is reported (i.e., multiple suspect products). Therefore, conservative measures are often necessary to ensure that due diligence is demonstrated during investigative processes. An example of this may involve the implementation of policies that require low threshold complaint volumes (product lot trends) to trigger routine inspection or testing/retesting of retained finished goods. Likewise, signal detection for pharmacovigilance programs should use statistical algorithms which are specific to low complaint volume thresholds and apply dataset conditions/constraints which are sensitive to minor changes in product or ingredient trends.

25.9 WAYS TO USE POSTMARKET DATA: SAFETY REVIEWS

Routine reviews of postmarket data are critical to ensuring the safe use of nutraceuticals in the general population. Safety reviews fall into two general approaches: incidence trending and signal detection (Evans et al., 2001).

Incidence trending is useful for broadly capturing the adverse event reporting profile of nutraceutical products over time. The cases and adverse events can be aggregated by the highest MedDRA coding level, the system organ class (SOC), and divided by the number of units shipped during the reporting period (International Conference on Harmonisation, 2015). Incidence calculations may also be normalized per 10,000 units shipped. For example, the incidence for 4 cases reporting gastrointestinal disorders involving a nutraceutical that shipped 50,000 units within the same timeframe would be calculated using Eq. (25.1) (Evans et al., 2001):

$$\frac{4\text{ cases}}{50{,}000\text{ units shipped}} \times 10{,}000\text{ units shipped}$$
$$= 0.8\text{ (gastrointestinal disorders) cases per } 10{,}000\text{ units shipped} \quad (25.1)$$

While incidence trending may focus on spontaneously reported adverse events, it can easily be expanded to encompass other report types.

Signal detection is a concept in pharmacovigilance widely utilized in the pharmaceutical industry (Barnes, 2003). In contrast to the broad overview given by incidence trending, signal detection can be used to focus on specific product–event relationships in an acute timeframe, such as a week or month. Events are typically aggregated at the MedDRA preferred term (PT) level, and the calculations do not rely on sales data. Rather, signal detection methods depend only on adverse events data.

25.10 SCRUTINY OF CURRENT GOOD MANUFACTURING PRACTICE FOR NUTRACEUTICALS

As indicated previously, adverse event monitoring, complaint reporting, and GMP requirements vary significantly around the world. Unfortunately, the inconsistent regulatory climate for GMP (as was described for pharmacovigilance) is another area of quality and safety oversight where nutraceutical product manufacturers receive scrutiny from those who are skeptical or not in support of this industry. In fact, very recently the US dietary supplement industry became the subject of review by the New York Attorney General's office to determine whether the methods which are commonly in place to identify product ingredients are sufficient to properly comply with cGMP, which includes having an effective analytical measure in place to detect adulteration and to properly characterize product ingredients in general to confirm that the labeled Supplement Facts panel is accurate (Harbaugh-Reynaud et al., 2015). Currently, the industry and the New York Attorney General's office are discussing the role of DNA analyses as a complement to existing testing standards for the industry's ingredient and allergen detection methods. However, industry trade associations such as the American Herbal Products Association, the American Botanical Council, the Council for Responsible Nutrition, and other field experts have expressed concerns regarding the proposal, including the fact that false positives may potentially be introduced when using these methods without proper context (Engredea News & Analysis, 2013; Harbaugh-Reynaud et al., 2015). For example, what industry has concluded and is sometimes misunderstood by the proponents of the DNA method is that botanical ingredients lose most if not all DNA material during the extraction process. Therefore, DNA identification and quantification results are likely to falsely show an absence of the ingredient or its marker constituents. Thus, the product would be incorrectly categorized as adulterated because the labeled ingredients would not be consistent with what is actually in the product (according to the analytical results). For this reason, existing validated methods for ingredient identification should be used in conjunction with new strategies such as DNA analyses. Many companies understand the importance of reaching a mutual agreement regarding this issue in an effort to enhance safety measures industry-wide and instill confidence at both the consumer and regulatory levels.

The DNA discussion is one of many that are currently occurring between FDA, industry, and other regulatory entities in the US. However, there are many different global or market-specific efforts often brought forth to further investigate matters which may stem from misconceptions of dietary supplement oversight/regulation, safety, and/or consumer risk assessments. However, these and other interactions between regulators and industry can present an opportunity to legitimately improve upon manufacturing standards for nutraceuticals as necessary.

25.11 WHERE SHOULD OUR NEXT BEST EFFORTS IN PRODUCT SAFETY BE DIRECTED: PREMARKET, POSTMARKET, OR BOTH?

After having discussed the various global pre- and postmarket provisions for nutraceuticals and the various safety measures which are expected of the manufacturer and/or regulatory authority, one may ask where the next best efforts should be directed in order to enhance the perception of consumer safety.

One question industry is currently pondering or being asked is, "How necessary is the premarket clinical trial process for nutraceuticals?" These substances often have well-established histories of safe use in the literature or compendium monographs, contain ingredients which are generally recognized as safe by authoritative bodies, are relatively prevalent in the food supply, are scientifically substantiated at the dose and frequency at which they are recommended, and/or have been vetted by other means to demonstrate safe use in the populations for which they are intended. "Additionally, it is important to acknowledge certain deficiencies which are common to any premarket analysis (for drugs or nutraceuticals) including sample populations which are not an accurate representation of the population at large and therefore, may not account for relevant factors which can affect the expectedness of certain events or outcomes such as medication use, preexisting medical conditions, metabolic differentiation, and/or lifestyle considerations (e.g., diet, tobacco or alcohol use, etc.). In fact, the frequency at which drug products are recalled postmarket for safety reasons despite extensive premarket research and clinical testing, exemplifies some of the limitations with premarket safety assessments. Several studies have been conducted to demonstrate this point and in 2001, a retrospective review of 150 drug withdrawals revealed the most frequent safety reasons for removal from the market which included several serious concerns such as hepatotoxicity (27.9%), cardiovascular toxicity (17.4%), nephrotoxicity (5.6%), neurotoxicity (6.3%), and carcinogenicity (6.3%)." Therefore, the confinements which are inherent to premarket product analyses and/or the lack of extensive premarket clinical testing required for nutraceuticals leave postmarket adverse event monitoring as the only practical way to evaluate the quality, safety, and efficacy of nutraceuticals after they are introduced into the marketplace (Fung et al., 2001; Johnson, 2012; LeDoux et al., 2015).

Consumer complaints are the most common source of postmarket adverse event data which a company receives. Therefore, it is also important that a company recognize other potential external recipients of adverse event complaints such as contract manufacturers or other third party associates. Additionally, adverse event reports may be received through clinics, hospitals, or other health care settings. These incidents may be reported by the recipient directly to the local regulator or the manufacturer may be contacted to provide further information and/or to take additional steps as necessary. In some cases, adverse event reports may be incorporated into a formal study and published in the literature. Continuous monitoring and follow-up with all of the aforementioned sources are critical in staying abreast of consumer feedback. Some examples of strategies to monitor nonconsumer sources of complaints may include: routine literature searches, regulatory requests for adverse event report information [e.g., US FDA Freedom of Information Act request (FOIA)], and regular searches in publicly accessible databases [e.g., US openFDA, Australia TGA Database of Adverse Event Notifications (DAEN), Health Canada MedEffect Canada Vigilance Adverse Reaction Online Database, and EMA EudraVigilance European database of suspected adverse drug reaction reports] (European Medicines Agency EudraVigilance, 2015; Food and Drug Administration, 2015; MedEffect Canada, 2015; Therapeutic Goods Administration, 2013).

FOIA data can prove to be a useful source of information in addition to spontaneous reports which are collected through a company's postmarket surveillance efforts. It is important to recognize that FOIA reports do not actually capture the majority of the information that is provided on a standard FDA adverse event report form for dietary supplements, drugs, and medical devices (MedWatch 3500A), which includes relevant details that are necessary in evaluating causality for any case report (e.g., description of the reported event, duration to onset, preexisting/pertinent medical history of the consumer, concomitant suspect products including those that are produced by another manufacturer and/or may include prescription or over-the-counter medications, etc.). However, for instances where multiple products are able to be queried and displayed simultaneously, such as a request for all applicable products from a specific manufacturer, the report does not differentiate which products may have been reported as suspect versus those that were reported as concomitant. For example, a consumer may report taking multiple supplements, but only allege one of the supplements as suspect in association with the reported event. In this case, all suspect and concomitant products reported would be listed without further specification as to the categorization of each. In addition, FOIA requests are limited to a product name rather than the ingredient, which significantly reduces the requestor's ability to properly construct trends for suspect ingredients.

While FOIA reports include the date on which the report was submitted to FDA, the adverse event onset date is omitted. This is significant because only manufacturers have the obligation to report adverse events within 15 business days. The absence of mandatory reporting timelines for all report sources may result in fluctuations in incidence trends which are not accurate reflections of current data trends amongst a specific consumer population.

The reporter of an adverse event is sometimes not a health care provider or other qualified professional who is able to make diagnostic determinations regarding their health-related experience or that which they are reporting on someone's behalf. For this reason, FDA also follows up with the individual who reports an adverse event to attempt to collect further details regarding the event, which helps to substantiate the causality analysis of each report. FDA's thorough investigation of reported data is significant because FDA also trends this data for the purpose of detecting "signals" or potential health concerns amongst products reported as suspect through the MedWatch system. Once a signal is detected, FDA follows up with the company directly to conduct further investigation of the matter and to determine whether any action is necessary.

The advantages of FOIA reports include consumer and health care accessibility for reporting potential health concerns to regulatory authorities. FOIA reports also can inform companies of adverse events which may not have been received through routine (internal) collection processes. Having access to a broader collection of adverse event reports enhances a company's signal detection and data trending capabilities. Furthermore, having insight to reports received directly by a regulatory agency allows the company to be proactive in assessing reports with the same perspective as the agency, with the ability to respond in a comprehensive and timely manner should the agency inquire about a specific product, case(s), or complaint type.

25.12 SOCIAL MEDIA: TWEETS, BLOGS, POSTS, OH MY!

An emerging concern for industry and challenge for regulators is adverse event complaints which arise from social media. Related matters include: product positioning through social media channels, which may put consumers at risk by promoting or indirectly suggesting misuse of products, inaccurate claims use, and/or failing to disclose all safety-related information within online marketing material. The nutraceutical industry is still waiting for guidance from regulators regarding social media and complaint collection/management. US FDA is the first regulatory authority to publish any social media guidance documents for industry, although the US FDA released two social media guidance documents for the drug/medical device industry. Although neither of the documents are specifically targeted to adverse event collection/management, the documents still give some insight to nutraceutical product manufacturers regarding due diligence and best practices for managing social media and limiting consumer risk through online interactions (Food and Drug Administration, 2014; Johnson, 2012). Furthermore, FDA is working on additional guidance documents related to social media and adverse event or product complaint guidance is anticipated. However, the Food, Drug, and Cosmetic Act as amended clearly states that a dietary supplement company is responsible for complaints which they become "aware of" alleging an association between product consumption and a health-related event (adverse event) and therefore, there is little ambiguity as to whether the company should make diligent efforts to follow-up with each case received though social media in accordance with these regulations (Food and Drug Administration, 2007).

25.13 POSTMARKET DATA COLLECTION AND REVIEW: WHAT TO DO WITH ADVERSE EVENT DATA ONCE YOU HAVE IT

As mentioned, the safety of a product once it is distributed to the population at large often cannot be fully quantified or qualified with premarket studies. Through routine postmarket data mining and assessment processes, however, significant trends can be established and compared against consumer exposure to determine incidence for these outcomes. Some common data endpoints that are of interest include the general prevalence of an event type, seriousness, product or ingredient, and/or specific population.

As part of general due diligence and best practices for compliance with pharmacovigilance regulations, a company is likely to implement strategic steps which ensure systematic adverse event collection and review. Strategies may include standardized adverse event collection forms to support data integrity and meaningful analyses, a robust safety database to support comprehensive trending and signal detection processes, and standardized terminology and coding procedures for adverse events by using the Medical Dictionary for Regulatory Authorities (MedDRA) (International Conference on Harmonisation, 2015). MedDRA is used by regulatory authorities worldwide and is therefore a useful tool for industry to communicate in common terms with regulators while having the advantage of standardizing and optimizing internal data trending and signal detection processes. Most importantly, a company should be able to show through thorough documentation practices that a case was sufficiently reviewed and investigated as necessary (the investigative process and resolution should also be documented with adequate details) and clearly demonstrate that there was no causal connection associated with the reported event when reviewed by a qualified health or other professional with relevant expertise, and/or explain that appropriate corrective and preventive action (CAPA) was implemented as necessary.

Trending and signal detection data can be utilized so that a manufacturer may collaborate internally with cross-functional teams. For example, safety trends may be indicative of a deviation from quality control such as targeted microbial or ingredient specifications. Such scenarios may include gastrointestinal-related signals which help to identify microbial contamination issues or a specific health condition which suggests ingredient excess or toxicity—that is, a particular ingredient trends in association with a medical condition which typically results from an excess of the ingredient.

Signal detection informs product development decisions. In the example mentioned earlier, an issue in quality control may result in product recall, reformulation, and/or relabeling of a product. Additionally, label warnings and contraindications may be added to a product label if adverse event data assessments suggest that a specific population is uniquely susceptible to a product or one or more of its ingredients because of an existing health condition(s), use of medication, or other significant predisposition(s).

Adverse event data trending may also identify certain consumer behaviors which potentially introduce risk with product usage. For example, consumers may misunderstand product use directions or indications/claims, resulting in excessive use or other patterns of consumption which may be harmful to some individuals. Additionally, misinterpretations of product claims or other indications have potential legal implications for a company if it is found that the product labeling or other marketing material was misleading or deceptive. Therefore, adverse event data may be used to launch consumer education initiatives or other risk mitigation strategies which target consumer behavior.

25.14 INCIDENCE TRENDING

Quarterly incidence trending of adverse events for nutraceuticals is a common endpoint for safety surveillance, with serious and nonserious adverse events analyzed separately. While this provides a general overview of the adverse event reporting for that particular time period, multiple time points are typically needed to fully understand the adverse event profiles of products and groups of related products. Furthermore, a single quarter's incidence is meaningless without a basis for comparison.

One method for establishing a basis is to calculate the 95% confidence interval (CI) around the average of previous time points—for example, compute the average and 95% CI for the previous 8 quarters, which equates to 2 years of data. Since all incidence calculations are normalized per 10,000 products shipped, the current quarter's incidence can easily be directly compared to the average incidence. As an example, if the current incidence exceeds the upper bound of the 95% CI around the average incidence, that may be considered a higher than normal incidence and indicate a potential safety issue requiring medical review of the pertinent cases.

The main challenge with incidence trending is that it requires data from multiple databases. Typically, safety data is stored separately from sales data, which may not be readily accessible to the safety department. The time spent gathering data from different sources creates delays in the ability to expedite incidence trending reports, which is critical to uncovering and resolving potential safety issues as quickly as possible (Evans et al., 2001).

Additionally, sales data may not be an adequate proxy for determining the exposed population. In other words, one unit of product shipped might not necessarily equate to one person exposed to the product. For example, a household of three persons may share one unit of a product, or multiple units could be purchased and used by a single person. Inflation of the sales units relative to the true number of persons exposed is of particular concern as it will artificially deflate the incidence calculations—that is, when the denominator of Eq. (1) is overestimated it yields a lower incidence estimation. A potential solution to inflated sales units is to calculate incidence based on unique purchasers; however, this metric may not be available or easily accessible in the sales database. Units shipped is typically preferred over other metrics, such as units ordered or units sold, as it is the most likely to be closest to the time when the product is consumed.

Though incidence trending and its reliance on sales units presents several challenges, it does provide a fundamental way of assessing product safety consistently over time. In the event that an increase in incidence is observed and suspected to be indicative of a potential safety issue, product trending serves as the starting point for further investigations. Subgroup analyses of specific consumer profiles may indicate that the spike in incidence is being driven mainly by adverse event reporting from a susceptible population. Serving size estimations are also critical components for assessing the true population exposure and the likelihood that the suspected safety issue is due to a specific product. Furthermore, the literature and publicly accessible regulatory databases are excellent resources for evaluating the potential safety issue and provide a broader context of the subject beyond the scope of the internal data.

25.15 SIGNAL DETECTION

Pharmacovigilance tools, particularly signal detection methods, have been developed and utilized by the pharmaceutical industry to ensure the safety of drugs in the postmarket setting. While pharmacovigilance methods were developed within the context of conventional medicines, the concepts can also be applied to monitoring the safety of herbal medicines/nutraceuticals (Barnes, 2003).

A signal is defined as "reported information on a possible causal relationship between an adverse event and a drug, the relationship being unknown or incompletely documented previously" (World Health Organization, 2004). Therefore, signal detection refers to methods used to uncover these adverse event–drug relationships with potential safety implications. The process begins with statistical analysis of the data and moves through the medical review process of the relevant cases. Adapting these pharmacovigilance approaches to nutraceutical vigilance denotes that signals and signal detection refer to adverse event–nutraceutical product relationships that indicate possible safety issues.

Common signal detection methods such as the proportional reporting ratio (PRR) and reporting odds ratio (ROR) are based on the concept of disproportionality. Disproportionality refers to adverse event–product combinations that are reported more frequently than would be expected based on all adverse event reports (van Puijenbroek et al., 2002). Hence, these combinations are often referred to as signals of disproportionate reporting or SDRs. The focus in this section will be the PRR, which has published guidelines on its use in the European Medicines Agency (EMA) EudraVigilance data analysis system (Eudravigilance Expert Working Group, 2006).

The PRR involves a relatively straightforward calculation, which can easily be interpreted as the strength of a SDR (Evans et al., 2001). It calculates the probability that an adverse event is reported given a specific product, divided by the probability that the adverse event is reported given all other products. Therefore, a PRR greater than the null or expected value of one implies that an adverse event is more frequently reported in association with a specific product than when compared to all other products in the database (Alvarez et al., 2010). As a result, the strength of the

SDR increases with higher PRR values—that is, the numerator and the denominator become more disproportionate (Evans et al., 2001). The probabilistic interpretation of the PRR can be written as Eq. (25.2) (Zorych et al., 2013):

$$\frac{\Pr(\text{adverse event}|\text{product})}{\Pr(\text{adverse event}|\text{-product})} \tag{25.2}$$

While the PRR is a simple yet powerful tool for signal detection, it is known to be a very sensitive method that may yield many false positive SDRs. Therefore, it is necessary to compute additional metrics and impose threshold criteria in order to define SDRs that may constitute true safety signals. One method used in the EudraVigilance data analysis system for their weekly and monthly reports is the 95% CI around the PRR. When the PRR is displayed with its 95% CI, an adverse event–product combination is defined as a SDR when the lower bound of the 95% CI is greater than or equal to one, and the number of individual cases with this adverse event–product combination is greater than or equal to three (Eudravigilance Expert Working Group, 2006).

There is currently no gold standard for signal detection in postmarket safety databases, and methods continue to be developed around the world. The World Health Organization Uppsala Monitoring Centre (WHO-UMC) and the Food and Drug Administration (FDA) currently use two Bayesian methods for signal detection: the Bayesian confidence propagation neural network (BCPNN) and the (multi) gamma Poisson shrinker (MGPS), respectively (Ahmed et al., 2010). Even the thresholds used for the PRR within the EudraVigilance system have been assessed for efficiency. Increasing the criterion of three or more individual case counts to five or more was found to decrease the generation of false positive SDRs by 25%. However, there was also a comparatively smaller decrease in the early detection of true signals. Conversely, it is possible that less stringent thresholds (e.g., two or more individual case count as opposed to three or more) may be adapted as part of a more cautious approach where more false positives are generated in order to not fail to uncover any true positives. The methods and thresholds used for signal detection will depend in large part on the resources available to the pharmacovigilance/postmarket safety surveillance department and the safety profile of the products being monitored (Slattery et al., 2013).

The main advantage of signal detection methods over incidence trending is the reliance on data from only one database. This eliminates any delays in receipt that may occur with the use of external data, such as sales or purchaser data (Evans et al., 2001). However, even within a single dataset and for a much shorter time frame than that used in incidence trending, the sheer number of possible adverse event–product combinations for which these metrics need to be computed is daunting in the absence of robust statistical programs. PhViD is a statistical package available in R that contains several pharmacovigilance signal detection methods, including those previously mentioned (Ahmed and Poncet, 2013; R Core Team, 2015). Its implementation has been well-documented in several published studies exploring different topics within pharmacovigilance, and it can serve as a valuable tool for nutraceutical vigilance (Gurulingappa, 2013; Johnson et al., 2012; Kajungu et al., 2014; Pulley et al., 2010). The PhViD package is able to quickly compute the selected signal detection metrics for every adverse event–product combination within a dataset, and its default settings can be adjusted for different thresholds and criteria, such as those used in the EudraVigilance system. This allows for expedited reporting of the SDRs and processing through medical review stage where the pertinent cases are evaluated more comprehensively within a clinical context and potential safety signals are identified (Eudravigilance Expert Working Group, 2006). The earlier potential safety signals are identified, the greater the ability to further investigate and rectify the situation (if necessary) without serious public health implications. As with incidence trending, it is important to utilize as many available resources, such as reports published in the literature and open-access regulatory databases, as possible to properly contextualize the potential and develop the most appropriate and effective solution.

25.16 RELEVANT ADAPTATIONS FOR NUTRACEUTICALS

The differences between conventional medicines and nutraceuticals, especially the challenges with postmarket safety surveillance, have been well documented (Barnes, 2003; World Health Organization, 2004). Recognizing these differences is fundamental to effectively applying and adapting the aforementioned pharmacovigilance methods to nutraceutical vigilance.

Nutraceuticals tend to have a much less established safety profile compared to conventional medicines since they are not required to undergo stringent testing or human clinical trials in most markets (Barnes, 2003). This relative ambiguity arguably warrants the more cautious approaches to signal detection, such as lowering the threshold for case counts or automatically processing SDRs containing a serious adverse event through the medical review stage irrespective of case count. Concurrent multiple product use is also a complicating factor. One resolution is to operate under the assumption that every suspect product is independently associated with every adverse event reported within the case. Another may be to perform different subanalyses, such as limiting the dataset to single product users if the number of cases is sufficiently large, or comparing cases involving a suspect product of interest to cases involving the same product of interest along with other suspect products that are relevant. This technique requires subject matter expertise to determine which product combinations are of interest and extracting or manipulating multiple datasets to be run in PhViD. Similar to multiple product use, nutraceuticals are complex products made up of several constituents (Barnes, 2003). Particular ingredients may be of interest and analyses can be performed in the context of substance exposure as opposed to product exposure. This method also requires subject matter expertise and manipulating a dataset to group products by their ingredients—whether that is plausible depends on the availability of the formulations and/or product labels. Despite the considerations that must be taken into account when monitoring the safety of nutraceuticals, the experiences of the pharmaceutical industry provide an excellent framework for adapting established pharmacovigilance methods to nutraceutical vigilance.

In the event that an SDR has been medically evaluated and a true safety signal is suspected in the database, there are several options for corrective and preventive action. Product labels can be modified to provide cautionary wording, such as, "Not recommended for use in" a certain population, or advisory statements to seek medical advice prior to using the product. Similarly, additional package inserts that provide material directed at product safety awareness concepts, like ingredient reading and patch testing for cosmetic products, are effective approaches for consumer education. It should be noted that these approaches may be subject to regulatory approval in certain markets. For suspected signals involving serious adverse events with potentially high public health risk, such as material contamination or other breaches in quality control, a more drastic and expeditious approach is warranted. Product recall is the process used to mitigate the public health risk by limiting further exposure to the suspect product or ingredient. Furthermore, additional quality tests can be performed on the recalled product(s) to identify a potential root cause in the manufacturing process that must be corrected. The corrective and preventive actions taken to address a suspect safety signal will depend on the scope of the issue and the public health threat that it poses, and these are only a few examples of strategies that have been used in the industry.

25.17 DETERMINING CAUSALITY: CAUSALITY ASSESSMENT CRITERIA

There are two major globally accepted criteria for assessing causality for adverse events. One of the causality assessment methods is the Roussel Uclaf Causality Assessment Method (RUCAM) for assessing case reports in which the product is suspected to be associated with hepatotoxicity, and the second is the World Health Organization (WHO) causality criteria which are applied to assess general causality for nonhepatic case reports (Arimone et al., 2005; National Library of Medicine, 2015). Both criteria assess the strength of the association between the reported event and the consumption of the reported product by assigning specific classifications such as: unlikely, unassessable/unclassifiable/excluded, possible, probable, and certain/definite. RUCAM uses an in-depth calculation to assign causality classifications while the WHO criteria use specific definitions for each classification.

Using standardized causality assessment criteria is advantageous in adverse event case reviews for several reasons, including: its application in determining regulatory requirements for case reportability, specific causality definitions aid in consistent case review and documentation for internal or external assessments, and causality criteria provide a means for noninvasive case assessments. However, the industry should consider whether the current "standard" processes, which are typically utilized in the drug industry to collect, manage, and assess adverse events (including the WHO and

RUCAM methods mentioned earlier), are sufficient to effectively and acutely monitor safety trends for nutraceuticals or other complex substances. Unfortunately, in the case of nutraceutical products, the disadvantages to causality assessment criteria, as it is standardized today for pharmaceutical products, outnumber, but do not necessarily outweigh, the advantages. This includes the fact that the criteria may only be accurately applied to one substance or active at a time. Many nutraceutical products are composed of several active ingredients and/or the consumer may take several products at once. Not only is it cumbersome to conduct causality assessments on each active ingredient present in each product reportedly associated with an adverse event, consumers who take complex substances like multiingredient nutraceuticals or consume multiple products at once make it very difficult to identify a suspect product, let alone a suspect ingredient regardless of whether the causality assessment is actually conducted. Additionally, synergies and other ingredient/ingredient, food/ingredient, or drug/ingredient interactions potentially associated with nutraceuticals are often not fully understood. This is also a challenge with nutraceutical products that are not standardized to one specific dose or even one specific ingredient source type (e.g., extraction type and/or plant part in the case of botanical ingredients may vary greatly and have different effects, safety profiles, etc., as a result). Finally, many of the details or definitions of case criteria are too general or subjective to accurately apply to adverse event cases associated with nutraceuticals. For example, the RUCAM causality worksheet contains a section that asks if the suspect agent is "known" to be associated with hepatotoxicity (National Library of Medicine, 2015; Vatistas and Samuels, 2012). The case reviewer answering this question might then ask, "known to whom?" or, "what does known actually mean?"—"a clinically proven or likely association for this product?" Considering that premarket clinical data is often lacking for nutraceuticals, ingredient variability can largely impact safety outcomes, formulary differentiation, and there is often an absence of other purely objective data to answer these questions. However, answering yes is often an inaccurate choice if the case being reviewed is, in and of itself, inconclusive, or other cases have been inconclusive, and/or it is the first case being assessed for such a reaction for a particular substance or its ingredients. That said, the answer to the "known" question can significantly impact the strength of the RUCAM causality outcome.

RUCAM criteria also contain a section for ruling out potential differential diagnoses for the reported event. However, the section is not exhaustive or current with all the common conditions which may be valid differentials associated with a given clinical presentation. As an example, the section includes certain viral etiologies and alcoholism but not other common causes for acute hepatitis such as Hepatitis E or even nonalcoholic fatty liver disease. Therefore, different clinicians may include or exclude certain risk factors for disease based on personal experience, knowledge, and expertise, epidemiological data specific to the origin where the adverse event reportedly occurred, etc. The scoring in this section can also significantly impact the RUCAM causality outcome.

Another example of inherent ambiguity within the causality assessment criteria is the potential default to insufficient data rendering the case unassessable (Arimone et al., 2005). The concept of there being insufficient data is highly subjective and open to interpretation by the reviewer. Furthermore, if the reviewer already suspects that an agent is associated with the reported adverse event, then very little substantiation may be necessary to determine that a causal association was possible or even probable.

As discussed earlier, even the more rigorous causality assessment criteria such as RUCAM may be easily misapplied and/or misinterpreted. There have been studies to determine the potential for these varying outcomes and the general reliability of standardized causality assessment criteria (even for the drug model for which it is intended), and it has been concluded that a significant amount of variability can exist when the same criteria are applied to the same case by different practitioners (all showing relevant expertise for the case being reviewed) (Teschke et al., 2013). In the case of nutraceutical products, ingredient and product variability, as well as inconsistent or inappropriate application of the causality assessment criteria, has resulted in mixed outcomes and significantly impacted the perceived safety of a product or ingredient in some cases. In fact, some nutraceuticals or ingredients have been deemed causally associated with an event regardless of the fact that some patients consuming these products present with a condition(s) which may provide a more likely etiology for the reported event (Teschke et al., 2013).

As discussed, there are obvious opportunities for improving the existing causality assessment criteria and the guidelines for using such criteria when applied to drug-related adverse event case reviews, as well as evaluations involving dietary supplement products. Revising these criteria and having more consistent standardized assessment processes are critical steps in having a better understanding of product safety, meaningful trending and signal detection outcomes, and determining the most appropriate corrective or preventive steps for cases deemed causally associated with a reported event.

25.18 CONCLUSIONS

In conclusion, monitoring and evaluating safety data associated with nutraceutical products (or any product) is a highly involved activity which ideally incorporates a manufacturer's regulatory or other legal obligations, standardized definitions or case review procedures, appropriate technology, organizational/internal collaboration, and due diligence in the best interest of consumer safety as the leading factors in case assessment and corrective or preventive action. The strategies and considerations presented in this document may be used to create a global pharmacovigilance or safety surveillance framework for the nutraceutical industry. However, there are several elements of each area listed previously which could be further enhanced, understood, or even harmonized to support more consistent and accurate outcomes, which will further improve postmarket safety surveillance activities and the most prudent application of postmarket data for optimal monitoring and regulation of these products globally.

The future of nutraceuticals will likely bring changes to the industry in regard to safety, as this is typically the topic of highest concern according to industry "skeptics" (e.g., select medical/academic communities, regulators, etc.). There are currently strategic discussions occurring in the US (amongst regulators as well as industry) to determine whether it is appropriate to establish additional regulatory requirements for premarket notifications and/or approval processes similar to drug products. Other opinions lean toward more of a compromise which may involve enhanced efforts in regulatory enforcement (e.g., routine inspection schedules and increased monitoring) to ultimately raise expectations for industry compliance (with premarket notifications, postmarket adverse event reporting, as well as GMPs). However, regulatory staffing and support is a major consideration for determining the next best steps. This would be relevant whether upcoming changes involve additional regulations to ensure more comprehensive premarket notifications and/or approval processes or enhancing enforcement activity for existing regulations. Currently, there are limited resources at FDA for reviewing incoming postmarket reports, conducting regular field inspections (GMP as well as pharmacovigilance), or monitoring product labeling and claims, which may also prompt FDA inquiries regarding product safety or quality. Therefore, it would be prudent to select a solution(s) which is compatible with this important consideration.

Similar to the US, there is concern for nutraceutical safety globally, and as discussed, regulations are being developed accordingly—whether it be establishing new regulations or revising existing regulations (for pre- or postmarket activities). Therefore, it is expected that further regulatory developments will occur which best meet the market-specific or regional needs, and will depend on whether those needs are perceived, precautionary/risk-based, quantified/qualified, or a combination of these approaches and potentially other factors. For example, in some countries where compliance with sanitation measures is a prevalent issue, establishing enhanced manufacturing/quality controls may take priority over safety surveillance.

Regardless of the market-specific or regional regulatory environment involved, there is likely a long road ahead for reaching the best solution between industry and regulators that adequately addresses nutraceutical product safety concerns. Ideally, industry and regulators will first come together to fully embrace existing mandates to ensure that these measures are regulating nutraceuticals as intended, which includes the ability of such regulations to monitor and control for issues that directly impact consumer safety.

REFERENCES

Ahmed, I., Dalmasso, C., Haramburu, F., Thiessard, F., Broët, P., Tubert Bitter, P., 2010. False discovery rate estimation for frequentist pharmacovigilance signal detection methods. Biometrics 66, 301–309.

Ahmed, I., Poncet, A., 2013. PhViD: an R package for PharmacoVigilance signal Detection. R package version 1.0.6.

Alvarez, Y., Hidalgo, A., Maignen, F., Slattery, J., 2010. Validation of statistical signal detection procedures in EudraVigilance post-authorization data. Drug Saf. 33, 475–487.

Arimone, Y., Bégaud, B., Miremont-Salamé, G., Fourrier-Réglat, A., Moore, N., Molimard, M., Haramburu, F., 2005. Agreement of expert judgment in causality assessment of adverse drug reactions. Eur. J. Clin. Pharmacol. 61, 169–173.
Barnes, J., 2003. Pharmacovigilance of herbal medicines. Drug Saf. 26, 829–851.
Cohen, P.A., 2014. Hazards of hindsight—monitoring the safety of nutritional supplements. N. Engl. J. Med. 370, 1277–1280.
Consumer Reports magazine, 2012. 10 surprising dangers of vitamins and supplements.
Electronic Code of Federal Regulations, 2015. Title 21 Food and Drugs Part 110 Current Good Manufacturing Practice in Manufacturing, Packing, or Holding Human Food.
Engredea News & Analysis, 2013. ABC: retract DNA barcode analysis article.
Eudravigilance Expert Working Group, 2006. Guideline on the use of statistical signal detection methods in the Eudravigilance data analysis system. European Medicines Agency.
European Medicines Agency, 2013. Guideline on good pharmacovigilance practices (GVP) module II – pharmacovigilance system master file (Rev. 1).
European Medicines Agency EudraVigilance, 2015. European database of suspected adverse drug reaction reports.
Evans, S., Waller, P.C., Davis, S., 2001. Use of proportional reporting ratios (PRRs) for signal generation from spontaneous adverse drug reaction reports. Pharmacoepidemiol. Drug Saf. 10, 483–486.
Food and Drug Administration, 1994a. Dietary Supplement Health and Education Act of 1994.
Food and Drug Administration, 1994b. Dietary Supplement Health and Education Act of 1994.
Food and Drug Administration, 2007. Guidance for Industry: Questions and Answers Regarding Adverse Event Reporting and Recordkeeping for Dietary Supplements as Required by the Dietary Supplement and Nonprescription Drug Consumer Protection Act.
Food and Drug Administration, 2011. Draft Guidance for Industry: Dietary Supplements: New Dietary Ingredient Notifications and Related Issues.
Food and Drug Administration, 2014. Guidance for Industry: Internet/Social Media Platforms: Correcting Independent Third-Party Misinformation about Prescription Drugs and Medical Devices. Draft Guidance.
Food and Drug Administration, 2015. Open FDA. 2015.
Fung, M., Thornton, A., Mybeck, K., Wu, J.H., Hornbuckle, K., Muniz, E., 2001. Evaluation of the characteristics of safety withdrawal of prescription drugs from worldwide pharmaceutical markets-1960 to 1999. Drug Inf. J. 35, 293–317.
Gurulingappa, H., 2013. Automatic detection of adverse events to predict drug label changes using text and data mining techniques. Pharmacoepidemiol. Drug Saf. 22, 1189–1194.
Harbaugh-Reynaud, D., Mishler, B., Neal-Kababick, J., Brown, P., 2015. The capabilities and limitations of DNA barcoding of botanical dietary supplements. Authentechnologies.
Health Canada, 2009. Compendium of Monographs. 2015.
Health Canada, 2011. Guidance Document for Industry—Reporting Adverse Reactions to Marketed Health Products.
Health Canada, 2013. Product Licensing Forms and Templates.
ICH Harmonised Tripartite, 2000. Good Manufacturing Practice Guide for Active Pharmaceutical Ingredients. Q7A, Current Step 4.
Instant GMP, 2012. What is GMP for Dietary Supplements? 2015.

International Alliance of Dietary Food Supplement Associations, 2011. Global Guide to Good Manufacturing Practice for Supplements.
International Conference on Harmonisation, 2015. Medical Dictionary for Regulatory Activities (MedDRA). 2015.
Johnson, D., 2012. Predicting drug safety: next generation solutions. J. Drug Metab. Toxicol. 3, e106.
Johnson, K., Guo, C., Gosink, M., Wang, V., Hauben, M., 2012. Multinomial modeling and an evaluation of common data-mining algorithms for identifying signals of disproportionate reporting in pharmacovigilance databases. Bioinformatics 28, 3123–3130.
Kajungu, D.K., Erhart, A., Talisuna, A.O., Bassat, Q., Karema, C., Nabasumba, C., Nambozi, M., Tinto, H., Kremsner, P., Meremikwu, M., 2014. Paediatric pharmacovigilance: use of pharmacovigilance data mining algorithms for signal detection in a safety dataset of a paediatric clinical study conducted in seven African countries. PLoS One 9, e96388.
Korean Food and Drug Administration, 2010. Health Functional Foods Act.
LeDoux, M.A., Appelhans, K.R., Braun, L.A., Dziedziczak, D., Jennings, S., Liu, L., Osiecki, H., Wyszumiala, E., Griffiths, J.C., 2015. A quality dietary supplement: before you start and after it's marketed—a conference report. Eur. J. Nutr. 54, 1–8.
MedEffect Canada, 2015. Canada Vigilance Adverse Reaction Online Database. 2015.
Medicines and Healthcare Products Regulatory Agency, 2011. Implementing Measures in Order to Harmonise the Performance of the Pharmacovigilance Activities Provided for in Directive 2001/83/EC and Regulation (EC) NO 726/2004.
Ministry of Health Malaysia, 2002. Malaysian Guidelines for the Reporting & Monitoring. March 2002.
National Library of Medicine, 2015. LiverTox: Clinical and Research Information on Drug-Induced Liver Injury. Roussel Uclaf Causality Assessment Method (RUCAM) in Drug Induced Liver Injury.
Pulley, J., Hassan, N.N., Bernard, G.R., Jirjis, J.N., Schildcrout, J., Robertson, D., Masys, D.R., Harris, P., 2010. Identifying unpredicted drug benefit through query of patient experiential knowledge: a proof of concept web-based system. Clin. Transl. Sci. 3, 98–103.
R Core Team, 2015. R: a language and environment for statistical computing. R Foundation for Statistical Computing.
Schultz, H., 2015. Continuing confusion over herbal supplement contents points to need for premarket notification, expert asserts. Nutraingredients.
Slattery, J., Alvarez, Y., Hidalgo, A., 2013. Choosing thresholds for statistical signal detection with the proportional reporting ratio. Drug Saf. 36, 687–692.
Srinivasan, V., 2014. USP and dietary supplements...a return to our roots. Quality Matters 2014.
Teschke, R., Schulze, J., Schwarzenboeck, A., Eickhoff, A., Frenzel, C., 2013. Herbal hepatotoxicity: suspected cases assessed for alternative causes. Eur. J. Gastroenterol. Hepatol. 25, 1093–1098.
Therapeutic Goods Administration, 2011. Australian Guidelines for Pharmacovigilance Responsibilities for Sponsors of Medicines.
Therapeutic Goods Administration, 2013. Database of Adverse Events Notifications. 2015.
van Puijenbroek, E.P., Bate, A., Leufkens, H.G., Lindquist, M., Orre, R., Egberts, A.C., 2002. A comparison of measures of disproportionality for signal detection in spontaneous reporting systems for adverse drug reactions. Pharmacoepidemiol. Drug Saf. 11, 3–10.

Vatistas, T.J., Samuels, J.G., 2012. The regulation of dietary supplements in the United States: advocating for a reasonable approach, protecting patient safety, and the role of nursing. Policy. Polit. Nurs. Pract. 13, 113–116.

World Health Organization, 2004. WHO Guidelines on Safety Monitoring of Herbal Medicines in Pharmacovigilance Systems.

Zorych, I., Madigan, D., Ryan, P., Bate, A., 2013. Disproportionality methods for pharmacovigilance in longitudinal observational databases. Stat. Methods Med. Res. 22, 39–56.

CHAPTER

Genetically modified products and GMO foods: a game of chance? 26

John Hudson

Department of Economics, University of Bath, Bath, United Kingdom

26.1 INTRODUCTION

This chapter is written from the perspective of an economist, albeit one who is trying to reach out to other disciplines, in both the sciences and social sciences. The focus is very much on the costs and the gains from genetically modified (GM) food, including all of the potential costs and gains and not just those limited to producers and consumers. In order to understand these costs and gains, we must first have some understanding of some of the science involved. GM foods are produced from organisms that have had changes introduced into their DNA using genetic engineering (GE) as opposed to more traditional methods. There are several ways of doing this, and many involve patented technology. Plants, animals, or micro-organisms that have been modified through GM are referred to as genetically modified organisms (GMOs). The GMO itself is likely to be patented, particularly if developed by a private company. In the United States the term GE tends to be favored over GM as a more accurate term. The World Health Organization defines GM foods as derived from organisms whose DNA has been modified in a way that does not occur naturally. In this chapter we shall use the terms GM and GE interchangeably. The fact that there are these definitional differences is reflective of the problems different countries and regulatory bodies have had in coordinating a common approach.

Almost all of the discussion has focused on transgenesis and there has been much less on cisgenesis. The former involves genes from species which cannot naturally interbreed and hence the evolution of the modified product is almost impossible to happen naturally. Cisgenesis can happen naturally, although often over a very long time period and involves genes which are derived from the gene pool of sexually compatible species, for example the introduction of resistance genes that exist naturally in wild/crab apples into commercially grown apple trees (Hudson et al., 2015). The evidence is, as we shall see later, that people are more apprehensive of transgenesis than of cisgenesis.

GM technology moves in waves or generations, often overlapping. With respect to crops, the first wave focused on introducing herbicide-tolerant and insect-resistant

Developing New Functional Food and Nutraceutical Products. http://dx.doi.org/10.1016/B978-0-12-802780-6.00027-4

traits into four main crops. The former increased crop tolerance to glyphosates, a group of low-toxicity herbicides, which allows farmers to control weeds more easily. The latter caused crop plants to produce the naturally occurring chemical *Bacillus thuringiensis* (Bt), which is toxic to common agricultural pests. This toxin repels pests without farmers' having to use chemicals. The next wave includes traits that would increase product shelf life, tolerance to drought, nutritional content, adaptability to climate change, and nitrogen fixation (Barrows et al., 2014). Improving the nutrient content of food crops through GM or conventional processes is known as biofortification. An example is Golden Rice, which contains significant amounts of provitamin A. This is present in the leaves of rice plants, but not the rice seeds. This could counter vitamin A deficiency, which is a severe health problem in many developing countries. A third wave is designed to produce substances for pharmaceutical or industrial purposes (Qaim, 2009). Most commercialized products, and thus most of the scientific literature, relate to the first wave.

The first commercial GM crops involved tomatoes and were planted in 1994. It was two years later before a significant area of crops, 1.66 million hectares, was planted. Hence this is still a relatively new technology. Once they got a foothold in the market they quickly expanded. Thus,GM traits accounted for 74% of soybean plantings in 2013 and 71% for cotton, 29% for maize/corn, and 22% for canola (Brookes and Barfoot, 2015). It is a technology which has grown rapidly, and by 2014, there were 181.5 million hectares in 28 countries worldwide.[a] But it has spatially not been an even growth. In 2013, two countries, the USA (38%) and Brazil (26%), dominated the production figures with the other main countries being Argentina, India, Canada, and China. Thus, large areas of the world are excluded from this process, including of course much of Europe. Even in countries such as China and India which have adopted the technology, there is some hesitancy on the part of governments and substantial hostility on the part of both consumers and farmers. Thus, adoption has been limited. Why is this, given that most of the scientific evidence to date is of substantial benefits and little proven adverse impact? Why is it that a technology which promises so many benefits, potential and achieved, has not been received enthusiastically around the world? Why too has the European Union adopted such a hard line regulatory position, when this has handed leadership of this technology to others, primarily biotech firms in the USA? At a time when innovation is critical to growth, this seems a slightly odd position. These are the questions we seek to, if not answer, at least put some light on in this chapter.

The structure of the chapter is as follows. In the next sections we will discuss the costs and benefits to farmers, the environment, and consumers. We then look at regulatory issues and different systems in different countries. In doing this we suggest firstly that these differences can in part be put down to a difference on where the burden of proof should lie: with those attempting to prove that GM foods offer a net gain to the world or those who argue that it leads to net losses. Much of the focus in this chapter is on GM crops. This is because there have been far fewer examples of

[a] <http://www.isaaa.org/kc/default.asp> (accessed 06.06.16).

commercialized GM animals as food, although this may be beginning to change and in the penultimate section we look at this issue. Finally we conclude the paper. Here we argue that we are in a sense in a game with nature: we make a move and nature then makes a counteracting move. We are not saying that hidden away somewhere there actually is this being called nature with whom we are playing a game. But economists are fond of working with "as if" assumptions and this one fits the facts as well as, and even better than, many other contexts where this assumption is applied. This game is something which some, although by no means all, of the analysis tends to ignore, resting on the often implicit assumption that current gains in production will last indefinitely into the future. There is one final point: nature often waits a long time to reveal its hand, by which time we are well and truly committed to our initial, and sometimes subsequent, moves.

26.2 THE BENEFITS AND COSTS TO THE FARMER

There is considerable data to suggest that GM seeds increase crop yields. However, acting against these gains, as we shall see later, prices should fall from the increase in aggregate supply and this will reduce farmers' income. Nonetheless the various estimates suggest that farmers should still gain. For example, in the USA, the estimated share of the gains accruing to farmers is between 5% and 40% (National Research Council, 2010). The advantages are diverse. Thus, GM seeds may increase the practice of double cropping, that is, growing two crops per season (Trigo and Cap, 2003). With conventional seeds, pest damage is often too great to facilitate profitable production throughout the potential growing season. GM seeds facilitate weed control, reducing problems prior to the crop appearing above ground, which tends to prolong the growing cycle. One factor to bear in mind is that unlike conventional seeds, there are often limits on reuse. For example, a "no saved seed" provision in a contract between the farmer and the biotech company can prohibit the farmer from saving seed and/or reusing GM seed from GM crops (Hamilton, 2001). Thus, the farmer will need to buy the seeds every year. The negative externalities of GM food also need to be considered. NonGM farmers may need to devote resources to protecting their crops from neighboring GM farmers. In addition nonGM farmers may eventually suffer from the evolution of "superweeds," and this may even force them to use GM seeds. This is important: in evaluating the benefits to farmers the focus is often on the GM farmers, while the costs to nonGM farmers are not so often taken into account. There is also relatively little empirical work looking at the socioeconomic outcomes of GM crops, particularly going beyond the farmers who grow GM crops.

There are claims in the literature that the greatest potential for gains is in developing countries, where pests are particularly prevalent and by implication it should be these countries which see the most substantial reductions in use of insecticides and toxic herbicides (Qaim and Zilberman 2003; Qaim, 2009). There is evidence on the gains, although to date herbicide-tolerant crops have not been widely adopted in the small farm sector, which characterizes many developing countries. But this is not the case for Bt cotton, especially in China and South Africa. Thus the adoption of Bt cotton resulted in a 50% increase in profit per hectare compared to an 18% increase in expenditures in India (Kathage and Qaim, 2012). There were also gains to smallholders in South Africa (Gouse et al., 2004). The evidence also suggests that the gains to small-scale farmers are similar to, or even greater than, those of larger farmers (Qaim, 2009). One problem with this argument is that being as the development of new seeds depends upon the motivation of private profit, there is a lack of incentive to develop traits for which the potential beneficiaries have low purchasing power. This affects some markets in developed countries, but also markets in developing countries. There is arguably the potential for GM seeds to help with many of the problems in developing countries including food safety and health, but whether the free market can deliver this potential is open to question.

26.3 THE BENEFITS AND COSTS TO THE ENVIRONMENT

There are both benefits and costs to the environment of GM crops. Insect-resistant seeds reduce the need for insecticide applications. They thus have the potential to reduce chemical use in agriculture and thus constrain the damage from chemical runoff. The greatest impact is, as we have seen, on cotton rather than food. Qaim (2009) has estimated that Bt cotton adoption reduced pesticide use by 65% in China, 47% in Argentina, and 36% in the United States. There are however still substantial benefits from GM foods. Herbicide-tolerant seeds do not reduce pesticide use, but it is argued that they facilitate the substitution of more benign alternatives, for example, some glyphosates, for the chemicals used on conventional crops (Cerdeira and Duke, 2006). The beneficiaries of this include mammals, birds, and fish. The reduction in tillage, used to control weeds, which GM seeds facilitates also benefits the environment. Tilling results in soil erosion, releases carbon from the soil, thus contributing to the buildup of CO_2 in the atmosphere, and increases agricultural runoff, thus polluting the seas (National Research Council, 2010). Both tillage and the use of insecticide require farm machinery and fuel to power that machinery. Hence the reduction in both of these also results in a reduction in energy consumption, again aiding efforts to reduce carbon emissions. Overall because of all these reasons, and others such as reduced pollution from the transportation and production of agrochemicals, Bennett et al. (2004) calculate GE crops involve about 10% of the toxicity of conventional crops. This of course is a very substantial gain.

It is possible that the introduction of a gene which strengthens a particular crop could also spread to other plants, for example, weeds, also strengthening them, making them more invasive. A particular problem relates to organic farmers who depend upon the purity of their produce. Protection tends to rely on buffers between the two types of crop (Beckmann et al., 2006). However, it is perhaps unlikely that over a sufficiently long period of time it will be possible to prevent such spread. Indeed there is some suggestion that this is already happening. For example, there is evidence from Mexico of locally produced maize containing part of the DNA found in glyphosate-tolerant and insect-resistant maize (Gilbert, 2013).

Barrows et al. (2014) argue that GM food has helped maintain crop diversity. In part this is through the reintroduction of seeds which pest damage had previously caused to be abandoned. GE also facilitates the production of different varieties with unique properties, thus it is suggested enhancing biodiversity. There is of course a second aspect to biodiversity and that is the impact on nontarget species (birds, snakes, nontarget arthropods, and soil macro and microfauna). Again the literature is substantial and there is relatively little evidence of a negative impact of GE crops (Nicolia et al., 2014). Of course one also has to consider the alternative, which tends to involve chemical use which frequently has a more adverse impact (Ammann, 2005).

26.4 THE BENEFITS AND COSTS TO THE CONSUMER

Consumers potentially gain from lower prices and in some parts of the world greater food security. The latter is of course of critical importance in many developing countries and could potentially on its own justify the use of GM foods. The impact of GM seed on farmers' production can be used to estimate the impact on prices, given a number of assumptions. Doing this Barrows et al. (2014) concluded that the effect in 2010 was to lower prices by between 13% and 27%. Cotton was expected to decline by 7–19% and soybeans by between 2% and 65%. Of course not all the gains accrue to the farmers or to developing countries as a whole; the biotech companies also benefit. Price et al. (2003) estimated that in the late 1990s Bt cotton generated a total annual economic surplus gain of approximately $164 million in the United States. The farmers got 37%, consumers 18%, and biotech companies 45%. The absolute gains as well as their distribution will differ from country to country and crop to crop. Thus, in Spain for Bt maize, Demont and Tollens (2004) estimate that 60% went to farmers and 40% to the seed companies. The biotech companies gain much less, 1.5% of the surplus, in China, due to weaker intellectual property rights (IPRs) (Pray et al., 2001). However, since then of course there has been a move globally to strengthen IPRs and hence it may be unlikely to be this low now.

One of the prime potential costs to consumers relates to health, both the impact on health and concerns about the possible impact. However, there are also health benefits to GM crops. Among the positive impacts, consumers experience reduced risk of digesting the toxic and carcinogenic chemicals produced by fungi that infest crops. They may also extend to farm workers through reduced exposure to pesticides (Huang et al., 2005). The benefits are particularly substantial in developing countries, due to poor pest control and protective equipment. Turning to the potential hazards, one concern relates to the possibility that genes inserted into the plants may prove toxic to humans in some manner. This is a very technical issue and in order to understand the economics of it, it is necessary to be at least familiar with some of the technology. This is particularly well handled in Nicolia et al. (2014) and in doing so they conclude that no significant hazard has been directly connected to the use of GM crops. In an agricultural context, gene flow can be defined as the movement of genes, gametes, individuals, or groups of individuals from one population to another, and occurs both spatially and temporally (Mallory-Smith and Sanchez-Olguin, 2011). For instance, gene flow may occur after several years as GE crop plants may be capable of surviving through seed or asexual propagules for years in the field. They may also be able to fertilize sexually compatible nonGE plants (nonGE crop or wild relative plants). The occurrence of gene flow may lead to the spread and persistence of transgenes into the environment or the market. There is also the possibility of horizontal gene transfer (HGT) to bacteria in the soil, which may lead to the development of "super pathogenic bacteria" resistant to antibiotics. At the moment the problems seem limited, but are the subject of ongoing scientific research.

Perhaps the issue which is of most direct concern to people is the possibility that by eating GM food, they may absorb part of the DNA, with adverse effects on their own health. In a normal diet we consume between 0.1 and 1 g of DNA/day (Parrott et al., 2010). This is partly digested, but there are concerns that transgenic DNA may resist this process, leading to HGT to live bacteria in the gastrointestinal tract, or accumulating in the body. The same worries concern livestock animals who are given GM feed. In general Nicolia et al. (2014) seem to conclude that for various reasons the causes for concern are small, with one qualification. Zhang et al. (2012) have reported the first evidence of HGT of a food-derived exogenous miRNA[b] naturally found in rice. The research is not specifically linked to GE as such, but demonstrates the potential for HGT from plants to humans. This is relatively new and research, including attempted replications, and discussion are ongoing. Even more recent is the work by Séralini et al. (2014) which documents long-term deleterious effects arising from consumption of a widely used GMO and its associated herbicide involving maize. The study was based on rodents and found evidence of, for example, kidney disturbances.[c] Nonetheless, bodies such as the OECD, the UN Food and Agriculture Organization, and national bodies such as the French Academies of Science find there is no evidence as to the unsafeness of the current wave of GM products (Paarlberg, 2010). At the same time the possibility is noted that the next generation could present new risks. The European Union has spent a considerable amount of money researching the health aspects of GM foods and, as with others, concluded that the use of GE plants as such does not involve greater risks than traditional approaches (European Commission, 2010).

---

[b]MicroRNAs, which play an important role in immune cells.

[c]It is worth noting that the paper was originally published in another journal, but subsequently withdrawn and a slightly changed version republished. This series of events, which is contentious in itself, is discussed in Fagan et al. (2015) who also analyze this issue and reach similar conclusions. Also see Casassus (2014).

26.5 PUBLIC CONCERNS AND THE REGULATORY ENVIRONMENT

Consumer and citizen approval for a new innovation is critical for its success in the marketplace. It is also important in even allowing it to reach the marketplace in the first place. Thus public hostility to GM foods has effectively prevented it from being sold and produced in the European Union. More surprisingly perhaps, public opinion has held up its development in China (Wang, 2015). In India too, widespread hostility from both farmers and consumers has also held up its development (Kumar, 2015). Thus despite being the world's fourth-biggest GM-crop producer, GM cotton is the only commercial GM crop which can be grown in India. Occasionally the literature reads as if governments should take a paternalistic attitude and ignore public opinion, proceeding on the basis of the scientific evidence, much of which says there are advantages to GM crops and the risks are relatively low. This is to misunderstand democracy. But this still leaves the question as to why, given the potential gains from GM foods, public opinion is so hostile?

Perceptions of gain and fear of risk have been shown to be critical, but not the only, factors in determining people's attitudes to GM foods. As well as relative risks and gains, perceptions of the naturalness and environmental impact[d] of the technology also influence people, as shown by Hudson et al. (2015) in an analysis of attitudes across the countries of the European Union using Eurobarometer data. They analyzed attitudes to both cisgenesis and transgenesis in the context of apple cultivation. They found that people were much less hostile to cisgenesis than to transgenesis. Thus, overall 57.1% approved of the former, but only 31.4% the latter. Such attitudes vary substantially between different groups of people. Women tend to be less favorable than men, in part perhaps because of greater fears of risk, as do less well-off people. This is consistent with much of the literature, for example, Burton et al. (2001). On the other hand those with a better education tend to be more favorable, particularly if they or their parents have a scientific background. As Ceccoli and Hixon (2012) point out, an understanding of science helps increase an individual's understanding of the scientific issues underlying GM foods, and it may be the corresponding reduction in uncertainty which helps to increase approval. Slightly more surprising perhaps was the influence of religion. For example, with respect to cisgenesis, Muslims, Catholics, and Orthodox Christians are significantly less approving than others. Most people perceive potential gains to GM technology, but equally they see risks. But it is not simply a risk–gain calculus, also relevant are concerns about the naturalness of the technology. This ties in with the view that the depth of the emotional and cultural affiliation maize has for the Mexican population impacts on their attitudes to GM maize (Gilbert, 2013).

In order to both protect consumers and also farmers, as well as allay their concerns, a good regulatory framework is important. The European Union's stance on

[d]This suggests that the latter two cannot be simply combined with risks and gains, but are to an extent separate dimensions.

regulation tends to rely on the precautionary principle, that potential risks should be avoided even if there is no scientific evidence to suggest there is harm (Foster et al., 2000). This can be seen as placing the onus on would-be developers of a new technology to prove there is no harm. The alternative principle, often couched in terms of there is no scientific proof of the damage done by ...., puts the onus of proof on those who would seek to prevent the development of the new technology. To a large extent this characterizes the position of the USA. Provided standard tests for known risks such as toxicity have been passed successfully, products can usually be commercially placed on the market (Paarlberg, 2010). The United States' approach is based on the concept of "substantial equivalence," comparing GM plants to existing ones. If any characteristic differs substantially from its conventional counterpart, then the possibilities of allergenicity, toxicity, or other unintended effects are investigated. Regulation of GM plants is the responsibility of the US Department of Agriculture (USDA), that is, they use an existing agency rather than creating a new one, suggesting it is no different to other foods. The process is not cheap; in this sense it is more akin to the pharmaceutical market than foods. Thus it is approximately 10 times more expensive to bring a GM crop to market than a nonGM one (Potrykus, 2010). This expense tends to limit the development of new products to the larger firms, although extensive and necessary patenting of the technology may in any case have had a similar effect. Qaim (2009) is unequivocal, calling it illogical to subject GM crops to a much higher degree of scrutiny than conventionally bred crops, further arguing that the regulatory complexity often observed appears linked to a politicized public debate and the lobbying of antibiotech interest groups. In doing this Qaim appears to particularly be attacking the European Union position and he is far from being alone in this.

The European Union's approach can also be seen as evaluating the process and the American approach as evaluating the product (Nicolia et al., 2014). This has further resulted in new legislation and regulatory authorities in the European Union, while in the United States there has been an integration into existing legislation and institutions. In general more countries have adopted stances closer to the European than to the American system (Qaim, 2009), which helps explain the limited geographical spread of GM crops. These different regulatory positions are limiting trade which, of course, is particularly irritating to the USA which has spent so much money developing the technologies and has sparked complaints to the World Trade Organization (WTO). Although of course a different regulatory position in the European Union would have encouraged more competition to the United States worldwide. Trade itself is governed by the Cartagena Protocol, which is based on the precautionary principle and allows countries to ban GMO imports if they feel there is not enough scientific evidence regarding the safety of the product. Most countries have ratified this, but not the United States or Canada who are among the leading producers of GM crops.

Regulation of new products onto the marketplace is one aspect of regulation, controlling their production once approved is another. One strategy, which is mandatory in the USA, are refuges, or areas of land planted with nonGM crops in among fields planted with GM crops. Pests focus on these areas. The refuge area provides the pest with adequate food, and thus it has no need to develop resistance to GM crops. The same logic carries through to weeds, although these are easier to control as they spread through seed dispersion, whereas insects may spread much more rapidly. A further aspect of postcommercialization regulation relates to the labeling of GM products. The products being deemed safe, safety is not seen as an issue per se, it is rather about informing consumers and allowing nonGM producers to differentiate their product from GM alternatives. In the European Union this is mandatory and again the European Union has come in for criticism from Qaim (2009). In the USA, it is not mandatory as such, but Vermont has introduced a law, which should come into effect in 2016, requiring the labeling of GM food. This is however, being looked at by the courts.

26.6 GM ANIMALS

There is also the potential for, and actual development of, GM livestock, fish, etc. We do not focus so much on these as, until very recently, none had been commercialized for sale as food, although there has been considerable research done on, for example, the development of a GE cow that produced allergy-free milk (Jabed et al., 2012). However, several GM animals have been developed for recreational purposes, for example a GM Zebra fish with a fluorescent gene which allows it to glow in the dark under UV light (Vàzquez-Salat et al., 2012). However, the United States Food and Drug Administration (FDA) has in Nov. 2015 approved a GM salmon as food. The FDA argue that the inserted genes remained stable over several generations of fish, that food from the GE salmon is safe to eat by humans and animals, that the GE is safe for the fish, and the salmon meets the sponsor's claim about faster growth. In addition, the FDA found that the approval would not have a significant impact on the environment of the United States, as the multiple containment measures the company will use in the land-based facilities make it extremely unlikely that the fish could escape and establish themselves in the wild. Other GM animals are in the regulatory pipeline.

Animals bring new factors into the calculations with respect to costs and benefits, as well as others, for example, the growth of superweeds, losing their relevance. Important in the former are considerations of animal welfare. As we have seen from a regulatory perspective, the United States has placed GM animals with the FDA rather than the USDA, although of course they are persisting with the approach that there is nothing sufficiently new as to justify a new agency. Vàzquez-Salat et al. (2012) suggest that placing control with the FDA is in response to awareness of the American consumers' apprehension with respect to GM animal applications and is designed to increase trust in the regulatory system. However, whether a simple change in federal regulatory agencies will achieve this remains to be seen.

26.7 CONCLUSIONS

Much of the literature seems predicated on the assumption of an unchanging environment, for example, present net benefits are representative of those in the future. This is unlikely both because the producers of GE seed may evolve their strategy to exploit a monopoly position and because the natural environment will evolve to counter GM crops. The latter is illustrated in a report by the National Research Council (2010) that three toxic resistant pest species had evolved to counter insect-resistant crops. This may not seem many, but this is a continually evolving situation (Bennett et al., 2013). The seed producers have responded to this by "stacking" multiple traits in insect-resistant seed (Barrows et al., 2014). These target pests differently, making it more difficult for them to build resistance against insect-resistant traits. In this sense we are in a game with nature, we make a move and nature counters with another move, and at this point in time it is difficult to forecast how the game will evolve. This is further illustrated by the growing resistance of weeds to glyphosates, which is leading farmers to increase dosage and resort more heavily to tilling operations, potentially impacting on soil erosion and water quality (Mueller et al., 2005).

Apart from anything else, GM crops will move on from the first generation which has focused on protecting crops from attack. As we saw, the next generation in part is focused on traits which change the food to increase product shelf life and nutritional content (Barrows et al., 2014). The costs and benefits of these will not be the same as with the first-generation crops. There will be fewer gains to farmers from reduced fertilizer use, double cropping, and so on, and thus also fewer potential gains to the environment. The risks too change. There may be fewer risks of nature moving to superweeds,and so on, but will this different focus increase, or pose new, health risks? On the other hand GE crops which increase tolerance to drought, adaptability to climate change, and nitrogen fixation (Barrows et al., 2014) may bring real benefits to the environment and those seeking to deal with climate change, including potentially at least the viability of whole countries. Third-generation crops, involving producing substances for pharmaceutical or industrial purposes, change the game again. The former has obvious health benefits, but also health risks, while the latter may shift the focus away from food-growing crops with adverse impacts on poverty and the ability of populations to feed themselves.

In many respects this is a catch 22 situation. One cannot statistically prove or disprove harm until the product is marketed. Under the position in the USA, if the technology does harm then it may be too late to stop it. Under the precautionary principle, technological development is stalled perhaps indefinitely. The United States' position is based on a belief in the efficacy of the free market, coupled with substantial influence of big business. The European Union's position is based on a more qualified view of the free market, together with a division of power and responsibility between the nation states and the European Union per se. With respect to GMOs which regulatory position is best? It depends upon the outcome, although we note that if there are negative effects which outweigh any benefits, it is perhaps unlikely that the European Union can avoid them as once established on the planet, they may well eventually spread to all or at least most countries.

The hostility of many to GM foods rests in part on a distrust of big business, of government, of the regulatory agencies, and even of science. In part such distrust rests on ignorance and can be reduced through an education process. But some distrust may well be justified. In the wake of various corporate scandals of recent years in banking and the automobile industry, it is hard to argue that mistrust in business is totally misplaced and solely based on ignorance. Hence governance and regulation need not just to focus on the technology, but on corporate governance and governance per se. Academics too need to engender trust. In some of the writing one can detect a possible bias toward one position or the other which tends to lead the academic away from their normal level of scientific detachment. This is a highly emotive subject which arouses strong views in many of us, but that should not get in the way of normal scientific rigor.

Of course there are more players in the game than just two, that is, humans and nature. The biotech companies have their own agenda, separate to both governments and the regulatory authorities, as well as of course to nature. It is possible that having established a monopoly position in the first stage of the game, they will then seek to exploit this more and more in later stages, increasing their own share of the surplus. But there is another concern which has not been so well explored. Frequently the biotech company does not just research and produce GM seeds, it also markets other products such as herbicides, and farmers who use their seeds must also use their herbicide against weeds. It is in effect a bundled product. The problem here is that there may be other herbicides which would be more efficiently combined with GM seeds, but because the firm has one on the market it is in their interests to develop this and not those others, thus giving them a monopoly position not just in the seeds but also in the herbicide. There are other examples of this in economics and there is an argument that GM seed developers should be restricted from developing herbicides and insecticides and vice versa.

Adam Smith too is a player in the game, in this case representing basic economic forces. Some of the arguments in favor of GM foods rest on the foods benefitting farmers whose incomes rise, helping to reduce poverty. But greater profits may under certain circumstances attract more farmers into the production process, thus driving down profits until new farmers are no longer attracted into the industry. In this case the gains to farmers may be short-lived, although of course the gains to consumers may be longer lasting and enhanced. Whether nonGM farmers can survive these lower prices in the absence of productivity gains will depend upon the extent they can sell their differentiated product on the marketplace.

Turning to IPRs, the monopoly position these give firms is theoretically limited by patent expiry. However this is of little help in biotech if the game is continually changing due to nature's reactions and the biotech companies are having to bring out new versions of the original GM seed to meet nature's latest challenges. This is particularly disturbing if those challenges make it harder for farmers to use conventional seed, forcing some to use GM seeds.[e] It is also of particular concern that the biotech companies are primarily in advanced economies with few in the poorer developing economies, particularly the smaller ones, leading to a transfer of income to the former. Yet it must be recognized that the motive of private profit has driven this revolution and if we had relied on public sector research many GE crop varieties would not have been developed (Graff et al., 2003). Nonetheless, should we have done so in order to keep IPRs out of an area with such large externalities? It is also a slightly contentious area, being based on patenting genes which are in fact nature's invention. There are particular problems because of the multiplicity of patents. A new GM crop may involve in excess of 40 patents and license agreements (Barrows et al., 2014). Similarly, the development of a GM crop may involve "dozens of patented intermediate Technologies" (Qaim, 2009). These are process technologies and are necessary, for example, in transmitting the gene to the original crop. This position on patents is in contrast to that in medical biotechnology. According to Qaim this particularly disadvantages public sector research, where returns on research tend to be more limited. It is worth emphasizing that this is not a debate in isolation, rather there is much discussion on patents per se.

Throughout this chapter we have sought to present a balanced view of the risks and gains to GM foods without letting our own views get too much in the way of the analysis. In part this is because, despite the large volume of evidence, we believe nature has not had sufficient time to reveal all of its moves, some of which we are still guessing at and others we are still evaluating. But at the end we come off the fence, just a little bit, with respect to regulation. If we are to proceed with GM foods we need to be careful and arguably go slowly. We need to be careful not just with respect to new crops seeking clearance to be sold on the market, but also to well-established crops with an eye continually on nature's latest moves. We should go slowly, particularly with new-generation crops, because that gives us more time to observe the adverse consequences which may characterize such crops and thus improve the regulatory framework for such crops. We need if you will "pioneer crops" on which we can gather information on other similar crops, to improve the regulatory environment. Now whether slowly means at the pace of, or even slower than, the European Union, at the pace of, or faster than, the USA, or somewhere in between we leave open to discussion. Finally, we should also comment that the argument against labeling, in that it may misinform ignorant consumers, seems a little dubious, particularly when our research shows that people are not just concerned about safety. First, much of the argument in favor of free markets rests on the assumption that consumers can make rational and correct decisions. But second, in all countries a large number of consumers do not want to consume GM products; to deny them the information to allow them to do so on the grounds that they are safe, and hence consumers should not have this preference, seems to us misguided.

[e]On the other hand it must be recognized that if the GM seed leads to a reduction in pests, rather than their evolution into more formidable threats, then nonGM farmers stand to benefit.

REFERENCES

Ammann, K., 2005. Effects of biotechnology on biodiversity: herbicide-tolerant and insect-resistant GM crops. Trends Biotechnol. 23, 388–394.

Barrows, G., Sexton, S., Zilberman, D., 2014. The impact of agricultural biotechnology on supply and land-use. Environ. Dev. Econ. 19, 676–703.

Beckmann, V., Soregaroli, C., Wesseler, J., 2006. Coexistence rules and regulations in the European Union. Am. J. Agr. Econ. 88, 1193–1199.

Bennett, A.B., Chi-Ham, C., Barrows, G., Sexton, S., Zilberman, D., 2013. Agricultural biotechnology: economics, environment, ethics, and the future. Annu. Rev. Environ. Resour. 38, 249–279.

Bennett, R., Phipps, R., Strange, A., Grey, P., 2004. Environmental and human health impacts of growing genetically modified herbicide-tolerant sugar beet: a life-cycle assessment. Plant Biotechnol. J. 2, 273–278.

Brookes, G., Barfoot, P., 2015. GM crops: global socio-economic and environmental impacts 1996–2013. PG Economics Ltd.

Burton, M., Rigby, D., Young, T., James, S., 2001. Consumer attitudes to genetically modified organisms in food in the UK. Eur. Rev. Agric. Econ. 28, 479–498.

Casassus, B., 2014. Paper claiming GM link with tumors republished. Nature News. http://www.nature.com/news/paper-claiming-gm-link-with-tumours-republished-1.15463

Ceccoli, S., Hixon, W., 2012. Explaining attitudes toward genetically modified foods in the European Union. Int. Polit. Sci. Rev. 33, 301–319.

Cerdeira, A.L., Duke, S.O., 2006. The current status and environmental impacts of glyphosate-resistant crops. J. Environ. Qual. 35, 1633–1658.

Demont, M., Tollens, E., 2004. First impact of biotechnology in the EU: Bt maize adoption in Spain. Ann. Appl. Biol. 145, 197–207.

European Commission, 2010. A decade of EU-funded GMO research. http://ec.europa.eu/research/biosociety/pdf/a_decade_of_eu-funded_gmo_research.pdf

Fagan, J., Terje Traavik, T., Thomas Bøhn, T., 2015. The Seralini affair: degeneration of science to re-science? Environ. Sci. Europe 27, 1–9.

Foster, K.R., Vecchia, P., Repacholi, M.H., 2000. Science and the precautionary principle. Science 288, 979–981.

Gilbert, N., 2013. A hard look at GM crops. Nature 497, 24–26.

Gouse, M., Pray, C., Schimmelpfennig, D., 2004. The distribution of benefits from Bt cotton adoption in South Africa. AgBioForum 7, 187–194.

Graff, G.D., Cullen, S.E., Bradford, K.J., Zilberman, D., Bennett, A.B., 2003. The public–private structure of intellectual property ownership in agricultural biotechnology. Nat. Biotechnol. 21, 989–995.

Hamilton, N.D., 2001. Legal issues shaping society's acceptance of biotechnology and genetically modified organisms. Drake J. Agricult. Law 6, 81–117.

Huang, J., Hu, R., Rozelle, S., Pray, C., 2005. Insect-resistant GM rice in farmers' fields: assessing productivity and health effects in China. Science 308, 688–690.

Hudson, J., Caplanova, A., Novak, M., 2015. Public attitudes to GM foods: the balancing of risks and gains. Appetite 92, 303–313.

Jabed, A., Wagner, S., McCracken, J., Wells, D.N., Laible, G., 2012. Targeted microRNA expression in dairy cattle directs production of β-lactoglobulin-free, high-casein milk. P. Natl. Acad. Sci. USA 109, 16811–16816.

Kathage, J., Qaim, M., 2012. Economic impacts and impact dynamics of Bt (*Bacillus thuringiensis*) cotton in India. P. Natl. Acad. Sci. USA 109, 11652–11656.
Kumar, S., 2015. India eases stance on GM crop trials. Nature 521, 138–139.
Mallory-Smith, C.A., Sanchez-Olguin, E., 2011. Gene flow from herbicide-resistant crops: it's not just for transgenes. J. Agr. Food Chem. 59, 5813–5818.
Mueller, T.C., Mitchell, P.D., Young, B.G., Culpepper, A.S., 2005. Proactive versus reactive management of glyphosate-resistant or -tolerant weeds. Weed Technol. 19, 924–933.
National Research Council, 2010. Impact of Genetically Engineered Crops on Farm Sustainability in the United States. National Academies Press, Washington, DC.
Nicolia, A., Manzo, A., Veronesi, F., Rosellini, D., 2014. An overview of the last 10 years of genetically engineered crop safety research. Crit. Rev. Biotechnol. 34, 77–88.
Paarlberg, R., 2010. GMO foods and crops: Africa's choice. New Biotechnol. 27, 609–613.
Parrott, W., Chassy, B., Ligon, J., et al., 2010. Application of food and feed safety assessment principles to evaluate transgenic approaches to gene modulation in crops. Food Chem. Toxicol. 48, 1773–1790.
Potrykus, I., 2010. Regulation must be revolutionized. Nature 466, 561.
Pray, C.E., Ma, D., Huang, J., Qiao, F., 2001. Impact of Bt cotton in China. World Dev. 29, 813–825.
Price, G.K., Lin, W., Falck-Zepeda, J.B., Fernandez-Cornejo, J., 2003. The size and distribution of market benefits from adopting agricultural biotechnology. Technical Bulletin, 1906, US Department of Agriculture, Washington, DC.
Qaim, M., 2009. The economics of genetically modified crops. Ann. Rev. Resource Econ. 1, 665–694.
Qaim, M., Zilberman, D., 2003. Yield effects of genetically modified crops in developing countries. Science 299, 900–902.
Séralini, G.-E., Clair, E., Mesnage, R., Gress, S., Defarge, N., Malatesta, M., Hennequin, D., de Vendômois, J.S., 2014. Republished study: long-term toxicity of a Roundup herbicide and a Roundup-tolerant genetically modified maize. Environ. Sci. Europe 26, 14.
Trigo, E.J., Cap, E.J., 2003. The impact of the introduction of transgenic crops in Argentinean agriculture. AgBioForum 6, 87–94.
Vàzquez-Salat, N., Salter, B., Smets, G., Houdebine, L.-M., 2012. The current state of GMO governance: are we ready for GM animals? Biotechnol. Adv. 30, 1336–1343.
Wang, Q., 2015. China's scientists must engage the public on GM. Nature 519, 7.
Zhang, L., Hou, D., Chen, X., et al., 2012. Exogenous plant MIR168a specifically targets mammalian LDLRAP1: evidence of crosskingdom regulation by microRNA. Cell Res. 22, 107–126.

Index

D

G

H

I

J

K

L

M

N

O

Q

R

Z

FSC
www.fsc.org
MIX
Papier | Fördert
gute Waldnutzung
FSC® C083411

Kolibri 360 GmbH
Ferdinand-Jühlke-Straße 7
99095 Erfurt, Deutschland
produktsicherheit@kolibri360.de